EQUINE NUTRITION
AND FEEDING

'1

EQUINE NUTRITION AND FEEDING

SECOND EDITION

DAVID FRAPE
PhD, CBiol, FIBiol, FRCPath

Blackwell
Science

© 1998 by Blackwell Science Ltd,
a Blackwell Publishing Company
Editorial Offices:
Osney Mead, Oxford OX2 0EL, UK
 Tel: +44 (0)1865 206206
Blackwell Science, Inc., 350 Main Street,
Malden, MA 02148-5018, USA
 Tel: +1 781 388 8250
Iowa State Press, a Blackwell Publishing
Company, 2121 State Avenue, Ames, Iowa
50014-8300, USA
 Tel: +1 515 292 0140
Blackwell Science Asia Pty, 54 University
Street, Carlton, Victoria 3053, Australia
 Tel: +61 (0)3 9347 0300
Blackwell Wissenschafts Verlag,
Kurfürstendamm 57, 10707 Berlin, Germany
 Tel: +49 (0)30 32 79 060

First edition published 1986 by Longman
 Group UK Ltd
Reprinted 1992, 1993, 1994
Second edition published by Blackwell Science
 1998
Reissued in paperback 1998
Reprinted 1999, 2000, 2001, 2002 (twice)

Library of Congress
Cataloging-in-Publication Data
Frape, David L. (David Lawrence), 1929–
 Equine nutrition and feeding / David
 Frape. — 2nd ed.
  p. cm.
 Includes bibliographical references (p.)
 and index.
 ISBN 0–632–04105–6 (hb)
 1. Horses—Feeding and feeds. I. Title
SF285.5.F73 1997
636.1'085—dc21   97–34539
                                          CIP

ISBN 0-632-05303-8

A catalogue record for this title is available
from the British Library

Set in 10/13 Times
by SNP Best-set Typesetter Ltd., Hong Kong
Printed and bound in Great Britain by
MPG Books Ltd, Bodmin, Cornwall

For further information on
Blackwell Science, visit our website:
www.blackwell-science.co.uk

Contents

Introduction to the Second Edition

About 12–13 years have passed since much of the first edition was written. During this period there has been an expansion in equine research in several countries. The nature of the studies has become increasingly complex and several metabolic problems related to nutrition have been uncovered. Moreover, the physiological implications of exercise have been explored in parallel with comparable human studies. Institut National de la Recherche Agronomique (INRA), the French research organization, has expounded and published proposals for estimating the nutrient and feed requirements which are distinct from those published in the USA by the National Research Council (NRC). One purpose of this second edition of *Equine Nutrition and Feeding* has been to integrate and discuss these developments in a palatable form. Some of this work has touched on scientific disciplines only briefly entertained in the previous edition, so the Glossary has been expanded to cover certain technical terms that should enable those unfamiliar with these subjects to follow and understand the arguments.

All the chapters have been rewritten to provide new information, without excluding satisfactory earlier evidence where this evidence meets the needs. Chapters 5, 6, 9, 10, 11 and 12 have all been extensively revised, owing to a greater recent output of research in those areas. More is known now about the nutrition of metabolic diseases, the effects of diet on exercise performance and pasture requirements peculiar to the horse. Possible solutions to a few unsolved scientific mysteries are given, but many remain to be resolved. Considerably more feed materials have been described in this edition, partly to rectify the paucity of information I provided previously and partly to assist in the solution of husbandry problems confronting an animal that desires space in a shrinking world.

David Frape

Acknowledgements

I would like to thank Pat Harris, MA, VetMB, PhD, for reading and commenting on Chapters 3 and 12 and for providing details of her experience with creatinine clearance methods, and Mr. J.C. Dickins, MA (Cantab), Dip. Math. Stats, C. Math., C. Stat., FIMA, for assistance with a description of the computational technique known as linear programming. Also, Sidney Ricketts, BSc, BVSc, DESM, FRCVS, for the use of photographs of selenium intoxication, and Tessa Milner for typing the manuscript.

List of Abbreviations

AAT	aspartate aminotransferase
acetyl-CoA	acetyl coenzyme A
ACTH	adrenocorticotropic hormone
ADAS	Agricultural Development and Advisory Service
ADP	adenosine diphosphate
a.i.	active ingredient
AI	artificial insemination
ALP	alkaline phosphatase
ALT	alanine aminotransferase
AMP	adenosine monophosphate
AN	adenine nucleotides
AP	alkaline phosphatase
AST	aspartate aminotransferase
ATP	adenosine triphosphate
BAL	bronchoalveolar lavage
BE	base excess
BFGF	basic fibroblast growth factor
BHA	butylated hydroxyanisole
BHT	butylated hydroxytoluene
BSP	bromsulphalein (sulphobromophthalein)
BW	bodyweight
CCO	cytochrome c oxidase
CFU	colony forming unit
CK	creatine kinase
COPD	chronic obstructive pulmonary disease
CP	crude protein
DCAB	dietary cation–anion balance
DCP	digestible crude protein
DDS	distiller's dark grains
DE	digestible energy

DM	dry matter
DMG	N,N-dimethylglycine
DMSO$_2$	dimethylsulphone
DOD	developmental orthopaedic disease
ECF	extracellular fluid
EDM	equine degenerative myeloencephalopathy
EMND	equine motor neuron disease
ERS	exertional rhabdomyolysis syndrome
EU	European Union
EVH-1/4	equine herpesvirus
FAD	flavin adenine dinucleotide
FE	fractional electrolyte excretion
FFA	free fatty acid
FTH	fast twitch, high oxidative
FT	fast twitch, low oxidative
FTU	fungal titre unit
GE	gross energy
GGT	gamma-glutamyltransferase
GI	gastrointestinal
GLC	gas-liquid chromatograph
GSH-Px	glutathione peroxidase
GSH	glutathione
Hb	haemoglobin
HI	heat increment
HPLC	high performance liquid chromatography
HPP	hyperkalaemic periodic paralysis
ICF	intracellular fluid
IMP	inosine monophosphate
INRA	Institut National de la Recherche Agronomique
iu	international unit
i.v.	intravenous
LBS	*Lactobacillus* selection
LDH	lactic dehydrogenase
LPL	lipoprotein lipase
LPS	lipopolysaccharides
MADC	matières azotées digestibles corrigées (*or* cheval)
MDA	malonyldialdehyde
ME	metabolizable energy
MSM	methyl sulphonyl methane
NAD	nicotinamide adenine dinucleotide
NADP	nicotinamide adenine dinucleotide phosphate
NE	net energy
NEFA	nonesterified fatty acid(s)
NFE	nitrogen free extractive

NIS	nutritionally improved straw
NPN	nonprotein N
NRC	National Research Council
NSHP	nutritional secondary hyperparathyroidism
OC	osteochondrosis
OCD	osteochondritis dissecans
OM	organic matter
PAF	platelet activating factor
PCV	packed cell volume
PCr	phosphocreatine
PDH	pyruvate dehydrogenase
PN	parenteral nutrition
PTH	parathyroid hormone
PUFA	polyunsaturated fatty acid
RDR	relative dose response
RER	recurrent exertional rhabdomyolysis
RES	reticuloendothelial system
RH	relative humidity
RQ	respiratory quotient
RVO	recovered vegetable oil
SAP	serum alkaline phosphatase
SDH	sorbitol dehydrogenase
SET	standardized exercise test
SG	specific gravity
SGOT	serum glutamic–oxaloacetic transaminase
SID	strong ion difference
SOD	superoxide dismutase
ST	slow twitch, high oxidative
STP	standard temperature and pressure
T_3	triiodothyronine
T_4	thyroxine
TAG	triacylglycerol
TB	Thoroughbred
TBA	thiobarbituric acid
TBAR	thiobarbituric acid reactive substance
TCA	tricarboxylic acid
TLV	threshold limiting value
TPN	total parenteral nutrition
TPP	thiamin pyrophosphate
TSH	thyroid-stimulating hormone
UDP	uridine diphosphate
UFC	unité fourragère cheval
UKASTA	United Kingdom Agricultural Supply Trade Association
VFA	volatile fatty acid
VLDL	very low density lipoprotein

Chapter 1
The Digestive System

A horse which is kept to dry meat will often slaver at the mouth. If he champs his hay and corn, and puts it out again, it arises from some fault in the grinders . . . there will sometimes be great holes cut with his grinders in the weaks of his mouth. First file his grinders quite smooth with a file made for the purpose.

Francis Clater, 1786

The domesticated horse consumes a variety of feeds ranging in physical form from forage with a high content of moisture to cereals with large amounts of starch, and from hay in the form of physically long fibrous stems to salt licks and water. In contrast, the wild horse has evolved and adapted to a grazing and browsing existence, in which it selects succulent forages containing relatively large amounts of water, soluble proteins, lipids, sugars and structural carbohydrates, but little starch. Short periods of feeding occur throughout most of the day and night, although generally these are of greater intensity in daylight. In domesticating the horse, man has generally restricted its feeding time and introduced unfamiliar materials, particularly starchy cereals, protein concentrates and dried forages. The art of feeding gained by long experience is to ensure that these materials meet the varied requirements of horses without causing digestive and metabolic upsets. Thus, an understanding of the form and function of the alimentary canal is fundamental to a discussion of feeding and nutrition of the horse.

THE MOUTH

Eating rates of horses, cattle and sheep

The lips, tongue and teeth of the horse are ideally suited for the prehension, ingestion and alteration of the physical form of feed to that suitable for propulsion through the gastrointestinal (GI) tract in a state that facilitates admixture with digestive juices. The upper lip is strong, mobile and sensitive and is used during grazing to place forage between the teeth; in the cow the tongue is used for this purpose. By contrast, the horse's tongue moves ingested material to the cheek teeth for grinding. The lips are also used as a funnel through which water is sucked.

As distinct from cattle, the horse has both upper and lower incisors enabling it to graze closely by shearing off forage. More intensive mastication by the horse means

1

that the ingestion rate of long hay, per kilogram of metabolic body weight (BW), is three to four times faster in cattle and sheep than it is in ponies and horses, although the number of chews per minute, according to published observations, is similar (73–92 for horses and 73–115 for sheep) for long hays. The dry matter (DM) intake per kilogram of metabolic BW for each chew is then 2.5 mg in horses (I calculate it to be even less) and 5.6–6.9 mg in sheep. Consequently, the horse needs longer daily periods of grazing than do sheep. The lateral and vertical movements of the horse's jaw, accompanied by profuse salivation, enable the cheek teeth to comminute long hay to a greater extent and the small particles coated with mucus are suitable for swallowing. Sound teeth generally reduce hay and grass particles to less than 1.6 mm in length. Two-thirds of hay particles in the horse's stomach are less than 1 mm across, according to work by Meyer and colleagues (Meyer *et al.* 1975b).

The number of chewing movements for roughage is considerably greater than that required for chewing concentrates. Horses make between 800 and 1200 chewing movements per 1 kg concentrates, whereas 1 kg long hay requires between 3000 and 3500 movements. In ponies, chewing is even more protracted – they require 5000–8000 chewing movements per 1 kg concentrates alone, and very many more for hay (Meyer *et al.* 1975b).

Dentition

As indicated above, teeth are vital to the wellbeing of horses. Diseased teeth are an encumbrance. Evidence has shown that abnormal, or diseased, teeth, can cause digestive disturbances and colic. Apparent fibre digestibility, the proportion of faecal short fibre particles and plasma free fatty acids were all increased after dental correction of mares. Consequently, diseased teeth and badly worn teeth, as in the geriatric horse, can limit the horse's ability to handle roughage and may compromise general health.

The normal horse has two sets of teeth. The first to appear, the deciduous, or temporary milk, teeth erupt during early life and are replaced during growth by the permanent teeth. The permanent incisors and permanent cheek teeth erupt continuously to compensate for wear and their changing form provides a basis for assessing the age of a horse. In the gap along the jaw between the incisors and the cheek teeth the male horse normally has a set of small canine teeth. The gap, by happy chance, securely locates the bit. The dental formula and configuration of both deciduous and permanent teeth are given in Fig. 1.1. The lower cheek teeth are implanted in the mandible in two straight rows that diverge towards the back. The space between the rows of teeth in the lower jaw is less than that separating the upper teeth (Fig. 1.1). This accommodates a sideways, or circular, movement of the jaw that effectively shears feed. The action leads to a distinctive pattern of wear of the biting surface of the exposed crown. This pattern results from the differences in hardness which characterize the three materials (cement, enamel and dentine) of which teeth are composed. The enamel, being the hardest, stands out in the form of sharp prominent ridges. It is estimated that the enamel ridges of an upper cheek tooth in a young

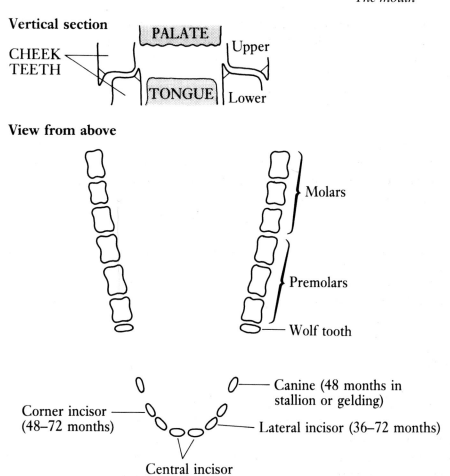

Fig. 1.1 Configuration of permanent teeth in the upper or lower jaw (the molars and premolars in the lower jaw are slightly closer to the midline). The deciduous teeth on each side of each jaw are: three incisors, one canine, three molars. The deciduous canines are vestigial and do not erupt. The wolf teeth (present in the upper jaw of about 30% of fillies and about 65% of colts) are often extracted as their sharp tips can injure cheeks when a snaffle bit is used. Months (in parentheses) are approximate ages at which permanent incisors and canines erupt, replacing the deciduous teeth.

adult horse, if straightened out, would form a line more than 30 cm (1 ft) long. This irregular surface provides a very efficient grinding organ.

Horses and ponies rely more on their teeth than we do. People might be labelled concentrate eaters; concentrates require much less chewing than does roughage. Even among herbivores, horses and ponies depend to a far greater extent on their teeth than do the domesticate ruminants – cattle, sheep and goats. Ruminants, as discussed in 'Eating rates of horses, cattle and sheep', swallow grass and hay with minimal chewing and then depend on the activity of bacteria in the rumen to disrupt the fibre. This is then much more readily fragmented during chewing the cud.

Saliva

The physical presence of feed material in the mouth stimulates the secretion of a copious amount of saliva. Some 10–12 l are secreted daily in a horse fed normally. This fluid seems to have no digestive enzyme activity, but its mucus content enables it to function as an efficient lubricant preventing 'choke'. Its bicarbonate content, amounting to some 50 mEq/l, provides it with a buffering capacity. The concentration of bicarbonate and sodium chloride in the saliva is, however, directly proportional to the rate of secretion and so increases during feeding. The continuous secretion of saliva during eating seems to buffer the digesta in the proximal region of the stomach, permitting some microbial fermentation with the production of lactate. This has important implications for the wellbeing of the horse (see Chapter 11).

THE STOMACH AND SMALL INTESTINE

Development of the gastrointestinal (GI) tract and associated organs

The GI tract tissue of the neonatal foal weighs only 35 g/kg BW, whereas the liver is large, nearly in the same proportion to BW, acting as a nutrient store for the early critical days. By 6 months of age the GI tract tissue has proportionately increased to 60 g/kg BW, whereas the liver has proportionately decreased to about 12–14 g/kg BW. By 12 months both these organs have stabilized at 45–50 g/kg BW for the GI tract and 10 g/kg BW for the liver. Organ size is also influenced by the activity of the horse. After a meal, the liver of mammals generally increases rapidly in weight, probably as a result of glycogen storage and blood flow. In the horse the consumption of hay has less impact on liver glycogen, so that following a meal of hay the liver weighs only three-quarters of that following mixed feed. Moreover, during and immediately after exercise the GI tract tissue weighs significantly less than in horses at rest, owing to the shunting of blood away from the mesenteric blood vessels to the muscles. At rest about 30% of the cardiac output flows through the hepatic portal system. More about these aspects is discussed in Chapter 9.

Surprisingly, the small intestine does not materially increase in length from 4 weeks of age, whereas the large intestine increases with age, the colon doing so until 20 years at least. The distal regions of the large intestine continue extension to a greater age than do the proximal regions. This development reflects the increasing reliance of the older animal on roughage. In an adult horse of 500 kg BW the small intestine is approximately 16 m in length, the caecum has a maximum length of about 0.8 m, the ascending colon 3 m, and the descending colon 2.8 m.

Transit of digesta through the GI tract

The residence time for ingesta in each section of the GI tract allows for its adequate admixture with GI secretions, for hydrolysis by digestive enzymes, for absorption of the resulting products, for fermentation of resistant material by bacteria and for the absorption of the products of that fermentation. Transit time through the GI tract is normally considered in three phases, owing to their entirely different characteristics. These phases are:

(1) expulsion rate from the stomach into the duodenum after a meal;
(2) rate of passage through the small intestine to the ileo–caecal orifice; and
(3) retention time in the large intestine.

The first of these will be considered below in relation to gastric disorders. Rate of passage of digesta through the small intestine varies with feed type. On pasture this rate is accelerated, although a previous feed of hay causes a decrease in the rate of the succeeding meal, with implications for exercise (see Chapter 9). Roughage is held in the large intestine for a considerable period that allows microbial fermentation time to break down structural carbohydrates. However, equine GI transit time of the residue of high fibre diets is less than that of low fibre diets of the same particle size, in common with the relationship found in other monogastric animals.

Digestive function of the stomach

The stomach of the adult horse is a small organ, its volume comprising about 10% of the GI tract (Fig. 1.2, Plate 1.1). In the suckling foal, however, the stomach capacity represents a larger proportion of the total alimentary tract. Most digesta are held in the stomach for a comparatively short time, but this organ is rarely completely empty and a significant portion of the digesta may remain in it for 2–6 h. Some digesta pass into the duodenum shortly after eating starts, when fresh ingesta enter the stomach. Expulsion into the duodenum is apparently arrested as soon as feeding stops. When a horse drinks, a high proportion of the water passes along the curvature of the stomach wall so that mixing with digesta and dilution of the digestive juices it contains are avoided. This process is particularly noticeable when digesta largely fill the stomach.

The entrance to the stomach is guarded by a powerful muscular valve called the cardiac sphincter. Although a horse may feel nauseous, it rarely vomits, partly because of the way this valve functions. This too has important consequences. Despite extreme abdominal pressure the cardiac sphincter is reluctant to relax in order to permit the regurgitation of feed, or gas. On the rare occasions when vomiting does occur, ingesta usually rush out through the nostrils, owing to the existence of a long soft palate. Such an act may indicate a ruptured stomach.

Gastric anatomy differentiates the equine stomach from that of other monogastrics. Apart from the considerable strengths of the cardiac and pyloric sphincters, almost half the mucosal surface is lined with squamous, instead of

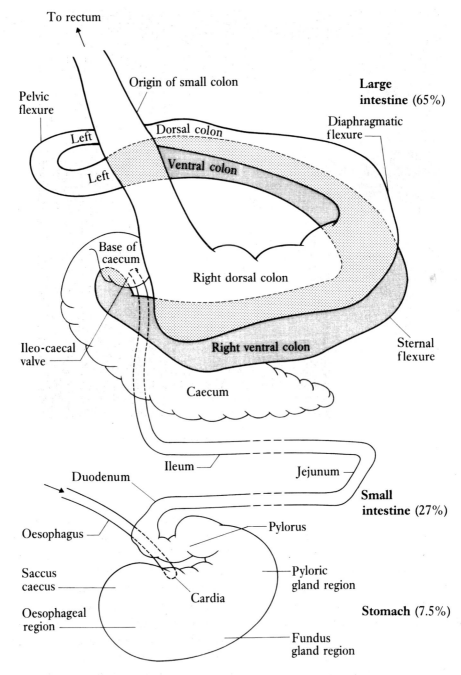

Fig. 1.2 GI tract of adult horse. (Relative volumes are given in parentheses.)

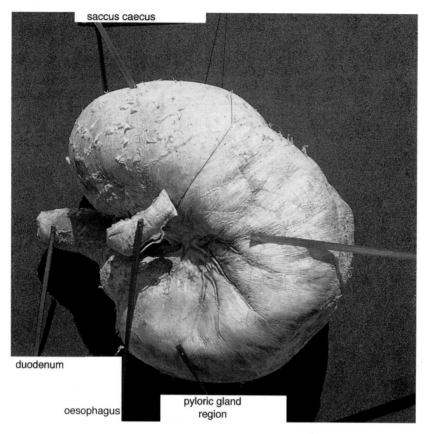

saccus caecus

duodenum

oesophagus

pyloric gland region

Plate 1.1 Stomach of a 550-kg TB mare, capacity 8.4l, measuring about 20 × 30 × 15 cm. Acid fermentation of stomach contents takes place in the saccus caecus (*top*).

glandular, epithelium. The glandular mucosa is divided into fundic and pyloric regions (Fig. 1.2). The fundic mucosa contains both parietal cells that secrete hydrochloric acid (HCl) and zymogen cells which secrete pepsin, while the polypeptide hormone gastrin is secreted into the blood plasma by the pyloric region. The hormone's secretion is triggered by a meal, and equine studies in Sweden show that a mechanism of the gastric phase of release seems to be distension of the stomach wall by feed, but not the sight of feed. Also in the horse gastrin does not seem to act as a stress hormone. The hormone strongly stimulates secretion of gastric acid and the daily secretion and release of gastric juice into the stomach amounts to some 10–30l. Secretion of gastric juice continues even during fasting, although the rate seems to vary from hour to hour. HCl secretion continues, but declines gradually at a variable rate when the stomach is nearly empty and hence at that time the pH is around 1.5–2.0. The pH rises rapidly during a subsequent meal, especially that of grain only, partly as a consequence of a delay in gastrin secretion, compared with the more rapid gastrin response to hay. The act of eating stimulates the flow of saliva –

a source of sodium, potassium, bicarbonate and chloride ions. Saliva's buffering power retards the rate at which the pH of the stomach contents decreases. This action, combined with a stratification of the ingesta, brings about marked differences in the pH of different regions (about 5.4 in the fundic region and 2.6 in the pyloric region).

Fermentation, primarily yielding lactic acid, occurs in the oesophageal and fundic regions of the stomach, but particularly in that part known as the saccus caecus, lined by the squamous cells. As digesta approach the pylorus at the distal end of the stomach, the gastric pH falls, owing to the secretion of the HCl, which potentiates the proteolytic activity of pepsin and arrests that of fermentation. The activity of pepsin in the pyloric region is some 15–20 times greater than in the fundic. Because of the stomach's small size and consequentially the relatively short dwell time, the degree of protein digestion is slight.

Gastric malfunction

Professor Meyer and his colleagues in Hanover (Meyer *et al.* 1975a) have made detailed investigations of the flow of ingesta and digesta through the GI tract of horses. In so far as the stomach is concerned their thesis is that abnormal gastric fermentation occurs when the postprandial dry matter content of the stomach is particularly high and a low pH is not achieved. There is, nevertheless, considerable layering and a differentiation in pH between the saccus caecus and pyloric region. Fermentation is therefore a normal characteristic of the region of higher pH and in that region the larger roughage particles tend to float. However, the dry matter content, generally, is considerably lower following a meal of roughage than it is following one of cereals. After meals of 1 kg loose hay and 1 kg pelleted cereals, the resulting gastric dry matters were, respectively, 211 and 291 g/kg contents.

The Hanover group compared long roughage with that which was chopped, ground or pelleted and observed that, as particle size of roughage was decreased, the gastric dry matter contents decreased from 186 to 132 g/kg contents and the rate of passage of ingesta through the stomach increased. The reason for this is probably that it is the finely divided material in a gastric slurry which passes first to the intestines. The slurry is forced into the duodenum by contractions termed 'antral systole' at the rate of about three per minute. Nevertheless, it should be recalled that particle size is generally small as a result of comminution by the molars. With larger meals of pelleted cereal, up to 2.5 kg/meal, the gastric dry matter content attained 400 g/kg, and the pH was 5.6–5.8, for as long as 2–3 h after consumption. The dry matter accumulated faster than it was ejected into the duodenum, and as cereals could be consumed more rapidly than hay, with a lower secretion of saliva, the dry matter of the stomach was higher following large meals of cereals. As much as 10–20% of a relatively small meal of concentrates (given at the rate of 0.4% BW) has been found to remain in the stomach 6 h after feeding ponies. A high dry matter acts as a potent buffer of the HCl in gastric juice and the glutinous nature of cereal ingesta inhibits the penetration of cereal ingesta by those juices.

Together with the delay in gastrin release during a cereal meal, these factors could account for the failure of the postprandial pH to fall to levels that inhibit further microbial growth and fermentation. Lactic acid producing bacteria (*Lactobacilli* and *Streptococci*) thrive (also see Probiotics, Chapter 5). Whereas *Streptococci* do not produce gas, some *Lactobacillus* species produce carbon dioxide, thrive at a pH of 5.5–6.0 and even grow in the range 4.0–6.8, some strains growing in conditions as acid as pH 3.5. The pH of the gastric contents will even increase to levels that permit non-lactic acid-producing, gas-producing bacteria to survive. Gas production at a rate greater than that at which it can be absorbed into the bloodstream causes gastric tympany, and even gastric rupture, and hence it is desirable that the post-prandial gastric pH falls sufficiently to arrest most bacterial growth and, in fact, to kill potential pathogens.

Gastric ulceration

The stratified squamous epithelial mucosa of the equine stomach exists in a poten-tially highly acidic environment and is susceptible to damage by HCl and pepsin. Routine post-mortem examination of 195 Thoroughbreds (TBs) in Hong Kong (Hammond *et al.* 1986) revealed that 66% had suffered gastric ulceration. In TBs taken directly from training the frequency was 80%, whereas it was only 52% among those that had been retired for a month or more. The lesions seem to be progressive during training, but to regress during retirement. These lesions are not restricted to adult horses. Neonatal foals are able to produce highly acidic gastric secretions as early as 2 days old, and the mean pH of the glandular mucosal surface and fluid contents of 18 foals at 20 days old were 2.1 and 1.8, respectively (Murray & Mahaffey 1993). Ulceration and erosion occur in the gastric squamous mucosa, particularly that adjacent to the margo plicatus, as the squamous epithelial mucosa lack the protective processes, especially the mucus–bicarbonate barrier, possessed by the glandular mucosa.

Observations by the research group in Hanover showed that clinical signs of periprandial colic and bruxism (grinding of teeth) were more pronounced in horses with the most severe gastric lesions of diffuse ulcerative gastritis. Their further evidence showed that ponies receiving hay only were free from lesions, whereas 14 out of 31 receiving concentrates had ulcerative lesions (see Chapter 11).

Although treatment with cimetidine, or ranitidine, is effective, one must wonder whether infection plays a part in the equine syndrome (as it frequently does in man, where the organisms shrewdly protect themselves from acid by urease secretion with an acid pH optimum), as periprandial microbial activity and pH of gastric contents are higher in concentrate-fed animals. Moreover, the pH is lowest during a fast. If this proposal is true then quite different prophylaxis and treatment should be chosen.

Digestion in the small intestine

The 450-kg horse has a relatively short small intestine, 21–25 m in length, through which transit of digesta is quite rapid, some appearing in the caecum within 45 min after a meal. Much of the digesta moves through the small intestine at the rate of nearly 30 cm/min. Motility of the small intestine is under both neural and hormonal control. Of a liquid marker instilled into the stomach of a pony, 50% reached the distal ileum in 1 h, and by 1.5 h after instillation 25% was present in the caecum (Merritt 1992 pers. comm.). The grazing horse has access to feed at all times and comparisons of quantities of feed consumed, where there is *ad libitum* access with similar quantities given following a 12-h fast, showed that the transit of feed from stomach to the caecum is much more rapid following the fast.

In consequence of the rapid transit of ingesta through the small intestine, it is surprising how much digestion and absorption apparently occur there. Although differences in the composition of digesta entering the large intestine can be detected with a change in diet, it is a considerably more uniform material than that entering the rumen of the cow. This fact has notable practical and physiological significance in the nutrition and wellbeing of the horse. The nature of the material leaving the small intestine is described as fibrous feed residues, undigested feed starch and protein, microorganisms, intestinal secretions and cell debris.

Digestive secretions

Large quantities of pancreatic juice are secreted as a result of the presence of food in the stomach in response to stimuli mediated by vagal nerve fibres, and by gastric HCl in the duodenum stimulating the release into the blood of the hormone secretin. In fact, although secretion is continuous, the rate of pancreatic juice secretion increases by some four to five times when feed is first given. This secretion, which enters the duodenum, has a low order of enzymatic activity, but provides large quantities of fluid and sodium, potassium, chloride and bicarbonate ions. Some active trypsin is, however, present. There is conflicting evidence for the presence of lipase in pancreatic secretion and bile, secreted by the liver, probably exerts a greater, but different, influence over fat digestion. The stimulation of pancreatic juice secretion does not increase its bicarbonate content, as occurs in other species. The bicarbonate content of digesta increases in the ileum, where it is secreted in exchange for chloride, so providing a buffer to large intestinal volatile fatty acids (VFA) (see 'Products of fermentation', this chapter).

The horse lacks a gall bladder, but stimulation of bile is also caused by the presence of gastric HCl in the duodenum. Secretion of pancreatic juice and bile ceases after a fast of 48 h. Bile is both an excretion and a digestive secretion. As a reservoir of alkali it helps preserve an optimal reaction in the intestine for the functioning of digestive enzymes secreted there. In the horse, the pH of the digesta leaving the stomach rapidly rises to slightly over 7.0.

Carbohydrates

A high proportion of the energy sources consumed by the working horse contains cereal starches. These consist of relatively long, branched chains, the links of which are α-D-glucose molecules joined as shown in Fig. 1.3. Absorption into the bloodstream depends on the disruption of the bonds linking the glucose molecules. This is contingent entirely upon enzymes secreted in the small intestine. These are held on the brush border of the villi in the form of α-amylase (secreted by the pancreas) and as α-glucosidases (secreted by the intestinal mucosa) (see Table 1.1). The concentration of α-amylase in the pancreatic juice of the horse is only 5–6% of that in the pig, whereas the concentration of α-glucosidase is comparable with that in many other domestic mammals. The α-glucosidases include sucrase, the disaccharidase present in concentrations five times that of glucoamylase and capable of digesting sucrose. Another important disaccharidase in the intestinal juice is the β-glucosidase, neutral β-galactosidase (neutral or brush-border lactase), which is necessary for the digestion of milk sugar in the foal. This enzyme has a pH optimum around 6.0 and its activity decreases as the horse matures, being absent from the brush border of horses more than 4 years old. In horses of this age, induction of neutral lactase by lactose feeding also seems impossible so that the sugar is fermented; large quantities of

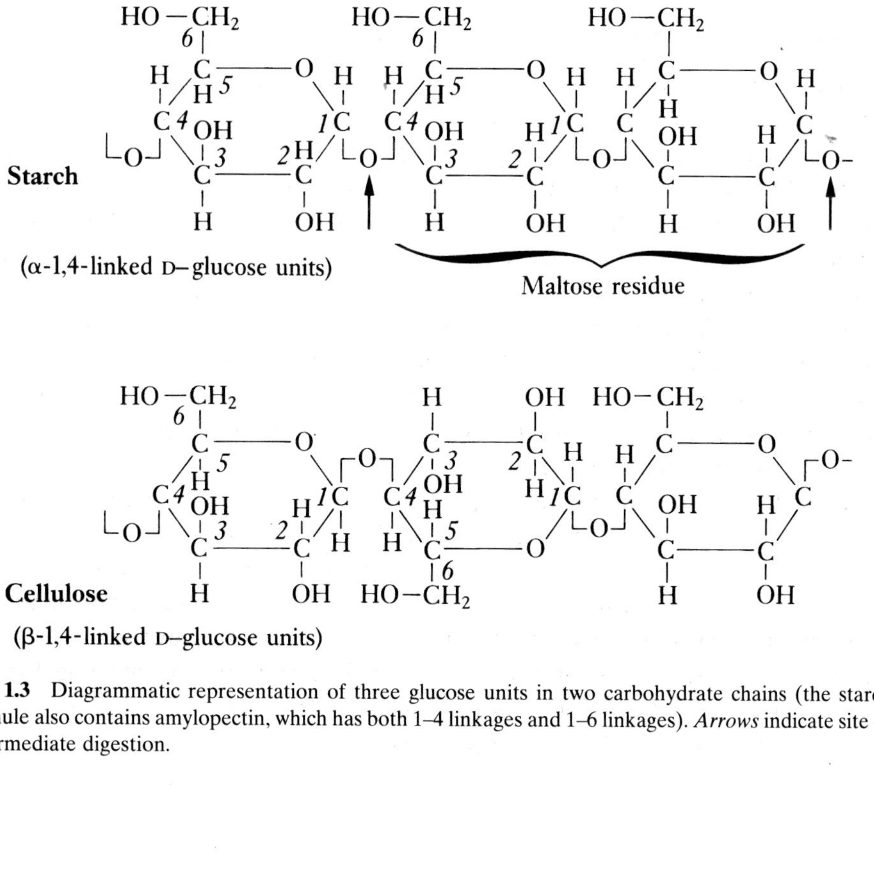

Fig. 1.3 Diagrammatic representation of three glucose units in two carbohydrate chains (the starch granule also contains amylopectin, which has both 1–4 linkages and 1–6 linkages). *Arrows* indicate site of intermediate digestion.

Table 1.1 Carbohydrate digestion in the small intestine.

Substrate	Enzyme	Product
Starch	α-Amylase	Limiting dextrins (about 34 glucose units)
Limit dextrins	α-Glucosidases (glucoamylase, maltase and isomaltase)	Glucose
Sucrose	Sucrase	Fructose and glucose
Lactose	Neutral-β-galactosidase	Glucose and galactose

lactose may thus cause digestive upsets. If a suckling foal, or one given cow's milk, lacks an active form of this enzyme, it suffers from diarrhoea.

Proteins

The amount of protein hydrolysed in the small intestine is about three times that in the stomach. Proteins are in the form of long folded chains, the links of which are represented by amino acid residues. For proteins to be digested and utilized by the horse, these amino acids must usually be freed, although the gut mucosal cells can absorb dipeptides. The enzymes responsible are amino-peptidases and carboxy-peptidases secreted by the wall of the small intestine.

Fats

The horse differs from the ruminant in that the composition of its body fat is influenced by the composition of dietary fat. This suggests that fats are digested and absorbed from the small intestine before they can be altered by the bacteria of the large intestine. The small intestine is the primary site for the absorption of dietary fat and long-chain fatty acids. Bile continuously draining from the liver facilitates this by promoting emulsification of fat, chiefly through the agency of bile salts. The emulsification increases the fat–water interface so that the enzyme lipase may more readily hydrolyse neutral fats to fatty acids and glycerol. These are readily absorbed, although it is possible that a considerable proportion of dietary fat, as finely emulsified particles of neutral fat (triacylglycerols, TAG), is absorbed into the lymphatic system and transported as a lipoprotein in chylomicrons. Many research workers have demonstrated that horses digest fat quite efficiently and that the addition of edible fat to their diet has merit, particularly in so far as endurance work, and also more intensive exercise, are concerned (see Chapters 5 and 9).

Feed modification to improve digestion

Considerable interest has developed over the past decade concerning the inclusion of commercially cooked cereals in feeds for horses. The processes used include infrared micronization of cereals and expansion or extrusion of products. The extent of cooking by the extrusion process varies considerably amongst the cookers used and the conditions of processing. Nevertheless, small intestinal digestibility is influenced by this cooking, even in adult horses; yet total digestibility is not improved (Table 1.2). That is, the digestibility of raw cereals and cooked cereals is similar when the values are derived from the difference between carbohydrate consumed and that lost in the faeces. Thus, the extent of precaecal digestion, or possibly preileal digestion, influences the proportion of cereal carbohydrate absorbed as glucose and that absorbed as VFA and lactic acid. Evidence from various sources indicates that somewhat more than 50% of the dietary starch is subject to preileal or precaecal digestion. The proportion so digested is influenced not only by cereal processing , but also by the amount fed. Lactate and other organic acid production is increased, and the pH is decreased in the ileum and caecum when undigested starch reaches those regions. In order to avoid starch 'overload', and therefore excessive starch fermentation in the large intestine, Potter *et al.* (1992a) concluded that starch intake in horses, given two to three meals daily, should be limited to 4 g/ kg BW per meal (also see Chapter 11). The Texas group (Gibbs *et al.* 1996) have also found that when N intake is less than 125 mg/kg BW, 75–80% of the truly digestible protein of soyabean meal is digested precaecally, and 20% is digested in the large intestine, while 10% is indigestible.

The preileal digestibility of oat starch exceeds that of maize starch or of barley starch (Meyer *et al.* 1995). When starch intake per meal is only 2 g/kg BW the preileal starch digestibility of ground oats may be over 95%, whereas at the other extreme that of whole, or broken, maize may be less than 30%. The grinding of cereals increases preileal digestibility compared with whole, rolled or cracked grain (note: the keeping quality, or shelf-life, of ground grain is, however, relatively

Table 1.2 Precaecal digestion of various sources of starch and digestion in the total GI tract of horses (Kienzle *et al.* 1992) and ponies (Potter *et al.* 1992a) (digested, g/kg intake).

	Starch intake, g/100 kg BW	Maize precaecal	Oats precaecal	Oats total	Sorghum precaecal	Sorghum total	Reference
Whole	200	289	835	—	—	—	Kienzle *et al.* 1992[1]
Rolled	200	299	852	—			Kienzle *et al.* 1992
Ground	200	706	980				Kienzle *et al.* 1992
Crimped[2]	264[CO] 295[CS]	—	480	944	360	940	Potter *et al.* 1992a
Micronized	237[MO] 283[MS]	—	623	938	590	945	Potter *et al.* 1992a

[1] Maize and oat digestibilities measured by these workers refer to preileal measurements.
[2] Dry rolled with corrugated rollers to crack the kernels. [CO], crimped oats; [CS], crimped sorghum; [MO], micronized oats; [MS], micronized sorghum.

short). Workers in Hanover found, in contrast to the results of the Texas workers, that in the jejunal chyme there is a much greater increase in the postprandial concentration of organic acids, including lactate, and in acidity, when oats rather than maize are fed. Whether this may be related to the putative heating effect of oats, compared with other cereals, is not established. The starch gelatinization of cooking enhances small intestinal digestion at moderate, or high, rates of intake.

Nitrogen utilization

At high rates of protein intake more nonprotein N (NPN) enters the GI tract in the form of urea. The N entering the caecum from the ileum is proportionally 25–40% NPN, varying with the feed type. Meyer (1983b) calculated that in a 500-kg horse between 6 and 12 g urea N pass daily through the ileo–caecal valve. The amount of N passing into the large intestine also varies with protein digestibility. At high intake rates of protein of low digestibility more N in total will flow into the large intestine, where it will be degraded to NH_3. From Meyer's evidence, about 10–20% of this total is urea N, as the daily range of total N flowing into the caecum is:

$$0.3–0.9 \, g \, N/kg \, BW^{0.75}$$

N also enters the large intestine by secretion there, although the amount seems to be less than that entering through the ileocaecal valve and net absorption nearly always takes place. Nevertheless, net secretion can occur with low protein, high fibre diets.

Utilization of the derived NH_3 by gut bacteria is between 80% and 100%. Excessive protein intake must increase the burden of unusable N, either in the form of inorganic N, or as relatively unusable bacterial protein. This burden is influenced by feeding sequence. The provision of a concentrate feed 2 h later than roughage, compared with simultaneous feeding, caused higher levels of plasma free and particularly of essential amino acids 6–9 h later (Cabrera *et al.* 1992; Frape 1994). Plasma urea did not rise with the dissociated, or separate, feeding, but rose continuously for 9 h after the simultaneous feeding of the roughage and concentrate. This indicates that mixed feeding led to a large flow of digesta N to the caecum with much poorer dietary protein economy; yet the separate feeding was in the reverse order to the standard practice of giving concentrates before roughage.

THE LARGE INTESTINE

Grazing herbivores have a wide variety of mechanisms and anatomical arrangements for making use of the chemical energy locked up in the structural carbohydrates of plants. A characteristic of all grazing and browsing animals is the enlargement of some part of the GI tract to accommodate fermentation of digesta by microorganisms. More than half the dry weight of faeces is bacteria and the bacterial cells in the digestive tract of the horse number more than ten times all the tissue cells in the body. No domestic mammal secretes enzymes capable of breaking down the complex molecules of cellulose, hemicellulose, pectin and lignin into their component parts suitable for absorption, but, with the exception of lignin, intestinal bacteria achieve this. The process is relatively slow in comparison to the digestion of starch and protein. This means that the flow of digesta has to be arrested for sufficient time to enable the process to reach a satisfactory conclusion from the point of view of the energy economy of the host animal.

During the weaning and postweaning of the foal and yearling, the large intestine grows faster than the remainder of the alimentary canal to accommodate a more fibrous and bulky diet. At the distal end of the ileum there is a large blind sack known as the caecum, which is about 1 m long in the adult horse and which has a capacity of 25–35 l. At one end there are two muscular valves in relatively close proximity to each other, one through which digesta enter from the ileum and the other through which passage from the caecum to the right ventral colon is facilitated. The right and left segments of the ventral colon and the left and right segments of the dorsal colon constitute the great colon, which is some 3–4 m long in the adult horse, having a capacity of more than double that of the caecum. The four parts of the great colon are connected by bends known as flexures. In sequence, these are the sternal, the pelvic and the diaphragmatic flexures. Their significance probably lies in changes in function and microbial population from region to region and in acting as foci of intestinal impactions. Digestion in the caecum and ventral colon depends almost entirely on the activity of their constituent bacteria and ciliate protozoa. In contrast to the small intestine, the walls of the large intestine contain only mucus-secreting glands, that is, they provide no digestive enzymes. However, high levels of alkaline phosphatase activity, known to be associated with high digestive and absorptive action, are found in the large intestine of the horse, unlike the large intestinal environment of the cat, dog and man.

The diameter of the great colon varies considerably from region to region but reaches a maximum in the right dorsal colon where it forms a large sacculation with a diameter of up to 500 mm. This structure is succeeded by a funnel-shaped part below the left kidney where the bore narrows to 70–100 mm as the digesta enter the small colon. The latter continues dorsally in the abdominal cavity for 3 m before the rectum, which is some 300 mm long, terminates in the anus (Fig. 1.2).

Contractions of the small and large intestine

The walls of the small and large intestine contain longitudinal and circular muscle fibres essential:

- for the contractions necessary in moving the digesta by the process of peristalsis in the ultimate direction of the anus;
- for allowing thorough admixture with digestive juices; and
- for bathing the absorptive surfaces of the wall with the products of digestion.

During abdominal pain these movements may stop so that the gases of fermentation accumulate.

Passage of digesta through the large intestine

Many digestive upsets are focused in the large intestine and therefore its function deserves discussion. The extent of intestinal contractions increases during feeding – large contractions of the caecum expel digesta into the ventral colon, but separate contractions expel gas, which is hurried through much of the colon. The reflux of digesta back into the caecum is largely prevented by the sigmoid configuration of the junction. Passage of digesta through the large intestine depends on gut motility, but is mainly a function of movement from one of the compartments to the next through a separating barrier. Considerable mixing occurs within each compartment, but there seems to be little retrograde flow between them. The barriers are:

- the ileo–caecal valve already referred to;
- the caecal–ventral colonic valve;
- the ventral–dorsal colonic flexure (pelvic flexure), which separates the ventral from the dorsal colon; and
- the dorsal small colonic junction at which the digesta enter the small colon.

Resistance to flow tends to increase in the same order, that is, the last of these barriers provides the greatest resistance (also see Chapter 11). This resistance is much greater for large food particles than for small particles. In fact delay in passage for particles of 2 cm length may be for more than a week. Normally the time taken for waste material to be voided after a meal is such that in ponies receiving a grain diet, 10% is voided after 24 h, 50% after 36 h and 95% after 65 h. Most digesta reach the caecum and ventral colon within 3 h of a meal, so that it is in the large intestine where unabsorbed material spends the greater proportion of time. The rate of passage in ruminants is somewhat slower, and this partly explains their greater efficiency in digesting fibre.

In the horse, passage time is influenced by physical form of the diet; for example, pelleted diets have a faster rate of passage than chopped or long hay, and fresh grass moves more rapidly than hay. Work at Edinburgh (Cuddeford *et al.* 1992) showed that fibre was digested more completely by the donkey than by the pony, which in turn digested it more effectively than the TB. These differences probably are owed

Table 1.3 Effect of diet on the pH, production of VFAs and lactate and on microbial growth in the caecum and ventral colon 7 h after the meal.

Diet	pH	FA (mmol/l)				Total bacteria per (ml × 10⁻⁷)
		Acetate	Propionate	Butyrate	Lactate	
Hay	6.90	43	10	3	1	500
Concentrate plus minimal hay	6.25	54	15	5	21	800
Fasted	7.15	10	1	0.5	0.1	5

Note: Values given are typical, but all except the pH show large variations.

in large measure to the relative sizes of the hindgut and, therefore, to the holding time of digesta. Although dietary fibre is not degraded as readily by the horse as by ruminants (Table 1.4), the horse utilizes the energy of soluble carbohydrates more efficiently by absorbing a greater proportion as sugars in the small intestine.

Pattern of large intestinal contractions

The caecum contracts in a ring some 12–15 cm from the caeco–colic junction, trapping ingesta in the caecal base and forcing some through the junction that in the meantime has relaxed. With a relaxation of the caecal muscles some reflux occurs, although there is a net movement of digesta into the ventral colon. Contractions of the colon are complex. There are bursts of contractile activity that propagate in an aboral direction, but some contractions propagate orally and some are isolated and do not propagate in either direction. Thus there is nonrhythmic haustral kneading and stronger rhythmic propulsive and retropulsive contractions. These contractions have the function of mixing the constituents, and promoting fermentation and absorption, as well as that of moving residues towards the rectum. The strong rhythmic contractions for the great colon begin at the pelvic flexure, where a variable site 'electrical pacemaker' exists. A major site of impactions is the left ventral colon, just orad (toward the mouth) to this pelvic flexure (Chapter 11). More detailed knowledge of this activity should ultimately help in the control of common causes of large gut malfunction and colic.

Microbial digestion (fermentation)

There are three main distinctions between microbial fermentation of feed and digestion brought about by the horse's own secretions:

(1)   The β-1,4-linked polymers of cellulose (Fig. 1.3) are degraded by the intestinal microflora but not by the horse's own secretions. The cell walls of plants contain several carbohydrates (including hemicellulose) that form up to half the fibre of the cell walls of grasses and a quarter of those of clover. These carbohydrates are also digested by microorganisms, but the extent depends on the structure and degree of encrustation with lignin, which is indigestible to both gut bacteria and horse secretions.
(2)   During their growth the microorganisms synthesize dietary indispensable (essential) amino acids.
(3)   They are net producers of water-soluble vitamins of the B group and of vitamin K_2.

Microbial numbers

In the relatively small fundic region of the stomach, where the pH is about 5.4, there are normally from 10^8 to 10^9 bacteria/g. The species present are those that can withstand moderate acidity, common types being lactobacilli, streptococci and *Veillonella gazogenes*. The jejunum and ileum support a flourishing population in which obligate anaerobic Gram-positive bacteria may predominate (10^8–10^9/g). In this region of the small intestine a cereal diet can influence the proportion of the population producing lactic acid, compared with that producing VFA as an end product, although the numbers of lactobacilli per gram of contents tend to be higher in the large intestine, where the pH is generally lower.

The flora of the caecum and colon are mainly bacteria which in fed animals number about 0.5×10^9 to 5×10^9/g contents. A characteristic difference between equine hindgut fermentation and that in the rumen is the lower starch content of the hindgut, which implies a generally lower rate of fermentation. Cellulolysis and proteolysis in the equine caecum have been studied, but less is known about the activity of hemicellulolytic and pectinolytic bacteria. In one pony study (Moore & Dehority 1993), the cellulolytic bacteria numbered 2–4% of the total. In addition, there were 2×10^2 to 25×10^2 fungal units/g, most of which were cellulolytic (also see probiotics, Chapter 5). In the horse, both caecal bacteria (which with fungi constitute the flora) and protozoa (fauna) participate in the decomposition of pectins and hemicellulose at an optimum pH of 5–6 (Bonhomme-Florentin 1988).

Fauna

Protozoa in the equine large intestine number about 10^{-4} of the bacterial population, that is, 0.5×10^5 to 1.5×10^5/ml contents. Although protozoa are individually very much larger than bacteria and they thus contribute a similar total mass to the large intestinal contents, their contribution to metabolism is less, as this is roughly proportional to the surface area. The species of the fauna differ somewhat from those in the rumen. Some 72 species of protozoa, primarily ciliates, have been described as normal inhabitants of the equine large intestine, with some tendency to species differences between compartments. Moore and Dehority (1993) found in ponies that the protozoa were from the following genera: *Buetschlia, Cycloposthium, Blepharocorys* and a few *Paraisotricha*. Removal of the protozoa (defaunation) caused only a slight decrease in DM digestibility, with no effect on numbers of bacteria, or on cellulose digestibility.

Flora

In the large intestine the bacterial populations are highest in the caecum and ventral colon. Here, the concentration of cellulose-digesting bacteria is six to seven times higher than in the terminal colon. About 20% of the bacteria in the large intestine can degrade protein.

Numbers of specific microorganisms may change by more than 100-fold during 24 h in domesticated horses being given, say, two discrete meals per day. These fluctuations reflect changes in the availability of nutrients (in particular, starch and protein) and consequentially changes in the pH of the medium. Thus, a change in the dietary ratio of cereal to hay not only will have large effects on the numbers of microorganisms but also will considerably influence the species distribution in the hindgut. Although frequency of feeding may have little impact on digestibility per se, it can have a large influence on the incidence of digestive disorders and metabolic upsets, which of course affect the wellbeing of the horse. These consequences result directly from the effect of diet and digesta upon the microbial populations (bacteria and protozoa).

Caecal bacteria from horses adapted to a grain diet are less efficient at digesting hay than are the microbes from hay-adapted horses. An analogous situation exists for hay-adapted caecal microbes when subjected to grain substrate. If such a dietary change is made abruptly in the horse, impactions may occur in the first of these situations and colic, laminitis or puffy swollen legs can result in the second (see Chapter 11). The caecal microorganisms in a pony or horse tend to be less efficient at digesting hay than are the ruminal microbes in cows. The digestibilities of organic matter and crude fibre in horses given a diet containing more than 15% crude fibre (a normal diet of concentrates and hay) are about 85 and 70–75%, respectively, of the ruminant values. This has been attributed to the combined effects of a more rapid rate of passage of residues in horses and differences in cellulolytic microbial species.

Recent work by Hyslop *et al*. (1997) has shown that under the conditions of their experiment the degradation of the acid detergent fibre and crude protein of sugar beet pulp, hay cubes, soya hulls and a 2:1 mixture of oat hulls:naked oats was no poorer in the pony caecum than in the rumen of the steer over incubation periods of 12–48 h. In fact, during incubation for 12 h the degradation of the beet pulp and the hay was marginally greater in the caecum. Thus, the equine hindgut microflora may not be inherently less efficient than are rumen microflora at feed degradation.

Products of fermentation

The microbial fermentation of dietary fibre, starch and protein yields large quantities of short-chain VFAs as byproducts, principally acetic, propionic and butyric acids (Table 1.3, Fig. 1.4). This fermentation and VFA absorption are promoted by:

- the buffering effect of bicarbonate and Na^+ derived from the ileum;
- an anaerobic environment; and
- normal motility to ensure adequate fermentation time and mixing.

Table 1.4 Proportion of VFAs in digesta to bodyweight (BW) in four herbivores (Elsden *et al*. 1946).

	g VFA/kg BW
Ox	1.5
Sheep	1.5
Horse	1.0
Rabbit	0.5

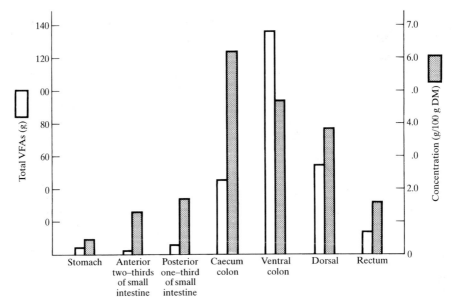

Fig. 1.4 VFAs (□) calculated as the total weight (g) of acid (as acetic acid) in the organ or as the concentration (g/100 g DM) (■) in the lumen (after Elsden *et al.* 1946).

Acetate and butyrate are major products of fibre digestion, whereas the proportion of propionate (and lactate, see Chapter 11) increases with increasing proportions of starch left undigested in the small intestine. In the pony, limited evidence indicates that 7% of total glucose production is derived from propionate produced in the caecum.

VFA, fluid and electrolyte absorption in the large intestine

The VFA produced during fermentation would soon pollute the medium, rapidly producing an environment unsuitable for continued microbial growth; however, an equable medium is maintained by the absorption of these acids into the bloodstream. In addition to this absorption there is the vital absorption of large amounts of water and electrolytes (sodium, potassium, chloride and phosphate).

Fluid absorption
The largest proportion of water that moves through the ileo–caecal junction is absorbed from the lumen of the caecum and the next largest is absorbed from the ventral colon. Fluid is also absorbed from the contents of the small colon, to the benefit of the water economy of the horse and with the formation of faecal balls. This aboral decline in water absorption is accompanied by a parallel decrease in sodium absorption. In the pony, 96% of the sodium and chloride and 75% of the soluble potassium and phosphate entering the large bowel from the ileum are absorbed into the bloodstream. Although phosphate is efficiently absorbed from both the small and large gut, calcium and magnesium are not, these being absorbed

mainly from the small intestine (Fig. 1.5). This phenomenon has been proffered as a reason why excess dietary calcium does not depress phosphate absorption, but excess phosphate can depress calcium absorption in the horse.

The water content of the small intestinal digesta amounts to some 87–93%, but the faeces of healthy horses contain only 58–62% water. The type of diet has a smaller effect on this than might be imagined. For instance, oats produce fairly dry faeces, but bran produces moist faeces, although in fact they contain only some 2 or 3 percentage units more of moisture.

VFA and lactic acid production and absorption
Microbial degradation seems to occur at a far faster rate in the caecum and ventral colon than in the dorsal colon (Fig. 1.4) and the rate is also faster when starches are degraded rather than structural carbohydrates. A change in the ratio of starch to fibre in the diet leads to a change in the proportions of the various acids yielded (Table 1.3); and these proportions also differ in the organs of the large intestine. Thus, proportionately more propionate is produced as a consequence of the consumption of a starch diet and the caecum and ventral colon yield more propionate than the dorsal colon does. Many bacteria have the capacity to degrade dietary protein, so yielding another blend of VFA.

An optimum pH of 6.5 exists for microbial activity that also promotes VFA absorption. VFAs are absorbed in the unionized form. As the pH moves closer to the pK of a particular VFA, more is absorbed. The H^+ ions required for this are probably derived from mucosal cells in exchange for Na^+. HCO_3^- buffer is secreted into the lumen in exchange for Cl^-. Thus, absorption of VFA is accompanied by a net absorption of NaCl. This in turn is a major determinant of water absorption. The ingestion of a large meal can cause a 15% reduction in plasma volume, ultimately resulting in renin-angiotensin, and then aldosterone, release. The increase in plasma aldosterone level causes an increased Na absorption, and with it water (see also

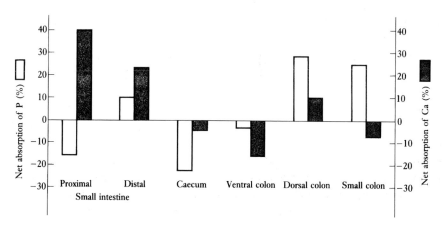

Fig. 1.5 Net fractional absorption of P (□) and Ca (■) from various regions of the small and large intestine (after Schryver *et al.* 1974a).

Chapter 9). However, whether a large meal, compared with continuous feeding, would increase the risk of impactions is unclear.

Whereas most ruminal butyrate is metabolized in the mucosa before entering the bloodstream, in horses all VFAs pass readily to the blood. Lactic acid produced in the stomach is apparently not well absorbed from the small intestine. On reaching the large intestine some is absorbed, along with that produced locally, but much is metabolized by bacteria to propionate.

Microbial activity inevitably produces gases – principally carbon dioxide, methane and small amounts of hydrogen – which are absorbed, ejected from the anus, or they participate in further metabolism. The gases can, however, be a severe burden, with critical consequences when production rate exceeds that of disposal.

Protein degradation in the large intestine and amino acid absorption

Microbial growth, and therefore the breakdown of dietary fibre, also depends on a readily available source of nitrogen. This is supplied as dietary proteins and as urea secreted into the lumen from the blood. Despite the proteolytic activity of microorganisms in the hindgut, protein breakdown per litre is about 40-fold greater in the ileum than in the caecum, or colon, through the activity of the horse's own digestive secretions in the small intestine.

The death and breakdown of microorganisms within the large intestine release proteins and amino acids. The extent to which nitrogen is absorbed from the large intestine in the form of amino acids and peptides, useful to the host, is still a hotly debated issue. Isotope studies indicate that microbial amino acid synthesis within the hindgut does not play a significant role in the host's amino acid status. Quantitative estimates obviously depend on the diet used and animal requirements, but a range of 1–12% of plasma amino acids may be of hindgut microbial origin. Absorption studies have shown that, whereas ammonia is readily absorbed by the proximal colon, significant basic amino acid absorption does not occur. S-amino acid absorption may occur to a small extent.

Horses differ from ruminants in absorbing a higher proportion of dietary nitrogen in the form of the amino acids present in dietary proteins, proportionately less being converted to microbial protein. As only a small proportion of the amino acids present in microbial protein is made available for direct utilization by the horse, young growing horses, in particular, respond to supplementation of poor-quality dietary protein with lysine, the principal limiting indispensable amino acid (Fig. 1.6).

Urea production

Urea is a principal end product of protein catabolism in mammals and much of it is excreted through the kidneys. It is a highly soluble, relatively innocuous compound and a reasonably high proportion of the urea produced in the liver is secreted into the ileum and conveyed to the large intestine (Table 1.5, showing total N, of which

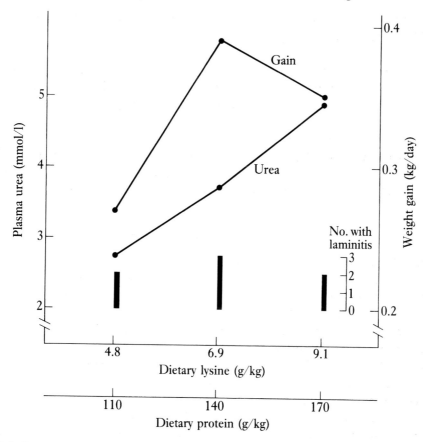

Fig. 1.6 Response of 41 6-month-old ponies over a 3-month period of diets containing different amounts of protein and lysine (initial weight 127 kg). In this experiment note that increased protein intake led to elevated protein catabolism and urea production without an increase in incidence of laminitis (Yoakam *et al.* 1978).

12–24 mg/kg BW is urea N) where most may be degraded to ammonia by bacteria. The possession by microorganisms of the enzyme urease, which does not occur in mammalian cells, makes this reaction possible. Most of the ammonia produced is re-utilized by the intestinal bacteria in protein synthesis. Some, however, diffuses into the blood, where levels are normally maintained very low by a healthy liver. If ammonia production greatly exceeds the capacity of the bacteria, and of the liver, to utilize it, ammonia toxicity can arise. The fate of any urea added to the diet is similar.

In summary, many studies have led to the conclusion that digestion and fermentation in, and absorption from, the large intestine, account, in net terms, for 30% of dietary protein, 15–30% of dietary soluble carbohydrate and 75–85% of dietary structural carbohydrate. The salient causes of variation in values for each of the principal components of the horse's diet are:

Table 1.5 Effect of diet on the flow of nitrogen from the ileum to the caecum (Schmidt *et al.* 1982).

Diet	Nitrogen flow daily (mg N/kg BW)
Concentrate, 3.75 kg daily (1%)*	62
Concentrate, 7.5 kg daily (2%)*	113
Hay	68
Straw	37

*Weight of concentrate given as percentage of BW.

- the degree of adaptation of the animal;
- the processing to which the feed is subjected; and
- the differences in digestibility among alternative feedstuffs.

Commercial enzyme and microorganism products

(See also Probiotics, Chapter 5.) A Directive from the EU (Council Directives 93/113/EC 93/114/EC) deals with the roles of supplementary enzymes and microbial cultures. There is a list of all permitted and properly identified products for use within the EU. The list includes products for use in feed, in drinking water and those given as a drench. The requirement for a complete taxonomic description of micro-organisms implies that bacterial cultures should be pure and that there is an identification of both the species and the culture collection strain type. The recommendations of the Nomenclature Committee of the International Union of Biochemistry and Molecular Biology are applied to enzyme nomenclature. The UK Medicines Act (1986) will apply if medicinal claims, such as growth promotion, are made; however, evidence of safety of all products will have been submitted.

Efficacy of products such as these may depend on conditions and length of storage, the loss of colony forming units of microbial cultures during storage, their ability to survive the low pH of the stomach, and certainly the retention of the activity of their enzymes. The efficacy of enzymes also must assume that their activity, as defined, applies to the region of the GI tract where they are expected to function.

STUDY QUESTIONS

(1) What are the advantages and disadvantages of a digestive system with a major microbial fermentation site in the hindgut only, compared to the fermentation system of the ruminant?

(2) The stomach of the horse is relatively smaller than that of the rat or human. What consequences do you draw from this?

FURTHER READING

Sissons, S. and Grossman, J.D. (1961) *The Anatomy of the Domestic Animals*. W.B. Saunders, Philadelphia and London.
Vernet, J., Vermorel, M. and Martin-Rosset, W. (1995) Energy cost of eating long hay, straw and pelleted food in sport horses. *Animal Science*, **61**, 581–8.

Chapter 2
Utilization of the Products of Dietary Energy and Protein

A horse whose work consists of travelling a stage of twenty miles three times a week, or twelve every day, should have one peck of good oats, and never more than eight pounds of good hay in twenty-four hours. The hay, as well as the corn, should, if possible, be divided into four portions.

J. White 1823

CARBOHYDRATE, FAT AND PROTEIN AS SOURCES OF ENERGY AND THE HORMONAL REGULATION OF ENERGY

Glucose, VFA and TAG clearance

Horse diets rarely contain more than 4% fat and 7–12% protein so that these represent relatively minor sources of energy in comparison to carbohydrate, which may constitute by weight two-thirds of the diet. Furthermore, protein is required primarily in the building and replacement of tissues and is an expensive source of energy. However, both dietary protein and fat can also contribute to those substrates used by the horse to meet its energy demands for work. Protein does so by the conversion of the carbon chain of amino acids to intermediary acids and of some of the carbon chains to glucose. Neutral fat does so following its hydrolysis to glycerol and fatty acids. The glycerol can be converted to glucose and the fatty acid chain can be broken down by a stepwise process called β oxidation in the mitochondria, yielding adenosine triphosphate (ATP) and acetate, or more strictly acetyl coenzyme A (acetyl-CoA), and requiring tissue oxygen (see Fig. 9.2).

Carbohydrate digestion and fermentation yield predominantly glucose and acetic, propionic and butyric VFAs. These nutrients are collected by the portal venous system draining the intestine and a proportion of them is removed from the blood as they pass through the liver. Both glucose and propionate contribute to liver starch (glycogen) reserves, and acetate and butyrate bolster the fat pool (Fig. 2.1) and also constitute primary energy sources for many tissues.

Sequence of feeding and amount fed

Studies in both France and Germany have shown that the sequence in which feeds are given can influence the metabolic outcome. When the concentrate was given to ponies 2h following roughage plasma urea concentration was significantly lower,

26

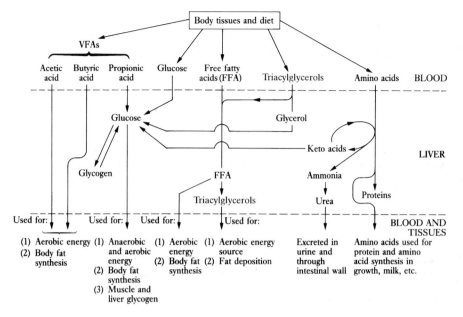

Fig. 2.1   Sources, metabolism and fate of major energy-yielding nutrients derived from body tissues and diet (after McDonald *et al.* 1981).

and plasma free amino acids higher, in the postprandial period, than when the concentrates and roughage were given simultaneously (Cabrera *et al.* 1992). This indicates that the conventional procedure of feeding the concentrate with, or before, the hay is likely to depress the potential net protein value of the diet. Concentrates appear to be better utilized if given after roughage consumption, probably as a result of retarded small intestinal passage of the concentrates so given. Larger feed allowances increase the rate of passage of ingesta through the GI tract, as occurs for horses commencing periods of greater work. Increased rates of passage slightly reduce roughage digestibility and account for somewhat poorer fibre utilization in ponies than in donkeys (Pearson & Merritt 1991).

Timing of feeds and appetite

(See 'Appetite', this chapter, for more detail.)

Feeding in the horse or pony causes mesenteric hyperaemia, that is, a diversion of blood to, and engorgement of, the blood vessels investing the GI tract. In a similar manner the imposition of exercise causes increased blood flow to muscles. Exercise within a few hours of feeding therefore increases the demands on the heart to supply blood for both activities. Even moderately strenuous exercise (75% of heart rate maximum) under these conditions leads to increased heart rate, cardiac output, stroke volume and arterial pressure, in comparison to the effects of exercise on fasted animals (Duren *et al.* 1992). The optimum timing of meals in relation to exercise is discussed in Chapter 9.

Blood glucose

Healthy horses and ponies maintain a blood plasma glucose concentration within certain defined limits. This is necessary as glucose represents the preferred source of energy for most tissues. In ponies, normal healthy resting levels may range between 2.8 and 3.3 mmol/l, but horse breeds may generally have higher resting levels, with TBs in the region of 4.4–4.7 mmol/l. The concentration in horses rises dramatically from the commencement of a meal to 6.5 mmol/l, or more, by 2 h (Fig. 2.2). A return to fasting concentrations is much slower than in the human and slower still in ponies. Nevertheless, the scale of plasma glucose response to a meal is greatly influenced by the intensity of any previous exercise, intense exercise greatly diminishing the response (Frape 1989). The plasma glucose response is measured generally as the area under the response curve to a known dose of carbohydrate. The more rapidly the plasma glucose is cleared, that is, the greater the tolerance, the smaller is this area (Figs 2.2 and 2.3). This clearance from the blood results from uptake particularly by liver and muscle cells, where it is converted to glycogen and also to fat, although the net conversion to fat in a fit athletic animal is small. Circulating glucose is used directly to meet immediate energy demands for muscular activity, nervous tissue activity, etc. The process of storage is stimulated by the hormone insulin,

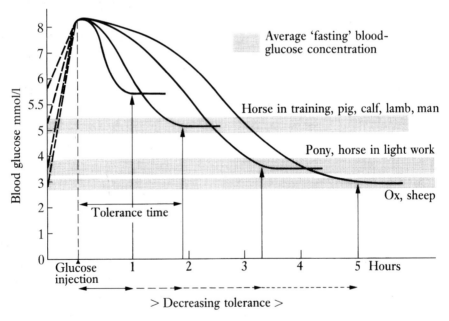

Fig. 2.2 Approximate glucose tolerance times (arrows) and normal 'fasting' blood-glucose concentrations (shaded). (Glucose is injected i.v. to allow comparisons between species with different digestive anatomy and mechanisms. By providing the glucose in the form of a starch meal, the peak is delayed 2–4 h in the horse. When oats are given as feed the maximum blood glucose concentration in TB horses occurs at about 2 h following the start of feeding.) The determination of 'tolerance time' has been largely superseded by the determination of the area under the response curve to a glucose dose, as this is determined with greater precision. The greater the area is, generally the longer is the tolerance time in an individual horse.

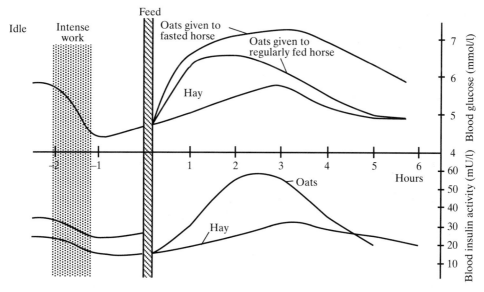

Fig. 2.3 Responses of blood glucose and insulin to feed.

which responds to a rise in blood glucose, and the insulin then also promotes fat clearance through activation of adipose tissue lipoprotein lipase.

Concomitant with the rise in plasma glucose concentration, the plasma nonesterified fatty acids (NEFA) level decreases. This indicates a reduction in the mobilization of fat stores, which results from insulin repressing the activity of intracellular, hormone-sensitive lipase (also see Chapter 9). Generally, plasma glucose and insulin concentrations are lower and plasma NEFA and urea concentrations higher in horses given very restricted rations. Stickler *et al.* (1996) found that these plasma changes occurred promptly in light horses, whereas plasma glucagon responds with a slightly slower rate of decrease and thyroxine with a much slower increase in concentration upon the imposition of the restriction.

Insulin response

Insulin is an anabolic hormone, the function of which is therefore primarily to switch on enzymes directed towards storage of blood glucose and fat. Blood insulin peaks shortly after that of blood glucose and concentrations may reach four- to eightfold fasting values 1–2h following a feed. Blood insulin may remain above fasting concentrations throughout the day, again unlike the human response to a single meal (Frape 1989). Horses and ponies have a lower glucose tolerance than do man and the pig, but a slightly greater one than do ruminants. TB and other hotblooded horses generally have a higher tolerance than do ponies; in other words, ponies tend to secrete less insulin and their tissues may be less sensitive to insulin, although there is considerable adaptation to diet. Ponies can therefore withstand fasting better than TBs, and TBs become more excitable after feeding.

Insulin prevents excess blood glucose from spilling out in the urine by increasing the uptake of tissues and so lowering blood levels. However, in order to avoid hypoglycaemia its effects are counterbalanced by that of other hormones (for example, glucagon, the glucocorticoids, and the catecholamines, epinephrine and norepinephrine). The system is thereby maintained in a state of dynamic equilibrium. The anabolic nature of insulin conflicts with the catabolic requirements of exercise, and although catecholamines, secreted during intense exercise, suppress further insulin secretion, elevated circulating postprandial insulin is undesirable if exercise is initiated. Both from this viewpoint and from that of blood redistribution exercise is to be discouraged during the postprandial interval.

Cases of hypoglycaemic seizures of horses have been reported, during which plasma glucose may fall to less than 2 mmol/l. The cause has been adenoma of pancreatic islet cell origin, with hyperplasia predominately of β-cells, causing hyperinsulinaemia.

Insulin resistance and hyperinsulinaemia

Insulin dependent diabetes is very rare in horses; however, the noninsulin dependent form, expressed as insulin resistance, does occur. Whether any of these occurrences involve a deficiency of dietary trivalent chromium in horses is unknown (see Chapter 3). Resistance may sometimes be incorrectly inferred. Horses given a diet high in starch content produced a similar glucose response to that of horses given a diet containing less starch, but the insulin response was higher in the first group (Ralston 1992). This may not indicate that there is a greater risk for horses given high cereal diets of developing insulin resistance. Insulin resistance is associated with an inadequate response of tissue receptors to the hormone; that is, a higher than normal local concentration of the hormone is required to elicit a normal tissue response. Plasma glucose and insulin responses to a meal, or to an oral glucose dose, are above normal. Ponies tend to be more intolerant of glucose than are horses, and fasted animals are more intolerant than fed. However, the response of ponies, and of Shetland ponies in particular, is a *reduced* insulin secretion in response to a carbohydrate load. This does not imply insulin resistance and it would explain a lesser decrease in plasma NEFA concentration in ponies following a glucose load. (The oral glucose tolerance test may suffer from influence by factors other than insulin, for instance, impaired gut function, and the intravenous loading route may be preferable; see Chapter 12.)

Plasma TAG concentration in ponies subjected to an extended fast is very much higher than that in horses; although this may not indicate insulin resistance. It may result from the raised plasma NEFA that is converted by the liver to TAGs, where it is mobilized and transported as very-low-density lipoproteins (glucose- and fat [TAG]-sensitive insulin tissue receptors exist in adipocytes. Insulin resistance retards fat deposition, and raised plasma TAG promotes insulin resistance, so it may not be coincidental that hyperlipaemia occurs more frequently in ponies than in horses).

Dietary causes of insulin resistance remain unresolved. The principal dietary and metabolic involvement in the causation of resistance is likely to be excessive feed energy intake, obesity, aging and inadequate exercise. Ponies previously suffering laminitis are also much more glucose intolerant (Jeffcott *et al.* 1986) than are those that have not had laminitis. Obesity in humans delays the plasma clearance of fat from a meal, owing to low insulin-sensitivity of receptors in adipocytes. It has been shown that 4 days of fasting can cause visibly lipaemic plasma in ponies but not in horses, and the same factors may be a cause of laminitis. Moderate regular exercise may prevent both laminitis and insulin resistance. In humans, raised postprandial plasma fat, following a fatty meal, causes a rise in plasma NEFA, which in turn may promote insulin resistance. Whether hyperlipaemia in ponies is a cause of insulin resistance, through an elevation in plasma NEFA in nonexercising ponies, has not been examined, but deserves to be so. Such a rise is unlikely to occur after a meal, as normal feeds are low in fat content.

ENERGY METABOLISM

Hard muscular work may require that energy is available for muscular contraction at a rate some 40 times that needed for normal resting activity. Thus, rapid changes in the supply of blood glucose could result unless the animal's system responds quickly. There are many changes to accommodate the altered circumstances, but our discussion at this point will relate to the supply of nutrients to the tissues.

During a gallop, pulmonary ventilation increases rapidly so that more oxygen (O_2) is available for transport by the blood to the skeletal and cardiac muscles for the oxidative release of energy. However, this process cannot keep pace with the demand for energy and glucose is therefore broken down to lactic acid, rapidly releasing energy in the absence of O_2. The fall in blood glucose stimulates the glucocorticoids and the other hormones that enhance glycogen breakdown so that blood glucose can rise during moderate exercise.

Repeated hard work (training, see Chapter 9) brings about several useful physiological adaptations to meet the energy demands of muscular work. First, the pulmonary volume and therefore the tidal volume of O_2 increase and the diffusion capacity for gases increases, so that carbon dioxide (CO_2) is disposed of more efficiently from the blood and O_2 is absorbed at a faster rate. This process is greatly assisted by changes in both numbers of red cells (erythrocytes) and the amount of haemoglobin in the blood. There is, therefore, a greater capacity for the oxidation of lactic acid and fatty acids to CO_2. Nevertheless, training is associated with a decrease in insulin secretion, possibly a higher glucocorticoid secretion, larger amounts of muscle glycogen and blood glucose and, because of the greater work capacity, higher concentrations of blood lactate. The glucocorticoids, and possibly epinephrine in the trained animal, stimulate a more efficient breakdown (lipolysis) and oxidation of body fat as a source of energy, so conserving glycogen and yielding higher concentrations of NEFA in the blood. The glycerol released during fat breakdown tends to accumulate during hard exercise, possibly because of the raised concentration of blood lactate, and only on completion of hard muscular work is it utilized for the regeneration of glucose (Fig. 2.1).

The energy requirements of extended work can be accommodated entirely within the aerobic breakdown of glucose and by the oxidation of body fat. Thus, no continued accumulation of lactate was observed in two horses subjected to an endurance ride (Fig. 2.4), and although body fat represents the primary source of energy, its relatively slower breakdown means that there is a gradual exhaustion of muscle and liver glycogen associated with a continuous decline in blood glucose (Fig. 2.5), despite elevated concentrations of NEFA in the blood. Exhaustion occurs when blood glucose reaches a lower tolerable limit. In a more general sense, hypoglycaemia (low blood glucose) contributes to a decrease in exercise tolerance. Therefore, horses and ponies conditioned to gluconeogenesis – that is, the production of glucose from noncarbohydrate sources through adaptation and training (Fig. 2.1) – may more readily withstand extended work. Hypoglycaemia may occur when extra-hard exercise coincides with a peak in insulin secretion, suggesting that horses and ponies conditioned to gluconeogenesis, through high-roughage diets, may more readily cope with sustained anorexia (persistent scarcity of food).

Glucose represents a much larger energy substrate in individuals given a high-grain diet, whereas VFA will do so in those subsisting on roughage. Horses and ponies accustomed to a diet rich in cereals will have, in a rhythmical fashion, greater peaks and lower troughs of blood glucose than those individuals maintained on a roughage diet, owing to differences in insulin secretion and the differences in rates

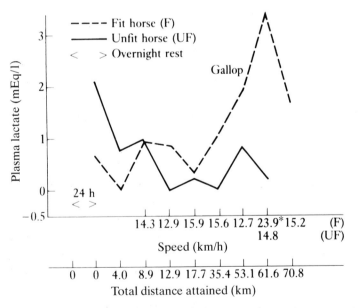

Fig. 2.4 Effect of speed but lack of effect of distance achieved on the concentration of plasma lactate. Horses separated at 53.1 km*. Note: only fit horse galloped between 53.1 and 61.6 km from start. Unfit horse retired after 61.6 km (Frape *et al.* 1979).

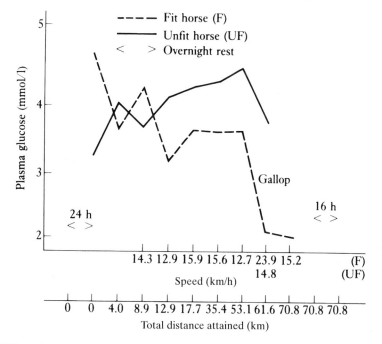

Fig. 2.5 Effect of speed and distance achieved on the concentration of plasma glucose in two horses (Frape *et al.* 1979).

of consumption of the two types of diet. The grain-fed horse at peak blood glucose is more spirited, and less so in the trough, but cannot necessarily sustain work better at the peak. The practical corollary of this is that individual horses and ponies accustomed to a diet rich in concentrates should be fed regularly and frequently in relatively small quantities, not only to prevent the occurrence of colic, but also to smooth out the cyclic changes in blood glucose (preparation for exercise is a different matter and is discussed in Chapter 9). In Fig. 2.6 the energy transfers of the young adult working horse are summarized.

Appetite

There is conflicting evidence about the factors that control appetite and hunger in horses and ponies. It is clear that amounts of NEFA in the blood are not significantly different between satiated individual animals and those with a normal hunger. It also seems that satiety is not directly associated with an elevation in blood glucose, although individuals with low concentrations of blood glucose tend to eat more and faster. Blood-glucose concentration in ponies may not influence the amount of food consumed in a meal, but it may influence the interval between voluntary feeds, without affecting the amount consumed when the pony goes to the feed trough. Supplementary corn oil seems to extend the interval before the next meal and reduce total feed intake 3–18h after administration. A trigger mechanism control-

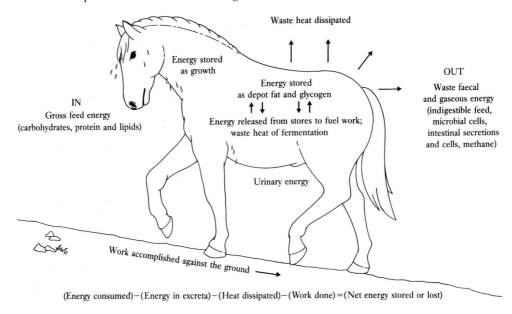

Fig. 2.6 Energy transfers in the young adult working horse.

ling the feeling of satiety, or hunger, in the horse or pony may be the amount of digestion products (especially glucose) in the intestine and VFA production in the caecum. That is, when these products in the intestinal lumen and mucosa attain certain concentrations, eating stops, and this may be mediated by afferent vagal nerve fibres. With access to feed, eating recommences when these concentrations have fallen below a certain threshold. The degree of fill in the stomach and the blood-glucose concentration, according to this evidence, have no influence on eating; but taste, visual contact between horses, energy density of feed, rate of eating and time of day all seem to influence feed intake. The practical interpretation of this for feeding management is considerable and will be discussed in Chapter 6.

AMINO ACIDS

Proteins consist of long chains of amino acids, each link constituting one amino acid residue. In all the natural proteins that have been examined, the links, or α-amino acids, are of about 20 different kinds. Animals do not have the metabolic capacity to synthesize the amino group contained in half the different kinds of amino acid. The horse and other animals can produce certain of them from others by transferring the amino group from one to another carbon skeleton in a process known as transamination. Ten or eleven of the different types cannot be synthesized at all, or cannot be synthesized sufficiently fast, by the horse to meet its requirements for protein in tissue growth, milk secretion, maintenance, etc. Plants and many microorganisms can synthesize all 25 of the amino acids. Thus, the horse and other animals must have plant material in their diet, or animal products originally derived from plant food, in order to meet all their needs for amino acids (i.e. they are unable to survive on an energy source and inorganic N). Whether or not microorganisms, chiefly in the horse's large intestine, synthesize proteins, the amino acids of which can be utilized directly by the horse in significant amounts, is still a contentious issue. The consensus of opinion is that although this source makes some contribution, probably in the small intestine, only small amounts of amino acid can be absorbed from the large intestine and by far the major part is voided as intact bacterial protein in the faeces.

During the digestion of dietary protein, the constituent amino acids are released and absorbed into the portal blood system. The amount of protein consumed by the horse may be in excess of immediate requirements and although there is some capacity for storing a little above those needs in the form of blood albumin, most excess amino acids, or those provided in excess of the energy available to utilize them in protein synthesis, are deaminated in the liver with the formation of urea. The concentration of urea rises in the horse's blood (see Fig. 1.6), although some of the amino-nitrogen may be utilized in the liver for the synthesis of dispensable amino acids (Fig. 2.7). An increase in the blood concentration of urea in endurance horses may simply reflect rapid tissue protein catabolism for gluconeogenesis in

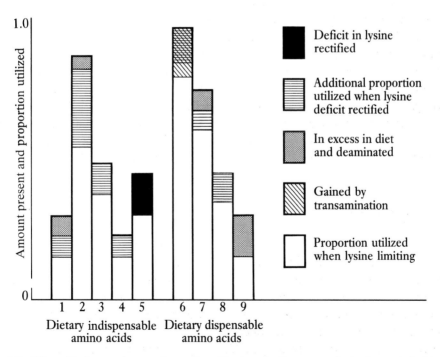

Fig. 2.7 Effect of lysine supplementation on the utilization of a dietary protein that is grossly limiting in its lysine (5) content. Only 9 of a possible 25 amino acids are shown. Amino acids 3 and 4 limit further utilization of a lysine-supplemented diet. Addition of supplementary 3 and 4 now would decrease deamination of 1, 2, 7 and 9.

glycogen depletion (Fig. 2.8) and, of course, the carbon skeleton of deaminated dietary glucogenic (e.g. glycine, alanine, glutamic acid, proline, methionine) and ketogenic (e.g. leucine and in part isoleucine, phenylalanine and tyrosine) amino acids is used as an energy source.

The extent to which dietary protein meets the present requirements of the horse depends on its quality as well as its quantity. The more closely the proportions of each of the different indispensable amino acids in the dietary protein conform with the proportions in the mixture required by the tissues, the higher is said to be the quality of the protein. If a protein, such as maize gluten, containing a low proportion of lysine is consumed and then digested, the amount of it which can be utilized in protein synthesis will be in proportion to its lysine content. As the lysine is 'limiting', little of it will be wasted, but, conversely, the other amino acids, both dispensable and indispensable, will be present in excessive quantities and so will be deaminated to an alarming extent.

Synthetic supplements

If the relative deficiency of lysine in the gluten is made good by supplementing the diet with either a good-quality protein, such as fishmeal, or with synthetic lysine,

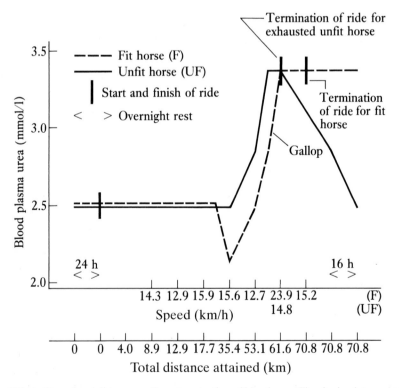

Fig. 2.8 Effect of speed and distance on the concentration of blood urea. The rise in plasma urea of the unfit horse resulted from a stress other than galloping (data from Frape *et al.* 1979).

Table 2.1 Treatment mean response (with SEs) of TB and quarter horse yearlings to amino acid supplementation of a concentrate mix (Graham *et al.* 1993).

	Gain per feed (g/kg DM)	Daily weight gain (g/day) (SE)	Girth gain (cm) (SE)	Urea N (mg/g serum) (protein) (SE)
Basal	71.0	570 (20)	9.7 (0.49)	2.6 (0.20)
Basal + lysine	77.2	640 (20)	10.1 (0.46)	2.5 (0.19)
Basal + lysine + threonine	78.7	670 (20)	11.3 (0.47)	2.0 (0.19)

then the amounts of each of the amino acids available in the blood plasma will more closely conform with the requirement, so that proportionately more of those amino acids can be used for protein synthesis (see Fig. 2.7). It has been shown that the proportions of amino acids in the common sources of feed proteins given to horses and ponies are such that lysine is the indispensable amino acid most likely to limit the tissue utilization of the protein and threonine the second most likely.

Several studies have been undertaken with growing TB and quarter horses to measure their growth response to the addition of lysine–HCl to conventional 12% crude protein concentrates containing maize, oats and soyabean meal, fed with hay. Graham *et al.* (1993) gave such a concentrate, to appetite, twice daily with coastal Bermuda grass hay (1 kg/100 kg BW) to yearlings for 112 days (Table 2.1). The concentrate was supplemented with 2 g/kg diet lysine or with 2 g/kg lysine plus 1 g/kg threonine, or neither. Amino acid supplementation increased the rate of body weight gain and the efficiency of gain, while decreasing serum urea content. This reduction would indicate that threonine improved the amino acid balance of the diet. Threonine increased the increment in muscle gain over the increment owing to lysine, as indicated by an absence of an increase in rump fat with the amino acid supplements.

Inadequate dietary protein

Inadequate dietary protein causes a fall in the concentration of plasma albumin, total protein and, according to Scandinavian evidence, in the concentrations of plasma free essential amino acids: isoleucine, leucine, lysine, phenylalanine, threonine and valine (see Chapter 8). These changes severely restrict the rate of protein synthesis.

Digestibility of protein

Another attribute of dietary protein that should not be neglected when alternative feeds are available is digestibility. For example, leather is a rich source of protein, but valueless to the horse because of its low digestibility. In practice, the problem arises only when byproducts are being considered. Most dietary proteins have an apparent digestibility coefficient of about 0.7–0.8. A related feature to that of digestibility is availability of amino acids and of lysine in particular. In practice, a reduction in lysine availability is encountered when skimmed milk, fishmeal and meatmeals are overheated during processing. The excessive heating in milk leads to a reaction between some of the lysine and both unsaturated fats and the reducing sugar, lactose. These reactions lead to products from which the animal's digestive system cannot recover the lysine.

In summary, the protein value of a diet is the product of the amount, quality and digestibility of its protein.

NONPROTEIN NITROGEN

Urea is synthesized in the liver from amino acids present in excess of need so that a rise in dietary protein above requirements is associated with a rise in plasma urea (see Figs 1.6 and 2.7). In ponies given diets containing from 6 to 18% protein, between 200 and 574 mg urea N/kg metabolic BW ($W^{0.75}$) daily are recycled and degraded in the intestinal tract. In a pony weighing about 150 kg, this range is equivalent to 54–154 g crude protein daily. While urea is within the tissues of a horse it cannot be degraded or otherwise utilized. However, when provided with an adequate source of dietary energy, microorganisms – chiefly in the large intestine – utilize it in protein synthesis, first degrading urea to ammonia by the action of bacterial urease. In the absence of an adequate supply of energy, which is normally present as fibre, starch and protein, a proportion of the ammonia, at a relatively neutral pH, diffuses back into the blood and may not be effectively utilized either by the horse or by its captive microorganisms (NH_3-nitrogen may be incorporated by the liver into nonessential amino acids during transamination reactions, but this would not necessarily increase net N-balance). A fine balance is required, for in the absence of sufficient nitrogen, microbial growth cannot occur at a maximum rate and therefore a maximal rate of fibre breakdown and utilization will not prevail. Whereas circulating urea is nontoxic to the horse, except when very high concentrations affect osmolality, the absorbed ammonia is highly toxic. A healthy liver copes adequately with low concentrations in the amination of keto acids, forming dispensable amino acids, and by urea synthesis. However, if liver failure occurs, and this is more frequent in older horses, ammonia intoxication can occur without any increase in blood urea (see Chapter 11).

Limited evidence (see Fig. 1.6) does not support the widely held view that excessive protein consumption per se predisposes horses to laminitis. The flow of urea and other nitrogen compounds into the large intestine from the ileum varies with the quantity of diet and its type (see Table 1.5). These digesta are relatively impoverished of nitrogen in horses receiving a diet of straw. The provision of nonprotein nitrogen (NPN) or, for that matter, of protein as a supplement to this diet, results in an increased flow of nitrogen and a stimulation of microbial growth in the large intestine. Urea, or more effectively biuret, added to low-protein diets in concentrations of 1.5–3%, has increased nitrogen retention in both adult and growing horses with functioning large intestines, and pregnant mares subsisting on poor pasture apparently benefit from the consumption of supplementary urea. Nevertheless, in most other circumstances the response to a urea supplement is poor and difficult to justify. Martin *et al.* (1996b) found that no nutritional benefit could be derived by mature horses from urea supplementation of a low-protein diet, in that N-balance was not increased. Where urea has been given to lactating mares the limiting factor has usually been energy intake. In this situation, feed intake and body weight have been depressed and plasma urea N has been increased, without raising blood ammonia concentration.

The large intestine of the horse can absorb small quantities of amino acids, including lysine, and the addition of urea or biuret to low-protein, poor-quality hay diets may increase DM and fibre digestion as well as N retention by stimulating microbial growth. These effects are, however, small from a practical viewpoint. Detailed studies with adult geldings, conducted by Martin *et al.* (1996b), failed to find any improvement in barley straw digestion, measured as dry matter organic matter or neutral detergent fibre digestibility from the addition of 20.3 g urea/kg dietary DM to a diet containing 4.4 g/kg N.

In summary, it would seem that horses and ponies with functioning large intestines and given diets containing less than 7–8% crude protein may make only minor use of supplementary NPN as an adjunct to that secreted back into the small intestine in digestive secretions and more directly from the blood. The reason for this is that bacterially synthesized amino acids are absorbed from the large intestine in only small amounts. In ruminants, large amounts of soluble N entering the rumen lead to a rapid production of ammonia and therefore to ammonia toxicity.

Treatment of ammonia toxicity

Ammonia toxicity, expressed as hyperammonaemia (blood levels exceeding 150 µmol/l; note that careful sample handling is required with rapid analysis), caused by excessive dietary nonprotein nitrogen, or protein, is less likely in the healthy horse with normal hepatic function, chiefly because much of the nitrogen is absorbed into the bloodstream before it reaches the regions of major microbial activity – the large intestine. Nevertheless, hyperammonaemia has been produced experimentally from the ingestion of large amounts of urea, but in these cases blood urea is also elevated. Where serum urea levels are normal (6–8 mmol/l) liver dysfunction is frequently the cause of hyperammonaemia with encephalopathic signs (ammonia readily crosses the blood–brain barrier to compete with K^+). Peek et al. (1997) reported evidence of hyperammonaemia associated with normal blood urea and liver enzyme levels, but with hyperglycaemia and acidaemia. Clinical signs included head pressing, symmetric ataxia, tachycardia and diarrhoea, and behaviour suggesting sudden blindness and abdominal pain.

Ammonia interferes with the citric acid cycle, oxidative phosphorylation and aerobic metabolism, resulting in lactic acidosis and hyperglycaemia. Treatment should therefore include administration of fluids, excluding dextrose, but including strong ions to counteract acidosis, given slowly intravenously. Where liver dysfunction has been eliminated as a cause, the origin of the ammonia is likely to be the large intestine. In this case, oral acidifying agents, such as lactulose, also should be given. They decrease ammonia absorption by converting it to the ammonium ion.

PROTEIN FOR MAINTENANCE AND GROWTH

Maintenance

Tissue proteins are broken down to amino acids and resynthesized during normal maintenance of adult or growing animals. This process is not fully efficient and, together with losses of protein in the sloughing of epithelial tissues and in various secretions, there is a continual need of dietary protein to make good the loss. However, these losses are relatively small in comparison with the protein synthesis of normal growth, or milk production, and proportionately less lysine is required. It follows that less protein, or protein of poorer quality, is needed for maintenance than is necessary for growth or milk secretion. Nevertheless, it has been shown that the protein needs of the adult horse for maintenance are less when good-quality protein is provided than when poor-quality protein is given. For example, adult TB mares were shown to remain in nitrogen balance when given 97 g fish protein/day, but they required 112 g for balance when the protein source was maize gluten.

The protein needed by the horse for body maintenance can be defined as the amount of protein required by an individual making no net gain or loss in body nitrogen and excluding any protein that may be secreted in milk. In these circumstances the animal must replace shed epithelial cells and hair, it must provide for various secretions and keep all cellular tissues in a state of dynamic equilibrium. The losses are a function of the lean mass of body tissues, depicted as a direct proportion of metabolic body size. For most purposes, the latter is considered to be the bodyweight (BW) raised to the power 0.75, and evidence suggests that horses daily require about 2.7 g digestible dietary protein/kg $BW^{0.75}$. A horse weighing 400 kg would therefore need daily about 240 g digestible protein, or 370 g dietary crude protein. This assumes that the protein has a reasonable balance of amino acids, although, as already pointed out, the lysine content of the protein for maintenance need not be as high as in that required for growth (discussed in more detail in Chapter 6).

Growth

A young horse with a mature weight of 450 kg normally gains 100 kg between 3 and 6 months of age at the rate of 1 kg daily. Growth rate in kilograms per day declines during the succeeding months and it therefore gains the next 100 kg between about 6 and 12 months of age and 75 kg between 12 and 18 months (Hintz 1980a). From a very young age the rate of gain per unit of bodyweight decreases continuously, while the daily maintenance requirement increases (Chapter 8). As the weaned foal grows, an increasing proportion of that daily gain is composed of fat and a decreasing proportion is lean. It is thus apparent that the dietary requirement for protein and the limiting amino acid lysine decline with increasing age in the growing horse. For colts aged 3 months, a maximum rate of gain has been achieved with diets containing 140–150 g protein/kg and 7.5 g lysine/kg. Diets may differ in the amount of digestible energy they provide per kilogram. For obvious reasons it is more accurate to state the protein requirements as a proportion of the digestible energy (DE) or net energy (NE) provided. Current evidence suggests that TB and quarter horse yearlings require 0.45 g lysine per MJ DE (Chapters 6 and 8). A compound stud nut for young growing horses may contain about 12–13 MJ DE/kg and oats about 11 MJ DE/kg. However, hard hay, containing 50–60 g protein/kg may provide 7.5–8 MJ DE/kg. If the yearling consumes a mixture (approximately 50:50) of concentrates and hay, the diet provides on average 10 MJ DE/kg and the minimum lysine requirement is 4.5 g/kg total diet (i.e. 0.45 g/MJ DE). Hay of 50–60 g/kg protein may contain only 2 g/kg digestible lysine and therefore the concentrate should contain at least 7 g lysine/kg in order to meet the minimum requirement. A yearling consuming 9 kg daily of total feed of this type would receive about 40 g lysine.

Much of the growth of horses may take place on pasture. Leafy grass protein of several species has been shown to contain 55–59 g lysine/kg. During the growing season the protein content in the dry matter of grass varies considerably from 110 to 260 g/kg in the leaf, whereas the flowering stem contains only 35–45 g/kg. Thus, the lysine content of the grazed material as a fraction of air-dry weight can vary from 5 to about 13 g/kg, and, if a leafy grass diet is supplemented with a concentrate mixture, the lysine and protein requirements may be met by cereals as a source of that protein. Because the quality of pasture varies so much, the use of cereals alone may mean that the protein and lysine requirements are not always met and, of course, the mixture may be inadequate as far as several other nutrients are concerned. Table 10.4 gives some analytical data for pasture in several months of the grazing season.

Laminitis and energy intake

Studies of the protein requirements of growing horses have failed to show any relationship between amounts of dietary protein and the incidence of laminitis (founder), but have shown that there is a greater incidence of the problem when starch intake is excessive. Restricted feeding of concentrates is therefore advocated and horses should be adjusted gradually to lush pasture by gradually increasing their time of access daily. The critical feature of dietary causes of laminitis is the rapid fermentation of dietary residues reaching the large intestine. These are principally starch and to a lesser extent protein, when total feed intake is excessive (a method for *ad libitum* feeding is outlined in Chapter 8 and laminitis is discussed in detail in Chapter 11).

STUDY QUESTIONS

(1)   The horse evolved as a browsing animal engaging in many small feeds each day. What impact has this had in respect of:
   (a)   food selection and metabolic responses;
   (b)   social habits; and
   (c)   safety in the wild?

(2)   What is meant by limiting dietary amino acid and what is the relation, if any, between this and (a) maintenance diets and (b) production diets? What is meant by nitrogen balance?

(3)   What factors should be considered when a horse has lost its appetite for sufficient feed to maintain body weight?

FURTHER READING

Frape, D.L. (1989) Nutrition and the growth and racing performance of thoroughbred horses. *Proceedings of the Nutrition Society*, **48**, 141–52.

Graham, P.M., Ott, E.A., Brendermuhl, J.H. and TenBroeck, S.H. (1993) The effect of supplemental lysine and threonine on growth and development of yearling horses. *Proceedings of the 13th Equine Nutrition and Physiology Society*, University of Florida, Gainesville, 21–23 January 1993, pp. 80–81.

McDonald, P., Edwards, R.A. and Greenhalgh, J.F.D. (1981) *Animal Nutrition*. Longman, London and New York.

Chapter 3
The Roles of Major Minerals and Trace Elements

Grass is the first nourishment of all colts after they are weaned. . . . Whereas when they are fed
with corn and hay, but especially with the first, . . . it exposes them to unspeakable injuries.

W. Gibson 1726

MAJOR MINERALS

Calcium and phosphorus

Function

The functions of calcium (Ca) and phosphorus (P) are considered together because
of their interdependent role as the main elements of the crystal apatite, which
provides the strength and rigidity of the skeleton. Bone has a Ca:P ratio of 2:1,
whereas in the whole body of the horse the ratio is approximately 1.7:1.0, because
of the P distribution in soft tissue. Bone acts as a reservoir of both elements, which
may be tapped when diet does not meet requirements. The elements of bone are in
a continual state of flux with Ca and P being removed and redeposited by a process
that facilitates the reservoir role and enables growth and remodelling of the skeleton to proceed during growth and development. The acute role of Ca relates to its
involvement in a soluble ionic form for nerve and muscle function. Consequently
$[Ca^{2+}]$ concentration in the blood plasma must be maintained within closely defined
limits.

Control of Ca and P metabolism

The flux and distribution of Ca and P in the body are regulated by two hormones in
particular, functioning antagonistically at the blood–bone interface, the intestinal
mucosa and the renal tubules (see also under vitamin D, Chapter 4). The horse
kidney seems to play a greater part in controlling concentrations of Ca in the
blood than does the intestinal tract and this may have practical significance for diet
and renal disease. The mean values and ranges for serum total Ca and P, among
others, are listed in Table 3.1. It should be noted that normal resting plasma
phosphate concentration decreases with increasing age and that ionized plasma
Ca concentration is approximately 1.7 mmol/l lower than the total values given in
the table.

Table 3.1 Mean values and ranges for serum total concentrations (mmol/l) of electrolytes in horses of different ages. (Modified from published tables of S.W. Ricketts, Beaufort Cottage Laboratories, Newmarket, Suffolk.)

		Birth to 36h	3 Weeks	Yearlings	Horses in training	Mares at stud
Ca	Mean	3.2	3.2	3.3	3.4	3.4
	Range	2.7–3.6	2.5–4.0	2.7–4	2.6–3.9	2.9–3.9
P	Mean	2.5	2.5	1.8	1.3	1.1
	Range	1.2–3.8	1.6–3.4	1.4–2.3	1.1–1.5	0.5–1.6
Na	Mean	136	137.5	138.5	138.5	138.5
	Range	126–146	130–144	134–143	134–143	134–143
K	Mean	4.8	4.5	4.3	4.3	4.3
	Range	3.7–5.4	3.6–5.4	3.3–5.3	3.3–5.3	3.3–5.3
Mg	Mean	0.83	0.81	0.78	0.78	0.78
	Range	0.57–1.10	0.66–1.10	0.62–1.10	0.62–1.10	0.62–1.10
Cl	Mean	—	—	—	—	—
	Range	Normal range for all ages 99–109				

Serum phosphate values vary without untoward physiological effects. For example, strenuous exercise can depress blood phosphate to half the 'resting value' for 2–2.5h. Nutritional secondary hyperparathyroidism (NSHP) is a diet-related clinical disorder of horses in which serum phosphate is slightly raised and serum $[Ca^{2+}]$ values are slightly depressed (Figs 3.1 and 3.2).

Ca and P in bone

Horses have no 'horse sense' when it comes to selecting a diet containing a balanced mixture of Ca and P – they prefer the palatability of a P-rich diet, whereas this is not available for selection in the natural grazing environment. Hence, a dietary Ca deficiency is not an infrequent occurrence among domestic horses. Inadequate dietary Ca and P for growing foals causes a delay in the closure times of the epiphyseal plates of long bones and contributes to developmental orthopaedic disease (DOD). In adult working horses it causes lameness and bone fractures. Failure of the osteoid, or young bone, to mineralize is called rickets in young and osteomalacia in adult horses (some authorities argue that true rickets does not occur in the foal). In extreme cases when mineral is being reabsorbed from bone, the outcome may be generalized osteodystrophia fibrosa in which fibrous tissue is substituted for hard bone and a characteristic enlargement of the facial bones (bighead) may occur. In the presence of vitamin D in each of these conditions the body, through the agency of parathyroid hormone (PTH), is endeavouring to maintain homeostasis of blood Ca by accelerating the removal of Ca from the bones and increasing the tubular reabsorption of Ca. Diets based on wheat-bran and cereals are rich in organic P and low in Ca, predisposing horses to these conditions.

The tendency to lower blood $[Ca^{2+}]$ leads to increased bone resorption, increased renal excretion of phosphate, an increased rate of bone–mineral exchange and to a

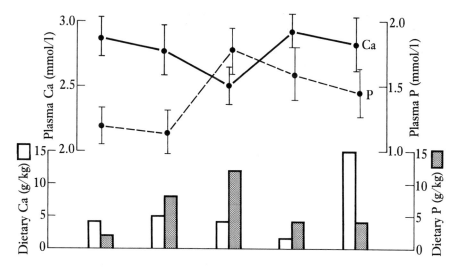

Fig. 3.1 Effects of dietary Ca (□) and P (■) on mean concentrations in blood plasma, ± SE (after Hintz & Schryver 1972).

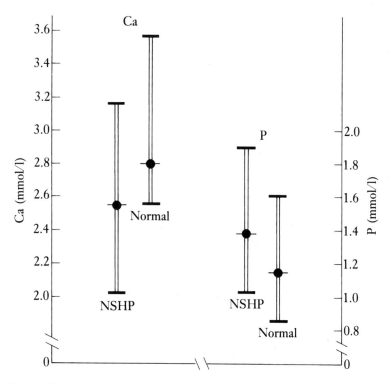

Fig. 3.2 Mean and range of serum Ca and P concentrations in spontaneous NSHP (Krook 1968). Notice how in cases of NSHP Ca is depressed and P elevated.

greater susceptibility of bones to fracture. Deposition of Ca salts in soft tissue, including the kidney (nephrocalcinosis), may also be apparent. A physiological example of rising serum PTH concentration occurs in periparturient mares during mammary secretion of Ca when there is a decrease in serum total and ionized Ca concentrations (Martin *et al.* 1996a). The hormone calcitonin, or thyrocalcitonin as it is secreted by the parafollicular C cells of the thyroid gland, opposes the effects of PTH. When plasma Ca concentration is elevated PTH secretion is decreased, reducing plasma Ca by decreasing the activity of osteoclasts and increasing that of osteoblasts.

Hypocalcaemia

(See also Chapter 9.) Ca is principally an extracellular cation existing in ionized, complexed and protein-bound forms. The ionized form is physiologically active and its concentration in blood plasma is precisely controlled, but the concentration is influenced by acid–base changes, increasing in acidosis and decreasing in alkalosis. Normal serum total Ca concentration in the horse is 3.2 mmol/l and ionized Ca concentration is about 1.5 mmol/l, although laboratory method influences the values obtained. Pronounced hypocalcaemia is unlikely to result from a dietary inadequacy of Ca, but rather from metabolic alkalosis. Thus, hypocalcaemia may occur in adult horses as a postexertional stress. The clinical signs are: muscle fasciculation and tetanic spasms, incoordination, synchronous diaphragmatic flutter, decreased gut sounds and even inability to stand. Extended work and overheating lead to a rise in blood pH that depresses the concentration of $[Ca^{2+}]$ in the blood. Moreover, elevated body temperature per se can bring about a loss of 350–500 mg Ca/h in sweat and this rate of loss may exceed the capacity of blood replenishment by bone mobilization.

The intravenous (i.v.) administration of a solution of 50–100 mmol $CaCl_2$, or Ca gluconate, in 1 l over 30 min should resolve the clinical signs of hypocalcaemia. The higher dose may cause cardiac arrhythmias and so administration should be accompanied by cardiac auscultation, decreasing the rate of dosing if necessary. Hypercalcaemia also induces cardiac arrhythmias (hypo- and hypercalcaemias, respectively, prolong and shorten the Q–T interval of the electrocardiogram).

Recent work in Spain has shown that the PTH concentration in normal horse plasma averages 20–30 pg/ml, determined by a radio-immune assay, employing an antibody against intact human PTH and 35–45 pg/ml with an antibody against the amino-terminal region of rat PTH. The higher values with the rat assay reflect the presence of amino-terminal fragments in plasma, which nevertheless possess biological activity. Normal equine plasma ionized Ca concentration is about 1.5 mmol/l, whereas PTH levels attain a maximum (>80 pg/ml) when the Ca concentration falls to approximately 1.25 mmol/l and they drop to a minimum (<12 pg/ml) when ionized Ca concentration rises to about 1.8 mmol/l.

Dietary levels of Ca and P in relation to requirenent

Excessive amounts of dietary Ca do not seem to initiate the kinds of dietary problems encountered in other domestic species. However, one experiment in which Shetland foals aged 4 months were given a diet containing 25 g Ca/kg, with a Ca:P ratio of 6:1, for 4 years (Jordan *et al.* 1975) resulted in a slight enlargement of the marrow region of the long bones and thinning of the cortical area, together with less bone mineral per unit of cortex. Lameness is frequently a characteristic of resorption of cortical bone through loss of support for tendons and ligaments. Excessive amounts of dietary Ca may make bones brittle, through abnormal bone storage of Ca. However, investigations in which diets containing from 7 to 27 g Ca/kg have been compared for up to 2 years show that differences are small; bone density is increased by high-Ca diets and the cortex of the long bones is slightly thinner (Schryver *et al.* 1970a, 1971b, 1974a, 1974b).

The equine kidney plays a vital part in Ca homeostasis, and daily urinary Ca excretion shows a direct relationship with intake. In many species the urinary loss of Ca is raised by increased intakes of sodium, disrupting Ca homeostasis. However, this relationship does not hold in the horse (see 'Sodium', this chapter). Diets rich in Ca yield urine containing a precipitate of Ca salts; the urinary loss of Ca in a 300-kg yearling given a diet containing 20 g Ca/kg was 20–30 g in 6–8 l of urine daily (that is, 0.36% Ca). The absence of calculus formation in the kidney demonstrates the horse's ability to deal with large amounts of Ca despite the low solubility of the element; conversely, a dietary deficiency of Ca yields urine almost devoid of the element. In contrast, endogenous loss of Ca in faeces, representing the minimal obligatory loss that must be replenished from dietary sources, is largely unaffected by the dietary amounts. Endogenous faecal Ca and P have been estimated, respectively, to be 36 mg/kg BW and 18 mg/kg BW daily in growing quarter horses (Cymbaluk *et al.* 1989b). Urinary losses of Ca decrease by 50–75% in extended work (Schryver *et al.* 1975, 1978a), whereas sweat losses increase. During 20 min of hard work, ranges of 80 to 145 mg Ca and 11 to 17 mg P have been found in the total yield of sweat (Schryver *et al.* 1978c). Over a full day's hard work this source represents a considerable loss of Ca. On the other hand, horses and ponies idle for long periods retain less Ca than those worked when the dietary P concentration is excessive. Following such an inactive period dietary Ca and P should be raised 20% above minimal requirement levels.

Horses must absorb about 2.5 g Ca/100 kg BW daily to balance the obligatory loss. Those given two to three times their Ca requirement absorb nearly half of it, whereas at the requirement level the absorption rate may lie between one-half and two-thirds of intake. Absorption of half the total ingested necessitates a daily intake of 5 g/100 kg BW for maintenance of the adult, or about 2.5 g/kg diet. Similarly, assuming average digestibility values the horse must be provided daily with about 2 g P/100 kg BW, or 1 g/kg diet for maintenance.

Intestinal absorption

The lack of impact of dietary Ca on the efficiency of P absorption in the horse may be related to the fact that Ca and P are absorbed from different regions of the intestine (see Fig. 1.5). However, vastly excessive intakes of Ca increase the faecal loss of P and dietary Ca can affect absorption of other elements. For example, excessive Ca can depress the absorption of magnesium, manganese and iron owing to competition at common absorption sites, or possibly to the formation of insoluble salts. Meyer and colleagues in Germany have reported that between 50 and 80% of dietary Ca and between 45 and 60% of magnesium are absorbed in the small intestine (Meyer *et al*. 1982c), whereas there is a net secretion of these elements into the large intestine. True Ca and P digestibilities are reported (Cymbaluk *et al*. 1989b) to decline from 71 to 42% and from 52 to 6%, respectively, between 6 and 24 months of age. Aged horses may be even less efficient in the absorption of Ca. Meyer's group further demonstrated that the site of P absorption varies with composition of diet. No P is absorbed in the upper small intestine of horses fed solely on roughage, whereas some is absorbed in the distal small intestine, especially in those given only concentrates. Large amounts of phosphate secreted into the caecum and ventral colon probably act as a buffer to VFAs produced there, and the dorsal colon and small colon are the major sites of absorption and reabsorption of phosphate (Fig. 1.5).

Availability

The net availability of Ca in a variety of feeds has been estimated to lie between 45 and 70%, except where signficant amounts of oxalates are present. The dietary level of phosphate influences Ca absorption. When dietary P, as inorganic phosphate, was raised from 2 to 12 g/kg, Ca absorption was decreased by more than 50% in young ponies receiving a diet otherwise adequate in Ca (4 g/kg diet) (Schryver *et al*. 1971a). Dicalcium phosphate-P or bone flour-P is digested to the extent of 45–50%. Rock phosphate and metaphosphates are poor sources of P and Ca. Phosphorus in salts of phytic acid, the predominant source in cereal and legume seeds, is only 30% available despite the presence of large numbers of intestinal bacteria secreting phytase. Phytate-P constitutes at least 75% of the total P in wheat grain and 54–82% of the P in beans.

Large amounts of vitamin D in the diet can increase the utilization of phytin-P, but as these amounts are almost toxic they cannot be recommended. Phytin-P use may be improved by the addition of yeast culture to the diet (see probiotics, Chapter 5). Experiments in pigs have shown that the addition of phytase, derived from *Aspergillus niger*, to the diet increased phytate digestion. Generally, most dietary P is of plant origin which has a lower availability than most sources of Ca (excluding those rich in oxalates). Therefore sometimes growing horses may receive inadequate available P in their diet for normal growth. Pagan (1989) has also argued that the endogenous P losses by young horses are about double those (20 cf. 10 mg/kg BW) used by the National Research Council (NRC) in estimating the dietary P requirement. Thus, with some diets, relying on their natural P content, the P requirement of young horses may not be met.

Oxalates and other dietary factors affecting Ca absorption

The bioavailability, or true digestibility, of dietary Ca varies considerably. The principal factors controlling bioavailability are:

- amount of dietary Ca (true digestibility is 0.7 at requirement intake, cf. 0.46 at several times the requirement);
- amount of dietary P (10 g P added/kg diet containing 4 g Ca/kg reduced true Ca digestibility from 0.68 to 0.43);
- vitamin D status (of less significance for absorption in the horse than in some other domestic species);
- dietary phytates and oxalates (phytates and oxalates bind Ca and so reduce Ca availability; Ca:oxalate < 0.5:1 causes NSHP; implicated grasses rich in oxalates include: napier, guineagrass, buffel [*Cenchrus ciliaris*], pangola [*Digitaria decumbens*], green panic, paragrass, kikuyu [*Pennisetum clandestinum*], setaria [*Setaria sphacelata*] and probably some species of millet grass. Lucerne (alfalfa) contains oxalic acid, see below); and
- age of animal (bioavailability may decline slightly with age, but the relationship is not pronounced in horses).

Oxalic acid binds divalent cations, such as Ca, in an unavailable form. Many tropical grasses are rich in oxalates and their feeding is associated with osteoporosis and lameness in horses. Ca deficiency ailments are noticeable in horses grazing pastures, or given hay, containing an abundance of these species of tropical and subtropical grass (see also Chapter 10). The bioavailability of Ca may, nevertheless, differ among plant sources containing oxalates. Lucerne contains these compounds, but its Ca has a high feeding value, and about three-quarters of the Ca is absorbed, although the Ca content relative to that of oxalic acid is much higher than that in the grasses listed above. Lucerne Ca is better utilized than that from timothy hay, which is not a significant source of oxalic acid. Evidence in laboratory animals shows that the feeding of Ca sources that have a low bioavailability compromises both the quantity and quality of bone, whereas inadequate quantities of highly available Ca reduce only the quantity of bone. Whether this is a contributory factor to equine lamenesss is unknown.

In summary

The maintenance requirement for the major minerals Ca and P is that necessary to balance losses in the faeces and urine, as well as unspecified 'dermal losses'. There is an additional need for growth and, in the breeding mare, for mineralization of the fetal skeleton and for lactation. Each kilogram of lean body tissue in the horse contains about 20 g Ca and 10 g P; the amounts required in the diet to allow for maintenance and growth are shown in Table 3.2. Mare's milk contains on average about 900 mg Ca and 350 mg P/kg (see Fig. 7.3). A 500-kg mare may produce a total of 2000 kg milk in a lactation extending over 5–6 months – a total lactation deficit of 1.8 kg Ca and 0.7 kg P derived from skeletal reserves and feed. For milk synthesis,

Table 3.2 Minimum daily requirements for Ca and P of growing horses (Schryver *et al.* 1974a).

Age (months)	BW (kg)	Weight gain (kg/day)	Ca (g/day)	P (g/day)
3	100	1	37	31
6	200	0.72	33	27
12	300	0.50	31	25
18	375	0.30	28	23
Mature	450	0	23	18

dietary Ca and P daily requirements, with average availabilities of 50 and 35%, respectively, are 10 g Ca and 5.5 g P to balance that secreted daily.

Limestone flour, dicalcium phosphate and bonemeal are reliable sources of Ca and the second two of P also, although bonemeal should be used only if it has been steamed and marketed in paper or plastic sacks (this advice may require revision once the aetiology of BSE has been fully resolved). The requirement of 10 g Ca is met by 28 g limestone or 40 g dicalcium phosphate, which also meets the P needs entirely.

Magnesium

Magnesium (Mg) is a vitally important ion in the blood; it forms an essential element of intercellular and intracellular fluids, it participates in muscular contraction and it is also a cofactor in several enzyme systems. Bone ash contains 8 g Mg/kg in addition to 360 g Ca/kg and 170 g P/kg. There is a small net absorption of Mg from the large intestine, but the majority of net absorption occurs from the lower half of the small intestine. The 'obligatory loss' of Mg secreted into the intestinal tract amounts to about 1.8 mg/kg BW daily; a further obligatory loss of about 2.8 mg/kg BW occurs in the urine, and the maintenance requirement to offset these losses is about 13 mg/kg BW daily, or about 2 g/kg diet. Homeostasis is achieved largely by a balance between gut absorption and renal excretion. Adrenal, thyroid and parathyroid hormones influence status, although plasma $[Mg^{2+}]$ has a less potent effect on PTH than does $[Ca^{2+}]$. PTH causes an increase in plasma Mg by increasing absorption from intestines and renal tubules and resorption from bone (PTH also requires Mg ions for activation of adenylate cyclase in bone and kidney). Aldosterone secretion (see 'Sodium', this chapter) causes a lowering of plasma Mg and an increase in urinary Mg excretion. Typical diets may not meet the horse's Mg need without supplementation.

A rarely observed frank dietary deficiency of Mg leads to hypomagnesaemia associated with loss of appetite, nervousness, sweating, muscular tremors, rapid breathing (hyperpnoea), convulsions, heart and skeletal muscle degeneration and, in chronic cases, mineralization of the pulmonary artery caused by deposition of Ca and P salts. Normal blood serum values are given in Table 3.1.

Mg of vegetable origin, naturally present in feed, is available in proportions ranging from 45 to 60%, the more 'digestible' sources being milk and possibly lucerne. Absorption of Mg from high-temperature-dried lucerne is greater than it is from timothy hay, according to evidence from Edinburgh (see also 'Oxalates and other dietary factors affecting Ca absorption'). Sugar beet pulp and beet molasses are also reasonably good sources of digestible Mg (both contain about 2.8 g Mg/kg DM). Large amounts of dietary P seem to depress absorption of Mg slightly, but not as effectively as do dietary oxalates. Bacterial phytase in the gut may assist absorption. The inorganic sources of Mg – Mg oxide (calcined magnesite), sulphate and carbonate – are all about 70% absorbed in the horse, although oxides from different countries of origin have differing availabilities. Generally, therefore, Mg carbonate is a more reliable source. An increase in the dietary level of Mg from 1.6 to 8.6 g/kg was shown to increase Ca absorption without an effect on P. Table 3.3 gives the daily requirements of Mg for horses weighing 400 and 500 kg.

Potassium

Potassium (K) is subject primarily to precaecal absorption, where 52–74% is absorbed. An intake of 46 mg K/kg BW/day during rest is adequate for a positive K balance in adult horses. Foals require more, perhaps as much as 150–200 mg/kg BW, that is, 7 g K/kg diet. Losses during sweating, or diarrhoea, increase the need considerably. Young foals may become deficient in K as a result of persistent diarrhoea and this in turn tends to precipitate acidosis. Spontaneous changes in plasma [K^+] can therefore result from strenuous exercise; this will be discussed in Chapter 9. The normal plasma K concentration is in the range 2.4–4.7 mmol/l (or mEq/l) (Table 3.4). Nevertheless, the plasma concentration increases during episodes of acidosis, as intracellular, red cell K (normal concentration 83–100 mmol/l) (Muylle *et al.* 1984b) exchanges for H^+ ions when cardiac arrhythmias can occur. Red cell K concentrations <81 mmol/l are accompanied by signs of skeletal muscle

Table 3.3 Daily requirements of Mg (g).

	Mature weight (kg)	
	400	500
Adult		
Rest	5.6	7.0
Medium	6.5	8.0
Last 90 days of gestation	6.5	8.0
Peak lactation	6.6	8.1
Growing: age (months)		
3	5.5	6.8
6	5.2	6.3
12	5.3	6.5
18	5.6	7.0

Table 3.4 Serum total concentrations (mmol/l) of electrolytes in TB foals (Sato *et al.* 1978).

Days from birth	Ca	P	Na	K	Cl	Mg
0	2.64	1.51	131.6	4.68	103.6	0.88
10	2.62	1.48	142.0	4.42	97.8	0.97
20	2.83	1.79	138.6	4.08	98.4	0.90
50	2.34	2.11	137.2	4.28	96.6	1.32*
90	2.50	1.74	140.0	3.96	99.8	—
120	2.23	1.63	136.4	4.12	101.4	—

*Thirty days of age.

weakness (Frape 1984b). The means of assessing K status and the major causes of K depletion and their therapy are discussed in Chapter 11.

Deficiency

A dietary deficiency of K may reduce appetite and depress growth rate; a reduction in plasma K (hypokalaemia) occurs and in extreme deficiency there may be clinical muscular dystrophy and stiffness of the joints. Hypokalaemia can occasionally result from persistent diarrhoea, or from excessive sodium bicarbonate administration. In depletion of ponies the K losses in urine did not fall below 20 mmol/l (Meyer *et al.* 1986) and the K content of sweat remained invariable at about 27 mmol/l (cf. sodium). Blood plasma K decreased from 3.5 to 2.3 mmol/l, whereas red cell K changed little. However, the K depletion of the skeleton was as much as 60%, whereas losses in muscles, vital organs and GI contents were only 9, 15 and 7%, respectively. Food and water intake decreased, the ponies were more excitable and exercise exhaustion occurred sooner.

Hyperkalaemic periodic paralysis

A syndrome of episodic weakness in horses, accompanied by elevated serum K concentrations – hyperkalaemic periodic paralysis (HPP) – has been described in horses (Naylor *et al.* 1993). This is not caused by a dietary K deficiency. It is accompanied by myotonia, facial muscle spasm and fasciculations and recumbency. Tracheotomy may be required if there is severe dyspnoea. Most episodes resolve spontaneously in 15–120 min, but may require parenteral administration of Ca gluconate, Na bicarbonate, dextrose, insulin and subsequently dietary management. This situation reflects the exchange of intracellular K^+ with H^+ in acidosis mentioned in 'Potassium', above.

Sources

Cereals are relatively poor sources of K, but hay contains between 15 and 25 g K/kg; thus, most diets should contain ample if at least one-third is in the form of good-quality roughage. Animals in heavy work generally consume more cereals, thus lowering dietary K when losses in sweat would normally be increasing. Lush pastures can contain large amounts of K in the dry matter and so theoretically may 'interfere' with Mg metabolism.

Sodium

Sodium (Na) is the principal determinant of the osmolarity of extracellular fluid and consequently of the volume of that fluid. Chloride concentration in the extracellular fluid is directly related to that of Na. Na is derived from the ingested food and its excretion via the kidney is controlled by the renin–angiotensin–aldosterone system. Hyponatraemia can occur from reduced intake of Na and it causes aldosterone secretion; however, Na deficits can also occur through excessive losses from the GI or urinary tract relative to water loss. These may indicate intestinal obstruction, enterocolitis or renal failure. Excessive sweating normally causes hypernatraemia. In some other species increased plasma levels of aldosterone result in increased urinary Mg excretion, although what occurs in the horse is unclear.

Na is reabsorbed by the large intestine to the extent of 95%. In an Na deficiency state the reabsorption may reach 99% and renal Na losses are reduced, thus conserving the tissue content, according to Meyer's group in Hanover (Meyer *et al.* 1982b). The resting horse, receiving an Na deficient diet, can conserve Na. The Na content of the sweat of working horses is, however, decreased only slightly, being partly replaced by K. Such Na depleted horses exhibit a licking habit and a craving for large amounts of sodium chloride, decreased cutaneous turgor, reduced feed intake and ultimately cessation of eating, muscular and nervous dysfunction (muscular tremor, gait and chewing incoordination). In advanced Na depletion, plasma Na and chloride concentrations fall to 120 and 70 mmol/l, respectively, with an increase in plasma K of up to 5.5 mmol/l and a decrease in total body water, largely through increased dry matter of the GI tract.

Although pasture grass may contain as much as 18 times more K than Na, supplementary Na in the form of common salt for grazing stock is normally unnecessary. Forages are a richer source of Na than are cereals and within normal ranges the one element tends to inhibit the loss of the other in the urine, by the action of aldosterone on the kidney tubules, conserving the body's resources of Na in the grazing stock. Diets providing 2–4 g Na/kg should adequately meet the requirement for Na, except during periods of excessive sweating in very hot weather, or as a result of diarrhoea. Diets containing 5–10 g common salt/kg will amply meet the normal Na requirement.

Chloride

Where the requirements for common salt (NaCl) are met, it is unlikely that a deficiency of chloride (Cl) will occur. The major source of loss, particularly in hot weather, is sweat where even at moderate rates of work horses may lose 100 g salt/day (60 g Cl).

TRACE ELEMENTS

Most stabled horses in the UK now receive supplements containing variable quantities of trace elements, and the horse seems able to cope with some measure of abnormal intake without showing clinical signs of toxicity or deficiency. Those trace elements of prime importance in the diet of horses are discussed here. Cobalt (Co) is considered in Chapter 4 under vitamin B_{12}. The variety of geological strata underlying UK soils yields grazing areas that cause clinically recognizable signs of specific deficiencies in cattle and sheep. There is biochemical evidence to show that horses and ponies in these areas may similarly reflect their nutritional environment.

Table 3.5 Values for four trace elements (µmol/l) in normal serum or plasma and milk of mares (Smith *et al.* 1975; Blackmore & Brobst 1981; Cape & Hintz 1982; Schryver *et al.* 1986; Lawrence *et al.* 1987b; Bridges and Harris 1988; Saastamoinen *et al.* 1990). (Also see Chapter 12 for Cu determinations.)

	Blood	Milk		
	Sérum or plasma	Parxtum	1–8 days	9–120 days
Cu				
TB, USA mainly stabled	24–35	16	10–12	4–6
TB	8–18			3[1]
TB, 7, 14, 28 days of age	6, 14, 25			
Quarter horse	5–31			
Horse/pony, gestation[2]	16.3			
Horse/pony, lactation[2]	14.7			3.8
Horse/pony, yearlings	15.3–26.1			
Horse/pony, 2–3 years old	21.3			
Fe				
TB	28 ± 7			
Quarter horse	28			
Arabian	23	24	17	12
Standardbred	30 ± 7			
Shetland	19			
Zn				
TB, USA, stabled	—		46	60
TB, UK pastured	17 ± 7	98	52	36
TB, UK stabled	26 ± 8			
TB, 7, 14, 28 days of age	17, 12, 12			
Horse/pony, gestation[2]	5.8			
Horse/pony, lactation[2]	7.6			31.3
Horse/pony, yearlings	10.7, 12.5			
Horse/pony, 2–3 years old	10.1			
Pb				
TB and standardbred:				
suckling to adult (mean)	0.8–1.9			
mixed breeding	7–8			

[1] Unpublished data of author.
[2] (Saastamoinen *et al.* 1990). Finnhorse breed.

Table 3.6 Normal blood characteristics of horses.[1] Ranges indicate approximately ±1 SD from mean.

Sample	Characteristic	Units per litre	Yearlings	Horses in training	Mares	Foals
Plasma	Albumin	g	27–28	34	27	25
Serum	AST	iu	140	150–400	140	70–120
Serum	CK	iu	43	50–70	43	53–57
Serum	SAP	iu	100–120	85–95	70–80	150–400
Serum	ALT	iu	1–6.7	1–6.7	1–6.7	NA
Serum	GGT	iu	18–30	20–30	18–19	13–16
Red cell	GSH-Px	u/ml[7]	15–25	15–25	—	—
Serum	LDH	iu	45–100	45–100	—	—
Serum	SDH	iu	0.8–1.2	0.8–1.2	—	—
Serum	Bilirubin, total[2]	µmol	34	34–39	26	38–55
Serum	Fasting glucose	mmol	3.5–5	3.5–5	3.5–5	3.1–4.2
Serum	T_4	nmol	5–39	5–35	3–56	10–150
Serum	T_3	nmol	1.5–2	1–2	0.9–1.4	2–7
Serum	Creatinine	mmol	140	170–185	140	150–190
Serum	Cu	µmol – stable	15.2[3]	12.4	15[4]	6 to 20[5]
Serum	Cu	µmol – grass	—	15.9	16–26	20[8]
Serum	I	µmol	—	—	0.6	0.8–0.85
Serum	Zn	µmol – stable	—	26	9–12	13
Serum	Zn	µmol – grass	—	17	9–12	—
Serum	Mo	µmol	—	0.31	—	—
Serum	Se	µmol	0.8–1.3	1.5[6]	0.8–1.2	0.6–1.1
Whole blood	Se	µmol	—	1.6[6]–3.0	0.8–1.1	0.5–0.7
Plasma	α-Tocopherol	mg	1–3	1.5–4.4	1.8–2.6	1.1–2.4
Plasma	α-Tocopherol	µmol	2.3–6.8	3.4–10	4.1–5.9	2.5–5.5
Plasma	Retinol	µg	180	180	150–300	150
Serum	25-Hydroxy vitamin D	µg	—	—	2.9–3.6	2.2–2.5
Whole blood	Cyanocobalamin	µg – stable	3.7–6.6	1.2–6.6	3.7–6.6	3.7–6.6
Whole blood	Cyanocobalamin	µg – grass	2.8–20	2.8–20	2.8–20	2.8–20
Serum	Folate	µg – stable	4.5–12	4.5–12	4.5–12	4.5–12
Serum	Folate	µg – grass	5.3–13.5	5.3–13.5	5.3–13.5	5.3–13.5
Whole blood	Thiamin	µg	28	30	33	24
Plasma	Ascorbic acid	mg	—	2.5–4.5	—	—

[1] Data are drawn mainly from TBs. See Table 3.1 for serum Ca, P, Mg, Na, K and Cl values.
[2] Mainly unconjugated.
[3] Ponies.
[4] Whole blood.
[5] Rising from 1 week of age to plateau at 4 weeks.
[6] Minimum adequate.
[7] One enzyme unit of GSH-Px activity = 1 µmol NADPH oxidized/min.
[8] Plasma.
Abbreviations: ALT, alanine amino-transferase; AST, aspartate amino-transferase; CK, creatine kinase; GGT, γ-glutamyl-transferase; GSH-Px, glutathione peroxidase; iu, international unit; LDH, lactic dehydrogenase; SAP, serum alkaline phosphatase; SDH, sorbitol dehydrogenase; T_3, triiodothyronine; T_4, thyroxine.

Tables 3.5 and 3.6 give average serum values for some trace elements. Abnormalities in leg growth and development of foals and yearlings have been reported to be associated with dietary deficiencies of copper (Cu), manganese (Mn) and selenium (Se), and toxicities of iodine (I) and lead (Pb).

The extent to which pasture plants extract trace metals from soil depends on the soil's pH and moisture content and the plant species. Effects may also be attributable to the root systems of plants as legumes and many herbs have deeper roots than grasses do. The levels of trace elements in herbage are clearly of importance, but the horse will also consume soil while grazing. Soil intake will depend on the density of the herbage. In certain conditions cattle and sheep may consume more than 10% of their daily intake of dry matter as soil.

Mixed supplements

Ott and Asquith (1995) measured the response of growing TB and quarter horses from 340 to 452 days of age to trace mineral supplements, when given concentrates to appetite and coastal Bermuda grass *(Cynodon dactylon)* hay at the rate of 1 kg/100 kg BW daily. The unsupplemented total diet contained 195 mg Fe, 36 mg Mn, 41 mg Zn and 4.8 mg Cu, each per kg DM. The supplements had no influence on growth and development, but mineral deposition in the third metacarpal was increased by the trace mineral mix, although not by Cu or by Cu plus Zn, excluding the other elements. The results suggested that the other trace minerals (Fe, Mn, Co or I) were more critical. However, previous work by the same group (Ott and Asquith 1989) indicated that bone mineral deposition in yearlings was increased when a trace mineral mix, containing Fe, Mn, Zn and Cu, was added to a natural diet containing lower concentrations of these elements than those recommended by the National Research Council (NRC) (1978).

Mineral content of horse hair

The mineral content of animal hair varies not only with mineral intake, but also with season, breed, age, hair colour and body condition. Nickel (Ni) intake is apparently correlated with Ni in hair, but other heavy metals, at least at subtoxic intakes, are not (toxic intakes of Pb cause elevated hair Pb). There is little simple relationship between minerals of nutritional importance and their concentration in horse hair.

Copper and molybdenum

Hypocupraemia occurs widely in grazing cattle, attributed frequently to an excess of molybdenum (Mo) derived from the underlying Mo-rich strata, particularly Lias clays and marine black shales of Jurassic and Carboniferous ages. The high levels of Mo associated with relatively low copper (Cu) lead to Cu:Mo ratios in the herbage narrower than 6:1, causing Cu deficiency on the so-called teart pastures in these areas. Hypocupraemia can also occur, owing to low Cu levels per se in the soils and herbage.

The horse is not as susceptible to clinical signs of Cu deficiency as are ruminants, but signs have been described, such as erosion of the articular cartilage of joints, and anaemia and haemorrhage in parturient mares. Moreover, Mo and sulphate do not seem to have the same impact on Cu status in horses as they have in cattle. Further work is required, because Cu injections in horses in alkaline areas of Ireland with raised Mo availability are said to benefit horses. It has been suggested that interference from Mo will cause anaemia and failure of osteoblastic function, which results in thinning of the cortex (shank) of long bones in the horse. Mo interference may be implicated in a widening of the growth plate and fractures of long bones in foals in Southern Ireland. Yet thiomolybdates, which bind Cu ($CuMoS_4$ is very insoluble), have not been detected in horses with dietary Mo concentrations of 10 mg/kg. Their formation may depend on the presence of a rumen and its microbial activity, that is, an extensive microbial activity in a region of the GI tract proximal to absorption sites for Cu.

Measurement of Cu status and dietary requirements

Excessive intakes of Cu are hazardous in sheep and to a lesser extent in cattle. Ruminants differ from nonruminants in the propensity of the former to store Cu in their livers at low dietary Cu intakes. On the other hand, very high dietary Cu levels (>4 mmol/kg DM) are required to increase liver Cu of ponies substantially (referred to by Suttle *et al.* 1996). (Note: any increase above the minimum dietary requirement seems to lead only to some additional liver Cu storage.) For this reason horses are less subject to intoxication by dietary Cu. There is a curvilinear relationship in ruminants between liver Cu and plasma Cu concentrations. In ruminant plasma, or serum, Cu values plateau and, excepting toxic crises, they rarely exceed 16 μmol/l, even with the liver at >800 μmol/kg fresh weight. In 44 horses Suttle *et al.* (1996) found a linear relationship:

$$Y = 7.00811 + 0.086019X \ (R^2, 0.168)$$

where Y is μmol Cu/l serum, X is μmol Cu/kg fresh liver (values were restricted to liver Cu <190 and serum Cu <29) and R^2 is the proportion of the variation, or sum of squares, attributable to the model (in this case a value of 0.168, i.e. 16.8%, of the variation is accounted for by the relationship that the equation describes). In 48 horses Suttle *et al.* (1996) found mean Cu concentrations of:

*   16.7 μmol/l serum, with 13.5 and 19.5 μmol/l as lower and upper quartile values;
*   113.7 μmol/kg fresh liver;
*   172.5 μmol/kg feed (11 mg/kg) with 61.1 and 233.8 μmol/kg as lower and upper quartile values.

The animal has a need to maintain a normal cellular Cu concentration, measurable as hepatic concentration. In the horse Suttle *et al.* (1996) concluded that adequate equine liver levels would be achieved by a dietary concentration of 20 mg/kg DM and that these liver and dietary concentrations correspond to 16 μmol/l serum. These authors propose 16 μmol/l as a threshold value to distinguish the normal from subnormal serum values and a serum level of 11.5 μmol/l to distinguish deficient from marginal liver Cu concentrations of 52.5 μmol/kg fresh weight. Unfortunately, as Suttle *et al*. point out, high serum Cu values occur following inflammation, infection or vaccination and liver samples are then preferable, but not readily available. (Plasma Cu may be unreliable as a guide to Cu status, as about 70% of the circulating Cu in the horse is present in the form of caeruloplasmin [EC 1.16.3.1], an acute phase protein.) Therefore an alternative cellular source may be preferred as a means of assessing Cu status.

Cu deficiency decreases the activity of Cu–Zn superoxide dismutase (EC 1.15.1.1) of leucocytes and platelets. The measurement of this activity is not easy. The author's colleagues (Williams *et al*. 1995) have routinely used mononucleated leukocyte and platelet Cu contents and platelet cytochrome c oxidase (EC 1.9.3.1) activities as guides to status. Mononuclear cell Cu shows good promise (Table 3.7). The Cu contents will be present mainly as enzyme cofactor Cu (see Chapter 12).

Cu is transferred to the fetal liver, which, like the neonatal liver, contains more Cu than that of older foals, or of their dams. Recent experiments in New Zealand have shown that when the dietary Cu of the grazing pregnant mare is increased from approximately 6 to 30 mg Cu/kg DM throughout the last 4–5 months of gestation then the Cu content of the foal's liver at birth can be increased by two-thirds. No clinical benefit was observed, yet there was evidence that the additional Cu reduced abnormalities of bone and cartilage development in the leg bones of these foals. In contrast, the supplementation of these foals from 3 weeks of age had no effect. Mare's milk may provide less Cu than the suckling foal requires daily. (Cu content of milk is about 3 μmol/l (0.19 mg/l), whereas pasture grass may contain 4–9 mg/kg DM and grass hay 10–12 mg/kg DM.) Therefore, with the object of increasing the Cu and Zn contents of the milk, and decreasing the risk of osteochondritis dissecans (OCD), Baucus *et al*. (1987) doubled the Cu and Zn contents of the mare's diet to 53 mg Zn/kg and 12 mg Cu/kg at parturition. Nevertheless, the Cu and Zn contents of the milk were unaffected. The Cu needs to tide the suckling foal over the period during which it eats little grass and dry food. These needs are normally met when the fetal liver stores about 300 μg Cu/g liver DM, or more. The stores should range between 300 and 600 μg/g when the pregnant mare has received adequate Cu in her diet. Although Hoyt *et al*. (1995b) found that miniature horses absorbed between 42.2 and 50.7% of the total Cu from diets containing 12 mg/kg, Ott and Asquith

Table 3.7   Range in leukocyte Cu and Zn contents in seven Shetland ponies receiving poor hay only, before and after supplementation with Cu and Zn for 50 days (author's unpublished data).

	Cu (μg 10^{-9}) in leukocyte cells	Zn (μg 10^{-9}) in leukocyte cells
Before supplementation	0.11–0.18	2.57–6.25
After supplementation	0.40–2.86	2.57–10.57

(1994) reported that serum Cu and Zn concentrations in foals were increased only when their mothers received chelated trace minerals during pregnancy, rather than inorganic sources.

The obligatory losses of Cu in the faeces of ponies amount to about 3.5 mg/100 kg BW daily in the presence of low levels of dietary Mo (Cymbaluk *et al.* 1981a,b). However, in order to allow for adverse interactions with other trace elements and to maximize iron (Fe) retention, a dietary intake of 15–20 mg/kg dry feed is recommended for growing horses. Foals need 25–30 mg/kg feed to reduce, but not eliminate, risk of cartilage erosion.

Cu and cartilage formation

(See also Chapter 8, growth of foals.) The bones and cartilage of Cu-deficient animals show increased defects and fragility, and contain an enhanced proportion of soluble collagen. This solubility is caused by a reduction in the cross-linking within the molecules of collagen and elastin that require Cu as the cofactor of lysyl oxidase (EC 2.3.2.3). Osteoblast function is inhibited by Cu deficiency, whereas osteoclast function is unaffected. Excess dietary Cu can interfere with bone metabolism causing inhibition of collagen synthesis and a loss of bone density. No observations of excess are available for the horse, although an excess of dietary Cu of up to 791 mg/kg diet in ponies for 6 months (Smith *et al.* 1975), leading to liver Cu concentrations of over 4000 mg/kg DM, apparently caused no liver damage and no adverse effect on fertility or other characteristic.

Zinc

A deficiency of dietary Zinc (Zn) in many domesticated animals, including the horse, depresses appetite and growth rate in the young, causes skin lesions and is associated with a depression of Zn concentrations in the blood. A deficiency of Zn in the rat and several other species causes abnormal development of ribs and vertebrae, cleft palate, micrognathia (undersized mandible) and agenesis (absence) of long bones, but there is little direct evidence of this in the horse. Excess Zn may exacerbate bony lesions induced by low Cu diets.

Zn as enzyme cofactor

Zn is a cofactor for over 200 enzymes in animals, either as part of the molecule or as an activating cofactor. The enzymes include alkaline phosphatase (EC 3.1.3.1), collagenase (EC 3.4.24.3) and carbonic anhydrase (EC 4.2.1.1), all required in bone formation. Alkaline phosphatase also requires Mg, and excess dietary Zn may inhibit the enzyme if Mg is displaced. A deficiency of Zn thus has fairly widespread physiological effects, but quite high dietary levels are required for toxicity.

An increase in dietary Zn from 26 to 100 mg/kg progressively increases serum Zn concentrations. However, the dietary requirement of the horse is probably less than

50 mg/kg and supplements normally used include zinc carbonate or sulphate; these inorganic salts possess a higher availability than do phytate salts of zinc in cereal grains and oilseed meals. The efficiency of absorption of Zn in all forms is probably affected more by diets rich in phytate than is the absorption of other trace elements, but, even so, high phytate concentrations are unlikely to depress the utilization of Zn by more than 30–40%. Amino acid chelation of trace elements avoids any interruption to Zn absorption by other metals and the competition for absorption is then between the amino acids. Experiments have shown that chelated mixtures of Mn, Zn and Cu, cf. inorganic mixtures, have increased the growth rate of hoof horn in yearlings given a pelleted concentrate containing 120 g protein/kg at the rate of 1 kg/100 kg BW with hay.

Zn is one of the less toxic of the essential trace metals, yet where there is industrial pollution of pastures, grazing animals may show signs of toxicity. Toxic dietary concentrations probably exceed 1000 mg/kg. Zinc intoxication is associated with reduced Ca absorption. A dietary level of 5.4 g Zn/kg causes anaemia, epiphyseal swelling, stiffness and lameness, including breaks in the skin around the hooves (Willoughby *et al.* 1972a,b).

Cu and Zn interactions

In man and most other domestic animals that have been investigated there is an antagonism between Zn and Cu. Excessive intakes of Zn, especially where dietary phytate and Ca are not abundant, can cause Cu deficiency if the dietary Cu level is marginal. Investigations in horses have not demonstrated similar effects. Hoyt *et al.* (1995b) found that dietary Zn concentrations of 73–580 mg/kg, provided as zinc oxide, had no influence on either Cu absorption or retention from diets containing 12 mg Cu/kg. However, as little as 100 mg Zn/kg diet for horses has been shown to increase the faecal loss of Cu and to lower blood Cu by about 10%. This change in blood concentration may not itself reflect any change in Cu absorption efficiency. Nevertheless, as Cu adequacy seems to be critical for proper bone development we should assume that dietary Zn at concentrations above 100 mg/kg are to be avoided.

Chelated minerals do not compete one with the other for intestinal absorption sites. Experiments with chelated Zn and Cu given to pregnant mares have led to an increase in plasma concentrations of Zn and Cu in their foals. It may be concluded that the trace element allowances of the pregnant mare are more important than those given during lactation for the welfare of the foal.

Manganese

Manganese (Mn) is required as a cofactor for glycosyltransferases, which catalyse the transfer of a sugar from a nucleotide-diphosphate to an acceptor, and so Mn is essential in several stages of glycosaminoglycan-chondroitin sulphate formation. Thus, epiphyseal cartilage and bone matrix formation are compromised by a deficiency of Mn. Mn is also required as a cofactor in Mn-containing superoxide dismutase.

A deficiency of Mn is thought to be a cause of enlarged hocks, and, by affecting the growth plate, to shorten legs with characteristic knuckling-over of joints. In the USA, excessively high concentrations of Ca in some samples of alfalfa have been said to precipitate a flexural deformity of the legs of growing horses that was rectified by Mn supplements. The young also seem to suffer lameness and incoordination of movement if they lack sufficient Mn, and Mn deficiency is a possible explanation of tiptoeing in situations where suckled foals are on pasture containing <20 mg Mn/kg DM. A severe deficiency can give rise to resorption *in utero*, or death at birth, and lesser deficiencies may provoke irregular oestrous cycles. Lawrence *et al.* (1987b) reported normal plasma Mn values in yearling ponies of 100–180 μmol Mn/l serum.

Iron

Most natural feeds, apart from milk, are fairly rich sources of iron (Fe), even when the availability may be questionable, and deficiences are unlikely unless the horse is anaemic through heavy parasitization. The foal is born with an adequate store of liver Fe and the foal's grazing activity is normally an adequate supplement to the mare's milk, which contains meagre amounts of most trace elements with the possible exception of selenium (Se) in mares supplemented with this element. The levels of Fe in Arabian mare's milk are shown in Table 3.5.

A deficiency of Fe causes anaemia, but evidence of dietary Fe deficiency anaemia in horses is rare. A dietary concentration of 50 mg/kg DM should be adequate for growing foals. Only 40 mg/kg is said to meet the maintenance requirements of adults. The basal diet used by Lawrence *et al.* (1987b) contained 436 mg Fe/kg (typical of many natural ingredient diets) and was composed of coastal Bermuda hay, maize, soya, alfalfa and minerals, including a supplement providing 40 mg Fe/kg basal diet. The addition to it of 500 or 1000 mg Fe/kg, as citrate, brought no benefit.

True dietary Fe deficiency may be confused with pseudo-Fe deficiency. Anaemia is not synonymous with Fe deficiency. Haemoglobin (Hb), which contains 67% of Fe stores in the horse, is synthesized preferentially to that of nonhaemoglobin Fe-containing proteins in Fe-depleted animals. On the other hand, blood packed cell volume (PCV) and Hb may be decreased by parasitism.

Fe status is assessed most reliably by serum ferritin concentration. Total serum Fe is an unreliable measure of status and may decline during microbial invasion, tissue injury, immunological reactions and in any inflammatory process causing an acute phase response, initiated by interleukin-1. Normal ferritin values fall between 100 and 200 μg/l serum. Deficiency may be assumed when values fall below 50 μg/l, and Fe overload may exist when values exceed 400–450 μg/l.

Adverse effects of Fe supplements

Fe-containing hematinics are widely used in the horse industry. Dietary Fe supplements and injectable Fe are used in the mistaken belief that PCV and Hb concentration can be regularly increased. The risk of Fe toxicity is a real one and antagonistic interactions with other trace elements may occur. Large, but natural, intakes of Fe in cattle can cause Cu deficiency. An induced Cu deficiency of this kind seems to be less likely in horses, although reductions of splenic Cu and Mn, associated with intakes of 1400 mg Fe/kg diet, were observed by Lawrence *et al.* (1987b). These workers also observed depressed serum and liver Zn with dietary Fe intakes of only 890 mg/kg. Interactions of this kind may arise from competition for binding by intestinal transferrin and/or other transport mechanisms.

Most Fe is stored in the liver and spleen, with no mechanism for disposal of excess, and hence excess can cause hepatitis and other forms of liver damage. The toxic level depends on several factors, including concurrent disease processes, previous liver damage and vitamin E and Se status. Fe is involved in oxidation–reduction reactions. When glutathione peroxidase activity is depressed and vitamin E status is reduced and there is an Fe overload, catalase activity increases with consequent liver damage, coagulopathy and raised mortality. Fe overload in the spleen may be the cause of the lymphopaenia frequently observed in neonatal cases.

Neonatal foals are particularly susceptible to iatrogenic Fe toxicity. Ferrous-Fe (divalent) is more soluble than ferric-Fe (trivalent) and therefore ferrous sulphate or fumarate is generally used in supplements and they are more likely to be a cause of toxicity. Ingestion of a single large dose of ferrous fumarate, shortly after birth, can cause death within 5 days. Daily administration of as little as 300 mg Fe in the ferrous form has been associated with signs of Fe toxicity in adult TBs. In extended Fe intoxication there is frequently:

(1) hepatic dysfunction, expressed as lethargy;
(2) yellow discolouration and petechial haemorrhages of mucous membranes; and
(3) thrombocytopaenia and elevated activities of both serum GGT (EC 2.3.2.2) and alkaline phosphatase (ALP).

These enzyme changes are indicative of cholestatic hepatopathy, with periportal bile ductule proliferation, as one origin of these enzymes is biliary epithelia.

Fluorine

Fluorine (F) readily substitutes for the hydroxyl ion in bone and teeth hydroxyapatite, creating a more stable crystal of Ca, P and Mg. It does not diffuse into formed bone, but becomes incorporated during bone formation. F increases osteoblast number by increasing osteoprogenitor cell proliferation. Risks of dietary excess (fluorosis) in grazing horses probably exceed risks of deficiency, as F has a narrow therapeutic index. Excess contamination of pastures, especially from brickworks, causes a softening, thickening and weakening of bones, through defects in mineralization, which are probably not prevented by Ca and vitamin D. A decrease in industrial effluent in many countries has ensured that very few cases occur today. The horse seems to excrete more F in its faeces than do cattle, but dietary concentrations should not exceed 50 mg/kg. A world shortage of sources of digestible P has led to an increase in the use of rock phosphates; as some of these are rich in F a careful scrutiny of their composition is essential before purchase and use.

Iodine

Iodine (I) is a relatively rare element in the earth's crust and it does not seem to be required by mono- and dicotyledonous plants, in which the concentration is low. In man and animals both I deficiency and toxicosis generally result in hypothyroidism (toxicosis may cause hyperthyroidism in some individuals). Goitre may then occur with hyperplasia and hypertrophy of the thyroid gland, induced by the elevated secretion of thyroid-stimulating hormone (TSH) originating in the anterior pituitary in the absence of feedback inhibition. This is not necessarily the outcome of I toxicosis in the pregnant mare (see below).

Insofar as the fetus is concerned the mare is at risk throughout gestation, and maybe even shortly before fertilization, in respect of both I deficiency and toxicity. When receiving insufficient dietary I mares may show no external signs, but may exhibit abnormal oestrous cycles and then produce hypothyroid foals. Serum triiodo-thyronine (T_3) and thyroxine (T_4) concentrations in the mare are low, generally below 1.3 and 19 nmol/l (0.7 and 15 µg/l), respectively (Table 3.8). The I concentration of deficient mare's milk is below 20 µg/l. A deficiency with the above signs is shown when mares graze inland pasture areas that are frequently deficient in I.

Deficiency signs in foals

Deficiency signs in foals are enlarged thyroid glands, weakness, persistent hypothermia, respiratory distress and high neonatal mortality. There is an increased susceptibility to infectious disease and respiratory infections are frequent. T_4 levels in foal serum will be low, whereas T_3 concentration may be normal. Plasma T_3 and T_4 levels are generally not too helpful in diagnosis and their correlation is poor. Nevertheless, T_4 concentration is depressed in hypothyroidism. Of greater value is

Table 3.8 Normal ranges for plasma T_3 and T_4 concentrations in TBs (nmol/l).

	T_3	T_4
Pregnant mares	0.6–2	3–56
Foals*	2–10	10–150
Adults in training	0.3–2	5–28

*T_3 levels increase to a maximum at 48 h of age and then decline.

inspection of feed and determination of I concentrations in feed samples and in plasma.

Toxicity signs in mares and foals

In mares given 300–400 mg I daily in the feed infertility and abortions occur, some mares developing hyperthyroidism, expressed as elevations in the plasma concentration of both T_3 and T_4 and suppressed plasma TSH, or they may develop hypothyroidism, expressed as normal T_3 and depressed T_4, although clinically they may appear euthyroid. At birth the foal is hypothyroid, frequently with colloid goitre, variably sized thyroid follicles, containing a single layer of low cuboidal cells and there are low concentrations of circulating T_3 and T_4. Interestingly the tracheal mucosa reveals squamous metaplasia, reminiscent of vitamin A deficiency. In man excess circulating iodide induces what is called the Wolff-Chaikoff block which inhibits I uptake, preventing the synthesis of T_4 from I. In many species this is a temporary phenomenon, but in the horse the block may depend on the precise level of I intake, although in the author's experience it may be only temporary. In the neonatal foal I toxicity blocks the release of T_3 and T_4 from the follicle, apparently by interfering with colloid proteolysis in the acini. Some pathologists give weight to the $T_3:T_4$ ratio, although in the author's experience the two hormones are poorly correlated in the healthy horse. Algae, such as red wrack seaweed, generally contain high concentrations of I and the feeding of dried kelp, a seaweed, to pregnant mares is frequently a cause of iodism in foals, with similar clinical signs to those of an I deficiency. There have also been cases where foals given excessive thyroprotein supplements develop focal depigmentation.

If the daily I intake of a pregnant 550-kg mare is between about 40 and 400 mg I, goitre may appear in her offspring. At the upper end of the range the mare may have goitre. Intakes by the mare of 30–50 mg I daily may cause only enlargement of the thyroid in her offspring, but the foal may be euthyroid and present no other abnormality, yet hirsute foals are seen. If the daily intake exceeds about 100 mg I for any length of time during pregnancy, her newborn foal is likely to show additional signs of hypothyroidism, expressed as weakness, lethargy, high neonatal mortality, poor muscular development and, most patently, osseous dysplasia of long bones. The author's observations indicate that these include angular deformity, tendon contraction and hyperextension of the lower limbs with poor ossification of the carpal and tarsal bones. This extension causes the foal to walk on its heels with its toes raised from the ground. Abnormal growth of the bones occurs both at the epiphysis and in the shank of long bones. These bones are thin and small with cortical thickening. This thickening appears to result from a greater reduction in osteoclasia than from osteogenesis. Observations of protruding jaws (mandibular prognathism) and parrot mouth (brachygnathia) have also been recorded.

The reason for these consequences is that I is concentrated in the placenta which delivers excessive amounts to the fetus, causing congenital malformations and concentrations of I in fetal blood two to three times that of the dam's blood (yet T_3 and

T_4 may be depressed). I is also concentrated in the mammary gland and is secreted in the milk. If the excessive I source given to pregnant mares is excluded from the latter's diet well before parturition the euthyroid state may be reimposed and normal foals may result. Involution of the thyroid, congenitally enlarged in young foals owing to I excess, can be brought about by removal of the I source(s). This includes removal of the mare's milk. Normal development may occur, as long as bony abnormalities have not already developed. The foal may be slightly undersize, as hypothyroidism is associated with decreased synthesis and release of growth hormone.

Goitrogens

I is required by animals for the synthesis of the hormones T_4 and the more potent T_3. I is absorbed from the gut principally as iodides. Some feeds (raw soyabeans, white clover and cabbage) contain goitrogens (see Chapter 5) which can inhibit the uptake of free I by the colloid proteins of the thyroid, preventing its incorporation into tyrosine. Thioglucosides, thiocyanates and perchlorates may be goitrogenic. The effects of these substances can be overcome by increasing the I intake. However, the goitrogenic effects of thiouracil-, thiourea- and methimazole-containing drugs cannot be overcome in this way, as they interfere with T_4 synthesis. (See 'Selenium and iodine interactions', this chapter.)

Supplements

Diets supplemented with between 0.1 and 0.2 mg I/kg should adequately meet the requirements of horses. The I is preferably added as potassium iodate rather than as potassium iodide. The reason for this is that iodides are readily oxidized to I which is volatile and so gradually lost from premixes and diets; however, iodates are more toxic to handlers in concentrated form and should be handled with care.

Treatment of hypothyroidism

Young foals that are diagnosed as having goitre should not be treated with iodide, or iodate, salts in the first instance. After receiving colostrum they should be removed from their mother's milk to a source known to contain normal amounts of I. These foals are first treated with thyroxine while other determinations are undertaken to determine the cause of the malady. The author's evidence is that the thyroxine rapidly raises blood T_4, whereas treatment with I salts has only a marginal initial influence on plasma T_4.

Selenium

Selenium (Se) has gained prominence since the realization that it forms an integral part of the GSH-Px (EC 1.11.1.9) molecule. This enzyme catalyses peroxide detoxification in bodily tissues during which reduced glutathione (GSH) is oxidized; it is closely involved with the activity of α-tocopherol (vitamin E), which protects polyunsaturated fatty acids from peroxidation. The requirement for α-tocopherol and Se is increased in the presence of high levels of dietary polyunsaturated fatty acids – cod liver oil, linseed and corn oils – and, in fact, pasture grass. A possible case for fish oils is put in Chapter 5, although they require higher levels of protection from vitamin E, owing to a greater degree of unsaturation than of vegetable oils.

Feed fats contain variable levels of polyunsaturated fatty acids; vegetable oils are generally much richer sources of these acids than are hard animal fats. Although linoleic acid (an n-6 fatty acid) is perhaps the most abundant polyunsaturated fatty acid in vegetable oils, some oils, including those present in grasses, are richer in α-linolenic acid (an n-3 fatty acid), which is particularly sensitive to peroxidation in tissues and which may therefore be more likely to cause problems than linoleic acid (however, α-linolenic acid may otherwise have benefits; see Chapter 5, Fat Supplements). The alkali treatment of roughage to improve its digestibility has attracted interest. It must, however, be appreciated that such treatment destroys α-tocopherol and β-carotene, and unless appropriate supplementation is given signs of myopathy may occur in animals consuming significant quantities of such roughage.

Horses require about 0.15 mg available Se/kg feed to meet their dietary requirement for this element. Se supplementation is normally given as either sodium selenite, or sodium selenate, and evidence in horses (Podoll *et al.* 1992) indicates there is no difference between them in potency. Measurement in laboratory animals, however, shows that organic plant sources of Se are more potent than inorganic, although barium selenate and amino acid chelated Se possess a high availability.

Deficiency

Se deficiency produces pale, weak muscles in foals and a yellowing of the depot fat; it is known that this form of muscular dystrophy in foals is related to a subnormal level of blood Se and a depressed activity of the enzyme GSH-Px. It is essential that pregnant mares receive adequate Se in their diet as their status affects that of their foals at parturition, whereas their milk provides only modest amounts (Lee *et al.* 1995). Serum Se values may fall to <0.3 μmol/l in foals and reduced amounts among TBs in the UK have been associated with poor racing performance (Blackmore *et al.* 1979, 1982). Only in extreme deficiencies, not normally seen in adult animals, is there sufficient muscle damage for extensive membrane leakage of enzymes such as aspartate amino transferase (AAT; EC 2.6.1.1) and creatine kinase (CK or CPK; EC 2.7.3.2) to be detected. The normal peroxidation occurring in muscles after exercise has, however, not been shown to be reduced by supplementary Se. Se deficiency *in vitro* changes neutrophil function and so may depress resistance to infection.

The Se content of herbage, cereals and other crops depends on the Se content of the soil in which they were grown. Areas of deficiency include New Zealand, and low soil concentrations occur in parts of Scotland and the eastern and western states of the USA (Scottish wheat 0.028 cf. Canadian wheat 0.518 µg Se/g DM; Barclay & MacPherson 1992). The great plains and central southern states of the USA have regions of high Se concentration in crops. These differences are reflected in blood Se concentration, in GSH-Px activity and in performance. In the western USA concentrations of Se in the blood show a negative correlation with the incidence of reproductive diseases in mares (Basler & Holtan 1981). In this particular study, blood concentrations ranged from 1.2 to 3.1 µmol/l and dietary concentrations ranged from 0.045 to 0.461 mg/kg, i.e. deficient to slightly excessive – normal blood Se is said to range from 0.8 to 2.8 µmol/l in adult mares (according to Blackmore & Brobst 1981), possibly suggesting that some 'normal' animals may have impaired reproductive capacity owing to Se inadequacy. Amounts of Se in the serum are closely correlated with whole-blood Se, so serum values in the horse seem, on present evidence, to be a good and reproducible measure of status. An increase in dietary Mn from 38 to 50 mg/kg increases both Se retention and blood Se, according to Spais *et al.* (1977).

Toxicity

Se is highly toxic to animals and also to persons handling the salt in highly concentrated forms. The minimum toxic dose of Se through continuous intake is 2–5 mg/kg feed and acute toxicity is caused in sheep given amounts equal to, or greater than, 0.4 mg Se/kg BW in a single dose.

Outside the UK there are areas where soils can contain in excess of 0.5 mg Se/kg and amounts of between 5 and 40 mg/kg DM are found in certain accumulator plants. Concentrations of up to several thousand mg/kg have been detected in species of milk vetch (*Astragalus*). Various species of woody aster (*Xylorhiza*) and goldenweed (*Oonopsis*), which grow in low-rainfall areas, are also indicator plants, containing relatively high levels of Se. Toxicity is thus more common in dry regions, but horses select grasses rather than these toxic weeds where there is adequate grazing, as the indicator plants are unpalatable. Where grass is sparse, animals suffer from 'alkali disease' in which excessive Se causes a loss of hair on the mane and tail, lameness, bone lesions, including twisted legs in foals, and sloughing of hooves. Plate 3.1 (S. Ricketts personal communication) depicts signs in Se-intoxicated adult horses. The diseased hooves were hollow, indicated by percussion, and there was greying of hair in the mane and tail, a characteristic not previously noted (Plate 3.1a–c). The abnormal hoof horn and hair contained much higher Se concentrations than did the healthy material. Doses of Se received by intoxicated horses are normally impossible to estimate accurately from field evidence owing to an accompanying loss of appetite. There is no simple remedy for this intoxication apart from removal of the animals from the region if destruction of the seleniferous plants is impractical.

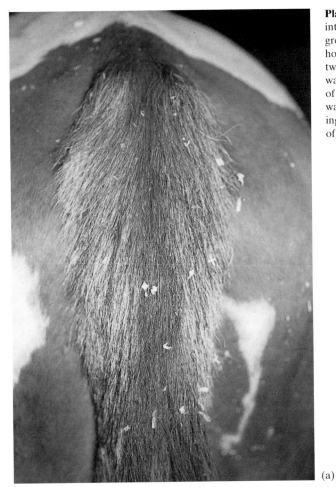

(a)

(b)

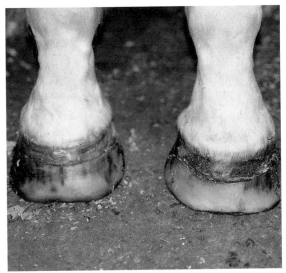

(c)

Plate 3.1 A selenium-intoxicated horse. (a) Loss and greying of tail hair; (b) crumbly hoof below coronary band; (c) two feet showing healthy hoof wall distal to (prior to) a period of intoxication during which the wall detached from the underlying tissue. (Photographs courtesy of Sidney Ricketts.)

Se and I interactions

Selenium functions not only as a component of glutathione peroxidase (GSH-Px; EC 1.11.1.9) but also as part of the enzyme 1,5′-iodothyronine deiodinase (EC 3.8.1.4), required in the conversion of T_4 to T_3. As T_3 is the more potent hormone, Se deficiency is said to be associated with human cretinism in northern Zaïre. There do not appear to be accounts of goitre in Se-deficient foals, so that the equine Se requirement for the functioning of this deiodinase may be less than that for normal function of GSH-Px.

Chromium

Chromium (Cr) is essential for normal carbohydrate metabolism as a potentiator of insulin action, and so it is found in insulin sensitive tissues where it stimulates glucose clearance. When given as an inorganic supplement, trivalent Cr (III) at dose levels of 300–500 µg Cr/kg diet accelerates glucose clearance. Trivalent Cr is relatively nontoxic, whereas hexavalent Cr, present in chromates and dichromates and acting as a pro-oxidant, is orally 10 to 100 times more toxic. Rats show no adverse effect from 100 mg CrIII/kg diet. The recommended safe daily intake of CrIII for humans ranges from 50 to 200 µg.

Cr appears to function as a specific complex, known as the glucose tolerance factor, containing niacin. Organic dietary forms, as found in brewer's yeast, may be more potent than inorganic. Barley bread has been traditionally given to individuals suffering from diabetes mellitus in Iraq. It appears that barley is a richer source of Cr than is wheat, samples typically containing over 6 mg Cr/kg. Some yeasts are very rich sources of organic Cr, and horses given 5 mg Cr, as yeast, in a natural basal diet providing 12 mg Cr/day, showed a decrease in both glucose and insulin responses. During exercise a lower plasma cortisol response and raised plasma TAG concentrations were observed with the high Cr yeast. As insulin inhibits the activity of intracellular hormone-sensitive lipase in adipocytes, the lower insulin response probably allowed more NEFA to be mobilized for both oxidation and return to the liver, where NEFA would be re-esterified, entering the circulation as TAG, contained in very-low-density lipoproteins. The Cr would seem to have reduced insulin resistance, reduced stress and facilitated energy metabolism in these horses.

Nickel

Nickel (Ni) is both essential for and toxic to the functioning of ruminal microorganisms, and so the dietary concentration may be critical for the horse. In the human subject toxicity from Ni, CrVI and Fe is mostly confined to pollution from metal industries.

Silicon

Although Silicon (Si) is the second most abundant element in the earth's crust, its absorption may be influenced by dietary Mo and aluminium (Al) and its retention by animals is sufficiently low for it to be classified as a trace element. Connective tissue Si is found in osteoblasts. It is a component of animal glycosaminoglycans, and their protein complexes, so it appears to be essential for bone matrix formation. It seems to have a structural function acting as a cross-linking element in the polysaccharide chains of the proteoglycans linked to the collagen (protein) of cartilage. Collagen synthesis also requires Si for the optimal activity of proline hydroxylase (EC 1.14.11.2). Si is probably also required for bone mineralization, as deficiencies in other species include abnormalities in the skull and leg bones. Independently of vitamin D-induced abnormalities, long bones have a reduced circumference, thinner cortex and reduced flexibility resulting from Si deficiency.

American evidence (Nielsen *et al.* 1993) indicates that the dietary supplementation of growing quarter horse foals from 6 to 18 months of age with sodium zeolite (an Al silicate) at the rate of 18.6 g/kg total diet increased subsequent plasma Si concentration and speed. It also extended work time before leg injury, reducing its frequency and delaying withdrawal from training. The critical period seems to be up to 1 year of age in quarter horses, as supplementation commencing at the yearling stage of growth had no physiologically important effects on running performance from 18 months of age. Al salts can also protect against toxic metal absorption. Whether this is a factor in the observation of Nielsen *et al.* (1993) is unknown. However, Al and Mo can inhibit tissue accumulation of Si and inhibit bone formation by reducing osteoblastic activity, osteoid mineralization and matrix formation. Whether the toxicity of Al is simply exercised by inhibiting Si use is not known, but it should be questioned whether an Al silicate is the most appropriate vehicle for providing Si. Further work in this area will undoubtedly occur.

Boron, gallium and vanadium

No equine information is available on the elements boron, gallium and vanadium.

STUDY QUESTIONS

(1) The use and function of calcium is now well understood. What factors affect calcium adequacy and use and why is it associated with a continuing problem in horses?
(2) Most feeds contain adequate potassium and dietary sources are generally well used. Why is potassium of any concern from nutritional and physiological viewpoints?
(3) If weanlings were said to have problems with (a) mineral status, or (b) trace element status, how would you set about resolving the position?

FURTHER READING

Cymbaluk, N.F., Christison, G.I. and Leach, D.H. (1989) Nutrient utilization by limit- and *ad libitum*-fed growing horses. *Journal of Animal Science*, **67**, 414–25.

Suttle, N.F., Gunn, R.G., Allen, W.M., Linklater, K.A. and Wiener, G. (eds) (1983) Trace elements in animal production and veterinary practice. *Proceedings of the Symposium of the British Society of Animal Production and British Veterinary Association*. BSAP Occasional Publ. No. 7, Edinburgh.

Underwood, E.J. (1977) *Trace Elements in Human and Animal Nutrition*, 4th edn. Academic Press, New York and London.

Williams, N.R., Rajput-Williams, J., West, J.A., Nigdikar, S.V., Foote, J.W. and Howard, A.N. (1995) Plasma, granulocyte and mononuclear cell copper and zinc in patients with diabetes mellitus. *Analyst*, **120**, 887–90.

Chapter 4
Vitamin and Water Requirements

The drink of all brute creatures being nothing but water, it is therefore the most simple . . . as it is the proper vehicle of all their food, and what dilates the blood and other juices, which without sufficient quantity of liquid, would soon grow thick and viscid.

W. Gibson 1726

VITAMIN REQUIREMENTS

Vitamins are nutrients that horses require in very small quantities, although the actual needs for each differ considerably. For example, the dietary requirement for niacin or for α-tocopherol (vitamin E) may be at least 1000 times that for either vitamin D or vitamin B_{12}. However, measurements of vitamin requirements lack precision; there is little direct evidence of the requirements for any of the vitamins in the horse and assertions are largely based on measurements in other domestic animals.

Like other mammals, horses require vitamins for normal bodily functions. These requirements will be met by vitamins naturally present in feed, supplementary sources, tissue synthesis and, in the case of vitamin K and the water-soluble B vitamins, additional amounts supplied from microbial synthesis in the intestinal tract. The tissue requirements are complicated by the synthesis of ascorbic acid from simple sugars in the horse's tissues, the production of vitamin D in the skin as a reaction to ultraviolet light, the tissue synthesis of niacin from the amino acid tryptophan, and the partial substitution of a need for choline by methionine and other sources of methyl groups.

Dietary requirements for specific vitamins are therefore affected by circumstance; for example, where horses are kept indoors or are maintained in very high northern latitudes, or indeed have highly pigmented skins and thick coats of hair, their dietary requirements for vitamin D will be greater. Young foals possess a poorly developed large intestine so that little dependence on it for B vitamin or vitamin K synthesis may be assumed. Foals grow fast and, in common with other domestic animals, one must assume that as their tissue requirements exceed those of adults, so their dietary needs are far greater. Tissue demands will also be larger for lactating mares than for barren mares, but as the former are likely to be eating more, this tends to lessen the difference per unit of feed. Adult horses are able to draw on much larger reserves of some vitamins to see them through periods of deprivation. For example, a good grazing season on high-quality grass can satisfy the mare's vitamin A requirement through about 2 months of the ensuing winter. In some instances, however, old mares or other horses have a diminished ability to assimilate nutrients, particularly the fat-soluble vitamins, through a decline in digestive efficiency with age and possibly through the debilitating damage of intestinal parasites, and there is some evidence that the fertility of old barren mares benefits from larger than normal doses of vitamin A. A role for β-carotene, other than as a precursor of vitamin A, is equivocal in the horse, and indeed nothing is directly known of possible functions of the hundreds of other carotenoids in equine metabolism, although some may be suspected.

Inferences drawn from other domestic animals cannot be used in estimating the effect of strenuous work on vitamin needs. It has been asserted, with some justifica-

Table 4.1 Adequate concentrations of available vitamins*/kg total diet (assuming 88% DM).

	Mature horse		Mares: last 90 days' gestation Stallions	Lactating mare	Weaned foal	Yearling
	Maintenance	Intense work				
Vitamin A (iu)	1600	1600	3500	3000	3000	2500
Vitamin D (iu)	500	500	700	600	800	700
Vitamin E (mg)	50	80	60	60	70	60
Thiamin (mg)	3	4	3	4	4	3
Riboflavin (mg)	2.5	3.5	3	3.5	3.5	3
Pyridoxine (mg)	4	6	5	6	6	5
Pantothenic acid (mg)	5	10	5	8	10	5
Biotin (µg)	200	200	200	200	200	200
Folic acid (mg)	0.5	1.5	1	1	1.5	0.5
Vitamin B_{12} (µg)	0	5	0	0	15	0

*There is no evidence of a dietary requirement for vitamin K, niacin or ascorbic acid in healthy horses.

Table 4.2 Signs of advanced vitamin deficiency in the horse and pony. The status should always be kept well above that leading to these signs to provide positive benefits.

Vitamin A	Anorexia, poor growth, night blindness, keratinization of skin and cornea, increased susceptibility to respiratory infections, infertility especially in older mares, lameness
Vitamin D	Reduced bone calcification, stiffness and abnormal gait, back pain, swollen joints, reduction in serum calcium and phosphate
Vitamin E	Pale areas of skeletal muscles and myocardium, red cell fragility, reduced phagocytic activity
Vitamin K	Extended blood-clotting time (prothrombin time), but this is rarely seen unless it is induced by drugs
Thiamin	Anorexia, incoordination, dilated and hypertrophied heart, low blood thiamin and elevated blood pyruvate
Folic acid	Poor growth, lowered blood folate
Biotin	Deterioration in the quality of the hoof horn expressed as dish-shaped walls that crumble at the lower edges so that shoe nails fail to hold

tion, that the dietary requirements for certain B vitamins involved in energy metabolism are increased for animals in heavy work, both in total and per unit of feed. This conclusion is reinforced by a frequent decline in appetite during extra hard work. The nutritional requirements of work should, however, not be confused with pharmacological responses. For example, thiamin given parenterally in single doses of 1000–2000 mg is said to have a marked sedative effect on nervous racehorses.

Recommended dietary allowances for vitamins are given in Tables 4.1 and 6.20 and a summary of deficiency signs is given in Table 4.2.

Fat-soluble vitamins

Vitamin A (retinol)

Grazing horses derive their vitamin A from the carotenoid pigments present in herbage. The principal one of these is β-carotene and fresh leafy herbage contains the equivalent of 100000–200000 iu vitamin A/kg DM for most domestic animals [1 iu equals 0.3 μg retinol (vitamin A alcohol)]. The horse, however, seems to be relatively inefficient in the conversion of β-carotene to vitamin A, and the carotene in good-quality grass or lucerne hay is estimated to possess only a fortieth of the value, weight for weight, of retinol (vitamin A). Although fresh pasture herbage would normally provide well in excess of the requirement, hay used for feeding horses in the UK provides meagre amounts of carotene and particularly where it is more than 6 months old should be considered to contribute none, unless it is visibly green. Greiwe-Crandell *et al.* (1995) found that reserves in horses deprived of pasture were depleted in about 2 months.

Various signs of vitamin-A deficiency have been recorded as it has several important functions, among them the integrity of epithelial tissue, normal bone development and night vision. One of the earliest signs of deficiency includes excessive lacrimation (tear production); a protracted deficiency may cause impaired endometrial function in the mare. Figure 4.1 indicates that these clinical signs and symptoms of deficiency occur under fairly extreme conditions of deprivation. As many horses are stabled for most of their time, when they consume little or no fresh herbage, the possibility of this deprived state exists. However, few cases of overt vitamin-A deficiency are recognized among stabled horses in western countries as most routinely receive supplementary synthetic sources. There is evidence of responses in several animal species to rates of intake above the minimum requirement level (Fig. 4.1) under the stress of certain chronic transmissible diseases. Some forms of infertility, particularly in elderly mares, may respond to vitamin-A therapy and responses among TBs in training suffering tendon strain and lameness have been noted (Abrams 1979).

Measurement of vitamin A status from blood values
Blood plasma retinol has a relatively low concentration in horses. Butler and Blackmore (1982) give a range of 60–300 μg/l for TBs in training in the UK. This

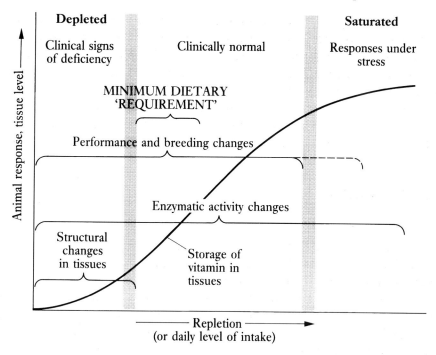

Fig. 4.1   General relationship of animal response to level of vitamin intake.

range is consistent with that given for other countries, but a marginal dietary deficiency causes little change in this. Plasma retinol is sustained by hepatic reserves and so varies only to a small extent with intake. The vitamin A status can be assessed by use of the relative dose response (RDR). For the horse this is determined in jugular blood as:

$$100(A_4 - A_0)/A_4 = RDR\%$$

where A_4 is plasma retinol concentration 4h following feeding 10000 iu vitamin A, and A_0 is fasting plasma retinol concentration (Jarrett & Schurg 1987). A deficiency is indicated where RDR% is greater than 20%.

The response at 4h, measured as total vitamin A, probably represents a combination of retinyl ester, in the process of absorption, plus the subsequent release of retinol from the liver. However, the object of the assay is to measure hepatic release of retinol, which peaks at 12–15h postdose of the ester. Thus, a more sensitive RDR test is obtained by measurement of plasma retinol only, by high performance liquid chromatography (HPLC), at 0h and at about 14–15h following the dose. The revised procedure employs an oral dose of 123.5mg retinyl palmitate (224152 iu) (Greiwe-Crandell *et al.* 1995):

$$100(A_{14} - A_0)/A_{14} = RDR\%$$

By this test a deficiency probably occurs when RDR% exceeds 10–12%.

Requirement

The NRC (1989) recommends 9–18 µg/kg BW retinol daily. This is equivalent to 15000 iu daily for a 500-kg horse (or, say, 1500 iu/kg total feed). Most commercial feeds are supplemented with at least 10000 iu/kg, so providing much more on a daily basis, where an allowance of, say, 4–7 kg is given. Donoghue *et al.* (1981), however, reported that the optimum vitamin A intake for normal growth rate of foals was 17 µg/kg BW daily, for adequate liver-secreted serum constituents 65 µg/kg BW daily and for normal haematopoiesis 120 µg/kg BW daily. As 120 µg/kg is equivalent to 15000 iu/kg feed the optimum intake is clearly not well established. Moreover, the presence of respiratory infections may increase the requirement, as these infections are frequently associated with a decrease in plasma vitamin A concentration. A dose rate of 120 µg/kg BW daily is equivalent to 150000–200000 iu daily for a horse, and, at an intake rate five or more times this, toxic reactions should be expected.

Preformed vitamin A and β-carotene, in common with all other fat-soluble vitamins, are unstable, being subject to oxidation, so that natural feed gradually loses its potency. Synthetic forms of vitamin A are stabilized and, when undiluted and stored in reasonable conditions, they retain more than three-quarters of their potency for several years. The grain/protein concentrate portion of the diet supplemented to the extent of 10000 iu vitamin A/kg (3 mg retinol/kg) should allow all foreseeable vitamin A demands to be met (with the possible exception of haematopoiesis in foals, as indicated above). In practice, deficiencies may arise from failure to supplement feed or from the provision of badly stored old feed.

Vitamin-A deficiency can also be induced in livestock by other dietary abnormalities. Evidence from several grazing species indicates that signs of deficiency can arise in stock subsisting on poor forage marginal in carotene, which, like much horse hay, contains less than about 7% of crude protein and is deficient in zinc. Authenticated evidence for such interactions in equids is unavailable, but they probably explain the observations of Jeremiah (14:6): 'The wild asses stand on the bare heights, they pant the air like jackals; their eyes fail because there is no herbage'. Thus, stock under range conditions should also be given adequate supplementary protein and trace elements.

β-Carotene

β-Carotene, a plant pigment, is the precursor of retinol (vitamin A); however, the pigment appears to function in the animal body independently of this precursor function. There is evidence that cows deprived of β-carotene respond to its supplementation by an improvement in fertility. It may simply function as an intracellular antioxidant, for in rats stressed with high intakes of corn oil the consumption of β-carotene lowered the activity of liver superoxide dismutase. However, the corpus luteum accumulates β-carotene and 1 g/mare daily has been shown to stimulate ovarian activity in both the follicular and corpus luteum phases and to increase circulating progesterone during dioestrus. Treated mares are said to display an enhanced oestrus, an increased pregnancy rate and reduced cycling disorders. This treatment of stabled breeding mares has, in some studies, led to blood β-carotene concentrations similar to those found in pastured mares. However, problem barren mares are said to accumulate lower amounts in the blood. 'Foal heat' diarrhoea of foals is said to be reduced by the elevated blood β-carotene concentrations of their dams.

A report from Edinburgh (Watson *et al.* 1996), on the other hand, indicates that water-soluble synthetic β-carotene is not absorbed by the horse as no increase in plasma β-carotene occurred with dietary supplements of either 1.8 or 10 mg/kg BW *per os* daily. Moreover, no change in cyclical ovarian activity, or in plasma progesterone concentration, was produced in that study. Pasture legumes are particularly rich sources of β-carotene and these natural sources are absorbed by horses, although the equine response is relatively low. Mixed grass–clover pasture contains at least 400 mg β-carotene/kg DM during the growing season, providing many times the equivalent of the vitamin A requirement. Fonnesbeck and Symons (1967) recorded plasma concentrations of 70 μg/l β-carotene from daily intakes of 650 mg lucerne β-carotene in comparison with 40 μg/l β-carotene from 170 mg β-carotene derived from fescue grass (also see lucerne, Chapter 5).

Breeding mares deprived of pasture during the winter can become depleted of the fat-soluble vitamins A, D and E, as grass hay, or silage, is generally not an adequate source of these vitamins. Blood concentrations tend to be lowest in late winter (Mäenpää *et al.* 1988a); moreover, mares may be depleted of vitamin A 2 months following removal from pasture, or by mid-winter in pastured mares (Greiwe-Crandell *et al.* 1995).

Vitamins D_2 *(ergocalciferol) and* D_3 *(cholecalciferol)*

Function
The widespread occurrence of bony abnormalities in horses and the misunderstandings concerning the interpretation of Ca and phosphate values in blood justifies a short summary of the functioning of vitamin D, which has been elegantly unravelled. Vitamin D is required for the maintenance of Ca and phosphate homeostasis, impairment of which produces the lesions in bone called rickets, or osteomalacia and the risks of lameness and bone fractures.

Metabolism
Under the influence of PTH secreted by the parathyroid gland in the neck, vitamin D is converted to its metabolically active form. This has two principal targets in many mammals – the small intestine and bone. The active form 1,25-$(OH)_2$-vitamin D is derived from both ergo- and cholecalciferol. In these forms the vitamin fits the classic model of a steroid hormone, but it is present in extremely low concentrations in plasma (human plasma 20–40 ng/l, 1 ng = 1/1000 μg). The intermediary metabolite 25-OH-vitamin D, formed in the liver, is present in much higher concentrations (in human plasma 15–38 μg/l) and is therefore more easily measured. Horse blood contains relatively low concentrations of the 25-OH metabolite. In Finland

Mäenpää *et al.* (1988b) reported winter mean concentrations of 2.14 µg/l and 8.05 µg/l, respectively, for 25-OH-vitamin D_2 and 25-OH-D_3. In summer the respective mean values were 2.16 and 16.6 µg/l. The relatively higher 25-OH-D_3 value reflects the synthesis of vitamin D_3 in response to solar ultraviolet irradiation of the skin in the summer, whereas the ergocalciferol vitamin D_2 is a poor source, derived from the action of ultraviolet light on cut leafy forage, i.e. hay.

The low equine concentration of even the 25-OH metabolite in blood, following vitamin D injections (Harmeyer *et al.* 1992), seems to indicate that the hormone has a lesser function in the horse than in other domestic animals, where it is critical for the intestinal absorption of Ca. The 25-OH metabolite is converted to the 1,25-$(OH)_2$ hormone by 1α-hydroxylase in the renal cortex, the activity of which was found to be nearly undetectable in the horse (Harmeyer *et al.* 1992). Nevertheless, blood plasma concentrations of Ca in the horse are higher than those in the gut lumen, so that active absorption must be taking place, presumably requiring a Ca-binding protein. In fact vitamin D-responsive Ca-binding protein has been identified in the equine duodenum, but it seems to be of less metabolic significance than in other domestic species investigated. It is possible that dietary vitamin D needs to be given in larger quantities than those recommended, as it is essential in the prevention of dyschondroplasia, as well as ensuring adequate bone ash. Nevertheless, excessive toxic doses cause signs similar to those of a deficiency.

In bone tissue, together with PTH, the 1,25-$(OH)_2$-vitamin D hormone serves to mobilize bone minerals. In the kidney tubules, PTH stimulates the reabsorption of Ca ions but blocks the reabsorption of phosphate. The vital objective of these two hormones, together with thyrocalcitonin, is to sustain a constant level of blood Ca. It is a fascinating fact that they modulate both Ca and P nutrition, but with different signals. When the diet is deficient in Ca, but adequate in P, then a fall in plasma Ca ions triggers the release of PTH from the parathyroid gland. This stimulates renal Ca reabsorption and the production of the vitamin D hormone. Intestinal absorption and bone mobilization of both Ca and phosphate are facilitated so that blood Ca and blood pH are returned to normal. Blood phosphate does not rise because of the PTH blocking effect on the renal reabsorption of phosphate.

Thyrocalcitonin counterbalances and modulates the effect of PTH by increasing the net deposition of Ca in bone stimulated by a raised serum Ca. By contrast, deficiency of dietary P depresses blood phosphate, which in turn directly raises ionized Ca in the blood, but also stimulates production of the vitamin D hormone. The combined effect of this is to suppress PTH production, increasing phosphate retention by the kidney (negating phosphate diuresis), stimulating Ca and phosphate absorption from the small intestine. Blood Ca, however, does not rise excessively as the lack of PTH increases the urinary loss of Ca (see also Chapter 11).

It is evident that when vitamin D nutrition is adequate for a given age of horse, but not so Ca or P, blood Ca is held within fairly well-defined limits and phosphate will be more variable. In the absence of vitamin D the efficiency of Ca absorption from the intestinal tract and the mobilization of bone Ca are depressed so that blood-Ca levels will fall. Some mobilization of bone Ca will continue, however, so that osteomalacia, or gradual bone decalcification, occurs in the adult horse and rickets, or reduced calcification of bones, is displayed by the young. On the other hand, some authorities dispute the existence of true rickets in growing horses and vitamin D is probably of less metabolic significance than in the young pig. Nevertheless, a loss of appetite, discomfort on standing, lameness, increased risk of bony fractures and a thinning of the cortex of long bones have been described in foals deprived of sunlight and dietary vitamin D. In young horses the growth plate (epiphyseal plate) of long bones is irregular, widened and poorly defined and the epiphyses are late in closing.

Dietary requirement

If the cereal/protein concentrate component of the diet is supplemented with 1000 iu vitamin D/kg (25 μg cholecalciferol or ergocalciferol, as 1 iu is equivalent to 0.025 μg cholecalciferol or ergocalciferol), then the daily requirement for vitamin D should be met. Moderately large doses of vitamin D can, to some extent, compensate for low dietary Ca by promoting further Ca absorption, particularly where dietary P is in excess. However, large doses of vitamin D (in excess of 2000–3000 iu/kg BW daily, or in excess of 60000–100000 iu/kg diet) will cause similar signs to those of the deficiency and eventual death, owing to the effect of vitamin D hormone on bone mineral mobilization.

Natural sources of toxicity

Several plant species, not found in the UK, actually synthesize this highly active hormone so that horses grazing areas where the plants exist will develop rickets and soft-tissue calcification (for example, *Cestrum diurnum*, a member of the potato family, sometimes incorrectly called wild jasmine, found in Florida and other subtropical states, including Texas and California, causes the condition).

Vitamin E (α-tocopherol)

Vitamin E potency is possessed by several tocopherol and tocotrienol isomers and undoubtedly all these have some antioxidant value in the feed and in the GI tract. However, analysis has shown that α-tocopherol is the only isomer to be found in significant amounts in equine tissues and so is the only form to have significant vitamin potency. As tocopherols have an antioxidant property that protects other substances in food, they are themselves destroyed by oxidation. This is accelerated by poor storage, mould damage and by ensilage of forage or the preservation of cereals in moist conditions. After the crushing of oats or grinding of cereals, the fats are more rapidly oxidized and vitamin E is gradually destroyed unless the material is pelleted. Fresh, green forage and the germ of cereal grains are rich sources of vitamin E, but feeds are frequently supplemented today with the relatively stable acetate ester of α-tocopherol.

Vitamin E status

Evaluation of the vitamin E status of horses is problematic, owing to a relatively low normal plasma α-tocopherol concentration. The normal range is from 1.5 to 5 mg/l. Adipose tissue of horses contains large quantities (10–60 μg/g) of α-tocopherol that are not prone to the short-term fluctuations characteristic of blood levels. The wide range indicated for these two tissues is intended to accommodate limited evidence for variation found among breeds. TBs tend to have plasma and adipose tissue concentrations at the lower end of these ranges and overall storage is lower than that for vitamin A.

Vitamin E and exercise

Vitamin E adequacy has been measured for many years as the dietary amount required to minimize erythrocyte haemolysis in the presence of dialuric acid, hydrogen peroxide or other haemolytic agents. Early studies (NRC 1978) indicated that the horse required only 10–15 mg/kg diet to ensure this. Limited experimental evidence (Lawrence *et al.* 1978) suggests that vitamin E supplements increase amounts of blood glucose and lactate in exercised horses and may help maintain the normal packed cell volume of the blood. A vitamin-E deficiency is known to reduce endurance in rats and the vitamin may be particularly important for extended work. Further evidence indicates that vitamin E supplements result in higher red cell GSH-Px activity following exercise (Ji *et al.* 1990).

Ronéus *et al.* (1986) found that to provide adequate standardbred tissue saturation with α-tocopherol the daily supplement of DL-, α-tocopheryl acetate (all-*rac-a*-tocopheryl acetate) should be 600–1800 mg. This is equivalent to 1.5–4.4 mg/kg BW. The question that arises then: is it necessary to saturate tissues with the vitamin? α-Tocopherol forms an integral part of cellular membranes, where it protects polyunsaturated fatty acids. The peroxidation of these π-6 fatty acids probably increases during exercise – frequently measured as an increase in thiobarbituric acid reactive substances (TBARs) – yielding n-pentane, a hydrocarbon gas that is excreted in the breath, although an assessment of breath pentane following exercise is complicated by an associated increase in breathing rate. McMeniman and Hintz (1992) recorded an increase in plasma ascorbic acid and a tendency to a decrease in plasma TBARs in exercised ponies supplemented with additional vitamin E. However, the additional amount was only 100 iu/day and very few animals were used. TB horses may have higher needs. Schubert (1990) reported that quite large supplementary levels of vitamin E improved the track performance of racehorses.

Fat supplementation

There is increasing interest in fat supplementation of diets for exercising horses (see Chapter 9). Many vegetable and fish oils, such as corn, soya or cod liver oil, are rich in polyunsaturated fatty acids and oil supplementation of other species is known to increase the vitamin E requirement. In the horse muscle TBARs have been shown to rise following this supplementation, despite the naturally high concentration of vitamin E in fresh corn or soya oil and the addition of antioxidants during manufacture of the oils. Clearly, stale, badly stored oils should not be used.

Breeding mares

Vitamin E functions by protecting unsaturated lipids in tissue from oxidation. In conditions where the intake of selenium and vitamin E is low, which can occur on pasture, mares give birth to foals suffering from myodegeneration. Pale areas in the myocardium and skeletal muscles are apparent on post-mortem examination and muscle-cell damage is seen histologically. If the foals survive, damage is said to be irreversible. Other symptoms include steatitis, or yellowing of the body fat, and general fat necrosis with multiple small haemorrhages in fatty tissues. As with other causes of muscle damage, the activity of blood CK (EC 2.7.3.2) and AAT (EC 2.6.1.1) rises and probably the fragility of red blood cells increases.

Immune function

Vitamin E is required for normal immune function. Baalsrud and Øvernes (1986) reported that a vitamin-E supplement given to oat-fed horses increased their humoral immune response to tetanus toxoid and equine influenza virus. An increase in the phagocytic activity of foal neutrophils has been induced by supplementation.

Vitamin E in prevention of equine degenerative myeloencephalopathy
and equine motor neuron disease

Two neurological disorders of horses have been recognized over the last decade to involve α-tocopherol status – equine degenerative myeloencephalopathy (EDM) and equine motor neuron disease (EMND). Oxidative injury of myelinated nerve fibres occurs in both diseases. The sheaths of these fibres are rich in unsaturated fat normally protected by α-tocopherol.

EMND is typified by weight loss, despite normal or increased appetite, and increased recumbency and trembling of proximal limb musculature (Mayhew 1994). The lifting of one thoracic limb from the ground for a few minutes may induce trembling, especially in proximal limb muscles of the opposing thoracic limb. Other signs include hyperaesthesia (hypersensitive response to stimulation) and low head and neck carriage. Affected horses move better than they stand. The clinical signs and neuronal lesions bear some resemblance to the clinical signs and lesions present in autonomic neurons of equine grass sickness (see Chapter 10), apart from the normal feed intake of EMND patients. In common with EDM, the disease is typically associated with an absence of access by horses to pasture, with the consumption of poor quality hay and with low concentrations of circulating levels of α-tocopherol. EMND seems to prevail among adult horses of a wide age distribution. Divers *et al.* (1994) recorded EMND in 28 horses, ranging in age from 3 to 18 years, with the highest frequency among 4–9 year-olds. It is also thought by Mayhew (1994) that EMND may require a neurotoxin to initiate oxidative neuronal degeneration in a pre-existing vitamin-E deficient state of the adult horse (also see Grass Sickness, Chapter 10).

EDM, on the other hand, usually develops in growing horses and is possibly more prevalent in some sire lines. EDM is a diffuse degenerative disease of the spinal cord and caudal medulla oblongata of equids. The disease may arise in foals where breeding horses have no access to pasture, or where large numbers of horses are crowded on poorly productive, often dry, pasture and are given poor sun-baked hay. The clinical signs include an abrupt, to insidious, onset of symmetric paresis, ataxia and dysmetria. Stiffness of the limbs of foals and yearlings is seen to persist to adulthood. The signs are present particularly in the pelvic limbs, but also in the thoracic. In the author's experience, EDM has been associated with absence of pasture and the presence of DOD (see Chapter 8), and signs were not expressed until 3–5 months of age. Older horses exhibit a striking hyporeflexia over the neck and trunk. Histologically neuroaxonal dystrophy is widespread, and neuronal atrophy, axonal swellings (spheroids), astrogliosis and lipofuscinlike pigment accumulation are prominent in older horses. Plasma α-tocopherol concentrations range from about 1 to 1.5 mg/l. The condition is vitamin-E responsive, where it has not progressed too far, and then plasma levels may rise to 2 mg/l. Selenium does not seem to be involved.

For breeding horses without access to pasture the daily vitamin E supplementation needs to be high, 2000 iu daily for breeding mares. For those presenting signs of ataxia, 6000 iu DL-α-tocopheryl acetate/horse daily should be mixed in 30 ml vegetable oil (see below) which should be added to 1 kg freshly ground cereal on a daily basis (not stored). Little response is likely for 3 weeks. Once improvement has been achieved the supplementary dose may be slowly reduced to 2000 iu daily. Proof of absorption should be sought by α-tocopherol determinations on blood samples.

The lipids of plant leaves are much richer in n-3 polyunsaturated fatty acids than are those of seeds. The author suggests that the cause of EDM may be a combination of deficiencies of natural antioxidants and of n-3 polyunsaturated fatty acids. This assertion would need testing before recommendations could be advanced. Fish oils are rich sources of the higher n-3 fatty acids. As an intermediary step, vegetable oils, such as rapeseed oil, richer in the lower members of the n-3 series than corn oil, might be used as the carrier for the vitamin E; but the dose of oil would have to be much greater than 30 ml – say 200 ml daily. Natural antioxidants include the carotenoids. The author has found that control of EDM is assisted by providing a source of these carotenoids, such as mould-free carrots, or dehydrated lucerne, and the riboflavin supplementation of a concentrate mix should be increased to 12 mg/kg (Table 4.3).

It is commented above that DOD seems to be present in some EDM cases. To this end, as discussed in Chapter 8, the author has found that the *ad libitum* feeding of growing foals and yearlings is helpful. Most published evidence suggests that this aggravates the condition; however, in several studies including Savage *et al.*(1993a, b) meal-fed and *ad libitum*-fed horses have been compared when given the *same concentrated feed* (this comparison was necessary for the needs of good experimental design), whereas the author's experience indicates that the feed for *ad libitum*

Table 4.3 Summary of proposed daily treatment for a horse presenting EDM.

Vitamin E	2000–6000 iu
Rapeseed oil[1]	30–200 ml
Clean, mould-free carrots	2 kg
Dried lucerne[2]	2 kg
Riboflavin	12 mg/kg concentrate
Coarse mix diluted with molassed chaff	To appetite, given *ad libitum* to growing weanlings

[1] If fish oil is used, the minimum vitamin E dose should be 4000 iu/day.
[2] A good source of riboflavin, although the additional 12 mg is an insurance.

feeding should be a coarse mix or a small pellet diluted with molassed chaff, so that the amount taken in at each 'sitting' provides a relatively small amount of energy, akin to nibbling.

'Tying-up'
Some benefit to 'tying-up' or myositis cases is said to accrue from treatment with selenium–vitamin E injections (see Chapter 11) seen, on occasion, after 1 or 2 days' rest in hard-worked horses.

Recommended dietary allowance for normally managed horses
Although more vitamin E may be needed when selenium is deficient, both are required nutrients and the amount of vitamin E that should be present in the diet rises in proportion to the level of dietary unsaturated fats (Agricultural Research Council 1981). Typical rations for horses should contain 75–80 iu vitamin E/kg (1 iu is equal to 1-mg DL-α-tocopheryl acetate, i.e. all-*rac-a*-tocopheryl acetate), although the requirement of very young foals may be slightly greater and that of idle adult horses somewhat less than this.

Vitamin K

Vitamin K_2, along with the B vitamins, is synthesized by functioning gut microorganisms in amounts that should normally meet the horse's requirements. However, this source may be inadequate during the first couple of postnatal weeks, or during extended treatment with sulphonamides. There is some body storage of vitamin K, and natural feeds, particularly leafy material, are fairly rich sources, so that no supplementary source is normally necessary. Vitamin K is essential for blood clotting. Airway haemorrhages in bleeders are an expression of blood vessel fragility and not one of a failure in the clotting mechanism and so may not be controlled by vitamin-K therapy. It is common for racehorses to present evidence of a mild form of haemorrhaging after races.

Water-soluble vitamins

Normal intestinal synthesis plus the quantities naturally present in typical horse feeds seem to meet the maintenance requirements for the B vitamins riboflavin, niacin, pantothenic acid and pyridoxine. Should there be a basic change in diet towards root vegetables and certain byproducts in the future, then an increase in supplementary needs might result. The needs of lactating mares and weanling foals should be met if good-quality pasture grass is provided. Additional nutrient demands of exercise are discussed in Chapter 9.

Thiamin

Signs of equine thiamin deficiency have been inferred, to a large extent, from studies in other species. However, a deficiency of this vitamin (Carroll *et al.* 1949) was shown to cause loss of appetite and weight, incoordination of the hindlegs, a dilated and hypertrophic heart, and a decline in blood concentration of thiamin and in the activity of enzymes requiring thiamin as a cofactor. Cymbaluk *et al.* (1978) made four standardbred horses thiamin-deficient by the daily oral administration of 400–800 mg amprolium/kg BW (amprolium is a structural analogue of thiamin). After 1–2 months thiamin deficiency signs of bradycardia and reduced heart rates, ataxia, muscular fasciculations and periodic hypothermia of extremities, blindness, diarrhoea and body weight loss were observed and erythrocyte transketolase activity was depressed.

Table 4.4 gives normal values for blood-thiamin levels. This vitamin has been used in the treatment of 'tying-up', but there is no corroboration that a deficiency of it is a cause. Grazing animals on heathland infested with bracken fern (*Pteridium aquilinum*) can exhibit signs of thiamin deficiency if they take to eating bracken. Treatment with thiamin is usually effective. About 25% of the free thiamin synthesized in the caecum is absorbed into the blood and a total dietary level of 3 mg/kg seems to meet the dietary requirement. Whether or not the requirement per kilogram of feed rises during periods of hard work has yet to be demonstrated.

Vitamin B$_{12}$ (cyanocobalamin)

The cyanocobalamin molecule contains the element cobalt (Co). Cattle and sheep grazing areas deficient in this element develop vitamin-B$_{12}$ deficiency as the rumen

Table 4.4 Normal blood-thiamin concentrations (µg/l) in standardbred horses (Loew & Bettany 1973).

Stallions	2.25 ± 0.16
Geldings	3.03 ± 0.13
Mares and fillies	3.36 ± 0.11
Less than 1 year	2.42 ± 0.26
1–4 years	2.81 ± 0.10
5–10 years	3.53 ± 0.14
10–20 years	3.35 ± 0.17

microorganisms are then unable to manufacture the vitamin. Co therapy rectifies the situation. Horses seem to be more resistant to Co deficiency, but undoubtedly they require Co at a minimum level of about 0.1 mg/kg diet for intestinal synthesis of the vitamin, as it is assumed, then, to be absorbed in adequate quantities. Synthesis in foals may be inadequate and, in fact, supplemental vitamin B_{12} has been shown to increase the blood concentration of the vitamin. It is required for cell replication; thus, a deficiency may cause anaemia and a reduction in the number of red blood cells. Macrocytic anaemia is common to both B_{12} and folic acid deficiencies through a limitation to DNA synthesis.

Although an overt deficiency has not been produced in adult horses, it has been suggested that a response, including a stimulation to appetite, can be obtained in some anaemic animals. Adult horses in training on high-grain rations may need dietary supplementation because a decline in appetite shown by such horses may reflect a buildup of blood propionate. This VFA is produced proportionately and absolutely in much greater quantities when diets of this type are consumed, and its metabolism to succinate requires methylmalonyl-CoA mutase (EC 5.4.99.2) that has adenosylcobalamin as a coenzyme. In fact, vitamin-B_{12} deficiency causes an elevation in urine of both methylmalonic acid and homocysteine which distinguishes it from folic-acid deficiency, where this does not occur. Early-weaned foals should receive a supplement of 10 μg vitamin B_{12}/kg dietary DM.

Folic acid

Natural folates exist as conjugates of *p*-aminobenzoic acid with mono- or polyglutamic acid. Both the stability and the availability to the horse of these compounds undoubtedly vary, but there is a scarcity of evidence relating to equine nutrition. Folic acid is closely associated with vitamin B_{12} in single-carbon metabolism and in some domesticated animals a deficiency causes macrocytic anaemia, through impairment of methionine synthetase (EC 2.1.1.13) function and so depressed DNA and RNA synthesis. In folic-acid deficient humans, elevated concentrations of homocysteine (but not of methylmalonic acid) occur in blood and urine. Folic acid is synthesized in the intestinal tract, but supplementation of stabled TBs receiving an inadequate diet produces an increase in serum folate concentration from about 4–9 μg/l and where blood haemoglobin is low the level may be raised. Australian work on folic acid in the horse supports the author's observations suggesting that there is an increased utilization of folic acid by horses in hard work. Green forage legumes are rich sources of the vitamin, but its availability in some sources is low. As horses required to partake in intensive exercise tend to receive less green forage it is recommended that they receive a supplementary folic acid source, or green forage in their diet. A daily supplement of 1 mg folic acid for foals and working horses is appropriate.

Biotin

Biotin is the only water-soluble vitamin to have brought about clinically observable responses with normal diets in otherwise healthy horses. This may have occurred *because* the biotin contained in wheat, barley and milo (sorghum) grains and in rice bran is almost completely unavailable for utilization. That contained in oats is only slightly more digestible. However, all the biotin in maize, yeast and soyabean is accessible, together with most of that in grass and clover foliage.

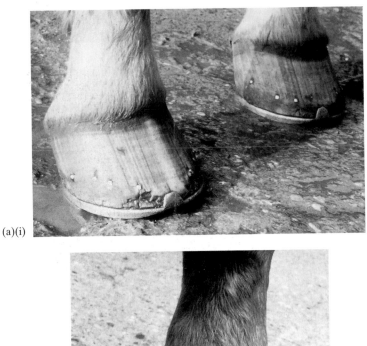

(a)(i)

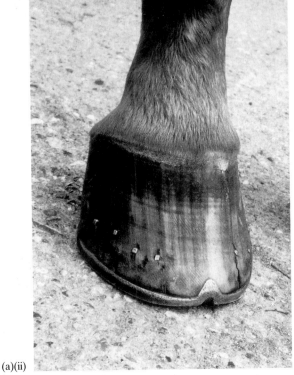

(a)(ii)

Plate 4.1   (a) Off-forehooves of 8-year-old Irish chestnut gelding, before (i) and after (ii) receiving 15 mg synthetic biotin/day orally for 13 months.

Hoof wall disease is common in horses. Slater and Hood (1997) reported that 28% of the horses in their Texas survey had some type of hoof wall problem, largely with an undetermined predisposing cause. Nevertheless many cases of weak, misshapen, cracked and crumbly hooves, that tend to separate from the sole, have responded to dietary supplementation with biotin and a greater response, as measured by horn hardness, tensile strength and possibly growth rate, has been evoked with 15 mg/horse daily than with 7.5 mg (Buffa *et al.* 1992) (Plate 4.1). These

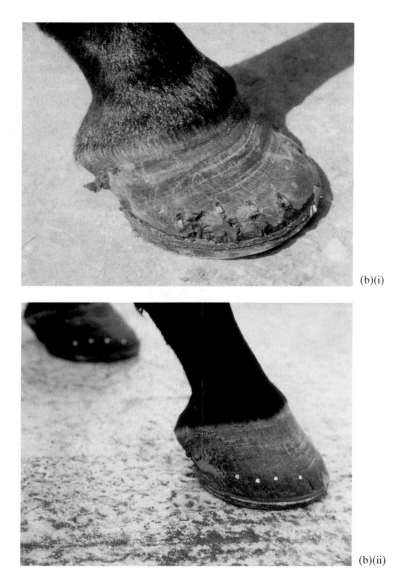

(b)(i)

(b)(ii)

Plate 4.1 (b) A 5-year-old TB gelding hack, before (i) and after (ii) receiving 15 mg synthetic biotin/day orally for 5 months. (Photographs 4.1(a) and (b) by kind permission of Norman Comben, MRCVS; Comben *et al.* 1984.)

amounts are considerably larger than should normally be adequate as a daily maintenance dose. It is also essential that the hooves are properly shaped and trimmed.
Long, unpared hooves exert excessive pressure on the heels and this restricts blood
flow and hence impedes adequate nutrition of the foot, leading to poor quality and
crumbly unsatisfactory growth of walls, sole and frog. The hoof wall grows linearly
from the coronary band at the rate of 8–10 mm/month, so that 9–12 months are
required for the wall of the toe to grow from the proximal to the distal border
(Pollitt 1990) (the sole and the frog are replaced every 2 months) and at least 10–12
months are required before a maintenance dose of about 2–5 mg supplementary
biotin/horse daily may be allowed.

Comprehensive recent evidence indicates that a dose level of 20 mg/day may be
required by large horses, and this for up to 3 years, for the maximum benefit to
be wrought (Josseck *et al.* 1995; Zenker *et al.* 1995). In Josseck *et al.* (1995) plasma
biotin concentration in untreated horses was 350 ng/l and in treated horses
1000 ng/l. Kempson (1987) reported that biotin did not overcome a second hoof
defect of poor attachment of horn keratin squames. Supplementary limestone (7.5 g/
day) with biotin improved this defect. Nevertheless, it is possible that this additional
response resulted from the more extended use of the biotin. Yet many diets are
deficient in Ca, a deficiency that should be rectified.

Horn is primarily composed of protein, rich in sulphur-amino acids, and many
proprietary supplements contain methionine, or methyl sulphonyl methane, as a
source of biologically available sulphur (see dietary supplements, Chapter 5), in
addition to biotin, which may be an asset where a low protein diet is given. Kempson
(1987) also noted infection of the keratin squames by *Bacteriodes nodosum* responsive to metrinidozole, although these organisms probably represent secondary
invaders.

Riboflavin, niacin, pantothenic acid, pyridoxine and lipoic acid

Each of the vitamins riboflavin, niacin, pantothenic acid, pyridoxine and lipoic acid
has metabolic functions in the horse; however, no evidence of a dietary deficiency
has been established, owing presumably to adequate biosynthesis, e.g. intestinal
synthesis by the gut flora. Animals are capable of lipoic acid biosynthesis. Whether
supplementation may be of benefit in special circumstances, e.g. ill-health or extreme exertion, has not been established. Thiamin, niacin, riboflavin and lipoic acid
are intimately involved in energy metabolism.

Thiamin as thiamin pyrophosphate (TPP), niacin as nicotinamide adenine
dinucleotide (NAD^+), riboflavin as flavoproteins, and lipoic acid all function in the
tricarboxylic acid (TCA) cycle for the aerobic combustion of acetate to CO_2 and
water with the production of energy, eventually provided to muscle cells as high
energy phosphate in ATP. Pantothenic acid functions as a carrier of acyl groups in
the form of CoA, which makes high energy thioester bonds with carboxylic acids,
the most important of which is acetic acid. Acetic acid is formed from the metabolic
catabolism of both fats and carbohydrates. Acetyl groups from these sources have to

be in the form of acetyl-CoA to be further metabolized in fat synthesis, or for energy production in the TCA cycle.

In carbohydrate metabolism acetate is derived from pyruvate. Pyruvate requires TPP, or cocarboxylase, the active form of thiamin, for the cleavage of pyruvate (an α-ketoacid), otherwise the formation of lactic acid from pyruvate will be accelerated. In intense exercise the build-up of lactic acid is a component of fatigue. It is possible therefore that supplementary pantothenic acid and thiamin may promote the aerobic metabolism of pyruvate to acetyl-CoA, reducing lactate production, although there is no evidence that this would occur in horses given normal feeds.

Vitamin B_{15} (pangamic acid)

Pangamic acid (vitamin B_{15}) is a term that has been used indiscriminately to describe several products. It is not a definable entity and no substantive data appear to have ever been presented to support claims of beneficial biological effects.

Vitamin C (ascorbic acid)

Ascorbic acid is synthesized in the tissues of horses from glucose. Dietary sources are very poorly absorbed, and in fact Löscher *et al.* (1984) concluded that where supplementation may be required after surgery and trauma, an i.v. dose of 10 g ascorbic acid was needed to raise the blood concentration. It has been employed by injection in post-traumatic wound infections, epistaxis, strangles and acute rhinopneumonia. Its acidic nature leads to local irritation following subcutaneous and intramuscular administration and so it is also of value by i.v. administration for removal of renal calculi. Snow *et al.* (1987) reported that single oral doses of 20 g had no effect on plasma concentration, whereas daily administration of either 4.5 or 20 g ascorbic acid resulted in an increased plasma level after 5–10 days. No benefit has been demonstrated from such administration to healthy horses.

WATER REQUIREMENTS AND FLUID LOSSES

Water constitutes some 65–75% of the bodyweight of an adult horse and 75–80% of a foal's. Water is vital to the life of the animal. The horse also needs to take in water with its food to act as a fluid medium for digestion and for propulsion of the digesta through the GI tract, for the useful products – milk and growth, and to make good losses through the lungs, skin and in the faeces and urine. In healthy adult horses undertaking light work one estimate showed that the losses of water were distributed such that 18% occurred in the urine, 51% occurred in the faeces and the remaining 31% represented insensible losses (Tasker 1967). Restricted water intake will depress appetite and reduce feed intake.

Equids differ in their ability to conserve body water and to withstand dehydration. Asses from the dry tropics can thwart extreme dehydration because they can conserve water more efficiently than horses. A rise in environmental temperature from 15 to 20°C increases the water requirement of horses by 15–20%. Work, depending on its severity, will raise requirements by 20–300% above the needs for maintenance through increased losses from the lungs and skin. For obvious reasons, peak lactation can lead to requirements twice those of maintenance (Table 7.4).

The horse obtains water for its metabolic needs from three sources – the consumption of fresh water, the water content of natural herbage and other foods and from metabolic water. Fresh, young, growing herbage may contain 75–80% water, so that under many circumstances additional fresh water may not be required, but a source should always be provided. In arid conditions herbage is very different, and horses will seek and consume poisonous shrubs and succulent plants unless water and feed are provided.

Metabolic water is that produced during the degradation of carbohydrates, fats and proteins in cellular respiration, e.g.:

$C_6H_{12}O_6 + 6\ O_2 \rightarrow 6\ CO_2 + 6\ H_2O$ for carbohydrate
$C_{57}H_{104}O_6 + 80\ O_2 \rightarrow 57\ CO_2 + 52\ H_2O$ for typical fat
$2\ C_3H_7O_2N + 6\ O_2 \rightarrow (NH_2)_2CO^* + 5\ CO_2 + 5\ H_2O$ for the amino acid alanine

where * indicates urea. Thus, per 100g glucose, average fat or average amino acid metabolized, there are, respectively, 60, 106 or 101g water produced. Per kilogram of feed ingested it amounts to the equivalent of 350–400g water, depending on the feed's digestibility. Nevertheless, in circumstances of choice, the water intake of horses is highly correlated with intake of DM and amounts to between 2–4l/kg DM in stabled horses worked moderately.

Water requirements

Maintenance

For the maintenance of adult horses in an equable environment, the total water requirement is probably less than 2l/kg DM intake (about 5l/100kg BW).

Working horses

Strenuous effort in hot climates increases the water requirement to as much as 5–6l/kg DM intake (12–15l/100kg BW) when there is an inevitable loss of relatively large amounts of sodium and potassium chloride in sweat. Excessive dehydration can be fatal. Certain breeds of horse and other species of *Equus* (e.g. *E. asinus*) (Maloiy 1970) can sustain extensive water loss without apparent discomfort, but horses of temperate breeds may succumb to water losses that amount to 12–15% of their bodyweight (Hinton 1978; Brobst & Bayly 1982). The estimates of the Hanover group (Meyer 1990) indicate that a 500-kg horse trotting at 3.5m/s in an ambient temperature of 27°C would require a minimum of 10–12l water/h to replace inevitable losses. Water repletion should be accompanied by balanced electrolytes, although electrolytes may often have to be given first in order to induce drinking when isotonic or hypotonic dehydration has occurred. Where the horse is fit it should be walked or allowed to graze so that it cools down gradually over an hour before being given substantial amounts of water. Excessive consumption of cold water by hot horses can precipitate colic or founder. During very cold weather, warm water of a temperature between 7 and 18°C should be provided and will be taken more readily than will very cold water. Decreased consumption of water may contribute to the incidence of impaction colic and of depressed performance in racehorses.

Foals and weaned horses

A high-yielding 500-kg mare may produce 12 kg water daily in her milk (see Table 7.4 for water requirements in the stud). However, a foal has a greater requirement than an adult in proportion to its size because it is less able to concentrate urine. A frequent cause of death in neonatal foals is rapid dehydration through persistent diarrhoea, which requires treatment with a physiological salt solution (see Chapter 11). The fluid intake of foals suckling grazing mares was measured (Martin *et al.* 1992) in Queensland, where the maximum ambient temperature averaged 30°C and the RH averaged 70%. The results are given in Table 4.5.

Weaned horses obtain their water needs by relatively brief periods of drinking, even when meal-fed. Sufit *et al.* (1985) found that ponies drank during 27 min daily, much of which occurred periprandially .

Water quality

Where it is feasible, water should be provided from the mains. If mains water is unavailable, then well water, or watercourses, must be free from pollution by sewage or fertilizer seepage. Ideally a new source should be first assessed by a competent analyst. Potentially dangerous microbiological contamination can occur. For instance, the urinary excretion of leptospira by rodents can pollute water, and river and sewage flooding can cause abortion in mares and death of foals. River water is normally safer than pond water. Nevertheless, in addition to risks incurred from any pollution derived upstream, a decreased rate of flow in summer droughts encourages the growth and accumulation of plant life. Whereas this may assist

Table 4.5 Daily fluid intake per foal (Martin *et al.* 1992).

Age (days)	Milk intake (kg)	Water intake (kg)	Total fluid intake (g/kg BW)	Milk consumed (kg/kg BWgain)
11–18	16.9	nil	246	12.8
30–44	—	3.9	202	15.7
60–74	18.1	5.5	172	16.4

oxygenation of water, blue–green algae cyanophyceae (Myxophyceae) can proliferate and can be held in position by other plant life. The species that has received most attention is *Anacystis cyanea* (Clarke & Clarke 1975), the consumption of which by horses can cause liver damage, icterus and photosensitization, or death. Microcystin, a cyclic decapeptide and an alkaloid hepatotoxin are said to be the toxic agents in this species. There are several other species of algae that contain a variety of toxins, and avoidance of water sources that may contain these species is to be recommended.

Water losses and deprivation

Renal losses

The nonexercised horse disposes of most excess water by excretion through the kidneys. This water is the vehicle for the excretion of excess salts of sodium and potassium, and much of the breakdown products of nitrogen metabolism. Whereas calcium salts may, in part, be excreted in a solid form there is a limit to which the horse can concentrate urea and the highly soluble salts of sodium and potassium. Thus, where diets are rich in salt, or in protein, more dietary water will be required. Based on evidence from other species, an increase in total dietary salt from 7.5 to 30 g/kg would be expected to increase the ratio of free-choice water to DM in the diet from 2:1 to 3.5:1, other things being equal.

Meyer (1990) reported that 73–89% of total water intake was lost by renal excretion in horses given concentrates, whereas <60% was lost by this route in horses given hay. Water restriction decreases renal losses, but does not affect sweat losses in normal horses. Where water restriction persists then a considerable stress is induced in exercised horses, with an increase in plasma protein and urea concentrations and increased breathing frequency during exercise. Urinary urea at rest, according to the data from Hanover (Meyer 1990), was 6–8 g/l at rest, but the amount increased to 15–50 g/l after exercise, reflecting the solubility of urea and the relative lack of effect on water conservation in some horses. The highest of the urea values occurred with high protein diets and extended water deprivation.

Table 4.6 Composition of sweat following 2 h of exercise (after Meyer 1990).

	g/l		mg/l
Na	2.77	Ca	123
K	1.42	Mg	52
Cl	5.33	Zn	11.4
		Fe	4.3
		Cu	0.27
		Mn	0.16
		Se	2–5 µg/l

Sweat

Horses lose a large amount of fluid as sweat during exercise. The amount increases greatly with a rise in environmental temperature. Meyer (1990) recorded losses of 1 l/100 kg BW/h at 18–20°C, but for each degree increase in environmental temperature (range 15–27°C) sweat production increased by 3%. In contrast to fluid losses from the lungs, sweat contains significant amounts of electrolytes and small amounts of trace elements (Table 4.6). During the 2 h of exercise (Table 4.6) the losses in sweat were calculated to be:

	mg/kg BW
Na	28–69
K	17–30
Cl	56–118

Thus, during this exercise a 500-kg horse could lose 50–90 g sodium chloride in sweat alone.

Evaporative losses from the lungs

The evaporation of sweat, or of water from the lung surface, absorbs heat and so cools the horse (the isothermal evaporation of 1 kg insensible water absorbs 2256 kJ). The water loss from the surface of the lungs increases greatly during exercise, owing to an increase in body tremperature of the horse and an increase in both respiratory frequency and tidal volume. Heat loss through the lungs at 20°C and 60% relative humidity (RH) was calculated to be 289 kJ/h at rest and 3059 kJ/h while trotting (Table 4.7).

Table 4.7 indicates that an increase in body temperature during exercise increases the moisture content of exhaled air by 15%, but this together with the increased respiratory volume during exercise increases the expiratory water loss nearly eight-fold. Water losses of exercising horses, in sweat and from the lungs, cannot be reduced during water deprivation if body temperature is to be contained within physiological limits, whereas urinary losses are decreased.

Determining water deprivation and response to thirst

Apart from various clinical signs of fluid loss, including skin turgor, the PCV of blood is not a guide to dehydration and water deprivation, because the shrinkage of

Table 4.7 Net water lost per unit volume (m³) of exhaled air and water lost in exhaled air per hour.

	Water lost during expiration (g/m³)	Expired water lost (g/h/horse)
Resting	38.3	172.5
Trotting 3.5 m/s	44.2	1356.0

Table 4.8 Mean fluid contents of the GI tract in ponies given 18 g DM/kg BW as hay or as concentrated feed (after Meyer 1990).

	Hay	Concentrates
3.5 h postfeeding		
DM (g/kg BW)	19.4	18.3
Water (ml/kg BW)	183	101
Na (mg/kg BW)	398	226
K (mg/kg BW)	289	220
After 1 h treadmill		
exercise at 3.5 h postfeeding		
Reduction in water		
content (ml/kg BW)	20	8
Reduction in Na		
content (mg/kg BW)	291	No significant change

red cells and the changes in the release of red cells from the spleen during dehydration and exercise make this an unreliable guide. Total plasma protein, however, may increase by 10–12 g/l (say, from 62 to 73 g/l) with a fluid loss causing a 12–15% decrease in bodyweight. Changes in plasma and urine electrolytes and urea depend on several associated factors. In one study (Brobst & Bayly 1982), fluid losses of this extent resulting from dehydration of TB geldings increased the concentration of serum and urinary urea by 68 and 130%, respectively. The specific gravity of urine reached at least 1.042 and urine osmolality increased 30% to 1310 mOsmole/kg when the osmolality ratio urine:serum increased to 4.14:1.00. Voluntary drinking has been found to start in ponies when plasma osmolality (mOsmole per litre) increased by 3% from normal. Drinking is also stimulated by inducing the formation of urine with a consequential decrease in plasma volume, measured as an increase in plasma protein concentration (Sufit *et al.* 1985).

Water reserves

See also Chapter 9, long distance work. Voluntary water intake is greater in horses fed hay than in those given concentrates, and the water content of the GI tract is considerably greater in those given a hay diet than in those given a concentrate mix only. Consequently the GI tract following hay feeding can act to some extent as a reservoir of water and sodium for metabolic needs (Table 4.8).

Sodium chloride dosing is harmful in states of water deprivation, but it will increase voluntary water intake prior to endurance work, increasing water retention. Maximum retention occurred at the 3rd–4th hour after providing salt in a feed (Meyer 1990) and so this may be the optimum time after a small meal for extended exercise during hot weather.

STUDY QUESTIONS

(1) Are there any circumstances in which you might expect a useful response of horses to a fat-soluble vitamin preparation? If there are, describe the circumstances and give reasons why these might exist?

(2) What factors should be considered in deciding on the adequacy of a water source for horses?

FURTHER READING

Josseck, H., Zenker, W. and Geyer, H. (1995) Hoof horn abnormalities in Lipizzaner horses and the effect of dietary biotin on macroscopic aspects of hoof horn quality. *Equine Veterinary Journal*, **27**, 175–82.

Martin, R.G., McMeniman, N.P. and Dowsett, K.F. (1992) Milk and water intakes of foals sucking grazing mares. *Equine Veterinary Journal*, **24**, 295–9.

Meyer, H. (1990) Contributions to water and mineral metabolism of the horse, In: *Advances in Animal Physiology and Animal Nutrition*, pp. 1–102. Supplements to *Journal of Animal Physiology and Animal Nutrition*, Paul Parey, Hamburg and Berlin.

Watson, E.D., Cuddeford, D. and Burger, I. (1996) Failure of β-carotene absorption negates any potential effect on ovarian function in mares. *Equine Veterinary Journal*, **28**, 233–6.

Zenker, W., Josseck, H. and Geyer, H. (1995) Histological and physical assessment of poor hoof horn quality in Lipizzaner horses and a therapeutic trial with biotin and a placebo. *Equine Veterinary Journal*, **27**, 182–91.

Chapter 5
Ingredients of Horse Feeds

Some parts of the kingdom produce no grain so much as oats which probably may be the reason why they have come to be used as our chief provender.

W. Gibson 1726

Some of the main chemical characteristics of the ingredients of horse feeds are given in Appendix C. Ingredients should be selected not only to provide the nutrients required, but also to be uniform in quality, to avoid harmful contaminants and dust and to balance dense energy-rich feeds with more bulky feeds. The rate of consumption of DE should not be excessive and the stomach contents should retain an 'open' physical texture.

ROUGHAGE

Loose hay

Grasses and forage legumes are cut for hay. Most common species of grass are suitable, but probably the more popular and productive ones include ryegrasses (*Lolium*), fescues (*Festuca*), timothy (*Phleum pratense*) and cocksfoot (*Dactylis glomerata*). Many species found in permanent pastures, for example, meadow grasses (*Poa*), bromes (*Bromus*), bent grass (*Agrostis*) and foxtails (*Alopecurus*), are also quite satisfactory. Among legumes, red, white, alsike and crimson clovers and trefoils (*Trifolium*), lucerne (*Medicago*) and sometimes sainfoin (*Onobrychis*) are used. Although the crude-fibre content of crimson clover (*Trifolium incarnatum*) hay may be similar to that of other clovers, the fibre tends to be less easily digested by the horse (forage legume fibre generally is more lignified than that of grass). For serving as horse hay or haylage crops, two reliable seed mixtures are:

(1) three perennial ryegrass strains – Melle 5 kg/ha, tetraploid Meltra 15 kg/ha, hybrid tetraploid Augusta 13 kg/ha – or as a two-year crop;
(2) tetraploid broad red clover – Hungaropoly 7 kg/ha – and tetraploid Italian ryegrasses – Wilo 15 kg/ha, Wisper 10 kg/ha.

The leaves of forage legumes and grasses are much richer in nutrients than are the stems, as stems contain about two-thirds of the energy and about half of the protein

96

and other nutrients found in the aerial parts. The leaves of legumes tend to shatter more readily than grass leaves so that care is necessary at haymaking to conserve the nutritional quality of legume hay. Even so, at the same stage of maturity, legume hay contains more DE, calcium, protein, β-carotene and some of the B vitamins, including folic acid, than does grass hay. Horses consuming hay composed predominantly of forage legume tend to produce more urine with a strong ammonia smell and containing deposits of calcium salts. These events are normal physiological responses in healthy animals.

As long as hay is composed of safe, nontoxic, nutritious plants, the stage of maturity of the crop at the time of cutting and the weather conditions and care to which the haymaking is subject are much more important characteristics than the species of plant present. As pasture herbage matures the yield of DM per hectare increases, the moisture content of the crop decreases and in the UK the weather becomes warmer. At Jealott's Hill Research Station, Bracknell, Berkshire (ICI Ltd) (now owned and operated as a research station by Zeneca CTL, Macclesfield, Cheshire) many years ago the average yield of DM from early hay crops was only 57% of that cut at a later date. Even when the aftermath was included, the total yield of the early hay amounted to only 71% of that produced by later cutting. Thus, there is a considerable commercial incentive to produce hay composed of grasses at the late flowering stage. Nevertheless, where hay of good nutritional quality is required for horses, mixtures of grass and clover should be cut before the grass is in full flower when the protein content of the crop lies between 9 and 10% DM and the crop contains high concentrations of calcium, phosphorus and other minerals. Hard, mature grass hays, however, frequently contain between 3.5 and 6% crude protein, lower concentrations of minerals and more crude fibre (Plate 5.1). Good-quality hay from pure stands of lucerne or sainfoin is difficult to make when natural drying is relied on because moisture loss from the thick juicy stems is relatively slow and mechanical turning and tedding can result in a considerable loss of leaf, which dries sooner and shatters more readily. For the best product, these legumes should be cut before flowering at the bud stage, because after first flowering the crude protein content declines at a rate of some 0.5% daily and the digestible energy declines by some 0.75% daily.

Horses should never be given mouldy hay, so the making of satisfactory leafy hays during inclement weather presents a considerable problem in the absence of a facility for artificial drying. Best-quality hay should be leafy and green, but free from mould dust, weeds and pockets of excess weathering. When ley mixtures are grown for hay, the first cut may contain more weeds, the second cut is generally produced from a faster growth and contains more stem, but the third cut may have the highest nutrient content and leaf, giving a small yield per hectare.

Haylage

(See also Chapter 10, particularly for grass species and safety aspects.) Haylage (grass cut between early silage and hay stages and normally containing, after pres-

Plate 5.1 Hay samples of various types. (a) Hard 'seeds' hay cut when the grass has formed seed heads. The material is clean, low in dust but of low nutritional value.

Plate 5.1 (b) Lucerne (alfalfa) hay, which is similarly stemmy and has been sun dried; bleaching destroys its vitamin-A potency but it adds some vitamin-D potency. Poor harvesting has led to the loss of most of the leaf, so depriving the hay of its most digestible component.

ervation, 40–65% DM) is being used increasingly, in place of hay, for horse feeding. However, the production of haylage for home consumption can be justified only if:

- there are a sufficient number of horses available to make use of the minimum quantity that could be produced in an efficient manner, and a number that could use an opened bale within a couple of days;
- there is adequate grassland, that has not been grazed that year by horses, and that can be set aside and fertilized; and
- there is manpower with the appropriate knowledge, equipment and space available for making, checking and storing the product.

Plate 5.1 (cont.) (c)
Good-quality lucerne hay,
which may have been
barn-dried. This is 'rich'
material and care should
be exercised to avoid loss
of leaf during feeding
(note the leaf particles at
the base of the sample).
Artificial drying deprives
the hay of vitamin-D
potency.

Plate 5.1 (d) Meadow
hay containing a
proportion of timothy
(*Phleum pratense*). This
sample is of reasonable
quality and is free from
significant moulding.

The most successful producers and users of haylage seem to find that a product
very high in DM content is the most successful, and acceptable to more of the horses
in a stable. At the Yorkshire Riding Centre, in England, meadow grass is cut from
mid-June, when the grass is commencing to flower, about 2 weeks before a hay crop
would be taken. The cut material is allowed to wilt in the field for a day and is then
square-baled (Plate 5.2) a day before it would be ready as hay, yielding a product
with 45–68% DM and 90–120 g crude protein/kg DM. Square bales are preferred as
'round' bales are more inclined to mould in the centre, where there may be a space.

Plate 5.2 Haylage. (a) A
stack of plastic-wrapped
bales: square bales are
less inclined to mould in
the centre than round
bales; (b) surface mould
on a bale, adjacent to a
puncture; (c) a cut section
of the bale, showing that
mould penetration of the
compressed haylage is
slight.

(a)

(b)

(c)

Table 5.1 Daily rations for horses using grass haylage of 50–60% DM.

	Haylage (kg/day)	Concentrates (kg/day)
Event horses	7–7.5	5–6
Riding horses for novices 1.68–1.73 m (16.2–17 hands) – summer	7–7.5	1.5–2.5
– winter	11–12	0–1
1.57–1.68 m (15.2–16.2 hands) – summer	5–6	1–1.5
– winter	7–7.5	0–0.5
1.42–1.57 m (14–15.2 hands) – summer	5–6	0.5–1
– winter	6.5–7	0–0.5

It is important to avoid puncturing the plastic bale cover, as moulding will occur at that point. Yet it is normally safe to pull away and discard only the portion that is mouldy, as the mycelium of *most* moulds is benign and does not penetrate well compressed bales (Plate 5.2). If there is secondary fermentation, occasioned by excessive heating in the centre, then the whole bale should be discarded. A well-made and packed bale may be left open for up to 4 days during cool weather, although for a lesser time in hot weather, before the residue should be discarded.

By weight, at 50–60% DM content, haylage is equivalent in energy value to the same weight of typical grass hay. At the Yorkshire Riding Centre 0.405 ha (1 acre) provides a year's supply of haylage for one horse. About 375 bales meet the requirements of 40 horses for a year, except for one or two that will not eat haylage. The droppings of most horses are looser when they are introduced to haylage, an effect similar to that which occurs when horses are put out to grass, yet no colic should be present. So horses 'coming off' grass adapt more easily to a regime that includes haylage. One bale daily is sufficient for about 35 horses in winter, when the horses receive a ration from it on the floor, two or three times during that day (5.5–7.5 kg for smaller horses and up to 12 kg for larger horses daily). This rationing scheme seems to keep the horses in a mental state that is suitable for novice riders. During the summer, when the horses are generally more active, more concentrates are given so that the haylage allowance is reduced by 20–25% and the horses are run on grass at night (Table 5.1). Moreover, competition horses receive less than other horses, in order to avoid an excessively large 'belly'.

'PROCESSED' FEEDS

Pelleting aids

Common pelleting aids include molasses, lignosulphonite and clay (bentonite, hydrated aluminium silicate, principally montmorillonite) pellet binders. Sodium bentonite is also used in feeds as a ruminal fluid buffer and calcium bentonite has been demonstrated to decrease some of the adverse effects in pigs of dietary aflatoxin. Clays have a large surface area and this characteristic, combined with their buffering capacity, is the reason bentonite is included in products to counter laminitis and gas colics of horses (see Chapter 11). It has been considered that clay may inhibit nutrient absorption, but the author's own studies show it does not interfere with vitamin A absorption from a retinyl palmitate source.

Pellets and wafers

Straw, chemically processed with sodium hydroxide or ammonia, will be discussed later in this chapter (see 'Sodium hydroxide-treated straw'). Hay is also occasionally processed; it may be ground and pelleted or chopped and wafered. During pelleting molasses and a binding agent are normally added to achieve a satisfactory product. Despite the additional costs of processing, pellets possess a number of advantages:

- The product is easier to weigh and ration.
- There is less waste during feeding and particularly with leafy legume or grass material the sifting out of small particles of leaf and their loss in bedding is avoided. This occurs regularly when leafy long hay is consumed.
- Less storage space is required than for long hay.
- Transport costs are lower.
- Horses particularly prone to respiratory allergies (heaves and broken wind) are less subject to dust irritation when given pelleted hay. Coughing is reduced in normal horses and bleeders are less prone to episodes of epistaxis.
- Older horses with poor teeth are inclined to masticate long hay incompletely, so occasionally precipitating colic through impaction. The introduction of pelleted hay should overcome this risk.
- Hay belly may be reduced.

Pellets do, however, have some disadvantages, the principal ones being as follows:

- Incorrectly pelleted material, or good pellets that are allowed to become wet, may be crumbly and soft so that fines are lost, and wet pellets mould within 18–24 h.
- It is visually difficult to assess the quality of pelleted hay.
- Horses may choke on pellets of about 12-mm (0.5-in) diameter. The problem is said to be more common when pellets are fed from the hand, but might be avoided by placing a large spherical rock (too large to be chewed) in the manger, forcing the horse to eat around it. The scale of this problem is probably exaggerated and episodes of apparent choking are normally overcome without intervention.
- Wood chewing and coprophagy (faeces eating) are more prevalent where pelleted hay is given without any long hay. The provision of 0.25–0.5 kg long hay/100 kg BW daily, or good-quality straw bedding, is normally sufficient to minimize these problems. Their incidence seems to be less, but is not eliminated, when wafered hay containing hay particles 4–5 cm (1.5–2 in) long is used. There

is evidence to suggest a relationship between the caecal environment and the incidence of wood chewing in horses given hay or grain. A higher proportion of propionate in the VFAs and a lower pH of the caecal fluid seem to be linked with a greater inclination to chew wood (Willard *et al.* 1977). A diet of all hay induces a higher proportion of acetate in the VFA of the caecal fluid (see Table 1.3).

- Although the grinding and pelleting of hay do not affect the digestibility of protein, the digestibility of both DM and crude fibre is decreased slightly, possibly because of a decrease in time taken to consume a given amount of feed. From a practical point of view the effect on digestibility is more than offset by the reduction in wastage.
- Grinding and pelleting can increase hay costs by up to 10%.

Feeding time can be influenced by the conditions and method of processing the hay. Researchers in Hanover (Meyer *et al.* 1975b) recorded that horses took 40 min to consume either 1 kg long hay or 1 kg hard, pressed, wafered hay. They took longer to consume chopped or ground hay but less time to eat soft pressed wafers. Hay of poor quality and high-fibre content took longer to eat than better quality hay. Highly digestible chopped maize silage was eaten much more rapidly than hay. Horses of between 450 and 500 kg made between 3000 and 3500 chewing movements in consuming 1 kg long roughage, but only between 800 and 1200 such movements in eating 1 kg concentrate. However, ponies of between 200 and 280 kg required twice as long to eat hay and a concentrate meal and between three and five times as long to eat whole oats or pellets. They made between 5000 and 8000 chewing movements in consuming 1 kg concentrate. The ingesta of horses given chopped hay, or ground hay, passed more rapidly through the stomach than did that of those given long hay and the former led to more fluid stomach contents.

Several other reports have clearly shown that the digesta of horses given ground hay passes more rapidly through the GI tract (Wolter *et al.* 1974), notwithstanding the evidence that horses masticate roughage to particles of less than 1.6 mm long before it is swallowed (Meyer *et al.* 1975b). In one experiment, the mean rate of passage of long meadow hay was 37 h compared with 26 and 31 h for ground meadow hay, and ground and pelleted meadow hay, respectively. The decrease in fibre digestibility experienced on grinding almost certainly is a function of rate of passsage through the GI tract. However, by the same token, the faster the rate of passage becomes the greater is the capacity of the horse for feed; but an extension of the time for each meal of pelleted roughage may improve digestibility of fibre.

Where horses are given a choice of loose hay, wafered hay and pelleted hay, more is consumed of the latter two than of the loose hay. Generally speaking, the horse is a reasonable judge of the quality of loose hay and, among grass hays, well-made ryegrass can be one of the best. Horses prefer grain to either chopped or long hay, and if given a chopped hay–grain mixture they are inclined to sort out the grain. Nevertheless, such a mixture frequently affords a useful function of depressing the rate of grain consumption by a greedy feeder. Some evidence suggests that the consumption of concentrate feed before hay, rather than after, causes a more intense mixing of ingesta and less variation in the concentration of VFA in the lumen of the large intestine (Muuss *et al.* 1982). This should be an advantage, but other evidence (Cabrera *et al.* 1992) indicates that the consumption of roughage, before concentrates, improves amino acid utilization from digested proteins.

For stabled horses, long hay is given on either a clean area of the floor in the corner of the box, in a hay rack or in a net. The latter should be placed sufficiently high to avoid the possibility of a horse entangling its hoof in an empty net. The amount of hay wasted may be greatest where it is placed on the floor, but this procedure leads to less atmospheric dust.

Dried grass nuts

Grass, clover, lucerne (alfalfa) and sainfoin crops are frequently cut when green and leafy, artificially dried, and preferably chopped and pelleted with a moisture content of about 120 g/kg. In the UK the product must have a crude protein content of at least 130 g/kg, on the assumption that the moisture content is 100 g/kg, to be designated 'dried grass'. High protein grass nuts contain approximately 160 g protein/kg. Dehydrated alfalfa manufactured in the USA contains 150–170 g protein/kg (90 g moisture/kg). These products contain little vitamin D_2, but are rich sources of high quality protein, β-carotene, vitamin E and minerals, well suited to horse feeding and of relatively balanced composition. However, where the product is rich in legume forage the Ca:P ratio is frequently too wide, and the protein content too high for it to form the entire diet. It should then be supplemented with a cereal product rich in P.

The artificial drying of green forage yields a product more valuable than hay, as the raw material to be dried is less mature, leaves are not shattered and lost, moulding is avoided and dustiness is minimal. The only disadvantages are the absence of long fibre, and the contents of β-carotene and α-tocopherol are variable and influenced particularly by length of storage of the product. Thus, as much as half the vitamin A potency (initially it may be equivalent to 30 000–40 000 iu vitamin A/ kg for horses) can be lost during the first 7 months of storage where facilities are not ideal (see Feed storage, this chapter).

FUNCTIONS OF HAY AND USE OF OTHER BULKY FEEDS

Haylage

Fibre and bulk are useful attributes for part of the horse's diet to have. By diluting more readily fermentable material, fibre suppresses a rapid fall in pH of the large intestinal contents and, by stimulating peristaltic contraction, feed with these characteristics probably aids the expulsion of accumulated bubbles of gas. There are many alternatives to hay as sources of fibre and for horses with sound teeth several are useful where reliable hay cannot be obtained. Best-quality silage and haylage free from moulds can be fed to most horses. Good-quality acidified grass silage with a high content of DM may replace between one-third and all horse feed; but success depends on its composition, freedom from abnormal fermentation and general quality, on the horse and the skill of the feeder. Compensation should be made for its deficiency, compared with grass, in potency of vitamins A and E. Horses suffering from respiratory allergy should benefit most by changing from hay to silage. Very acid silages should be avoided. Silage with low amounts (less than 25%) of dry matter and baled or bagged material with a higher content of dry matter and with a pH of around 6 may lead to a greater risk of abnormal clostridial fermentation (see Chapter 10), or may very occasionally precipitate explosive intestinal fermentation and colic if improperly fed. The reason for this may be that the rate of intake of highly fermentable DM is much greater in this form than it is in the form of long hay.

Wasteproducts

Good-quality spring barley or wheat straw in small quantities acts as a source of fibre for horses with sound teeth but is deficient in most nutrients. The inclusion of various wasteproducts in complete diets has been examined, particularly in France and the USA. These products include dried citrus pulp, which is quite satisfactory (Ott *et al.* 1979b), and such unusual materials as sunflower hulls, almond hulls, corrugated-paper boxes and computer paper. Digestibilities for the last two seem to be about 90 and 97%, respectively, but are much lower for the first two because they are heavily lignified. The characteristics of some other bulky feeds are discussed elsewhere in this chapter. With increasing competition between domestic animals for feeds and an expanding world population, undoubtedly the search will continue for satisfactory and safe means of sustaining a healthy population of domesticated herbivores by the greater use of wasteproducts of human activities.

Alkali treatment of roughage

Treating straw with sodium hydroxide – nutritionally improved straw (NIS) – increases its digestibility to the horse (Mundt 1978) and, with dietary adjustments to its sodium content, the product is an important supplier of dietary fibre. Ammonia-treated straw also has promise (Slagsvold *et al.* 1979), but results in cattle and horses have been mixed. The potential digestibility of poor-quality roughages is difficult to predict by chemical analysis, as the factors that inhibit the complete digestion of plant cell-wall polysaccharides include a difference in structural organization, as well as differences in chemical composition of those structures.

Group feeding

Feeding habits and hunger of stabled horses can vary enormously and succulent roughages are sometimes used to stimulate animals with flagging appetites. One

French study (Doreau 1978) showed that the intake *ad libitum* of a dry feed by a group of stabled horses varied from 8.1 to 19.2 kg daily and the time spent eating ranged from 6 h 40 min to 15 h 50 min. The horses ate several large meals and some small diurnal and nocturnal meals. The night meals represented 30% of the total intake. Several factors may contribute to the fastidiousness of finicky eaters, such as environmental stress and nervousness, unpalatability and monotony of ration, nutritional deficiencies, poor health and teeth, lack of exercise and peck order (hierarchy) among group-fed horses.

Succulents

Many succulent vegetables and fruits (for example, sugarbeet roots, carrots, apples, pears, peaches and plums) are satisfactory as treats for horses. Peaches and plums should be stoned, and hard root vegetables should be sliced into strips to avoid choking, and then mixed with compounded nuts or grain. Carrots contain over 100 mg carotene/kg and care should be exercised in the quantities used (not more than 0.5 kg fresh material/100 kg BW daily) if other large supplements of vitamin A are being given. A similar attitude should apply to all other treats as they represent an unbalanced feed and in large quantities (more than 20% of the total dry-matter intake) can do more harm than good. It should also be realized that succulents, including both root vegtables and fruit, contain 80–90% water, and on a dry-matter basis they may therefore be a very expensive source of energy and protein. Only if they are relatively cheap can succulents be justified and bulkiness restricts their role to that of a supplement to normal rations. Of all the main flavour groups present in feed, the horse is deterred by sour tastes and attracted by sweet flavours in moderation.

COMPOUNDED NUTS

Several ingredients in a ground form are generally incorporated, among them the common cereal grains, oilseed meals, milling, brewing and distilling byproducts, dried grass and lucerne, fishmeal, and mineral, trace-element and vitamin supplements. Their principal role in this form is to provide a balanced source of all nutrients, but they have to be supplemented with loose hay as a source of long fibre, with water and sometimes with common salt. Nuts rich in nutrients and with high digestibility can be supplied to young foals, high-energy nuts can be given to horses in hard work, and low-energy nuts can be provided for adult horses engaged only in light work. The advantages of nuts thus include standardized diets for particular purposes, constant quality, extended shelf life, freedom from dust, palatability, and uniform physical characteristics and density, all of which facilitate routine feeding.

Compounded nuts, particularly high-energy, nutrient-dense formulations, should be introduced gradually to give the horse and its microbial flora time to adapt to the new regime. A too-rapid introduction of nuts, or for that matter of oats, sometimes leads to slightly loose droppings during the first 2–3 weeks, 'filled-legs' and even to colic. Complete nuts are also manufactured for feeding to horses in the absence of hay, but, generally, these should be used only for a greater control of dust where horses are subject to respiratory allergies. In the absence of loose hay, wood chewing and some other vices, including coprophagy, may be more prevalent. The preparation of feed in nut form may have the disadvantage that the user is unable to recognize good-quality from poor-quality ingredients. Products from reputable compounders should therefore always be used for feeding horses, but some indication of the chemical nature of the product can be conjectured by reference to declared analyses required by law in the EU and found on a ticket attached to the bag. The Statutory Statement included on the ticket should give the following information for complementary and complete feeding stuffs:

- the name or trade name, the price, country of origin and address of the person responsible for the particulars of the Schedule. The net quantity and minimum storage life (or batch number), and the moisture content of compounds if it exceeds 14%;
- whether a permitted antioxidant, colorant or preservative is included;
- the active levels of vitamins A, D or E, if included, and the shortest period over which the activities apply;
- the name of any Cu additive and the total level of Cu (naturally present plus added);
- bentonite and montmorillonite, the name of the additive; and (other) binding agents present;
- details of any enzymes or microorganisms added (see 'Probiotics', this chapter);
- information may be included on total levels of other trace elements and the total levels of other vitamins, provitamins and vitamin-like substances, including the minimum period over which the activity applies;
- ingredients in descending order by weight;
- the percentage by weight of crude oil (lipids extractable with light petroleum, 40/60°C boiling point without prior hydolysis, except in the case of milk products);
- the percentage by weight of crude protein (the nitrogen content multiplied by 6.25);
- the percentage by weight of crude fibre (principally organic substances remaining insoluble following alkali and acid treatment);
- the percentage by weight of total acid-insoluble ash;
- levels of lysine, methionine, cystine, threonine and tryptophan may be included;
- levels of starch, Ca, Na, P, Mg and K may be included (levels of Ca, if over 4.9%, and P, if over 1.9%, must be included).

Table 5.2 gives recommended declarations and some chemical values for compounded horse feeds.

Table 5.2 Recommended declarations and chemical values for compounded nuts and coarse mixtures (assuming 88% DM).

	Crude oil (%)	Crude fibre (%)	Crude* protein (%)	Total lysine (%)	Digestible energy (MJ/kg)	Total ash (%)
Foals						
1 month before weaning to 10 months old	4–4.5	6.5–7.5	17–18	0.9	13	7–9
11–20 months old	3–3.5	8.5	15–16	0.75	11	7–9
Adults						
Strenuous work	3.5–4	8.5	12–13	0.55	12	8–10
Light to moderate work, barren mares and stallions	3	14–15	10–10.5	0.45	9	9–10
Last quarter of pregnancy, lactation and working stallions	3	9–10	13–15	0.65	11	8–10

* Actual protein concentrations are less important than are the total lysine contents. Note: lysine and DE are not normally declared.

A range of sizes of compounded nuts has been found suitable for feeding to growing and adult horses. However, the optimum seems to be a diameter of 6–8 mm and a length of 12 mm. For very young foals being given a milk substitute nut, a diameter of 4–5 mm and a length of 6–7 mm is probably more suitable. South African work (van der Merwe 1975) indicates that acceptability is not affected measurably by hardness of nut, although most horses dislike nuts that crumble too readily, and those that are excessively hard may occasionally be bolted without mastication. This work revealed that smaller nuts are chewed more slowly and require more time for a given amount to be consumed – a decided advantage.

Where horses are in especially hard work, up to 80% by weight of the total ration can be provided in the form of nuts or grain and supplements, with the remaining 20% composed of hay. Regimens of this nature require considerable skill, 4–5 feeds per day and regular exercise every day. A much more typical regimen for stabled horses in strenuous work is a ration of 50–60% by weight of nuts or concentrate and 40–50% hay. As the amount of work is reduced, so the proportion of nuts can be decreased, or nuts of lower energy can be used. In stables where horses are given their concentrate measured in terms of the number of bowls per day, recognition should be given to the differences in bulk density and energy density of feeds. For example, a unit volume of barley is about three times the weight of the same unit volume of wheat bran. Furthermore, the common cereals have different amounts of digestible energy per unit weight. The combined effects of energy density and bulkiness imply that, for example, a unit volume of maize contains nearly double the digestible energy of the same volume of oats (Table 5.3 gives appropriate conversion values).

Table 5.3 Weights of common cereal grains and soyabean meal and average DE and UFC (MJ/10l) values per unit volume.

	Weights		DE	UFC
	lb/bushel	kg/10l		
Oats	27–45	3.4–5.6	51.7	3.9
Barley	36–55	4.5–6.9	73.0	5.7
Wheat	50–62	6.2–7.7	98.0	7.5
Milo (sorghum)	51–59	6.4–7.4	89.7	6.4*
Maize	46–60	5.7–7.5	93.7	7.4
Soyabean meal (solvent extracted, 44%)	47–53	5.9–6.6	83.1	5.7
Wheatbran	17–21	2.1–2.6	25.4	1.8

*NE value of milo estimated.

COARSE MIXES

Some horses in strenuous work have poor appetites and are more likely to 'eat up' when given a coarse mixture (sweetfeed) than when given nuts in similarly large amounts. Coarse mixtures have thus gained in popularity, although it is evident that poor appetite for nuts is due in part to similar volumes and therefore larger weights of nuts being offered, when refusals would be expected. Coarse mixes should be complete, to be supplemented only with loose hay, water and sometimes with common salt. They tend to be more expensive to produce than compounded nuts, but have an advantage in being less dense, and normally contain a proportion of cooked, flaked cereals and oil seeds and expeller oil-seed cakes. Their shelf life is less than that attributed to compounded nuts and their bulkiness demands proportionately more storage space. Storage of these mixtures and of all feeds should be in dry, cool, ventilated conditions where there is little variation in temperature, otherwise moulding can occur.

CEREALS

Whereas water is probably the most critical nutrient for the horse's immediate survival, fatness and lack of exercise are its worst enemies. The adequate control of energy intake is the most difficult aspect of optimum and reliable feeding. Cereals (Plate 5.3) are the principal source of energy in the diet of hard-worked horses and therefore a brief discussion of the characteristics of the common cereal grains and their byproducts is appropriate.

Cereal grains contain from 12 to more than 16 MJ DE/kg DM compared with about 9 MJ/kg in average hay. Cereal grains embody three main tissues: the husk and aleurone layer, endosperm and the embryo. The endosperm is a rich store of starch required as a source of energy during the early growth of the plant. Of the

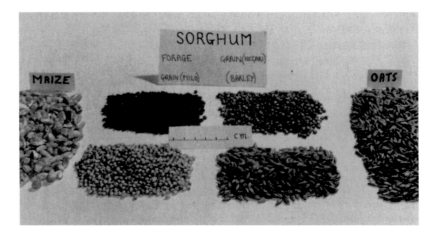

Plate 5.3   Cereal grains. Maize (corn) grains are the largest and may be given whole to horses with sound teeth, but as they can be very hard, cracking is often worthwhile. Barley is smaller and relatively hard and the grains should be crimped or rolled lightly. Oat grains are relatively light and bulky, and crimping or rolling is required only for young horses or for older horses with poor teeth. By comparison, sorghum grains are small, as they are 'naked'. Sorghum is grown in hot dry countries and white varieties are quite satisfactory for horses when coarsely ground, cracked, rolled or cooked. The brown varieties contain large quantities of tannins and are unsuitable for horses.

nitrogen compounds of cereal grains 85–90% are proteins; these are found in each of the tissue regions but are in higher concentrations in the embryo and aleurone layer. Cereal proteins are not as nutritionally useful as oil-seed and animal proteins because they are relatively deficient in lysine, threonine and methionine. The quality of protein (as distinct from the amount) decreases in the order oats, rice, barley, maize, milo, wheat and millet. Oat protein contains slightly more lysine than does the protein of other cereals. The oil content of cereal grains varies from about 15 to 50 g/kg, with oats containing slightly more than maize, which in turn contains more than barley or wheat. This oil is rich in polyunsaturated fatty acids, of which the principal one is linoleic acid which generally constitutes about half the fatty acid composition by weight in the oil. Unsaturated oils, such as these, are prone to rancidity subsequent to the grinding of cereals, unless the meal is compressed into pellets or otherwise stabilized. Cereal grains are deficient in calcium, because they contain less than 1.5 g/kg, but they have three to five times as much phosphorus, principally in the form of phytate salts. These salts tend to reduce the availability of calcium, zinc and probably magnesium in the intestinal tract; those of oat are said to have a greater immobilizing effect than do the phytates of other common cereals.

Palatability

Although different samples of one cereal species can vary considerably in quality, when given a choice ponies prefer oats to other common cereals. Comparisons made at Cornell (Hawkes *et al.* 1985) showed that the order of preference was oats, cracked maize, barley, rye and cracked wheat. Nevertheless, there was little depression in total intake when only the less palatable cereals were given.

Oats (*Avena sativa*)

Under traditional systems of feeding, in which a single species of cereal grain is given, grains of oat are safer to feed than are the other cereals as their low density and high fibre content make them more difficult to overfeed and the grain size is more appropriate for chewing. They need, therefore, no crimping or rolling for horses aged over 1 year if the teeth are sound. A greater quantity of oats than of the other cereals must be consumed to produce founder or other digestive problems. However, they tend to be more expensive per unit of energy than the others, as between 23 and 35% of the grain consists of the hull. Figure 5.1 gives a cross-sectional view of oats and in Appendix C their chemical characteristics are listed for comparison with the other common cereals.

Naked oats (varieties produced by crossing *Avena nuda* with *A. sativa*)

The groat of *Avena sativa* is enclosed in a husk that constitutes 250 g/kg of the grain's weight. The husk of the new varieties is loose and falls off during threshing, hence the energy content of naked oats is considerably higher than is that of the grain of *Avena sativa*. Naked oats typically contain 120 g/kg protein and 6 g/kg lysine. The P content is high (3.5 g/kg) but is principally present as phytates and the oil is prone to lipolysis during storage. At present it is advisable to restrict naked oats to 100–200 g/kg of the cereal mix of horses.

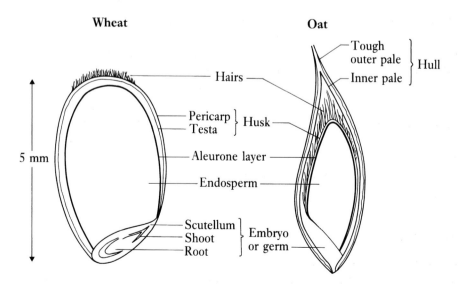

Fig. 5.1   Cross-sectional view through wheat and oat grains.

Barley (*Hordeum vulgare*)

Oats and barley differ from wheat, maize and grain sorghum in being invested in a hull (botanically known as the inner and outer pales), which the other three cereals have lost during harvesting. All the cereal grains are, however, encased tightly in the thin membrane composed of the less-fibrous fused testa and pericarp (Fig. 5.1).

The hull of barley constitutes 10–14% of the total grain weight and the hull is relatively smaller and more tightly apposed to the grain than the hull of the smaller grain of oats. Thus, for feeding to horses, barley grain should be crimped or lightly rolled to rupture the case shortly before feeding, but it may be fed as the only cereal after a period of gradual adaptation. This period is necessitated by the higher starch content and bushel weight of barley in comparison to oats. Processes that gelatinize starch grains, such as steam rolling or micronization, are discussed later in this chapter (see '(1) Cooking: expansion' and '(2) Cooking: micronization'). Barley protein is of slightly lower nutritional quality than that of oats, being relatively deficient in lysine, and the oil content is quite low, being generally less than 20 g/kg.

Some varieties of naked or hull-less barley, very low in fibre, have been bred. They are comparable with naked oats or oat groats from which the hull has been removed by processing, again yielding a starchy, high-energy, low-fibre product, but their price rarely justifies their use in horse feeding.

Wheat (*Triticum aestivum*)

The grains of wheat (Fig. 5.1) are free of hull and relatively small so that they may escape mastication if fed whole. Wheat should therefore be cracked, coarsely ground or steam flaked before use. The two endosperm proteins (known collectively as gluten) are deficient in lysine and can form a pasty impenetrable mass (similarly for rye grain) for digestive juices, especially when wheat is finely ground. In the uncooked state wheat should form less than half the grain fraction of the diet. Moreover, as the starchy endosperm constitutes 85% of the grain, excessive intakes can cause digestive disturbances, particularly if the adaptation period has been short. The bran and germ make up about 13 and 25% of the grain, respectively.

Triticale

Triticale is a hybrid resulting from crossing wheat *(Triticum)* and rye *(Secale)*. It contains more protein than barley, but in variable amounts (100–200 g/kg) which are deficient in the amino acid tryptophan (1.5 g/kg cereal), and richer in lysine than wheat protein. It is said to contain higher concentrations of trypsin inhibitors and alkyl resorcinols than wheat or barley. In ground form it should have a feeding value for horses slightly greater than that of barley. Some varieties are subject to ergot infection and samples should be examined to ensure there are no significant amounts present.

Maize (corn) (*Zea mays*)

Maize is the largest of the cereal grains and is acceptable in any form for feeding to horses. Frequently, however, when the grains are very hard they should cracked, especially for horses with poor teeth. Maize contains twice the energy per unit volume of oats. However, a considerable portion of maize starch is fermented in the hindgut when large amounts are given, unless the grain is cooked (see Chapter 1 and Fig. 6.10). The so-called heating of cereals generally results from the rapid assimilation of their products of digestion and their rapid fermentation by intestinal micro-organisms (with a fall in caecal pH). This causes an abrupt increase in the heat of fermentation and a rise in the concentrations of glucose and VFA in the blood, stimulating metabolic rate to a greater extent than occurs after a meal of hay. Precooking of cereals diminishes the fermentative component of this effect. The grains of maize contain about 650 g starch/kg, but only 80–100 g crude protein. The endosperm protein, zein, is deficient in tryptophan and to a lesser extent in lysine, but the protein of the germ in common with the germ of other cereals is of better nutritional quality.

Yellow and white maize varieties are produced, but the types for stock feeding are predominantly yellow and contain the xanthophyll pigments cryptoxanthin, a precursor of vitamin A and zeaxanthin. Infrequently some maize imported into the UK is more than 2 years old, when it is a poor source of vitamin A.

Grain sorghums (*Sorghum vulgare subglabrescens*)

Sorghums are the main food grain of Africa and parts of India and China where they are grown on land too dry for maize. They are also a major stock feed in dry areas of mid-west USA. The kernel is naked, like that of maize and wheat, but more spherical in shape and smaller than that of wheat. The protein content of the grain is variable (80–200 g/kg). The lysine content of the protein is rather low, and the oil content of the grain is less than in maize. Owing to its size the grain should always be rolled, cracked, coarsely ground or steam flaked before feeding to horses. This disrupts the waxy bran covering the endosperm. If finely ground it may become pasty and rolling is preferable. It is a high-energy cereal and, therefore, to avoid digestive disturbances, it should preferably form only a portion of the cereal intake.

Several sorghum varieties are grown, including some for forage use. The grains range in colour from white to deep brown. Varieties of grain sorghum that contain only low concentrations of tannins are widely used as a horse feed. Those varieties that are brown, or purple, contain significant amounts of tannins and the white, or milo, grain varieties are the types that should be used (Plate. 5.3). Tannins can cause colic in horses.

Rice (*Oryza sativa*)

The grain of rice is invested in a thick fibrous hull, which is easily removed but which constitutes about 20% of the total weight. The hull is rich in silica and when freed from the grain it is unsuitable on its own for feeding to horses, because the sharp edges may cause irritation. Rough rice, that is the grain before the removal of the hull, is more suitable as a horse feed. Rice protein is a reasonable source of lysine.

Millet (*Setaria* spp., *Panicum miliaceum*)

The name millet is applied to a range of grass species. The seeds have a high energy content, are the palatable staple diet of many people and typically contain 110 g/kg protein of low lysine content, 50–90 g/kg crude fibre and 25–35 g/kg oil. They are free from toxins except for the unripe seed husks of *Paspalum scrobiculatum*. However, millet is not readily available in Europe and parcels that are may have been rejected for human consumption for reasons of mould or admixture. The seed is small and the hull is not removed during normal harvesting. It requires coarse grinding, or crushing, for feeding to horses and its feeding value is somewhat similar to that of oats.

Processing of cereal grains

Specific procedures for processing each species of cereal grain have already been sketched. Where cereals are mechanically rolled the process should be one of kibbling, or bruising, rather than the complete rolling of uncooked material, otherwise the chemical stability of the product is jeopardized, no further increase in digestibility is achieved and greater processing costs are incurred. A similar argument can be advanced for coarse in contrast to fine grinding. Moreover, finely ground cereal endosperm is floury, unpalatable, dusty, less stable and may lead to digestive disturbances. Some other feeds such as bran may occasionally ball up and block the oesophagus if fed dry and therefore are normally dampened or mixed with cut or chopped hay, or fed damp with oats. Cereal grains, or high-energy nuts, should be distributed among as many daily feeds as possible to minimize the risk of colic.

(1) Cooking: expansion

Cereals should be cooked only in the presence of water in order to minimize the risk of heat damage to proteins and oil. Steam pelleting and expansion procedures achieve this objective and, for high-energy materials in particular, have been shown to improve the digestibility of dry matter, organic matter, starch and the nitrogen-free extract of cereals and nuts without interfering with the digestibility of crude protein (see Chapter 1). The digestibility of the crude-fibre fraction, however, is either not affected or is only slightly depressed if the food material is ground beforehand. This fraction represents only a relatively small portion of high-energy products. The process of expansion, or popping, relies on the cooking effects of superheated steam injected into a slurry compressed against a die face by a revolving worm and the subsequent rapid fall in pressure during extrusion. Material is subjected to a temperature of around 120°C for about a minute. Processing costs, which constitute more than 10% of the value of the product, are difficult to defend except for young stock and horses under competition rules. It is hard to quantify some of the indirect advantages, which, depending on the process, may include reduced storage space, increased stability of product, improved palatability, destruction of natural toxins, insect pests and bacterial pathogens, and the avoidance of high-starch concentrations in the large intestine. The last may be of prime import.

(2) Cooking: Micronization

Other cooking procedures include the traditional steam flaking of maize during which the grain is passed through heated rollers, the roasting of oil seeds during the industrial extraction of oil, and the micronization of cereals and vegetable protein seeds. For micronization a moving belt carries a thin, even layer of cereal grain horizontally beneath a series of ceramic burners that emit infrared irradiation in the 2–6-µm waveband. This results in a rapid internal heating of the grain and a rise in water vapour pressure, during which the starch grains swell, fracture and gelatinize. The product is usually then passed through helically cut rollers and from a cooler to a cyclone. The raw material achieves temperatures ranging from 150 to 185°C for between 30 and 70 s – for each specific raw material there are optimum values within these ranges. These products are frequently included in coarse mixtures for horses; the process increases digestibility and, for instance, in the case of soya beans, will inactivate antiprotease and other toxic factors.

Acidification

The alkali treatment of roughages has been briefly described already (see 'Alkali treatment of roughage', this chapter). Treatment with propionic acid of high moisture cereal grains has achieved a certain measure of popularity during harvesting in inclement weather. The acid acts as a mould inhibitor and preservative. Grain treated in this way is only marginally suitable for feeding to horses, owing partly to its acidity and more especially to the frequent presence of mouldy patches in the silo. The grain may become infected by the fungus *Fusarium*, which produces the toxin zearalenone known to cause 'poor doing' in all animals and infertility in breeding animals (also see Grass Sickness, Chapter 10). Furthermore, high-moisture cereals are deficient in α-tocopherol so that supplementation with synthetic forms of vitamin E at a level of about 30 mg/kg feed is essential.

Cereal byproducts

The industrial use of cereal grains leads to the production of two major types of byproduct: (1) those derived from the milling industries (the seed coats and germ) and (2) those derived from the brewing and distilling industries (principally spent grains, the residues of germinated grains and dried yeast).

Wheat, oat and rye milling byproducts

There are three byproducts of oats: the hull, dust consisting of oat hairs lying between the grain and the hull, and meal seeds composed of hull and the endosperm of small seeds. Oat hull has a crude fibre content of 330–360 g/kg with a digestibility little better than that of oat straw. A combination of oat hulls and dust in the approximate ratio of 4:1 gives oatfeed, which should by legal definition contain no more than 270 g crude fibre/kg. Each of these byproducts may be fed to horses when appropriately processed and included in balanced feeds in proportions of up to 20% of low-energy diets.

Undoubtedly the major milling byproducts fed to horses in Western countries are those derived from wheat milling. The offals of wheat consist of the germ, bran, coarse middlings and fine middlings, which comprise about 28% of the total weight of the grain and collectively are known as wheatfeed, although in some products a proportion of the germ is marketed separately. The germ contains 220–320 g crude protein/kg and is a rich source of α-tocopherol and thiamin. This particular byproduct is too expensive for general use but can be of value to sick animals. Bran is derived from the pericarp, testa and aleurone layers surrounding the endosperm, with some of the latter attached. It normally contains between 85 and 110 g crude fibre/kg and between 140 and 160 g crude protein. It is sold either as giant, broad or fine bran according to size, or as entire fraction 'straight-run bran'. These grades are similar in chemical composition, although the larger flaked varieties may contain slightly more water. Bran is typically expensive for the nutrients it provides, but it can form, as a mash, a palatable vehicle for oral administration of drugs and it has the capacity to absorb much more than its weight of water. Thus, it has a laxative action on the intestinal tract. Bran in particular, but also other wheat milling byproducts, are rich sources of organic phosphorus, as bran contains approximately 10 g/kg, or slightly more, of which 90% is in the form of phytate salts. As bran is deficient in calcium and as phytate depresses the utilization of dietary calcium and zinc, the use of large quantities accelerates the onset of bony abnormalities in young and adult horses.

Coarse middlings are similar to bran but contain somewhat more endosperm and therefore chemically contain only 60–85 g crude fibre/kg and about the same amount of crude protein. Fine middlings contain even more endosperm than the coarse, and consequently only 25–60 g crude fibre. When adjustments are made for the imbalance in minerals, wheat byproducts are safe feeds as supplements to horse and pony rations, besides being relatively rich sources of some of the water-soluble vitamins.

Oatfeed

During the commercial preparation of oatmeal for human consumption the husks, hairs and meal seeds (the husks and part of the endosperm of small grains) are removed. These husks and hairs are combined in a ratio of 4:1 to form oatfeed which in the UK should not, by legal definition, contain more than 270 g crude fibre/ kg. Oatfeed is a safe feed used in mixtures to reduce the energy concentration of starch-rich cereals in diets for nonworking horses on the 'easy list'. Oat hulls are rich but variable in lignin content. One study showed this to vary from 9 to 61 g/kg among worldwide varieties. The raw material is bulky and does not flow easily, so is frequently purchased in a pelleted form by manufacturers.

Rye bran
Rye bran is of limited acceptability to horses.

Maize (corn) byproducts

The byproducts resulting from the industrial production of glucose and starch derived from maize include the protein gluten, a small amount of bran and the germ. These byproducts are similar to the analogous byproducts of wheat and frequently all three are combined for sale as maize gluten feed. Although the protein is of poor nutritional quality, the feed is quite suitable as a supplement in horse rations as it is a good source of some of the water-soluble vitamins.

Maize germ meal
The maize germ is removed from the endosperm during the process of corn starch extraction. Much of the corn oil is then removed from the germ leaving the extracted maize germ to which may be added other maize offal products, bran and gluten. The composition is therefore variable, depending on the details of the manufacturing procedure. The residual oil is polyunsaturated and thus subject to peroxidation. It is a palatable feed, the protein has a reasonably good amino acid balance and there are no toxic factors present in well manufactured and stored product.

Maize gluten feed
Maize gluten is another byproduct of the maize starch industry. It is the maize protein, gluten, together with maize bran, extracted maize germ and dried residues of the steeping liquors. The product is rich in xanthophylls, approximately 20 mg/kg, giving it a bright yellow colour. The gluten, free from other byproducts, contains over 600 g protein/kg and is available as a branded product, Prairie Meal; but typical material contains 180–230 g protein/kg, 70–80 g crude fibre/kg and only 6–7 g lysine/kg, depending on the amounts of bran and germ remaining. The feed is palatable, free from toxins and a useful constituent of horse feed mixtures.

Rice bran

The quality of rice bran depends on how efficiently the indigestible, siliceous hulls have been removed before the husk is detached from the grain. The bran from the first stage milling consists principally of the husk (pericarp and testa), the germ and part of the aleurone layer. These components are incompletely removed and the remaining byproduct (second stage milling) constitutes rice polishings, which also contain some endosperm. The first and second stage milling byproducts are sometimes combined to form rice pollards.

Large quantities of rice meal, or rice bran, are produced globally. This byproduct consists of the husk, aleurone layer, germ and some of the endosperm of the rice grain. Inevitably some hull will be present, but this should be at a minimum in the preferred material. The bran has a crude composition of 90–210 g crude fibre/kg, 100–180 g crude protein/kg and 110–170 g lipid material composed of a very unsaturated fat. This fat becomes rancid rapidly and is therefore extracted, leaving a product of much better keeping quality. Extracted rice bran is widely available and is a good supplementary feed for horses when used as a component of a mixed ration. The extracted byproduct has a composition of about 15 g oil, 130 g crude protein and 120 g crude fibre/kg. However, frequently as much as 60 g silica/kg are present and the ash content of the bran is variable and ranges from 100 to 240 g/kg, reflecting the amount of hull that still remains. Extracted rice bran is also a very rich and variable source of phosphorus (11–22 g/kg), much of which is phytate-P. Care should be taken to ensure that rations in which it is used are appropriately balanced for minerals. Where good quality rice bran is obtained 150 g/kg may be included in balanced concentrate feeds.

Brewing byproducts

Three major byproducts are derived from brewing: malt culms, brewer's grains and brewer's yeast. When barley is sprouted for the purpose of hydrolysing the starch, the resulting malt sprouts, which include the embryonic radicle (root) and plumule (stem), remain after the malting process. These are removed and dried to form the malt culms. The remainder of the material is mashed to remove sugars, leaving the grains which may be disposed of as a wet byproduct or dried and sold as dried brewer's grains.

Malt culms (malt sprouts)

Malt culms (malt sprouts) are the dried shoots and radicle of germinated barley grain. The material contains 12–30 g oil and 140 g crude fibre/kg. The crude protein content is typically 240 g/kg, but very variable and represented by a proportion of nonprotein N. The lysine content is typically 12 g/kg. The byproduct is not very palatable, is dusty, and swells on wetting, stimulating peristalsis, but can contain moulds if not fully kiln-dried. An inclusion rate for good material of up to 50 g/kg in horse feeds is satisfactory, but it should not be fed dry on its own.

Dried brewer's grains

The residual grains after removal of the wort may include maize and rice residues in addition to those of barley, the main constituent. The dried byproduct contains 180–250 g crude protein and 140–170 g crude fibre/kg and it therefore forms a useful adjunct to mixed horse feeds.

Dried brewer's yeast

The most coveted and expensive byproduct of brewing is, of course, yeast, which in dry form contains 420 g high-quality protein/kg and is a rich source of a range of water-soluble vitamins and of phosphorus. This yeast is frequently fed to horses in poor condition at the rate of 30–50 g daily, but is too expensive for regular feeding (also see 'Probiotics', this chapter).

Distilling byproducts

The principal residues from the whisky distilling industry are the grains and the solubles. The grains in the malt-whisky industry consist solely of barley residues, whereas grain whisky residues may in addition include those of maize, wheat, oats and rice. A proportion of the grains is sold wet, but significant quantities are dried.

Distiller's dried, or light, grains
Distiller's dried, or light, grain is the fibrous residue of barley and of other grains (blended whisky), remaining after washing out the sugars derived from hydrolysis of the starch and used for fermentation to alcohol. For whisky production the alcohol is distilled off, leaving liquid pot ale from malt whisky production and spent wash from grain whisky production. This byproduct is suitable for inclusion in horse feeds.

Distiller's solubles
After distillation of the alcohol, the spent liquor is spray-dried to yield a light-brown powder known as distiller's solubles, quite suitable for use in mixed feeds. Frequently the dried solubles are added back to the dried grains and marketed as dried distiller's grains with solubles, known also as dark grains.

Distiller's dark grains
The liquid pot ale (or spent wash from maize and other grain distillation) contains unfermentable carbohydrates and products of yeast metabolism, such as protein and vitamins. Where maize is used for alcohol production this liquid contains maize oil and after evaporation of water and addition of lime the residue is spray-dried. With the grains the residue produces distiller's dark grains (DDS), or distiller's dried grains with solubles. This byproduct in small quantities is a valuable supplement to mixed horse feed and American evidence suggests that dried maize grains with solubles stimulate the digestion of cellulose by microorganisms in the horse caecum. Malt whisky DDS from barley contains about 270 g digestible carbohydrate/kg and 70 g ether extract/kg, whereas grain DDS from maize contains about 180 g digestible carbohydrate/kg and 110 g ether extract/kg. The crude fibre content of DDS is 100–130 g/kg, lipid 100–120 g/kg, crude protein 260–280 g/kg and the lysine content about 8 g/kg. The byproduct from grain whisky is generally more digestible, but both DDS byproducts are low in sodium, potassium and calcium. Both are free

from toxic constituents, and rich but variable in copper, malt DDS containing about 40 mg/kg, and grain DDS 80 mg/kg, and both are suitable for inclusion in horse feeds.

Cereal grain screenings

Cereal grain screenings are residues from the preparation, storage and treatment of cereal grains (barley, wheat, maize, sorghums and soyabeans) and their products. They are extremely variable in quality and include broken grains, chaff, weed seeds and dust, and therefore moulds. Mycotoxins may be present. Each batch may differ and their use should be in the pelleted form and strictly limited. The ash content varies from 20 to 180 g/kg. Higher proportions of ash indicate greater soil contamination, and the presence of broken grains contributes to an accelerated rate of rancid cereal oil formation, particularly important in respect of maize.

Chaff and molassed chaff

Chaff is composed of the husk, or glumes, awns and other fibrous waste material derived from the threshing of grain. It is highly lignified, and therefore not well fermented by intestinal microorganisms, but oat chaff, which is the best, is better utilized than is straw. That derived from barley threshing contains large amounts of barley awns, which have serrated edges and may cause some irritation. Nevertheless, it is a safe material to dilute energy-dense cereal grains and is frequently marketed mixed with molasses at a concentration of about 500 g/kg. This product is palatable to horses and overcomes any dustiness of the chaff. Some molassed products also include about 20 g/kg limestone, as a source of calcium; however, it should be borne in mind that these products are deficient in phosphorus.

Sodium hydroxide-treated straw

Sodium hydroxide (NaOH)-treated straw is an adjunct to mixed horse feeds and provides a useful source of palatable fibre. It should be recognized that NaOH-treatment of cereal grains, or of other materials that are a source of vitamin E, will destroy the vitamin. Where the feed is also deficient in selenium, nutritional degenerative myopathy can result.

Hydroponics

The practice of germinating barley seeds in lighted trays under humid conditions produces a high-quality feed as rapidly growing young plants. However, the cost per unit of dry matter in particular, owing to the inclusion of a realistic rate of depreciation for capital equipment and the high labour commitment, make the practice difficult to justify on economic grounds. The barley grains should not have been through a grain drier, which would severely damage their capacity for germination,

Table 5.4 Typical analytical values for hydroponic barley (g/kg DM).

Crude protein	120–170
Crude fibre	80–155
Ash	27–42
Digestible carbohydrate	550–740
Calcium	1.1–1.3
Magnesium	1.5–2
Potassium	4.6–4.8
Phosphorus	4.4–5.5

nor should they have been treated with mercurial seed dressings. They should be bright and free from broken grains. The time interval from germination to consumption should be minimized by establishing optimum conditions for growth of 20 h light per 24 h and an ambient temperature of 19–20°C. Slow growth increases the likelihood of moulding and its attendant risks. A buildup of mould spores in the room must be avoided by routine hygiene. About three-quarters of the product is moisture and therefore the DE content is only 2.5 MJ/kg despite its high digestibility (Table 5.4). In comparison with pasture grass hydroponic barley has a very low content of calcium and potassium, whereas the phosphorus contents are similar. The Ca:P ratios are therefore quite different from those of grass.

OTHER LESSER INGREDIENTS AND BYPRODUCTS

Carob seed meals (Ceratomia siliqua) *and locust beans* (Parkia filicoidea)

The evergreen carob tree *(Ceratomia siliqua)* grows in Mediterranean countries. Its green pods fall from the tree in the autumn and are harvested. The hard seeds are embedded in the thick fleshy maturing pods that are unsuitable for grinding, owing to their high moisture content. They are kibbled, or crimped, and the seeds fall out. These are used in the confectionery industry, leaving the pieces of pod that are distributed to animal feed manufacturers. These dark-coloured pieces are very sweet, containing 400–450 g sucrose, 35 g crude protein, about 70 g crude fibre and 160–240 g moisture/kg. It is not advisable to use batches containing more than 200 g moisture/kg. The pods also contain some tannic acid, but the best sources are a useful feed for inclusion in coarse mixes. Major sources in Europe are Spain, Portugal and Cyprus and the best quality may be derived from the latter source.

The West African locust bean pods, from a spreading tree *(Parkia filicoidea)*, contain dark-brown, sweet, fibrous seeds in fibrous pods that are free from undesirable chemicals. The pod and seed, containing about 300 g sucrose/kg, are drier than that from the Mediterranean area and so can be ground, bulk handled and used in ruminant and horse rations.

Biscuit meal

Bakery wastes are variable and high in energy content, owing to a wide variation in fat content (5–280 g/kg) and high energy digestibility. In addition to wheat and other flours the wastes contain sugar and salt, which is of no disadvantage to horses. Rancidity and moulding are typical problems and a perennial issue for the Rules of Racing and of Competition is the inclusion of chocolate products in the wastes. This has frequently been the source of theobromine detected in urine.

Molasses

The crystallization and separation of sucrose from the water extracts of sugarbeet (*Beta vulgaris saccharifera*) and sugarcane (*Saccharum officinarum*) leave a thick black liquid termed molasses, which contains about 750 g DM/kg, of which about 500 g consists of sugars. The crude protein in molasses is almost entirely nonprotein nitrogen and of minimal value in feeding horses. In beet molasses, a proportion of this is in the form of the harmless amine betaine (a substitute for choline), which is responsible for the somewhat unpleasant fishy aroma associated with that form of molasses, but cane molasses has a very pleasant smell. The sweet taste of both forms is attractive to horses when used in mixed feeds up to a level of 100 g/kg feed, and in these proportions molasses can act as a relatively effective binding agent in the manufacture of nuts. Cane molasses contains between 5 and 11 g calcium/kg and the potassium content ranges from 20 to 40 g/kg in cane molasses and from 55 to 65 g/kg in beet molasses. Cane molasses is reasonably rich in pantothenic acid and both contain around 16 mg niacin/kg.

Molassed sugar beet pulp (*Beta vulgaris* var. *saccharifera*)

The shredded residue resulting from the extraction of the sucrose contained in the root of sugar beet is a wet product containing 750–900 g water/kg. This could be ensiled for horse feeding, but it is rarely used in this way. Most beet pulp is dried and marketed as dried sugar beet pulp, or mixed with molasses to form molassed sugar beet pulp, providing sugar, pectin and betaine, and sold normally in a cubed form. The dry cubes should not be given to horses on their own, without soaking, as in the dry form they can cause choke in some animals (see Chapter 11) and if dried beet pulp constitutes a high proportion of the concentrate mix then large quantities of free fluid are absorbed by the mass in the stomach and colic may result. The carbohydrate it contains is rich in pectins, but also cellulose and hemicellose that are fermented by the large intestinal flora of the horse. The residual sucrose present is digested in the small intestine. Mixed with other feeds the dried form is used successfully in coarse mixtures and larger quantities, soaked, act as a useful cereal replacer. It has a uniform composition, is free from toxins and much less likely to cause laminitis than equal quantities of cereals. It contains more calcium than do cereals, on average 6 g calcium/kg, and <1 g phosphorus/kg; but it is a poor source of some vitamins, cf. cereals. The protein content and quality are similar to those of cereals. Urea is sometimes added to beet pulp. This is of no value to horses, but causes no harm either in those with fully functioning kidneys. There are no undesirable natural chemicals in beet pulp.

Dried lucerne (*dehydrated alfalfa*)

Although not a byproduct but a useful forage, dried lucerne is mentioned here as it has been attributed with several indirect effects, possibly caused by unidentified factors. Its meal stimulates cellulose digestion by equine microorganisms and enhances gross-energy digestion of feed. An Eastern European report suggests that lucerne hay may have a protective value in the development of glandular inflammation and may encourage white cell (lymphocyte) and red cell (erythrocyte) production in foals (Romić 1978). Pelleted lucerne meal is a better source of nutrients than grass hay or sun-cured lucerne (except for vitamin D_2), and the oxalates it contains do not hamper its calcium or magnesium digestibility. The drying takes place at very high temperatures, sufficient to inactivate mould spores. This will extend shelf-life, inhibit mould growth during storage and so reduce the likelihood of adverse respiratory responses. Pelleting of lucerne leads to increased consumption and its relative bulkiness prevents a rapid accumulation of highly fermentable starch in the stomach, while allowing the penetration of it by gastric juices, ensuring a fall in gastric pH. Horses fed dried lucerne seem to produce better hoof horn than those receiving grass hay. The explanation may be that lucerne provides considerably more sulphur-containing amino acids and calcium than is provided by grass hay.

Dried lucerne leaves contain several green pigments in addition to chlorophyll. One of these, pheophorbide-α, has been shown to cause photosensitization, expressed as skin lesions, of albino rats. Exposure to light in the visible range is sufficient to cause lesions in rats, and other white animals, given a lucerne leaf-protein concentrate. The pheophorbide-α is formed by breakdown of chlorophyll under the influence of chlorophyllase, during processing. There is a higher activity of this enzyme in legume forages than in grass, which may account for an absence of dermatitis associated with green grass products.

Carrots (*Daucus carota*)

Cairot roots are very palatable to horses and contain no undesirable chemicals, as long as they have not been moulded. Horses not accustomed to their consumption may bolt them, developing choke, and in that case it is desirable they should be sliced. They have a dry matter content of 110–130 g/kg and an energy value rather similar to that of oats per unit DM. The orange-coloured varieties are rich in β-carotene, containing 100–140 mg carotene/kg, 85% of which is present as the β-isomer. This is partially converted to vitamin A by the horse. Large intakes of carrots, and therefore of the provitamin, can cause intoxication in the human. β-

Carotene has been demonstrated by some workers to improve fertility in mares deprived of a dietary source (Ahlswede & Konermann 1980; Ferraro & Cote 1984; van der Holst *et al.* 1984) (see Chapter 4).

Potatoes (*Solanum tuberosum*)

Small (chat) and damaged potatoes are sometimes fed to livestock. However, green potatoes and sprouted potatoes contain the alkaloid solanin, which is extremely hazardous to horses. Damaged potatoes and those commencing to decompose are equally dangerous and there have been several reported fatalities. Small potatoes may also cause choke. It is not recommended that waste potato feeding be practised with horses. Moreover, Meyer *et al.* (1995) reported that the preileal digestibility of potatoes or cassava was less than 10% compared with 80–90% for oat starch.

Citrus pulp (*Citrus* spp. and *Ananassa sativa*)

Juice extraction from oranges, lemons, tangerines, limes and pineapples (*Ananassa sativa*) leaves a residue of pulp (peel, pith and seed) that is dried and pelleted, frequently after adding limestone to neutralize the acid and to aid the removal of moisture (average calcium content then is 12 g/kg). The product is low in contents of protein and phosphorus, rich in fermentable fibre (mostly pectins), clean and a useful addition to horse feeds, although the palatability may be variable. The content of oil may vary from 10 to 70 g/kg and that of crude fibre averages 130 g/kg. A high fibre level tends to indicate the presence of citrus seeds that contain relatively high concentrations of limonin, a metabolically interesting compound in human medicine that is toxic in high concentrations to pigs and poultry. The toxic threshold in horses has not been determined, but inclusion rates of the pulp in concentrates of up to 50 g/kg should be quite safe where seeds are present, and of up to 150 g/kg where they are scarce or absent. Ammoniated pulp is not recommended.

Olive pulp (*Olea europaea, O. sativa*)

The pulp and skin of olive fruits, following oil expression or extraction, is dried and pelleted. Where expression is used the pulp may contain 260 g crude fibre/kg and 100–180 g oil/kg which can be rancid and lead to increased vitamin E demands. For extraction, carbon disulphide, trichlorethylene or benzene is used and unsatisfactory solvent residue may be present. The product is equivalent to a poor roughage. Pressed cake containing the seeds causes digestive problems, and the digestibility generally is only moderate. The calcium and phosphorus contents average 10 and 2 g/kg, respectively; the protein (100 g/kg) is of poor quality and low digestibility. Some batches have been of recognizable value for horses when included in the diet up to 100 g/kg, but the variability of the product does not lend it to any general recommendation.

Cassava (manioc) (*Manihot esculenta*)

Cassava is a woody herbaceous shrub of tropical and subtropical areas. The tuberous roots provide a human and livestock food that is inferior to cereals and low in protein. The roots contain a glucoside, linamarin, which liberates hydrocyanic acid (HCN) by enzymic hydrolysis. A dried root meal, from which some of the starch has usually been extracted, is widely available in many western countries. This should contain a minimum of 620 g starch/kg, a maximum of 50 g fibre/kg and 30 g silica/kg (mostly from adherent soil). It also contains a residue of the glycoside and HCN so that its value for horses is limited.

FAT SUPPLEMENTS

There is increasing interest in supplementing horse feeds with edible fats or edible oils (see Chapter 9). Good sources of well digested and edible oils and fats have a calming effect on excitable horses. This effect may be imparted by the lecithin that oils contain (Holland *et al.* 1996). Choline, which is found in lecithin, is a component of the neurotransmitter acetylcholine found in parasympathetic and voluntary nerves to skeletal muscles.

Where oils and fats are byproducts of human food production, or of industrial processes, the quality of the fat or oil can be extremely variable. In order to establish the feeding value of a parcel, the fatty acid composition is frequently sought. This is generally inadequate and therefore it is opportune at this point to give some description of fats and of analytical procedures used in their definition. Crude fat consists of triacylglycerols (three fatty acid residues combined with glycerol, see Fig. 5.2), long chain (aliphatic) alcohols, as found in waxes, choline, phosphate, sterols (cyclic alcohols), such as cholesterol, moisture, and oxidized, polymerized and cyclized fatty acids. A typical feed fat (material soluble in a fat-solvent, e.g. ethanol, petroleum and diethyl ether) may have the composition given in Table 5.5. Table 5.6 gives characteristics of the gas-liquid chromatographic (GLC) fatty acid profile of fats suitable as horse feed supplements.

Table 5.5 Composition of typical feed fat and desirable limits (g/kg).

	Typical	Desirable maximum
Moisture and other impurities	8	10
Unsaponifiables	20	30
Glycerol	25	
Total fatty acid	870	
Free fatty acid	60	60
Oxidized fatty acid	20	25
Oxidized, polymerized fatty acid	77	0
Polythene	—	0.2
Total	1080	

Table 5.6   Desirable GLC profile of fatty acids in fats used as supplements (g/kg fat).

Saturated fatty acids	300
stearic acid (C18:0)	100 maximum
palmitic acid (C16:0)	200
Monounsaturated fatty acids	500–600
palmitoleic acid, oleic acid (C16:1, C18:1)	500–600
Polyunsaturated fatty acids	250
linoleic acid, linolenic acid (C18:2, C18:3)	200 maximum
arachidonic acid plus others, e.g. eicosapentaenoic acid (all in animal fats) (C20.4+)	50 maximum
Percentage total fatty acid content by internal marker	800 minimum

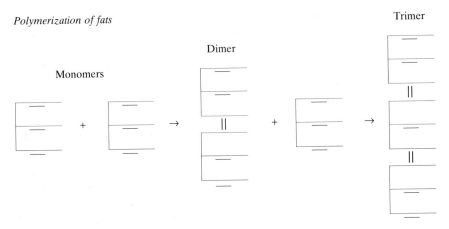

Glycerol   Three fatty acid residues

$$CH_2 - OOC - (CH_2)_n - CH = CH - (CH_2)_n - CH_3$$
$$|$$
$$CH \ \ - OOC - (CH_2)_n - CH = CH - (CH_2)_n - CH_3$$
$$|$$
$$CH_2 - OOC - (CH_2)_n - CH = CH - (CH_2)_n - CH_3$$

Under severe heating of fats, commonly with unsaturated fatty acids present, these monomers may combine by rearranging the linkages, forming new carbon–carbon double bonds, resulting in indigestible dimers and trimers.

Polymerization of fats

Monomers   Dimer   Trimer

Fig. 5.2   A triacylglycerol molecule (simplified to show one unsaturated bond in each fatty acid).

Only the glycerol and fatty acids have a reliable energy value. To determine the fatty acid composition of fat it is hydrolysed by saponification, yielding the sodium or potassium salts of fatty acids and free glycerol. The unsaponifiables (insoluble in water, but soluble in fat-solvents) include hydrocarbons, aliphatic alcohols, cholesterol and phytosterols. These unsaponifiables together with the high molecular weight polymerized and cyclized fatty acids are not eluted from a column with the fatty acids during GLC analysis and so *they are not recorded in the standard fatty acid profile of the fat*. The value for each fatty acid in this analysis is given as a percentage of the sum total of the fatty acids eluted and the values can therefore appear quite normal for fats containing damaged nonelutable fatty acids, as the total fatty acids will summate to 100%. Note that this is the percentage of the total eluted fatty acids and not of the total fat supplement purchased. The author has examined feed fats containing less than 400 g utilizable fatty acids/kg fat (<200 g/kg fat supplement, for most fats are sold on an inert base). If an internal marker of an unusual fatty acid, e.g. an odd-numbered carbon fatty acid, such as C17, is included in the analysis, an estimate may be made of the true proportion of the utilizable fatty acids in the total fat.

Polymerization and cyclization of fatty acids occurs during excessive heating of fat (thermal decomposition) (see Fig. 5.2). Fat that is heated in air also undergoes oxidative decomposition, and the extent of this tends to be correlated with the extent of polymerization. An indication of the existence of oxidation can sometimes be obtained by measuring the peroxide value of the fat. This is a measure of the oxidation of the double bonds in unsaturated fatty acids and it is likely to be more extensive in polyunsaturated oils unprotected by antioxidants. Values in excess of 20 mEq/kg oil must be suspicious. Low peroxide values, on the other hand, do not indicate a lack of damage, as peroxidation is a primary stage in the deterioration of fat quality. Secondary decomposition includes the formation of carbonyl-, hydroxy- and epoxy-derivatives, which are also valueless. The fat may have become hydro- lysed liberating free fatty acids, hence the value of determining free fatty acids in raw material that is not intended to contain such acids. (Hydrolysed edible fat containing 50% free fatty acids is also marketed. This is a very satisfactory product.) Saturated fatty acids can become rancid only following hydrolysis, as in butter, indicating deterioration by ketone formation.

The importance of using an internal marker in fatty acid analysis should be apparent. If such an analysis gives a value of *18%* for, say, the C16 fatty acid as a percentage of eluted fatty acids in a product, which is sold as 50% fat on a nonfat base, then the actual amount of the C16 fatty acid present in the product is only *3.6%*, where there are only 400 g utilizable fatty acids/kg in the product referred to above. Typical bases used for preparing fat premixes include palm kernel extrac- tions, ground straw, wood flour and vermiculite, a hydrous silicate.

Damaged fatty acids in fats may have a negative value, precipitating a vitamin E deficiency and may cause digestive upsets and other problems. Waste oils that are notorious for causing problems of these sorts are acid oil distillation residues, especially fish acid distillation residues. Other undesirable byproducts are 'roller oil' from the steel and tin plating industry, oxidized cod oil obtained from tanned hides and recovered vegetable oil (RVO) from food frying outlets. Waste fats can be contaminated with heavy metals, pesticides or polythene. The rate of deterioration of a fat will be accelerated by contamination with, for instance, copper and retarded by the presence of antioxidants (synthetic or natural). Most seed oils are highly unsaturated, but if fresh are generally rich sources of the antioxidant α-tocopherol (e.g. soyabean oil). Marine oils will deteriorate more rapidly than, for example, coconut oil, as a consequence of the difference in degree of unsaturation.

High quality fats, whether highly saturated or polyunsaturated, have a valuable role to play in the nutrition of stabled horses and it is imperative that the quality of the fat supplement is assured for this value to be realized. Fat increases the energy value of feeds, eases pelleting of fibrous feedstuffs and reduces product dustiness.

Polyunsaturated fatty acids

The constituent oils of most seeds and cereal grains are rich in n-6 polyunsaturated fatty acids, whereas only a few, including linseed oil and rapeseed oil, contain any significant amounts of n-3 polyunsaturated fatty acids. These two series are differentiated by the position of the first double bond in the carbon chain, which in the case of the n-6 series is six carbons from the terminal methyl group. Fish oils are rich in the higher (i.e. longer and more unsaturated chains) members of the n-3 group. Linoleic acid (C18 n-2) of the n-6 series in seed oils is elongated and desaturated during metabolism to arachidonic acid (C20:4 n-6) which is the precursor of the inflammatory eicosanoid prostaglandins. Horses in hard work receive a large cereal-based diet, that is, one in which the fat is rich in n-6 polyunsaturates. On the other hand, the grazing and browsing horse receives a diet in which the oils are derived from leaves, which by happy chance contain a much higher proportion of the lower members of the n-3 series. Competition is afforded by these n-3 nutrients during the elongation, desaturation and cyclooxygenase metabolism of the n-6 series, and thus the synthesis of inflammatory intermediates is partially suppressed. In fact there is much evidence in the human field that cod liver oil reduces inflammatory reactions in joint disease. Comparable evidence is coming forward that similar responses may be expected in horses. Working horses are subject to inflammatory ailments and so supplementation of their diet with oils rich in n-3 fatty acids may reduce the extent of painful leg ailments. A word of caution, however: fish oil n-3 fatty acids are more unsaturated than oils of the n-6 series and any increased use of the n-3 sources should be accompanied by an increase in vitamin E supplementation.

PROTEIN CONCENTRATES

Vegetable proteins

The richest sources of vegetable protein fed to horses and ponies are oil-seed residues, but other sources include peas, beans, yeast and in the future possibly new sources of microbial protein and, finally, high-quality dried forages, particularly lucerne meal (alfalfa). Soyabeans, linseed, cottonseed and, to some extent, sunflower seed after processing are widely used. Groundnuts cannot be recommended because of their frequent contamination with a toxin (aflatoxin) of the mould *Aspergillus flavus* to which the horse is relatively sensitive.

Two alternative procedures are adopted for the extraction of oil from oil seeds, both of which may be preceded by the removal of a thick coat by a process known as decortication, as practised for cottonseed and sunflower. Undecorticated meals contain very much more fibre. Where oil is removed by pressure, this is preceded by cooking at up to 104°C for 15–20min, after which the temperature is raised briefly to 110–115°C. Then pressure is achieved by passing the seeds through a horizontal perforated cylinder, in which a screw revolves and the oil is partially pressed out, leaving a residue containing perhaps 35g fat/kg. Expeller cakes, therefore, have the advantage of incorporating more fat than meals derived from the more-efficient chemical extraction process. However, the temperatures achieved during compression can damage the protein, which generally has a lower biological value (see Glossary) than that resulting from solvent extraction. In this latter process only material with less than 350g oil/kg is suitable so that the feeds are first subjected to a modified screw press, less extreme in its effects than in the expeller process. The seeds are then flaked and the solvent, usually hexane, is allowed to percolate through, effectively removing the oil. The solvent residues are evaporated by heating or toasting, which also benefits some meals by destroying natural toxins.

Oil-seed meals are much richer sources of protein than are cereals and their balance of amino acids is superior. However, linseed meal is a poorer source of lysine than is soya, considered the best quality of these proteins. Sunflower seeds are rich in the sulphur amino acids, cystine and methionine, although it is rare for horse diets to be limiting in respect of these amino acids. Oil-seed meals are also relatively reliable sources of some of the B vitamins and of phosphorus, but contain little calcium.

Soyabean meal (**Glycine max**)

Raw soyabeans contain allergenic, goitrogenic and anticoagulant factors in addition to protease inhibitors. The correct toasting and cooking of the beans, as in micronization and well-regulated oil-extraction procedures, destroy these factors without detracting from the protein quality (in fact amino acid value is radically influenced by variation in heat treatment, see 'Cottonseed meal', below). Reliably cooked products, therefore, may be used as the sole source of a supplementary protein in horse feeding.

Standard hexane-extracted soyabean meal contains 440g crude protein/kg. Dehulled meals of uniformly high quality containing 480–490g crude protein are also of general commercial availability. Both these meals contain less than 10g oil/kg. Full-fat soya flour and cooked soya flakes are much more costly, but the latter is widely used in coarse mixes and both contain 180–190g fat and 360–400g of crude protein/kg. The precise composition varies with the crude-fibre levels, which range from 15 to 55g/kg.

Linseeds and linseed meal (**Linum usitatissimum**)

Linseeds are unique in so far as they contain a relatively indigestible mucilage at concentrations of between 30 and 100g/kg. This can absorb large amounts of water, producing a thick soup during the traditional cooking of linseed and its lubricating action regulates faecal excretion and sometimes overcomes constipation without causing looseness. The cooking of linseed also destroys the enzyme linase, which, after soaking, would otherwise release HCN from a glycoside in the seeds, in the presence of water, so poisoning the horse. This action implies that the seeds should be added to boiling water rather than to cold water and then boiled, otherwise some enzymatic activity may be initiated. However, HCN is volatile and a proportion of any already present will be driven off by subsequent boiling. Cooked linseed mashes are highly regarded to improve coat condition of horses for sale. Linseeds should not be fed dry because of their water-absorbing propensity, although the contained linase would be rapidly inactivated by the stomach's secretions.

The low-temperature removal of oil during the production of linseed meal implies that the product may be toxic if fed as a gruel, linase may not be inactivated and HCN can be evolved. In comparison, oil removal by the expeller process normally yields sufficient heat to make a safe cake, whether it be fed wet or dry. UK laws stipulate that linseed cake or meal must contain less than 350 mg HCN/kg, although this takes no account of any linase activity that may be present.

Coconut meals *(copra)* (Cocos nucifera)

The dried flesh of the coconut contains variable amounts of oil (50–70 g/kg) depending on the efficiency of extraction. The oil is composed of highly saturated, medium chain fatty acids and so is less subject to rancid fat development than many other vegetable oils. The meal is prone to absorb moisture and mould unless lengthy storage is avoided and an environment of low humidity is achieved. It contains a similar amount of protein to that of peas, but of lower biological value (see Glossary). The meal typically contains 220 g protein, only 6 g lysine and 125 g crude fibre/kg. It is palatable and contains no toxic factors.

Cottonseed meal (Gossypium *spp.)*

Cottonseed meal is a cake byproduct from cotton and oil production and it tends to be dry and dusty, having a somewhat costive (constipating) action. Undecorticated, the cake contains 200–250 g crude fibre/kg. If the hulls are separated from the kernels the fibre content is halved. The oil is removed either by hydraulic expeller or solvent extraction methods. Glands within the seed embryo contain a toxic yellow pigment, gossypol, which may be at a concentration in raw cottonseeds of between 0.3 and 20 g/kg DM. Heating during processing partly inactivates the toxin in the raw material, but at excessive temperatures gossypol binds lysine, depressing protein quality. The binding process partially inactivates the toxin. However, even the bound form is reported to reduce intestinal iron absorption, partly counteracted by further iron supplementation. Free gossypol may be at a concentration of 1 g/kg in expeller meals and 5 g/kg in solvent-extracted meals. Mixed feeds containing more than 60 mg of free gossypol/kg are unsatisfactory for horses. Therefore only products of glandless varieties are suitable for horse feeding and these are scarce in Europe. Cyclopropene fatty acids are also present in cotton seeds, but at lower concentrations in glandless varieties, although the amount is proportional to the oil residue in the meal. Thus, extracted meals of glandless varieties should be sought, notwithstanding the limitation placed by the Feeding Stuffs Regulations on the gossypol content of meals sold in the UK to 1.2 g/kg. Good quality cottonseed meals are palatable and can contain 40% protein and 15 g lysine/kg.

Moreover, Gibbs *et al.* (1996) reported that the precaecal N digestibility of a sample of solvent-extracted cottonseed meal they examined amounted to 81.2% in comparison with only 57.1% for their sample of soyabean meal. The difference may have relied upon excessive, or inadequate, heat treatment of the soya. Although the N digestibilities over the entire GI tract were similar for the two samples, the amino acid value of dietary proteins partly depends on the extent to which precaecal digestion occurs (see Chapter 1).

Sunflower seed meal (Helianthus annuus)

Sunflowers are grown for the oil in their seeds. The oil content of the meal residue is greater where the oil has been removed by hydraulic pressure rather than by extraction. However, the oil in the residue is unstable. The fibre is poorly utilized and the fibre content of the meal is much lower where the seeds have been partially dehulled (decorticated) and so it ranges from 50 to as much as 350 g/kg. The protein is of high quality, rich in sulphur-containing amino acids (double that of soya protein), but variable in amount (from 240 g protein/kg in expeller cake to 500 g/kg in some samples of dehulled extracted meal). The methionine and lysine contents of the meal are 5–14 and 9–22 g/kg, respectively, and the meal is a good protein source for horses, containing no significant undesirable substances, but it may be un-palatable in large quantities.

Palm kernel meal (Elaeis guineensis)

Expeller and extracted meals are available, but they are not widely used for horse feeding owing to low palatability. The protein and crude fibre contents are typically both 160 g/kg; the protein is relatively impoverished of lysine (40 g/kg protein). Nevertheless, the meals are free from undesirable chemicals.

Groundnut meal (peanut meal) (Arachis hypogaea)

Groundnut or peanut meal is produced following oil removal by expeller and extraction processes. The former process leaves a larger residue of the polyunsaturated oil. The protein content of the meal is slightly higher than that of standard 44% soyabean meal and averages 470–480 g/kg. Groundnut meal is very palatable and contains no undesirable substances apart from frequent contamination with the hepatotoxin, aflatoxin, derived from the mould *Aspergillus flavus*. The horse is subject to severe aflatoxicosis, and owing to a history of serious damage among livestock the use of this byproduct in Europe has been severely curtailed by EU legislation. If this problem is resolved then groundnut meal could make a useful contribution to equine nutrition.

Rape seed meal (Brassica napus, B. campestris)

Rape is a member of the genus *Brassica* in the family Cruciferae and it is grown for the oil in its seeds. There is a large world production of rapeseed; varieties from two species are grown – *Brassica napus* and *B. campestris*. Western European production has increased rapidly owing to encouragement by the EU. A drawback to rapeseed in the past was the content of erucic acid in the residual oil and of glucosinolates in the meal. Glucosinolates, present in the unheated rapeseed, and widely distributed among the Cruciferae, are goitrogenic when hydrolysed during digestion by the enzyme myrosinase, present both in unheated rapeseed and in gut microorganisms. The active goitrogens released are isothiocyanates and oxazolidinethiones (goitrin) (see 'Goitrogens', this chapter). Although the protein quality of rapeseed meal is good and although heat treatment decreases the hazard by destroying the myrosinase, the intestinal enzyme can still release quantities of the thyroactive substances and so only small amounts of meal of unknown origin are suitable for feeding to horses and many other animals. This predicament led Canadian plant breeders to select varieties of *B. campestris* low in both erucic acid and glucosinolates (less than 3 g/kg seed), known as 'double-low' varieties and sold as canola meal. Comparable varieties are widely available in Europe, leading to their routine cultivation. The better varieties are a good source of protein for horses, but use should be restricted to 200 g/kg concentrates. The protein content of the meal ranges from 340 to 390 g/kg and the protein contains 60–64 g lysine/kg, and so the meal may replace soya on an equivalent protein basis. Several rapeseed varieties contain tannins, or polyphenols (averaging 30 g/kg seed) that reduce digestibility, limiting their usefulness and possibly contributing to a slightly lower protein value for the meal in comparison with that of soya, according to several reports (reviewed by Aherne & Kennelly 1983). A further discussion of these toxins is given later in this chapter (see 'Condensed tannins').

Lupin seed meal (Lupinus albus, L. angustifolius, L. luteus)

Lupin seed meal has two disadvantages for horse feeding. There are three species grown that have white, blue and yellow flowers and a number of varieties within each species. These are variably sweet or bitter. The bitter varieties contain significant amounts of toxic alkaloids and the ether extract content is variable (40–120 g/kg) and prone to rancidity. The crude fibre content is high and variable (80–160 g/kg), but relatively well utilized by the horse, and the protein and lysine contents vary from 250 to over 400 g/kg and from 14 to 23 g/kg, respectively. The seeds are small and must be ground, or crushed, for use. Only the sweet varieties should be used and limited to 50 g/kg concentrate.

Sesame seed meal (Sesamum indicum, S. orientale)

The seed oil of sesame is a source of unstable polyunsaturated fatty acids and so only extracted, and not expeller, meal should be used, which contains 440 g crude protein/ kg, rich in methionine but deficient in lysine. The meal is also rich in P, mainly present as phytates. The husks contain oxalates so that only decorticated meal is suitable.

Peas (Pisum sativum, P. arvense), *white and purple flowered*

Peas contain less protein than do field beans, but the biological value (see Glossary) to horses is equivalent to that of soya protein. Legume seeds generally contain antinutritive factors, although the content in modern pea varieties is low. Some older varieties may contain trypsin inhibitor and phytohaemagglutinins, which are partially inactivated by heating. Autoclaving for 5 min at 121°C completely destroys pea trypsin inhibitors and haemagglutinins. Trypsin inhibitors are stable at 80°C or below. Soaking for 18 h removes 65% of the haemagglutinin activity. The trypsin inhibitor content of peas is similar to that in field beans, but only 10% of that in soyabeans. Peas also contain chymotrypsin and amylase inhibitors, although the content of the latter is very low and destruction by heating is akin to that for trypsin inhibitor. The oxalate content of peas is about 7 g/kg, phytic acid 5–8 g/kg and tannins (polyphenolic compounds) 0.2–0.4 g/kg. Tannins, mostly in the testa (seed coat), bind to enzymes and other proteins forming insoluble complexes, reducing digestibility. Extended cooking is a good insurance (see 'Natural and contaminant toxicants in feeds', this chapter). The expansion (the extent of cooking varies among machine types) and micronization processes adopted for peas included in coarse mixtures and other compounded feeds are adequate to achieve a significant improvement in nutritive value. Nevertheless, peas are a very useful protein source for horses.

Field beans (Vicia faba), *winter beans (horse) and spring beans (tick and horse)*

Within the bean family (Leguminosae or Fabaceae) two genera *Vicia* and *Phaseolus* grow throughout the world and many are important food crops. Winter and spring varieties of field (horse) beans grown in the UK are members of the species *Vicia faba*, all safe for feeding to horses, especially after cooking. The winter field (horse) bean contains on average 230 g crude protein and 78 g crude fibre/kg, whereas the spring bean contains 270 g crude protein and 68 g crude fibre/kg. Amounts of fat in both varieties are low – about 13 g/kg; like most other seeds they are rich in phosphorus but poor in calcium and manganese. Field-bean protein is of high quality as it is a valuable source of lysine. The bean is normally cracked, kibbled or coarsely ground, but may be given whole to adult horses with sound teeth. The water-soluble polyphenolic condensed tannins in the hulls interact with both dietary and endogenous proteins in the intestines, increasing the faecal losses of both these protein sources. High dietary concentrations of condensed tannins depress appetite; however, the extent to which this occurs in the horse has not been determined.

Phaseolus *and some other legume species*

There are importations of beans belonging to the genus *Phaseolus*, especially the lima (navy) bean (*P. lunatus*) and the kidney (haricot, pinto and yelloweye) bean, (*P. vulgaris*). Kidney beans are normally refused by horses unless cooked; if force-fed they will cause colic. These beans must all be cooked (wet heat) before feeding because they contain several toxic factors, including antiproteases and lectins, which will cause diarrhoea, and many beans of the genus *Phaseolus* also contain a cyanogenetic glycoside identical to that in linseed. Other beans available in parts of the world include the hyacinth bean or lablab (*Dolichos lablab*), horse gram (*D. biflorus*), green and black grams (*P. aureus, P. mungo*) and chick pea (*Cicer arietinum*), which are widely used in Asia, and lentils (*Lens* spp.) for which India is the chief producer. One lentil is said to induce staleness in racehorses and hunters if given in excess. The Indian or grass pea (*Lathyrus sativus*), at one time imported as an animal feed, and some other members of the same genus cause lathyrism (see 'Lathyrism caused by $\beta(\gamma\text{-L-glutamyl})$ aminopropionitrile and L-α, γ-diamino butyric acid', this chapter). The horse is particularly, and characteristically, affected. If beans and peas of unknown origin are used to feed horses, extended cooking will provide a measure of safety from many of the toxins.

Lentils, split peas and red dahl (Lens esculenta, L. culinaris)

Lentils, split peas and red dahl are important and palatable legume crops, producing seeds containing protein of good quality (protein 260 g/kg, lysine 70 g/kg protein), for human consumption. They may contain small amounts of trypsin inhibitors and haemagglutinins and as they are in demand for human use those batches available as livestock feed may be contaminated with moulds and mycotoxins. Such batches should be avoided. Lentil bran may be obtained as a byproduct of lentil preparation for human consumption.

Animal proteins

There are only two high-quality animal protein sources suitable as horse feeds – white fishmeal and milk-protein products. They are reserved almost entirely for foals, either in creep supplements or as milk replacers. Small amounts are occasionally given to adult horses in poor condition, but large amounts of dried skimmed milk may cause diarrhoea, owing to the presence of lactose (see Chapter 1).

Fishmeals

Two types of fishmeal are recognized under British law, of which the first is a product from the drying and grinding of fish, or fish waste, of a variety of species.

The second, marketed as white fishmeal, is a product containing not more than 4% salt and obtained by drying and grinding white fish or the waste of white fish to which no other matter has been added. This is a high-quality protein source because it contains abundant lysine, and it is suitable for, but not essential in, the diet of young foals. It is rich in minerals (about 80 g calcium and 35 g phosphorus/kg), trace elements (especially manganese, iron and iodine) and several water-soluble vitamins, including vitamin B_{12}. This vitamin is found naturally only in animal products and bacteria. The dietary requirement of the young weaned foal for the vitamin can be met by the inclusion of fishmeal, or synthetic sources, in its diet. The suckling foal should, however, receive ample in the dam's milk. About 5 or 10% of white fishmeal in a creep feed, or milk replacer, is quite satisfactory for foals. During processing the fish waste is dried by one of two procedures. The first and more desirable one is steam drying, either under reduced pressure or with no vacuum applied. The other procedure is based on flame drying, when the temperatures achieved may decrease the digestibility of the protein and decrease the content of available lysine.

Fishmeals of unknown origin may be contaminated with pathogenic organisms, in particular with *Salmonella* species or other enteric organisms that cause diarrhoea. Good-quality white fishmeal, however, should be a safe feed. Meat meals, meat and bone meals and unsterilized bone flour should on no account be used for feeding to horses because many samples are contaminated with *Salmonella*, other transmissible pathogens, or are shipped in contaminated bags. The problem of ridding stock of infection, once contracted, is considerable.

Cow's milk

Where liquid milk is used for feeding to orphan foals, it should be diluted with 15–20% of clean water and given in small amounts in as many meals as is practically convenient. Liquid cow's milk, on average, contains 125 g DM, 37 g fat, 33 g protein and 47 g lactose/kg. It contains little magnesium and is deficient in iron, a source of which should be provided for the young foal (at birth the foal's liver acts as a reservoir of iron for the neonatal foal). Whole milk is rich in vitamin A and provides useful quantities of vitamin B_{12}, thiamin and riboflavin. Milk proteins contain abundant lysine.

Dried skimmed milk is widely available and sold commercially as a component of milk replacers and horse supplements. As its name implies, it contains very little fat and therefore practically none of the fat-soluble vitamins. However, the protein quality approaches that of the liquid product if drying has been carried out by the spray process. Roller drying subjects milk to higher temperatures which results in some loss of lysine availability, and in large quantities this product can cause diarrhoea. Any significant quantity of dried skimmed milk should not be fed to horses more than 3 years old, owing to their deficiency in the enzyme (β-galactosidase) that digests lactose.

Spray-dried skimmed milk is a useful supplement to feeds of young foals where the mother is providing inadequate quantities of milk to sustain normal growth.

Concentrations of 10–15% in the dry diet have proved satisfactory. On the other hand, its use in creep feed for foals approaching weaning may be less satisfactory if the main objective is to encourage the development of a faculty for the digestion of horse feeds to be given after weaning. Thus, satisfactory creep feeds for use in normal circumstances can be provided as nutrient-rich stud nuts.

Single-cell proteins

Yeasts (Saccharomyces cerevisiae, S. carlsbergensis)

Yeasts contain protein of good amino acid balance. However, a greater proportion of the N in single-cell organisms (bacteria in particular and yeasts) compared with plant cells is composed of nucleic acids (50–120 g/kg DM in yeasts and 80–160 g/kg DM in bacteria). While some of the purine and pyrimidine bases in these acids can be used for nucleic acid biosynthesis, large amounts of uric acid, the end product of nucleic acid catabolism, are excreted in the urine. Dietary inclusion rates of up to 50–75 g/kg feed are economically feasible only for foals.

The crude protein and fat contents range, respectively, from 400 to 450 g/kg and 25 to 55 g/kg. Yeasts are readily digested and are a rich source of B group vitamins with the exception of vitamin B_{12}. This vitamin is synthesized almost exclusively only by bacteria. The live yeast culture has been demonstrated to promote fibre digestion and growth rate in young horses when included in the diet at a concentration of only 1 kg/t (see 'Probiotics' below).

Bacterial cultures

The crude protein and fat contents of bacterial cultures range, respectively, from 340 to 720 g/kg and 20 to 210 g/kg, but their use as a dietary protein source is not generally justified (see 'Probiotics' below).

PROBIOTICS

Over the past two decades there has been increasing interest in the inclusion of probiotics in equine diets. Probiotics, or substances 'for life', may be defined as live organisms and their products that contribute to intestinal microbial health (as viewed from the aspect of the host). Under EU legislation only registered live culture strains may be used for feeding to animals. The microorganism strain must be identified according to a recognized international code, including: the deposit number of the strain; the number of colony-forming units (expressed as CFU/kg, measured by an acceptable method); the period during which the CFUs will remain present; and the characteristics of the microorganism that may have arisen during manufacture.

Fungi

The organisms that are predominately used in probiotic cultures are certain species of gram-positive lactic acid-producing bacteria and several fungal species. The fungi are commonly strains of *Saccharomyces cerevisiae* (baker's and brewer's yeast), *Aspergillus oryzae* and *Torulopsis* spp. Studies by Glade and Sist (1990) in which *A. oryzae* has been added to *in vitro* equine caecal cultures indicate that it does not alter fermentation products or pH. The consumption of live cultures of *S. cerevisiae* (Collection No. NCYC 1026) (10–20 g/day per horse of a culture of 10^9 CFU/g), especially by young growing horses, has been demonstrated to stimulate hindgut microbial growth, to improve dietary fibre fermentation (and so apparent digestibilities of pectins, hemicellulose and cellulose), to increase apparent crude protein digestibility, to reduce endogenous (metabolic) faecal N loss, to increase N-retention and to increase plasma concentrations in foals of arginine, glutamine, glycine, isoleucine, leucine, methionine and valine, while concentrations of ammonia, hydroxyproline and 3-methyl-histidine were decreased (Glade & Sist 1990). The mechanism by which all this is achieved is still unknown and progress in this area must await some deeper understanding. One critical aspect of the subject is whether killed cultures of fungi, or bacteria, retain some of the useful attributes of live cultures. Evidence from ruminant research indicates that *S. cerevisiae* (Collection No. NCYC 1026) protects anaerobic rumen bacteria by increasing the rate of oxygen disappearance, indicating that viable cultures are necessary for this respiratory activity. Under EU legislation only registered culture strains may be used for live feeding to animals.

Bacteria

The bacterial species employed are gram-positive organisms, and so they resist lysozyme. They ferment glucose and a variable number of other sugars with the formation of organic acids, principally lactic acid and acetic acid, and therefore they withstand a relatively low ambient pH. These species are generally in the genera *Lactobacillus* and *Streptococcus*, although there are some related species used, including *Pediococcus* spp. Certain other species employed may have disadvantages, including low viability in store or in the GI tract. The lactobacilli are probably all benign, whereas many *Streptococcus* species are parasitic in man and animals and some are highly pathogenic. Thus, cultures of the selected species must be pure.

The viability of species within the genera commonly used varies enormously among strains. Cultures should be preserved first by freeze-drying, the preferred method, in which water is removed by sublimation. With the object of extending shelf-life manufacturers frequently encapsulate organisms so that they may more readily withstand aggressive environments. The strain-type selected for the horse should have been demonstrated to survive and multiply in the equine GI tract. This assumes that viability is essential and, at this point, it is appropriate to remind ourselves of the characteristics that a useful culture of a strain should probably possess. It:

- should be incapable of causing disease;
- should contain 10^8–10^9 viable organisms /g after 12 months' storage;
- should have the correct host specificity, contain useful enzymes and certain antibacterial substances, and, if necessary, the strain should be capable of displacing lactic acid bacteria occurring naturally in the intestine;
- should be gram-positive and so resistant to gastric juice, both gastric juice's low pH and the enzymes it contains (lactobacilli are generally resistant to the acidity encountered in the stomach, although the degree of resistance varies from strain to strain. One of the enzymes, against which organisms must be resistant, is lysozyme. Gram-positive lactic acid bacteria have a greater resitance to lysis by this enzyme than, for example, do gram-negative coliforms. Gram-positive organisms are also more resistant to the adverse effects of freezing and freeze-drying);
- should be bile tolerant and tested on *Lactobacillus* selection (LBS) agar containing 0.15% oxgall. (Many bacteria entering the duodenum are destroyed by the bile secreted there. Bile salts lower surface tension and probably emulsify the lipids in the bacterial cell membrane.) Most strains of *L. bulgaricus* and *L. lactis* possess poor bile tolerance, despite their frequent inclusion in probiotic cultures.

Lactobacillus plantarum and *Bifidobacterium bifidum* are two species frequently used, but most strains fail to adequately meet the above criteria.

The possible functions of probiotics are:

- to act as a source of useful enzymes in the GI tract. (Yogurt can increase lactose digestion in humans, even though the bacteria may not survive and/or grow in the intestines. The lactobacilli contain an intracellular β-galactosidase, which is thus protected during its passage through the stomach, but bile salts alter the membrane permeability of the bacterial cell, allowing ingress of lactose and access to the enzyme. If the bacterium grows in the intestine it will produce large additional amounts of the enzyme);
- to inhibit the growth of intestinal pathogens. (Many lactic acid bacteria have the ability to produce bacteriocins, antibiotic-like compounds that are active against closely related species. Enteric pathogens require their attachment to intestinal epithelial cells for pathogenicity. Precolonization and attachment to those cells by probiotic bacteria reduces this pathogenic potential which can occur at times of stress); and
- to assimilate cholesterol.

How these possible functions can lead to increased performance of the horse is unknown. The objective selection of fungal and bacterial species and strains in the future will depend on knowledge of how probiotics work.

Enzymes (bacterial and fungal)

Reference to enzymes was made under 'Bacteria' above. Enzymes are proteins and, therefore, can be denatured. Those contained within microbial cells may be protected from destruction in storage, or, more particularly, in the case of those that function at a neutral or alkaline pH, from destruction by the acid environment of the stomach. As enzymes possess pH optima for activity, it is essential that the optimum for a particular enzyme coincides with that in the gut location of its anticipated activity. Adult horses are unlikely to respond to supplementation with active enzymes, whereas response in the young of several species has been reported.

Under EU legislation and in Part X of Schedule 4 of the Feeding Stuffs Regulations 1995 of the UK, only 3-phytase (EC 3.1.3.8), having a minimum phytase activity of 5000 FTU (fungal titre units)/g, produced by *Aspergillus niger* (CBS 114.94), is listed as a permitted enzyme. Note: it has been demonstrated that adding phytase to a phosphorus-deficient diet for turkey poults improves their bone characteristics. The Statutory Statement on the animal feeding stuff product container shall include the following information, where these are not listed in Schedule 4:

- the names of the active constituents according to their enzymatic actvities;
- the International Union of Biochemistry identification number, i.e. EU number;
- the activity units (expressed as activity units per kilogram, or activity units per litre), if these can be measured by an acceptable method, and
- an indication of the period during which the activity units will remain present.

DIETARY VITAMIN AND MINERAL SUPPLEMENTS

A simple mixture of cereals, a concentrate protein source such as field beans or soyameal, and hay may be adequate in terms of energy and amino acids to meet the daily needs of a horse, but it is likely to be inadequate in certain of the minerals, trace elements and vitamins for optimum performance, particularly in the longer term. Some raw materials are relatively rich sources, especially of water-soluble vitamins, which makes supplementation of these unnecessary, but such supplementation is unlikely to be hazardous. The natural mineral and trace element contents of roughages and of concentrate feedstuffs tend to be variable and to depend on source and quality. Optimum supplementation is therefore difficult to achieve without chemical analysis.

Where supplements are provided the total daily intake of each nutrient should fall within the limits imposed for each in Chapters 3 and 4. Where supplements are given in addition to a compounded feed the amounts from both sources should, of course, be summated. The risks from excess are real for the fat-soluble vitamins, with the exception of α-tocopherol, and for several of the trace elements. Those elements that have notoriously caused problems are selenium and iodine. The reason for this is that the minimum adequate levels and the toxic threshold concentrations are relatively close and some natural feeds are rich sources. Moreover, for iodine the clinical signs of excess are similar to those of deficiency (see Chapters 3 and 12 for diagnostic methods). Excessive consumption of the major minerals is rarely lethal, but inappropriate rates of intake can certainly contribute to poor performance.

Vitamin-like substances, metabolic enhancers and aids to GI health

Many supplements contain items that are not strictly nutrients, but that have some physiological value to the horse. Care should be taken, where it is essential, that none of these contravene rules of competition and racing.

L-Carnitine (β-hydroxy-γ-trimethylaminobutyric acid)

L-Carnitine is a conditionally essential nutrient, present in substantial amounts in diets composed of animal products, but scarce in feeds derived from plants, and so scarce in horse feeds. Carnitine facilitates the transport of long-chain fatty acids across inner mitochondrial membranes and it may regulate acetyl-CoA:CoA by buffering excess acetyl units during intense exercise. Supplements of 10 g L-carnitine given twice daily for 2 months have doubled plasma carnitine concentration of TBs, but there was neither increased content in nor loss of total carnitine from middle gluteal muscle, associated with intense exercise. However, a supplement of this amount given to broodmares increased the carnitine content of their plasma and that of their foals (Benamou & Harris 1993). The young foal, like the neonatal human, may have a reduced capacity for carnitine biosynthesis, accounting for normal plasma concentrations of about a third those in adults. The effect of dietary fat supplementation on carnitine need has not been examined, but it is doubtful whether the function of carnitine can be enhanced, except in rare individuals that have low biosynthetic ability. Healthy adult horses probably synthesize adequate amounts from lysine and S-adenosyl-methionine.

Carnosine (β-alanyl-L-histidine)

Carnosine is found in high concentrations in equine muscle and is a major physiochemical buffer. The carnosine content of muscle is positively correlated with the proportion of fast twitch glycolytic fibres, and attempts to increase the amount of carnosine present by dietary histidine supplementation have been unsuccessful.

Glucose

The horse has the ability to store large quantities of glycogen in muscle. Concentrations in well-trained, well-fed horses are 600–700 mmol glucosyl units/kg muscle DM. Attempts have been made to accelerate glycogen repletion rates after extended exercise by use of i.v., or oral, glucose and glucose polymers, but without effect.

N,N-dimethylglycine

N,N-dimethylglycine (DMG) is an intermediate compound in the metabolism of betaine to sarcosine. In the process there is methyl group transfer to homocysteine with methionine formation. There is evidence that DMG supplementation may delay the onset of fatigue induced by lactic acid. Mature, conditioned, exercised horses supplemented with 1.6 mg DMG/kg BW failed to show typical increases in blood lactic acid concentration (see Chapter 9). Further evidence with trimethylglycine has also shown that postexercise lactate oxidation is accelerated in untrained horses, but not in trained horses. However, there is a need for more corroborative evidence.

Dimethylsulphone, methyl sulphonyl methane

Dimethylsulphone, methyl sulphonyl methane ($DMSO_2$, MSM) is included in several horse feed supplements. It is a naturally occurring sulphur compound in both flora and fauna, which has been shown to provide biologically available sulphur in the metabolism of S-amino acids and it may be an important component of the natural sulphur cycle. MSM is the oxidation product of several natural and synthetic compounds and is an important odoriferous compound in milk. It has the formula: $2(CH_3)SO_2$. In laboratory animals with spontaneous chronic arthritis MSM has been demonstrated to lessen destructive changes in joints when given in relatively large, but quite safe, doses.

Hesperidine

A very large group of compounds (several thousand) referred to as bioflavonoids exists in plants with red, blue or yellow pigments, other than carotenoids. They are polyphenolic and the flavonoids include catechins, proanthocyanidins (condensed tannins) and flavanones (reduced flavonones). Undoubtedly much will be learnt about the fascinating significance of some of these compounds during the next decade. Some potentiate the antiscorbutic activity of ascorbic acid. Hesperidine is an aglycone flavonoid that has been reported to reduce capillary fragility and/or permeability. It may 'spare' vitamin C, possibly by chelating divalent metal cations (Cu^{++}, Fe^{++}), serving an antioxidant function. However, the function is speculative, and whether, for example, hesperidine would reduce the frequency of postrace bleeding has not, to the author's knowledge, been tested.

Fuller's earth (sodium montmorillonite, sodium bentonite)

Fuller's earth, or bentonite, is a fine greyish-white powder that consists mainly of montmorillonite, a native hydrated aluminium silicate, with which finely divided calcite (calcium carbonate), magnesium and iron may be associated. Fuller's earth is an adsorbent that can take up gases produced in the GI tract. It adsorbs water, swelling to about twelve times the volume of the dry powder. A mixture of 20 g/l water has a pH of 9–10 and it has some buffering capacity in the GI tract, reducing a rapid decrease in pH during starch overload (see Chapter 11). It can also act as a pellet binder. Bentonite is normally available in the feed industry as the sodium salt.

Ergonomic aids

Sodium bicarbonate

(See also Chapter 9.) Lactic acid accumulates in skeletal muscles and body fluids during and following high-intensity exercise. At the pH of normal muscle tissue the lactic acid is highly dissociated into lactate and H^+ ions. These ions accumulate, lowering the pH and reducing the activity of glycolytic enzymes, probably impairing the contraction process of working muscles, expressed as fatigue. H^+ ions are buffered by the bicarbonate buffering system:

$$H^+ + HCO_3^- \rightarrow H_2CO_3 \rightarrow CO_2 + H_2O$$

and the CO_2 is exhaled, disposing of the H^+ ions. Supplementation of horses with sodium bicarbonate at the rate of about 0.4 g/kg BW in 1 l water approximately 3 h before a race of more than 2 min duration (<1500 m) has been shown to raise venous blood pH and lactate, indicating an improved buffering capacity during metabolic acidosis.

FEED STORAGE

Some organic nutrients and nonnutrients in forages, cereals and compounded feeds deteriorate during storage. Labile, readily oxidized pigments, unsaturated fats and fat-soluble substances are destroyed at differing rates depending on their degree of protection, the environmental conditions, their propensity to oxidation and the presence or absence of accelerating substances. The immediate effects include a reduction in the acceptability of feed to the horse, which is perhaps one of the most discerning and perspicacious of domestic animals over its feed selection. All the fat-soluble vitamins present naturally in feed – that is, vitamins A, D, E and K – are subject to oxidation, together with the unsaturated and polyunsaturated fatty acids. Rancidity of the latter depresses acceptance, although some stability is imparted by natural and permitted synthetic antioxidants, which are respectively present in, or are used in, mixed feed. Added synthetic sources of vitamins A and E are much more stable than are their natural counterparts, but they contribute very little antioxidant activity. The critical water-soluble B vitamins are fairly resistant to destruction during normal storage, although riboflavin in feed will be lost where it is exposed to light. Advice given on labels attached to proprietary supplements and feeds should be followed.

Several factors are essential attributes of good feed stores and grain silos. These are a low and uniform ambient temperature, low humidity and good ventilation, absence of direct sunlight, and freedom from rodent, bird, insect and mite infestation. These characteristics imply that feed stores and grain stores should be insulated and without windows, but should be well ventilated and both clean and cleanable. Construction materials should be rodent-proof, stacked feed should be raised from the floor and accessible from all sides, and roofs should be free from leaks. Galvanized bins are generally preferred to plastic bins, which can be gnawed by rodents, but metal bins may be more subject to moisture condensation on the inner surface if the feeds they contain have excessive moisture contents (a maximum of 120 g moisture/kg feed should be achieved). Thus, the choice of store should rest on the level of general tidiness, whether rat infestation is likely, whether all sides of proposed plastic bins can be reached and whether a uniform temperature can be attained over 24 h. Uniformity in the latter reduces the risk of temperatures reaching the dew point during its fluctuation.

An Irish study showed that 14% of both Irish and Canadian oat samples were badly contaminated with fungi, although the Irish samples contained slightly more moisture (MacCarthy *et al.* 1976). These fungi grow during the maturation of the crop in the field and normally have a minor role in feed stability. However, high concentrations can affect acceptability to the horse and fungal invasion does detract from the stability of cereals during storage. *Fusarium* species and a few others may produce toxins that subsequently affect fertility or other aspects of animal health.

Storage fungi are another matter; these species can grow in environments with relatively low moisture and high osmotic pressure, known technically as conditions of low water activity. Such environments cause heating, mustiness, caking, lack of acceptability and eventually the decay of stored grains, oilseeds and mixed feeds. Feed stability and nutritional value decline, and fungal toxins may be produced. Species of *Aspergillus* and *Penicillium* are the main culprits and all feed stored, where there are moisture contents that give water activities of 0.73–0.78 at temperatures between 5 and 40°C, may be invaded by *A. glaucus*. However, *Chrysosporium inops* will spread at moisture contents as low as 150–160 g/kg. Most storage fungi have minimum temperatures for growth of 0–5°C, grow optimally at 25–30°C and do not grow at temperatures above 40–45°C. However, *A. candidus* and *A. flavus* (producing aflatoxin) grow vigorously at 50–55°C and *Penicillium* grows slowly at temperatures down to –2°C.

Uninsulated bins and waterproof sacks subject to variations in environmental temperature are particularly prone to moisture condensation on the inner surfaces even when the average moisture content of the product is low, but, of course, the probability of this occurring increases with greater average moisture levels. Once mould growth is initiated, this generates metabolic water and a vicious circle is established. Insects, for example, grain weevils and beetles, and flour mites not only accelerate deterioration of feed and grain, but also generate both heat and metabolic water, and are vectors of fungal spores. Dirty, badly stored grain will be ripe for the hatching of eggs and the multiplication of these insects and mites. Mites will multiply at moisture levels as low as 125 g/kg and a temperature of 4°C. Insects require slightly higher combinations of temperature and moisture. Cleaning and fumigation of long-stored feed and feed stores are, therefore, desirable. Hygiene and the storage of feed and grain at low temperatures and in a dry condition, without pockets of high moisture, are the greatest assurance for the maintenance of feed quality in the long term.

Weevils and beetles can be seen with the naked eye. Mites are extremely small but their presence can be detected in meals by observation for a minute, during which movement of fine feed particles should be apparent. There is a characteristic sour smell from mite infestation, whereas with moulding, discoloration of the grain, dust and a fungal smell can be readily detected.

Rodent infestation not only causes a loss of feed, but also both rat and mouse droppings introduce to horses the risk of enteric disease, principally salmonellosis, but also other pathogens, including Tyzzer's bacillus (*Clostridium piliformis*). This bacterium is carried in the GI tract of many wild species, e.g. rodents and lagomorphs (see Chapter 11).

Mould inhibitors, such as calcium propionate, propionic acid, sorbic acid and hydroxyquinoline, have been recommended, but they are really effective only for coating grain and are relatively ineffectual when included in mixed feed. Propionic acid is more effective than the calcium salt.

NATURAL AND CONTAMINANT TOXICANTS IN FEEDS

(See also Chapter 10.) Feed is sometimes naively considered to be a parcel of nutrients, both essential and nonessential. A consideration closer to the truth accepts that natural feeds also contain materials thought to either be inert or to influence the metabolism of other dietary constituents, and substances with nutritional value but which may be present in toxic concentrations. Many natural feedstuffs also contain substances in toxic concentrations with no known nutritive value. Many of these potentially hazardous substances are produced naturally, either by the plants themselves or by organisms infecting them and their products. Finally, there are contaminants that result from human intervention and activity. Table 5.7 gives an arbitrary classification to indicate the extent of the problem, yet the groupings and distinctions drawn are by no means absolute.

Toxicants produced by plants

Those toxicants likely to be consumed by browsing horses will be discussed in Chapter 10. Here our concern is with substances present in seeds used as feeds.

Condensed tannins

Condensed tannins comprise a diverse group of water-soluble polyphenolic compounds. They are contained in sorghum grains, lentils and many other legume seeds,

Table 5.7 Classification of naturally occurring toxicants and toxic contaminants of horse feed.

(1) *Naturally occurring toxicants*
  (a) Protein or amino acid derivatives
      Lectin (haemagglutinins)
      Trypsin inhibitors
      Lathyrogens
      Nitrates/nitrites

  (b) Glycosides
      Goitrogens
      Cyanogens

  (c) Miscellaneous
      Tannins (polyphenols), saponins
      Gossypol
      Phytin, oxalic acid
      Several antivitamin factors

  (d) Poisonous shrubs and weeds (principally alkaloids)

(2) *Moulds and pathogenic bacteria developing owing to bad harvesting, handling and storage*
  (a) Spores causing respiratory allergies
      Aspergillus fumigatus
      Micropolyspora faeni

  (b) Mycotoxins produced by moulds
      Hepatotoxins
      Hormonal
      Other

  (c) Pathogenic bacteria, or their toxins
      Pathogenic *Salmonella* spp.
      Clostridium botulinum toxin

  (d) Endophytic fungi in grasses
      Acremonium coenophialum
      A. lolii

(3) *Dietary allergens absorbed from intestines*

(4) *Dietary contamination during manufacture*
    Toxic antibiotics
    Dopes
    Pesticide and herbicide residues
    Nutrients with narrow margin of safety
    Heavy metals unlikely to have nutritional value
    Breakdown products of feed constituents

e.g. peas and field beans tend to contain significant amounts of condensed tannins, especially where the seed coat is black (not always a reliable guide) or where field beans are coloured-flowering. They are not toxic but depress appetite and react with, and lower the digestibility of, proteins and carbohydrates in a pH-dependent manner, if they are present in a sufficient concentration. Autoclaving or pressure-cooking destroys these tannins, but prolonged treatment is required at lower temperatures. The reaction in the intestines occurs with both dietary and endogenous proteins (for example, slightly lowering trypsin activity in the ileum), increasing their faecal loss. The soaking of some beans before cooking may lower digestibility further, as it seems the tannins may diffuse from the seed coat into the kernel where they react with the proteins. If these polyphenols are absorbed their subsequent detoxification involves methylation. This may lower the protein value of a foal diet, for example, if it is limiting in methionine content.

This enormous group of compounds is under intense scrutiny at present as the spatial configuration of the phenolic hydroxyl groups in some isoflavonoids may confer oestrogenic activity affecting fertility of domestic animals and man. Moreover, the antioxidant activity of the polyphenols in green tea and red grapes is thought to reduce atherosclerotic development.

Lipoxygenase activity

Ground lentils and other ground pulses rapidly become rancid following disruption. The polyunsaturated fatty acids are particularly susceptible to oxidation and the enzyme responsible is lipoxygenase. Heat treatment prior to grinding to destroy the enzyme is an effective preventative measure.

Antiproteases (trypsin, or protease, inhibitors) and lectins

Two widely distributed groups of compounds are known as digestive enzyme inhibitors and lectins (previously known as haemagglutinins). The specific compound, its toxicity and susceptibility to destruction by heat vary among the species of plant within which it is found. Plants producing trypsin inhibitors and lectins include field or horse beans, black grams and kidney, haricot or navy beans. Horse grams, moth bean (*Phaseolus aconitifolius*), certain pulses (also containing amylase inhibitors), groundnuts or peanuts (*Arachis hypogaea*), soyabeans and rice germ also contain these substances. Most rice bran fed to horses has had the germ removed, although some residual activity is normally found.

Trypsin inhibitors depress protein digestion, but lectins in commonly consumed legume seeds are considered to be more harmful because they disrupt the brush borders of the small intestinal villi, hamper absorption of nutrients, but apparently allow the absorption of certain toxic substances. They stimulate hypertrophy and hyperplasia of the pancreas and from evidence by McGuinness *et al.* (1980) can eventually cause adenomatous nodules and pancreatic cancer of exocrine glands. At high concentration they induce rapid depletion of muscle lipid and glycogen in laboratory animals. Tissue catabolism and urinary nitrogen are increased, and thus growth in young stock can be depressed. The activity of both these groups of substance is destroyed by steam heat treatment. For example, the trypsin inhibitor activity of field beans is reduced by 80–85% during steam heating at 100°C for 2 min and by about 90% during treatment for 5 min. However, the trypsin inhibitor and lectin activities of kidney beans are very stable because treatment for 2 h at 93°C is necessary for adequate destruction. Kidney beans are therefore generally unsuitable for nonindustrial processing and should not normally be fed to horses.

Lathyrism caused by β(γ-L-*glutamyl*) *aminopropionitrile and* L-α,γ-*diaminobutyric acid*

β(γ-L-glutamyl) aminopropionitrile and L-α, γ-diaminobutyric acid are present in various species of *Lathyrus*. The horse is particularly subject to intoxication. The Indian or grass pea, after long periods of feeding, causes a condition known as lathyrism, which is exemplified in the horse as a sudden and transient paralysis of the larynx with near suffocation brought on by exercise. This is associated with a degenerative change in the nerves and muscles of the region and profound inflammation of the liver and spleen. Other closely related species, including sweet pea (*L. odoratus*), wild winter pea (*L. hirsutus*), singletary pea (*L. pusillus*) and everlasting pea (*L. sylvestris*), can also cause lathyrism. Although the whole plant contains the toxin, the seeds appear to be the most potent source and it is only partially destroyed by heat.

Goitrogens

Goitrogenic activity is caused by goitrins, which are derived from glucosinolates found in many members of the Cruciferae family, including cabbages, rape and mustard. Goitrins are released by enzymes contained within the plant and the destruction of these enzymes by heat treatment to a large extent eliminates the potential hazard. The effect of goitrins is not counteracted by additional dietary iodine, but further enzymatic metabolism of certain goitrins can release isothiocyanates and thiocyanates. The antithyroid effect of these substances on young horses, in particular, can be overcome by dietary iodine (see Chapter 3). The enzymes are destroyed by adequate heat treatment. The slight antithyroid effect of uncooked soyabeans is said to be overcome by additional iodine.

Cyanogenic glycosides and hydrocyanic acid

Cyanogenic glycosides release hydrocyanic acid (HCN) on soaking the seed. Feeds containing these HCN-glycosides should be given in small amounts and dry. Larger amounts require preheating to a sufficient temperature (adding to boiling water) to destroy the enzyme that releases the HCN. The glucoside in *Phaseolus lunatus* remains stable to cooking. Vicine is the pyrimidine glucoside in *Vicia faba* responsible for the haemolytic anaemia of favism in subjects deficient in glucose-6-phosphate dehydrogenase and is generally unrecognized in horses. Glycosides, derivatives of α-hydroxynitriles, present in lima beans, sorghum leaves, linseed and cassava (tapioca) (*Manihot esculenta*), generate HCN when acted on by the specific enzymes the plants contain. HCN can cause respiratory failure by inhibiting cytochrome c oxidase (EC 1.9.3.1). Again, as the poison is released by enzyme activity, heat treatment will ensure safety, so long as prolonged storage of, for example, moist seeds or cassava roots has not led to some accumulation of HCN. The suppression of enzymatic activity is another reason for the importance of dry storage of certain uncooked feedstuffs. HCN can also react with any thiosulphate present, producing thiocyanate, which is itself responsible for thyroid enlargement after prolonged feeding.

Gossypol

Gossypol occurring in many strains of cottonseed is the reason the meal is not widely used in horse feeds. In both its bound and free forms, the pigment gossypol reacts, incompletely, with cottonseed protein to depress appetite and protein digestibility and therefore to reduce the efficiency of amino acid utilization; but its toxicity can result in death caused apparently by circulatory failure. The pigment also reacts with dietary iron, precipitating it within the intestines. Fairly large additions of supplementary iron to the diet will then promote further precipitation, which partially suppresses the adverse effects of gossypol.

Antivitamins

Several antivitamin factors are present in animal and vegetable feeds, but most are of little significance to horses. A thiaminase present in the bracken fern (*Pteridium aquilinum*) is a cause of bracken poisoning, which is counteracted with large doses of thiamin, and the antivitamin-E factor present in raw kidney beans is partly destroyed by cooking. Sulphonamides have an antimicrobial action and depress intestinal vitamin K synthesis.

Phytins and oxalates

High levels of phytic acid, or its salts, present in many vegetable seed products when consumed in large quantities will interfere with the availability of calcium and several trace metals, particularly zinc. Oxalates detectable in many plants and present in high concentrations in certain tropical species of grass have been reported to kill cattle and cause lameness in horses, owing to precipitation of calcium (see Chapter 10).

Nitrates

Green fodder containing more than 5 g nitrate/kg DM fed alone may cause digestive disturbances and respiratory and circulatory abnormalities. The rapid growth of pasture after high rainfall and excessive use of nitrogen fertilizers can lead to high concentrations of nitrates in the herbage, and contamination of water supplies through the leaching of soils. Although nitrates are only slightly toxic, they can be reduced to nitrites before or after consumption. High levels of nitrites may accumulate in plants after herbicide treatment and during the making of oat hay, owing to nitrate reduction encouraged by inclement weather. In the body, nitrites convert blood haemoglobin to methaemoglobin, which is unable to act as an oxygen carrier.

Large intakes therefore cause death. Pigs are probably more susceptible than are horses, which appear to react similarly to ruminants.

Alkaloids

In South Africa, America and Australia numerous species of the legume *Crotalaria* (one species of which causes Kimberley horse disease) have proved very poisonous in horses; lesions are induced in the liver by pyrrolizidine alkaloids similar to those found in ragwort (*Senecio*). Alkaloids are considered in Chapter 10 under grazing management. Other details are found in Table 10.17. Green potatoes (*Solanum tuberosum*) contain the alkaloid solanine. Horses are killed by eating quantities of potatoes that would not affect ruminants, even when the tubers are not apparently green.

Mould development

The effects of moulds are of two types:

(1) Certain mould spores in large numbers in badly harvested and stored roughages and in cereals can cause a respiratory reaction (allergy) when inhaled by sensitive horses (see Chapter 11).
(2) Many mould species in the appropriate conditions of temperature and humidity produce toxins which have a variety of metabolic effects.

Mycotoxicosis

Fusarium moniliforme and *F. proliferatum* are fungi found on maize, sorghum and other grains throughout the world. They produce mycotoxins called fumonisins which have been shown to cause leucoencephalomalacia and hepatotoxicity in horses. An early sign in the horse is a rise in plasma sphinganine concentration, resulting from a disruption to sphingolipid metabolism. This effect occurs before plasma levels of liver enzymes are noticeably elevated (Wang *et al.* 1992).

Colic and mycotoxins
Several mycotoxins have been examined for their propensity to cause colic. At the concentrations found in contaminated feedstuffs only T_2 toxin, at concentrations >50 μg/kg feed, and zearalenone, at concentrations >70 μg/kg feed, have caused colic in horses. No colic has been observed with aflatoxin, deoxynivalenol or with fumonisin B_1.

Ergotamine, ergometrine

There are numerous historical references to the consequences of consuming ergot of rye, namely abortion and other effects of blood vessel constriction. The mould concerned, *Claviceps purpurea*, also infects ryegrasses and some other grasses, and so in high concentrations can be a hazard in pasture and hay.

Aflatoxin

Several reports, especially from Thailand and the USA, have recorded deaths or brain, heart and particularly liver lesions, and haemolytic enteritis in horses receiving aflatoxin produced by *Aspergillus flavus* at levels of less than 1 mg/kg contaminated cereals, peanuts and even hay. Horses and ponies seem to be more susceptible to acute aflatoxicosis than are pigs, sheep and calves, as daily intakes of 0.075 and 0.15 mg aflatoxin B_1/kg BW are lethal for ponies in 36–39 days and 25–32 days, respectively (Cysewski *et al.* 1982).

Zearalenone

The toxin zearalenone produced by *Fusarium* species causes vulvovaginitis and reproductive failure in females of several domestic species. Whereas aflatoxin usually develops during storage, this toxin may develop preharvest. Although no reports of effects in horses are known (cf. grass sickness, Chapter 10), it could well cause breeding irregularity in them. Many other fungal toxins with a wide variety of effects and significance also exist.

Dietary allergens (other than mould spores)

Dietary allergens are not contaminants, but certain horses can react to normal nutritional constituents of specific feeds (Plate 5.4). These are probably protein in nature and the effects, which include respiratory and skin lesions, are normally overcome, according to the author's experience, by removal of the offending source from the diet. Problems of cross-reactivity in which related sources of proteins yield similar reactions can, however, pose problems of interpretation.

Heavy-metal and mineral contamination from pastures

(See also Frape 1996.) The approximate maximum tolerable dietary concentrations of heavy metals (these concentrations are influenced by the concentrations of essential minerals and trace elements) are:

	mg/kg feed DM
Arsenic (As)	2
Cadmium (Cd)	10
Fluorine (F)	50
Iodine (I)	1
Lead (Pb)	20
Mercury (Hg)	0.2
Molybdenum (Mo)	200 (excluding any possible interaction with Cu availability)
Selenium (Se)	2

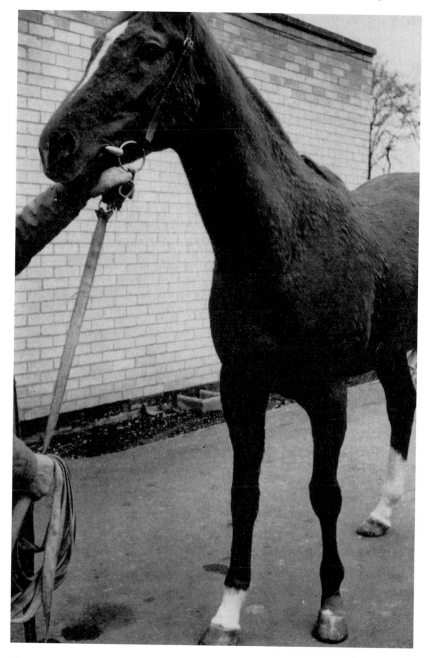

Plate 5.4 A 2-year old TB gelding with widespread 'bumps' on the head, neck, shoulders, ribs and flanks. An allergic reaction to bran and oats was detected in the blood serum. The horse recovered over several months when its diet consisted of a high-fibre cubed diet, low in cereal and with water *ad libitum*.

Lead is one of the commonest causes of poisoning in cattle, sheep and horses. Signs of toxicity are more frequent in young horses and include lack of appetite, muscular stiffness and weakness, diarrhoea and, in an acute form, pharyngeal paralysis and regurgitation of food and water. Lead accumulates in the bones and as little as 80 mg/kg diet may eventually cause toxic signs, which are sometimes precipitated by other stresses. Natural feeds with 1–5 mg lead/kg cause no problems. The acute lethal dose is 1–1.8 g/kg BW as lead acetate or carbonate. The chronic lethal dose depends on many factors but is said on average to be about 12 mg of lead/kg BW daily for 300 days.

The contamination of pasture with lead, cadmium and arsenic – derived from mine workings, dumping or sewage sludge, aerial dust and water erosion, even from car batteries and lead shot – are local risks. Where the pasture is dense, undoubtedly the greater problem arises from surface contamination of the plants, but where it is sparse, soil, either rich in these heavy metals or contaminated by them, can be consumed in sufficient amounts to cause problems. Lead shot is somewhat hazardous to horses, but, because they close graze, consumption can be greater than might be generally appreciated, followed by some solution in the stomach. Where grass is ensiled, entrapped shot partly dissolves during fermentation and highly toxic levels of 3800 mg soluble lead/kg DM have been detected in England by the author (Frape & Pringle 1984). Whereas lead seems mainly to contaminate the surfaces of plants, cadmium is readily absorbed and is accumulated from soils rich in this element. Of common pasture species, the daisy (*Bellis*) accumulates 60–80 mg cadmium/kg (30 times as much as in grasses) from contaminated soils (Matthews & Thornton 1982).

Pastures to the leeward of steel and brick works may amass abnormally high concentrations of fluorine. The horse is probably less subject to fluorosis than are cattle and sheep, but damage to its bones and teeth has been induced by this element. However, it will tolerate 50 mg/kg feed for extended periods. Mercury poisoning expressed as colic and diarrhoea has arisen in horses as a consequence of the mistaken use of dressed seed as a feed. Chance exposure from other sources is unlikely.

The required trace elements – iodine, selenium and molybdenum – may be consumed in toxic quantities following natural accumulation in vegetable materials. Seaweed can be a source of excessive iodine, and certain accumulator plants store large quantities of selenium from selenium-rich soils. When these plants die it is said that they in turn deposit selenium in a form readily absorbed by neighbouring plants. Many selenium-rich areas are sparsely covered and the consumption of soil rich in the element is another source of risk. Molybdenum is readily absorbed by most plants from soils containing excessive amounts. High concentrations of this element in herbage have been shown by the author to depress serum copper in horses, but less strikingly than in cattle, and the ingestion of soil rich in iron and sulphur is also known to reduce copper absorption in grazing animals. Season and the extent of soil drainage can influence the accumulation of several metals in herbage. Concentrations frequently tend to be higher during the winter months (see also Chapters 3 and 10).

Pesticide residues

Many normal feeds contain trace amounts of pesticide residues but, excepting gross contamination through negligence, the amounts normally detected are insufficient to cause any problem to horses (after herbicide treatment, pastures should be rested for 2 weeks before grazing is permitted). The rodenticide zinc phosphide (now rarely used) has been known to be consumed in lethal amounts by horses when it may release and eject from the stomach poisonous phosphine (PH_3). If this poison is suspected, a nasogastric tube should only be used in the open air. The acute lethal dose in horses is 20–40 mg/kg BW, or about 15 g for a horse (about 5–10 g for a pony).

FEED ADDITIVES

Domestic animal feed-additive drugs

Several drugs are used in the feed of farm animals to promote growth, to counteract diarrhoea and parasitic infection and to influence the carcass. Most of these drugs have little, if any, ill effect on horses when present in the diet at normal feed levels, or when horses are mistakenly given feed containing antibiotics intended for other species. Higher dosages are a different matter. Although framycetin sulphate is sometimes useful in cases of flatulence, or fermentative colic, the persistent use of some antibiotics, especially oxytetracycline, may cause a severe upset to the intestinal flora, possibly including a fungal overgrowth, precipitating acute and intractable diarrhoea, lethargy and lack of appetite. Two other drugs, and one of them in particular, can have severe toxic effects in horses when given at normal feed rates.

Ionophore antibiotics

Ionophores are polyether carboxylic antibiotics given to poultry for the control of coccidiosis and to ruminants to improve feed utilization. There are presently six of these carboxylic ionophores: monensin, lasalocid, salinomycin, narasin, maduramicin and laidlomycin. Whereas the horse is subject to intoxication from each of these at feed levels, there is a dietary threshold concentration below which no adverse effect has been observed. Above this level severe colic, sweating, trembling and occasionally haematuria can occur. Reliable and quantitative equine data, however, are not available for all of these chemicals.

Monensin sodium

Monensin is fed to beef cattle for promoting growth and to poultry as a coccidiostat. Poultry feed containing 100 mg monensin/kg, the normal feed level, has severe toxic effects when fed to horses. At a level of 30 mg/kg in the feed, horses experience a reduced appetite and uneasiness, although Matsuoka *et al.* (1996) indicate that horses can tolerate the highest usage rate for cattle of 33 mg/kg feed; at a level of

Table 5.8   Ionophore antibiotic toxicity.

Antibiotic	Active chemical usage rate (mg/kg feed DM)	Equine lethal oral dose (LD$_{50}$) (mg/kg BW)[1]
Salinomycin	60	approx. 0.6
Narasin	70	approx. 0.7
Monensin	100–120	1.38–3[2]
Lasalocid	30–50	21.5

[1] Lethal dose in mice is about ten times higher.
[2] Matsuoka *et al.* (1996) found an LD$_{50}$ of 1.38 mg/kg BW for a single dose by gavage using mycelial monensin.

100 mg/kg (about 2.5 mg/kg BW) in a diet fed continuously, it is lethal in a matter of 2–4 days to about half the individuals. Horses present signs, in the author's experience, of anorexia, posterior weakness, profuse sweating, tachycardia, occasionally muscular tremors, polyuria, myoglobinuria (dark-brown urine), elevated urinary potassium, elevated serum levels of muscle enzymes, progressive ataxia and recumbency. Post-mortem examination shows myocardial degeneration and monensin can normally be confirmed by analysis of stomach contents. In the early stages of toxicity, recovery can be achieved frequently by removing the offending feed and dosing the horse with mineral oil, although it may suffer permanent heart damage with increased risk during hard physical exertion.

Lincomycin

Lincomycin is an antibacterial drug sometimes included in pig feed. It is less toxic than monensin in horses, but above dose levels of 80 mg/kg BW daily (5 mg/kg total diet fed continuously) metabolic signs of toxicity and evidence of liver damage have been observed by the author.

PROHIBITED SUBSTANCES

The list of proscribed drugs of prohibited substances embraces a very high proportion of the drugs permitted in livestock feeds by EU legislation and the detection of any of them, or their recognizable metabolites, in urine, blood, saliva or sweat will lead to the disqualification of a horse subject to the Rules of Racing. Thus, apart from drugs that might be used directly in horses, any antibiotic, growth-promoter or other drug used for feeding to poultry, pigs or ruminants must not be detected in any of the above fluids. Of the proscribed drugs, most are unlikely to be present in feed. However, some drugs acting on the cardiovascular system, some antibiotics and one or two anabolic agents have been detected in contaminated feed ingredients – oats, soyabean meal, bran – and in feed additives, or they may be present in feeds for other classes of stock mistakenly fed to horses. In practice, those causing chief concern are the xanthine alkaloids – theobromine, caffeine and its metabolite theophylline.

Caffeine is present in tea, coffee, coffee byproduct, cola nuts, cacao and its hull, which is available as a byproduct, and in maté leaves. Tea dust contains as much as 1.5–3.5% caffeine, whereas coffee byproduct, as normally available, contains only some 200 mg caffeine/kg. Small amounts of theophylline are found in tea, but as much as 1.5–3% theobromine is typically present in cacao beans, and its wasteproduct, the hull, contains as much as 0.7–1.2%. The widespread international traffic in coffee and cocoa beans, and in their byproducts, constitutes a formidable risk through their contamination of the means of transport, from ships at one extreme to hemp sacks at the other. These means, in their turn, put in jeopardy cereals, pulses and other raw materials moved from one place to another, leading to infringements of the Rules of Racing through the inadvertent consumption of contaminated batches of these feeds. Gross contamination and its control in animal feedstuffs has been discussed in a code of practice published by the United Kingdom Agricultural Supply Trade Association (UKASTA) (1984).

After an oral dose with caffeine, about 1% appears unchanged in the urine, the excretion of which is almost complete after 3 days. About 60% of the caffeine is excreted in the urine as metabolites, including theophylline and theobromine. Traces of the latter may continue to be excreted for up to 10 days, whereas theophylline excretion is virtually complete after 4–5 days. Thus, the unintentional use of these alkaloids, or of coffee or cocoa wastes containing them, can result in their detection in the urine for up to 10 days and caffeine is, moreover, demonstrable in urine within 1 h of an oral dose. Great care must therefore be exercised in the feeding of race and competition horses during such a period. The extent of the excretion curves in urine of various other drugs has been determined, but certainly not those of all antibiotics similarly excreted.

Both caffeine and theobromine are readily absorbed from the intestinal tract and soon impart their effects of cardiac and respiratory stimulation and of diuresis. However, tests have shown that their effects on the speed of horses only occur when they are used in high doses. Experiments have shown that theobromine can be detected in the urine of TB geldings when they receive as little as 1 mg theobromine (in the form of cocoa husk)/kg racehorce cubes and 7 kg of those cubes are given daily, divided between two feeds. The threshold level for theobromine in the urine, accepted by the Stewards of the Jockey Club, is 2 μg/ml urine, when HPLC is used in the analysis (Haywood *et al.* 1990).

Growth hormone

There is increasing use of exogenous bovine growth hormone in TB studs. This increases the extent of 'double muscling' and rate of growth with little effect on ultimate height. Some of the author's evidence suggests it may also increase the risk of DOD.

STUDY QUESTIONS

(1)   How would you set about changing an existing stable over from hay to haylage?

(2)   What factors should be considered when it is proposed to introduce fat into a rationing system?

(3)   A number of feed processing techniques have been introduced during the last two to three decades. Have any of these been of particular value to horse feeds? If so, which and why?

(4)   If probiotics are to be introduced, what checks should be made concerning: (a) the product and (b) the feed system for their use?

FURTHER READING

Haywood, P.E., Teale, P. and Moss, M.S. (1990) The excretion of theobromine in Thoroughbred race-horses after feeding compounded cubes containing cocoa husk – establishment of a threshold value in horse urine. *Equine Veterinary Journal*, **22**, 244–6.

McDonald, P., Edwards, R.A. and Greenhalgh, J.F.D. (1981) *Animal Nutrition*, Longman, London and New York.

Moss, M.S. and Haywood, P.E. (1984) Survey of positive results from race-course antidoping samples received at Racecourse Security Services Laboratories. *Equine Veterinary Journal*, **16**, 39–42.

United Kingdom Agricultural Supply Trade Association (1984) *Code of Practice for Cross-contamination in Animal Feeding Stuffs Manufacture.* Amended Code, June 1984. UKASTA, London.

Chapter 6
Estimating Nutrient Requirements

What good receipt have you for a horse, that hath taken a surfeit of provender. This comes commonly to such horses as are insatiable feeders and therefore it is requisite that they be dieted, especially if they have too much rest, and too little exercise.

T. De Gray 1639

The formulation of adequate diets requires knowledge of three types of information:

(1) the requirement of the horse for each of the nutrients and energy, affected principally by the horse's size and function;
(2) the nutrient composition of each of the appropriate feeds available, and
(3) the capacity of the horse for feed.

Tables have been derived to provide the information necessary for (1) and (2), and some discussion is appropriate first concerning (3).

RELATIONSHIP OF CAPACITY FOR FEED TO BODYWEIGHT

The daily requirements of horses and ponies have been estimated in terms of the amounts of each nutrient – minerals, trace elements, vitamins and amino acids (or, more realistically, protein) – required per day for the various functions of maintenance, growth, lactation, and so on. The normal vehicle for these nutrients is the daily feed, and if a particular horse were to consume twice as much feed as another horse fulfilling the same tasks then it might be reasonable to suppose that the nutrient concentration in the diet of the first horse need be only half that in the diet of the second. Thus, in order to make useful statements about dietary composition and to facilitate calculation of adequate diets, it is necessary to predict reliably the appetite of a horse, or group of horses, for feed, or more specifically, to predict appetite for dry matter. The appetite and capacity of horses for acceptable feed are regulated by five dominant and related factors:

(1) the volume of different parts of the intestinal tract;
(2) the rate of passage of the digesta;
(3) the concentration of certain digestion products in the intestine;

(4)   the energy demands of the horse; and
(5)   the energy density and its chemical form in feed.

(3) seems to control meal size and (2) will be modulated by the physical form of the feed. (1) is controlled by the body size of the animal, but to some extent is modified by breed and adaptation.

   The most common easily corrected deficiencies in home-prepared feed mixtures are those of calcium, phosphorus, protein, salt and possibly vitamin A. However, in addition to water, the fundamental, immediate and long-term need of the horse is for a digestible source of dietary energy. Ideally, in situations of moderate work or productivity, the energy demands should just be met by the appetite and capacity of the horse for feed. Not only capacity, but also energy requirement for a variety of functions, is closely allied to bodyweight, although this weight varies from day to day according to the amount of gut fill. Therefore, a means of estimating weight is fundamental to any rationing system.

   In the absence of facilities for weighing horses, the most reliable predictions include a measurement of the heart girth, following respiratory expiration (Fig. 6.1). As conformation changes with age and differs among breeds, the measurement of girth alone is bound to yield only an approximate estimate. Some improvement on this is achieved by inclusion of the length of the horse from the point of the shoulder to the point of the buttocks (Fig. 6.1). If one wishes to include length, then the equation below (Carroll & Huntington 1988) gives an appropriate relationship:

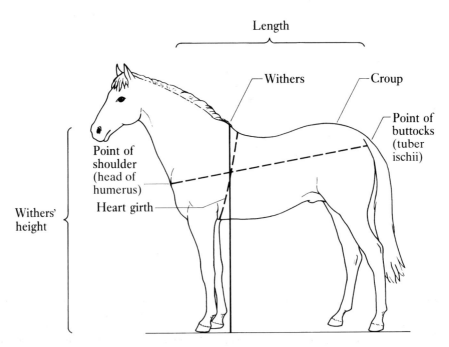

Fig. 6.1   Linear body measurements used in estimating bodyweight of horses and ponies.

$$\text{Bodyweight (kg)} = \frac{\text{heart girth (cm)}^2 \times \text{length (cm)}}{11877}$$

or

$$\text{Bodyweight (lb)} = \frac{\text{heart girth (in)}^2 \times \text{length (in)}}{328.8}$$

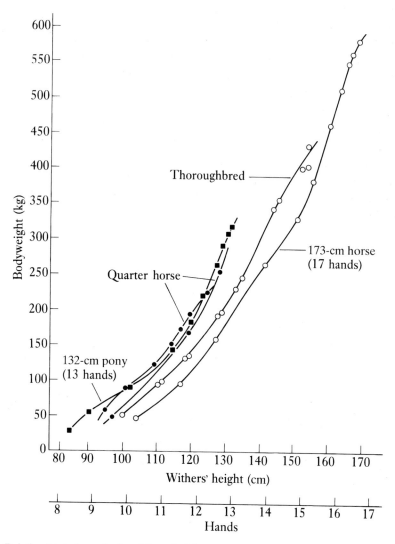

Fig. 6.2 Relationship between bodyweight and withers' height in normally growing horses and ponies (1 hand = 10.16 cm). Note that horses and ponies achieve mature height before mature weight so the curves must be concave upwards as maturity is approached (*upper end of curve*). Furthermore, as the curves are not coincident, withers' height is generally not a good predictor of bodyweight. (Data from Green 1961, 1969; Hintz 1980a; Knight & Tyznik 1985; R. W. W. Ellis & R. A. Jones, personal communications 1984.)

For many, withers' height will be a more familiar index of size; Fig. 6.2 shows its approximate relationship to bodyweight for several types of horse and pony and Fig. 6.3 delineates the change in withers' height with age during normal growth. Suggested average daily allowances for horses of different liveweights are indicated in Fig. 6.4. The allowances given to idle horses would be lower than those shown, whereas lactating mares will consume more feed. Furthermore, hard-worked animals such as TBs in advanced training for racing will be entitled to consume amounts near their capacity, although their appetites may decline when vigorous exercise is practised routinely. Observations in the USA showed that among seven racing stables the average daily intake of concentrate by 3–4-year-olds was 6.16 kg (4.9–7.5 kg) and that of hay 9.37 kg (6.4–11.9 kg) (Glade 1983a). Comparable observations among 2–4-year-olds in Newmarket by the author

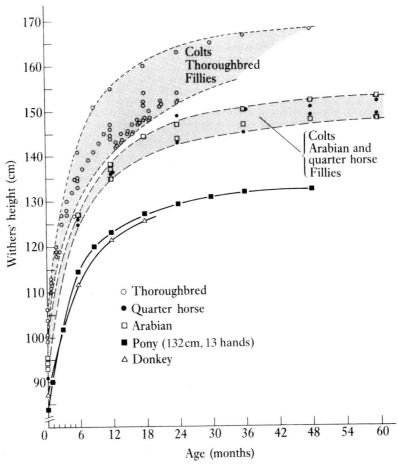

Fig. 6.3 Expected withers' height at various ages of normally growing horses and ponies (1 hand = 10.16 cm). (Data from Green 1961, 1969; Hintz 1980a; Knight & Tyznik 1985; R. W. W. Ellis & R. A. Jones, personal communications 1984.)

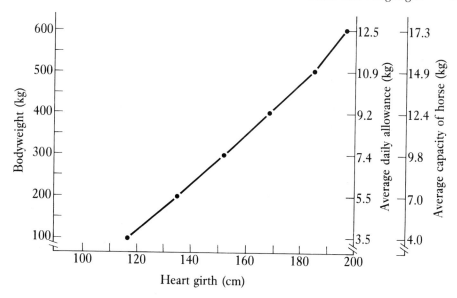

Fig. 6.4 Relationship for horses and ponies between heart girth and average liveweight, average daily capacity for concentrates and forage, and average allowance of concentrates and forage (12% moisture).

showed that the concentrate intake averaged 8.15 kg and that of roughage 5.5 kg daily per horse. The lower intake of roughage in the UK may reflect the generally poorer nutritional quality of this feed as supplied to horses. In the American study, the horses were estimated to average 496 kg bodyweight whereas those in the UK were about 480 kg.

CONCENTRATES AND ROUGHAGES

All horses should receive some form of long roughage, such as fresh grass, hay or silage. A proportion of this may be replaced by succulent green or root vegetables and soaked sugarbeet pulp. Hay or grass may comprise the total ration of idle horses and usually forms at least half the ration. The concentrate portion of the daily feed may therefore vary from nothing to 50%, and only in exceptional circumstances, or in the hands of experienced feeders, should it rise to the proportions of three-quarters of the daily allowance of dry feed.

Concentrated feeds, such as cereals, cereal byproducts, oilseed meals and the like, are traditionally fed by the bowl, that is, by volume. The energy and nutrients that these feeds provide are, of course, much more closely associated with their weight than with their volume and feeding containers should thus be calibrated to show the volume occupied by a unit weight of each type of feed. Table 5.3 contains average conversion values for cereals, although it will be appreciated that the bushel weight of cereals varies from season to season and from crop to crop, according to how well they were grown. The energy content per kilogram of each type of concentrate also differs. Ideally, therefore, feed bowls should be calibrated to indicate the volume, giving multiples of 2 MJ DE (or alternatively 1 MJ NE) for each type in use.

FEED ENERGY

The gross energy of a feed is the heat evolved when it is subjected to complete combustion in an atmosphere of oxygen. Obviously all this energy, measured as heat, is not available to the animal because a portion of the feed remains undigested and is voided in the faeces. In addition, a relatively unknown quantity is lost from the horse as the gases methane and hydrogen, in the main by passage out through the anus, but also by absorption into the blood and exhalation. Of the products of digestion and fermentation that are absorbed, a proportion of the amino acids is deaminated and the nitrogen incorporated in urea. Much of this is excreted in the urine. The gross energy of a feed, less the energy content of the faeces attributable to it, is the digestible energy (DE), and less the energy content of combustible waste gases voided, and urine excreted, leaves the metabolizable energy (ME). This is the residue of feed energy that is available to the body for its various processes of tissue repair, the functioning of organs, the physical work of skeletal muscles, growth and milk production. The efficiency of ME utilization depends on the precise chemical form of the nutrients derived from the diet and on which of these functions is performed. The efficiency is measured either as the amount of useful product or from the quantity of waste heat dissipated. The ME less this heat increment attributable to the feed is the net energy (NE). (Heat increment is the heat loss of a nourished animal in excess of that lost by a fasting animal.) The scheme is summarized in Fig. 6.5. NE is used for maintenance, growth, work, reproduction, etc.

When energy demands are great, concentrate feeds, such as cereal grains, must form part of the diet if those demands are to be met, simply because the horse can consume larger quantities of dry matter daily when cereals are included and they contain more ME per kilogram of dry matter. Conversely, an idle horse obviously has relatively low energy requirements, yet its appetite should be satisfied. As the horse can consume daily lesser quantities by weight of bulky fibrous feeds than of concentrates, then its appetite is more likely to be satisfied with lower intakes of energy when fibrous feeds are used. This idle horse is a stabled, normally active, nonworking animal, described as having energy requirements for maintenance only, that is, those leading to a zero change in bodyweight, or more accurately a zero change in body energy content.

NRC (1989) assessment of maintenance needs of adult horses

The energy requirements for maintenance per 100 kg bodyweight decline slightly with increasing bodyweight, so that relative to body size larger horses require slightly less food for maintenance than do ponies in similar conditions. This may account for the first term in the equations below, and, to compensate for it, ponies

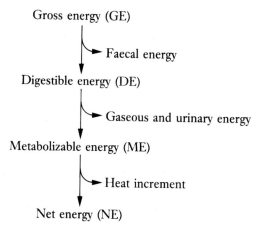

Gross energy (GE)

→ Faecal energy

Digestible energy (DE)

→ Gaseous and urinary energy

Metabolizable energy (ME)

→ Heat increment

Net energy (NE)

Fig. 6.5 Partition of dietary energy.

may develop a greater barrel or appear more pot-bellied. The relationship between bodyweight (BW) and the DE requirements for maintenance is depicted in Fig. 6.6. Two of these curves are derived from the identities formulated by the NRC (1989):

$$DE\ (MJ/day) = 5.9 + 0.13\,BW \tag{1}$$

where BW is the bodyweight (kg) of a normal nonworking horse weighing 600 kg or less. Work at the Texas Agricultural Experiment Station indicated that cutting horses, working in a hot environment, expend 10–20% more energy than would be predicted (Webb *et al.*, 1990).

Texas workers (Potter *et al.*, 1987) found that the energy requirement of heavy (675–839-kg) Belgian and Percheron horses was 10–20% lower than predicted by Pagan and Hintz (1986a). The difference was attributed to the lower activity of heavy horses and the slower rates of acceleration and deceleration during voluntary work. Hence, the NRC (1989) made an adjustment for heavy horses:

$$DE\ (MJ/day) = 7.61 + 0.1602\,BW - 0.000063\,BW^2 \tag{2}$$

where BW is the bodyweight (kg) of horses in excess of 600 kg, as these engage in somewhat less voluntary activity (partly accounting for the negative quadratic term in the equation).

Maintenance needs of growing horses

The daily maintenance requirements of growing horses, determined by extrapolation of growth data to zero gain, were found to be 158 kJ DE/kg BW and 148 kJ DE/kg BW for limit- and for *ad libitum*-fed horses, respectively (Cymbaluk *et al.* 1989a). (In the same study 24–83 g BW gain/MJ DE was achieved above maintenance.) This

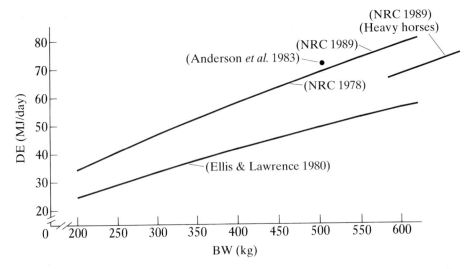

Fig. 6.6   Approximate relationship of kg BW to daily DE requirements for maintenance according to four sources: for weights 600 kg BW or less, DE (MJ/day) = 5.9 + 0.13 BW (NRC 1989), for weight >600 kg BW, DE (MJ/day) = 7.61 + 0.1602 BW − 0.000063 (BW)^2 (NRC 1989); DE (kJ/day) = 649 BW$^{0.75}$ ≡ 0.649 BW$^{0.75}$ MJ/day (NRC 1978); DE (kJ/day) = 465 BW$^{0.75}$ ≡ 0.465 BW$^{0.75}$ MJ/day (Ellis & Lawrence 1980).

is about one-quarter of the efficiency of poultry and pigs. The values of 148–158 kJ agree with Eq. (1) for animals of 200–300 kg BW, but clearly can be applied only to animals within the BW range studied by Cymbaluk and colleagues. Individual horses differ in their needs about all these means; some will eventually become fat when subjected to a regime under which others will lose condition.

Heat production and efficiency of ME use

An inactive horse at the maintenance level of energy intake and expenditure does essentially no work on its surroundings so that NE expended in maintenance (m) is ultimately degraded to heat:

$$ME_m = NE + HI = \text{heat production at maintenance} \qquad (3)$$

where HI is heat increment, or waste heat. The fact that the temperature of the horse's body is normally greater than that of the surroundings, to which heat is continuously being lost, is the expression of this situation. Exposure to a cold or a wet and windy climate accelerates metabolic rate so that the rate of heat production keeps pace with the rate of heat loss in order to maintain a steady body temperature, that is the energy requirements for maintenance rise. Conversely, in hot climates, where the environmental temperature is higher than that of the horse, the heat produced must still be dissipated. This is done primarily by evaporation of sweat and of water from the lungs, but also by a rise in body temperature. A physiological stress is induced. Thus, in one environment heat production is a boon and in the other a hindrance. Work at the Texas Agricultural Experiment Station (reviewed by

Hiney & Potter 1996) indicated that cutting horses, working in a hot environment, expend 10–20% more energy than would be predicted by the equations of Pagan and Hintz (1986b). Their rectal temperatures were often 41°C, and one may assume that, as metabolic rate is a function of body temperature, this was a cause of the greater need.

Can the heat production be manipulated to the horse's advantage? Waste heat (HI) is a measure of the efficiency of utilization of the ME of feed and it is known to vary between types of feed. If the NE available represents 80% of the ME (NE/ME = 0.8), then the remaining 20% is HI. When feeds are selected for use, their difference in HI should ideally be considered in the context of the climate and the purpose for which the horse is kept. Allowance for these differences is the basis of the justification for the French (INRA) NE system, discussed later in this chapter (see 'The NE system introduced in France by INRA 1984, updated 1990').

Some estimates of the likely efficiency of ME utilization by the horse are given in Table 6.1 and Fig. 6.7. The efficiency values (*k* values) in Table 6.1 subtracted from 1 show the proportion of energy lost as waste heat when the feed is used for maintenance or for fat deposition. Thus, 30% of the energy of meadow hay would be lost as waste heat by horses at maintenance, whereas only 15% of the ME of barley would be similarly lost (note that the utilized energy is ultimately degraded to heat also, but more hay would be required for maintenance). During winter ample meadow hay may be a more appropriate feed than in the summer, or than barley, as the greater HI of hay may contribute to the maintenance of body temperature when the weather is cold. The partition of the GE of four feeds is described in the histograms in Fig. 6.7. It should be clear that *k* represents, in the main, efficiency of glycogen and of depot fat formation. Efficiency of utilization of these sources by muscles is approximately 0.35–0.45. The concept of work measured for flat racing, etc. is illusory as the true efficiency can be measured only as a difference in energy expended between exercise on the level and that on a gradient not an inclined moving belt. When a horse moves up a gradient work is done against the force of gravity, whereas on an inclined moving belt the horse does not rise. It remains at the same level. However, it is known that, for other reasons, a horse expends more

Table 6.1 Estimated efficiency of utilization of the ME (NE/ME), or *k*, for various energy sources by the horse.

	For maintenance* (k_m)	For fat deposition (k_f)
Mixed proteins	0.70	0.60
Meadow hay	0.70	0.32
Lucerne hay	0.82	0.58
Oats	0.83	0.68
Barley	0.85	0.77
Fat	0.97	0.85

*These values are higher than those for fattening mainly because the use of these nutrients for that purpose spares the breakdown of body fat.

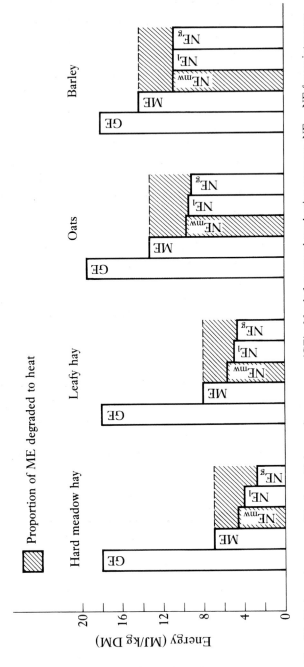

Fig. 6.7   Estimated average utilization efficiency of gross energy (GE) of feeds for several production purposes. NE_{mw}, NE for maintenance, or for work, which may in practice be approximately the same; NE_l, NE for lactation; NE_g, NE for growth.

Efficiency of utilization of GE as a source of ME = ME/GE = q
Efficiency of utilization of ME as a source of NE = NE/ME = k
$$NE/GE = qk$$

energy running on an inclined belt than when on a horizontal belt moving at the same speed. The *k* value includes an allowance for the energy costs of both ingestion and digestion, that is the energy expended in eating, digesting and fermenting feed, in addition to differences amongst nutrients in the systemic efficiency of their metabolism in ATP, tissue, milk etc. synthesis (see Appendix C for maintenance km calculation). Note: nearly two-and-a-half times as much hard hay is required for maintenance as would be required of barley so that nearly 25% more heat is produced at maintenance on the hay diet.

The ribs of both breeding and working horses in optimum condition cannot be seen, but can be felt with little fat between the skin and ribs. Acclimatization to cold weather does not necessitate excessive fat deposition but should allow sufficient time for the coat to grow. Horses should therefore be provided with a shelter protecting them from rain, snow and the worst of the wind. In other words, three sides and a roof provide sufficient protection in all seasons for properly fed adult animals. A long hair coat, if dry, and a modicum of subcutaneous fat are an excellent insulation for horses given an ample roughage diet, so that the rate of waste heat production without shivering, a function of $(1-k)$, equates with the rate of heat loss. In the spring when horses are brought in, daily grooming and 57–114 g oil added to the ration each day should accelerate the shedding of the winter coat.

Production needs

The measurements of Anderson *et al.* (1983) given in Table 6.2 indicate the energy needs for maintenance plus endurance work. Figure 6.8 gives the DE requirements of horses of various weights at maintenance and when engaged in work of a range of intensities, strenuous work causing a large increase in energy demand. Excessive HI, or waste heat, in working horses is an encumbrance and a contributory

Table 6.2   DE demands for maintenance plus work on a slope of 9°[1] at endurance rates (135 heart beast/min, 155 m/min).[2]

Bodyweight (kg)[3]	400	500	600
Distance travelled (km)	DE per day (MJ)		
1	68	79	90
2	76	88	100
4	89	103	117
6	99	114	127
8	107	121	133
Approximate appetite	108–113	125–130	140–145

[1] Work output on an inclined moving belt is in practice greater than that on the level.
[2] Based upon a quadratic equation relating energy requirements to bodyweight and work in quarter horses (Anderson, C.E. *et al.* 1981, 1983).
[3] Average bodyweight 503 kg.

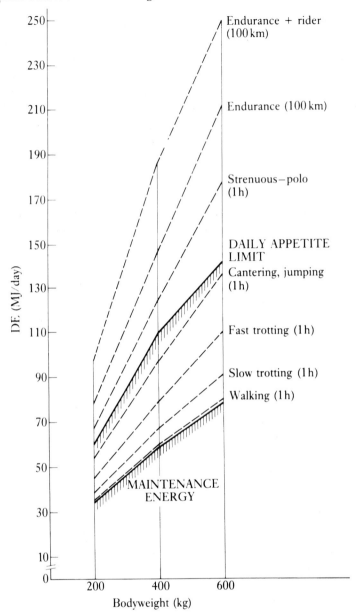

Fig. 6.8 DE demands of daily maintenance and work at a constant elevation in relation to appetite of horses of three bodyweights. (Effect of a 67-kg rider on 400- and 600-kg horses and 33-kg rider on 200-kg horse given for endurance rides only.)

cause of unnecessary sweating, indicating an important attribute of concentrate feeds, in addition to that of meeting the high energy demands of working animals. These demands of high productivity, also exemplified by peak lactation, are met from two main sources: (1) the breakdown of body fat and (2) increased feed energy.

The data in Table 6.1 show that in growing/fattening animals 68% of the ME of meadow hay and 40% of the ME of mixed proteins are lost as waste heat, whereas only 23% of barley grain and 15% of fat ME are similarly lost in fattening. The so-called heating effect of cereals and other concentrates reflects a more rapid rise in blood glucose and metabolic rate after a large meal and the associated feeling of vigour in 'hot-blooded' breeds (see also calming effects of dietary fats, Chapters 5 and 9). In Fig. 6.9 the interactions among the HI of feeds, environmental temperature, and production of body heat and critical temperature are depicted.

Partition of feed energy

Visual examination of feed reveals nothing about its ME content, but the feed can be weighed. Fortunately, the gross energy (GE) content of most horse feeds is just over 18 MJ/kg DM. This statement is untrue for feeds containing much more than 80 g ash/kg or 35 g oil/kg. For example, oats on average may contain 45 g oil and 19.4 MJ GE/kg DM. Figure 6.7 shows how the GE of samples of four different feeds might be utilized for growth or for hard extended work and the data give a revealing and objective comparison of roughages with cereals. The coefficient q represents the approximate efficiency by which the GE of each feed is utilized as a source of ME and as the coefficient k represents the efficiency by which this ME is

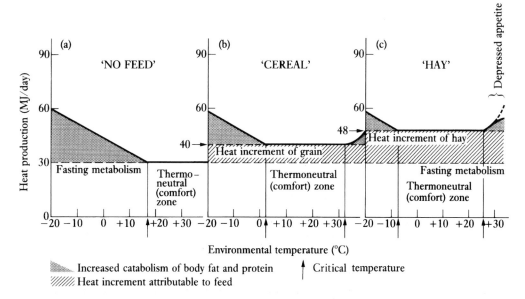

Fig. 6.9 Interactions among the heat increment of feeds, environmental temperature, heat production and critical temperature. Data are approximate; they assume minimal air movement and are not based on direct experimentation. (a) Fasted horse (also with slightly less subcutaneous fat); (b) horse fed 3.5–4 kg grain providing 40 MJ ME and 30 MJ NE (maintenance level); (c) horse fed 5–6 kg hay providing 48 MJ of ME and 30 MJ of NE (maintenance level).

utilized for the functions of maintenance, growth, etc. $q \times k = NE/GE$, or the overall efficiency of utilization of the 18 MJ for the productive function. Note that $q \times k$ of the hard meadow hay for growth (mainly fat deposition) is 0.12, whereas the equivalent for barley is 0.59, a value 4.9×0.12. The NE in hard meadow hay for growth, and probably for extended work, is only a quarter of that found in the two cereals, despite a similarity of their GEs. In energy terms, that is, MJ/kg feed, the losses in the utilization of ME from hay and cereals are not very different (Fig. 6.7), but the k values are (Fig. 6.10), for the reason that the ME values (MJ/kg) differ widely.

It is recognized that roughages are required by horses and ponies, particularly in a long form, in order to maintain general metabolic health and a feeling of wellbeing. However, are there lessons to be learnt from the above calculations? First, poor-quality roughages can be an expensive buy if they are to form a major part of the ration of growing or hard-worked animals. Second, hard-worked animals can lose condition if poor-quality roughages form a major portion of their diet. Finally, idle horses may put on unwanted fat if too much cereal is included in their diet. The data available on the ME content of horse feeds and on their NE for various functions is limited (but estimated values for maintenance are provided by INRA and selected data are given in Appendix C expressed as horse feed units, UFC). The large difference in the efficiency of DE utilization for productive activity between roughages and concentrated feeds led Martin-Rosset and colleagues (1994) to develop the UFC NE system (see 'The NE system introduced in France by INRA 1984, updated 1990'). For proper use of the DE system it is necessary to follow certain rules (see 'Ration formulation using the DE and NE systems', this chapter) and to use DE data provided in Appendix C.

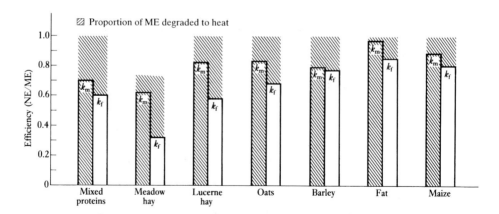

Fig. 6.10 Estimated efficiency of utilization of ME (NE/ME), or k, for various energy sources. $k_m = NE/ME$ for maintenance; $k_f = NE/ME$ for fat deposition (later growth). (Maintenance values are higher than those for fattening, mainly because the breakdown of body fat is spared by the use of these feeds for maintenance.)

DE, PROTEIN AND MINERAL REQUIREMENTS BASED ON NRC RECOMMENDATIONS

Reproduction and lactation

The dietary requirements of the breeding mare can be arbitrarily divided into those for: (1) the first 8 months of gestation; (2) the last 3 months of gestation; and (3) lactation (this may coincide with 0–4 months of gestation). Gestation length for most TBs is in the range 335–345 days and for other breeds 322–345 days. Lactations of 110–130 days are typical of many husbandry systems, although the nonpregnant mare would produce milk for much longer if given the opportunity.

Energy

The first 8 months of gestation have no practical impact on the nutrient needs – that is, they do not raise requirements above those for the barren mare, nor do they increase the already high requirements of the lactating mare. Thus, after weaning, this mare's energy requirements approximate those of maintenance until 8 months of gestation have been completed. Most of the fetal growth occurs during the last 90 days of gestation; even so, the nutrient drain incurred then to sustain normal fetal and placental growth is much less than that for lactation. The approximate energy contents of the fetus and the other products of conception at term, compared with the energy content of the mare's milk over a 4-month lactation, are given in Table 6.3. By assuming that all fetal growth occurs during the last 90 days, the approximate DE required daily above maintenance to meet these needs can then be calculated (Table 6.4) for comparison with the much greater demands of milk production.

Table 6.3 Approximate energy contents of the fetus and other products of conception at term compared with the energy content of mare's milk over a 4-month lactation.

Mare's weight (kg)	200	400	500	600
Products of conception at term (MJ)	110	200	240	270
Lactation of 17 weeks (MJ)	1700	2840	3400	3900

Table 6.4 DE required daily to meet the needs of fetal growth and lactation, but excluding maintenance requirements for energy of the mare.

Mare's weight* (kg)	400	500
Products of conception at term (MJ)	5.0	6.0
Average milk production (MJ)	40	50

*The mare's weight should increase by 15% during gestation so that her maintenance requirements rise proportionately.

As the fetus occupies an increasing proportion of the mare's abdominal cavity, her capacity for bulky feed declines during the period in which nutrient requirements increase. This may correspond to an increase in the quality of grazing (see Chapter 10), but where mares are given hay and concentrates the quality of the forage should be improved during the last 3 months of gestation. The energy and nutrient requirements of the breeding mare (Table 6.5) increase from period (1) to (2) and from (2) to (3) of the cycle given above. (Note that if the mare is lactating during part of period (1), her requirements then will exceed those of period (2).) However, the values are averages and the amounts of feed and therefore the DE given to individual pregnant mares should be adjusted to avoid either obesity or poor condition (see also Chapter 7). During the 11th month the DE requirement is equivalent to a 0.33:0.66 mixture of oats and hay of nearly 10 kg daily. Part of the demand in the 11th month is to sustain udder development. On average, these values differ little from the flat-rate recommendations of Table 6.5. Feeding studies have not established whether birthweight of the foal is generally influenced by deviations from those rates. One investigation (Goater *et al.* 1981) demonstrated an increase of both 1.5 kg in birthweight and 0.24 kg in daily weight gain during the first 30 days of life as a result of providing the mare with 120% of the NRC (1978) gestation rates. Other experiments in which quarter horse and TB mares were restricted to 55% and Arabian mares to 85% of the rates led to weight losses by the pregnant mares without affecting birthweight of the foal in comparison with mares receiving the recommended rates. Clearly, healthy mares possess the capacity to adapt without the foal incurring any significant handicap.

Protein and minerals

The most critical nutrients for breeding mares given traditional feeds are protein, Ca and P. Mares kept during the last 90 days of gestation entirely on reasonably good-quality pasture or high-quality conserved forage containing some 30–40% leafy clover, lucerne or sainfoin require no other source of Ca, and if the forage contains 10% protein per unit of dry matter, no supplementary protein. An increase in the physiological demand for Ca leads to a decrease in serum total and ionized Ca concentrations and a stimulation to parathormone secretion. This occurs in periparturient mares during mammary Ca secretion when serum total Ca has been shown to decrease from 3.1 to 2.7 mmol/l (12.5–11 mg/dl) (Martin *et al.* 1996a). The evidence indicated that the dietary requirement of the mares averaging 510 kg BW in late pregnancy was closer to 5.5 g Ca/kg dietary DM (45 g Ca/day) than to 3.5 g Ca/kg DM when the Ca was derived from bluegrass–clover pasture and concentrates, a third of the daily Ca being derived from the pasture. The P requirements, however, amount to 3 g/kg (0.3%) of the total dry diet. This would be provided by good pasture (see Table 10.2), but as both grass and legume hays given to horses in the UK normally contain less than 2 g P/kg (0.2%), a supplement of dicalcium phosphate, steamed bone meal or wheat bran will be required. The discrepancy would be met daily by a supplement of 60 g dicalcium phosphate or 1.5 kg bran for horses and

Table 6.5 Daily DE requirements of horses for various functions and amounts of hay and concentrates needed to provide the energy (based on NRC 1989 recommendations).

Mature bodyweight	200 kg			400 kg			500 kg			600 kg		
	DE (MJ/kg)	Hay[1] (kg)	Concentrate mixture[2] (kg)	DE (MJ/kg)	Hay[1] (kg)	Concentrate mixture[2] (kg)	DE (MJ/kg)	Hay[1] (kg)	Concentrate mixture[2] (kg)	DE (MJ/kg)	Hay[1] (kg)	Concentrate mixture[2] (kg)
Mature horse[3] maintenance	3.10	3.8	—	56.1	6.8	—	68.6	8.4	—	81.2	9.9	—
Mares last 90 days of gestation	35.6	2.9	1.1	64.9	5.1	2.0	79.5	6.1	2.6	94.5	7.6	2.8
Lactating mare first 3 months	64.0 (12)[4]	2.6	3.8	101.2 (18)[4]	4.7	5.5	122.2 (21)[4]	6.0	6.4	141.0 (23)[4]	7.2	7.2
Lactating mare 3 months to weaning	51.0	3.4	2.0	82.4	5.9	3.0	102.0	7.5	3.6	120.9	8.7	4.3
Stallion breeding	38.9	2.2	1.9	70.3	3.4	3.7	85.8	4.3	4.4	101.7	5.2	5.2
nonbreeding	35.0	2.8	1.1	62.0	4.7	2.0	75.0	5.8	2.4	89.0	6.9	2.9
Weanling (6 months old)	35.0	1.0	2.4	57.3	1.8	3.7	67.4	2.2	4.3	75.7	2.3	5.0
Yearling (12 months old)	39.7	2.2	1.9	68.2	3.6	3.4	84.1	4.4	4.2	100.0	5.1	5.1
Long yearling (18 months old)	37.5	3.1	1.1	69.0	5.0	2.5	87.9	6.2	3.2	104.6	7.5	3.8
Two-year-old excluding work	33.0	2.7	1.0	64.0	4.6	2.3	78.7	5.8	2.8	98.3	6.9	3.7
maintenance plus 1h moderate work	47.7	2.5	2.4	90.0	4.6	4.6	110.1	4.8	6.2	135.1	5.3	8.0

[1] Hay containing 8.2 MJ DE/kg and 88% DM.
[2] Concentrate mixture containing 11.4 MJ DE/kg and 88% DM. Quantities of concentrate up to 1.5% BW daily may be fed if a minimum roughage allowance of 1 kg/100 kg BW is given.
[3] 2.3 kg extra feed daily should produce 0.5–0.6 kg gain.
[4] Assumed peak daily milk yield (kg).

40 g or 1 kg, respectively, for ponies. Where grass hays are used, as assumed in Table 6.5, supplementary protein will be required as well. If the hay contains 7% of protein and it constitutes 70% of the diet, whereas the diet as a whole should contain 10% protein, then the concentrate must contain 16–17% of protein (Appendix A shows the type of calculation needed). This is the level found in most commercially prepared stud nuts that would also provide the necessary P, vitamins and trace elements.

Abundant good-quality pasture will meet the energy, protein, Ca and P needs of lactation, even though the minimum dietary protein requirement will have risen to 125 g/kg dry feed (12.5%). Responses in milk yield have been obtained from quarter horse mares by providing mixed feeds containing up to 170 g protein kg. However, grass and clover proteins are of high quality and it is unlikely that economic responses would be obtained by raising the protein level of the spring-grazing diet. If the stocking density is high, or good-quality pasture is otherwise moderately limited, supplementation can be provided by lower protein pony nuts (see Table 7.1), or a mixture of these and cereals. If pasture is more scarce, a stud nut or an equivalent mix containing 16–17% of protein should be given to lactating mares. Any conserved forage provided should be leafy hay containing a mixture of clover and grass, or should be well-conserved haylage. Typically in the UK grass hays of only moderate quality are fed when grazing is limited. These contain only 40–80 g crude protein/ kg and so do not meet immediate needs.

It is unlikely that one could overfeed lactating mares with roughage, except that large quantities of poor roughage would limit their capacity for concentrates, leading to a decrease in milk yield. When typical grass hays are given, a satisfactory milk yield is obtained only if at least 50% of the dry feed is provided as a stud nut or equivalent 16–17% protein mix. This mix may be based on oats or barley and soyabean meal, or an equivalent proprietary protein concentrate containing 440 g protein/kg. Horse mares would require 1.75 kg of soya daily (12–14% of the total grass hay and cereal-based ration). The reason why a 16–17% protein mix is sufficient for lactating horses, as well as for late pregnancy, despite a higher protein requirement in lactation, is that the mix forms over half the ration in lactation, to meet the greater energy needs, whereas it forms only 30% of the prefoaling diet.

This grass hay, cereal, soya diet must be supplemented with a mineral mix composed of 35 g dicalcium phosphate, 65 g limestone and 70 g sodium chloride when the total daily intake of dry foods is 14 kg. Proportionately less will be required for smaller rations. A proprietary mixture of trace elements and vitamins should also be given. The latter should include vitamin A for horses with no access to pasture. Where large amounts of silage or haylage are used, supplementary vitamins D and E will also be necessary at levels, respectively, of 7000 iu and 250 mg daily. Whereas the trace-element content of mare's milk (but see trace elements, Chapter 3), and therefore the adequacy of the foal's diet, is affected by the supplementation of the dam's diet, a deficiency of water, energy, protein, Ca or P will ultimately bring about a decrease in milk output, without altering its composition.

Table 6.6 Nutrient concentration in diets for horses and ponies expressed on the basis of 90% DM (based on NRC 1989).

	Crude protein (g/kg)	Ca (g/kg)	P (g/kg)
Mature horses and ponies at maintenance	72	3.2	2.0
Mare, last 90 days of gestation	94	5.5	3.0
Lactating mare, first 3 months	120	5.5	3.0
Lactating mare, 3 months to weaning	100	4.0	2.5
Creep feed	160	8.0	5.5
Foal (3 months old)	160	8.0	5.5
Weanling (6 months old)	135	6.0	4.5
Yearling (12 months old)	115	5.0	3.5
Long yearling (18 months old)	105	4.0	3.0
Two-year-old, light training	95	4.0	3.0
Mature working horse, light to intense work	95	3.2	2.0

Growth of the foal

As horses grow they do not simply increase in weight and size, they also display what is termed development. Various tissues and organs of the body grow at different rates. In proportion to body size the rate of weight gain of the body as a whole, if permitted by the feed allowance, is very much greater in the younger than in the older animal. In fact, from the suckling period onwards the rate of gain per 100 kg BW declines continuously, but the rate of growth of long bone and muscle declines at an even faster rate. An increasing proportion of the gain constitutes fat, which has much higher demands for feed energy. These trends are fundamental to a formulation of requirements for protein, Ca and P in particular, which decline fairly rapidly as a proportion of total diet with increasing age of the foal and yearling (see Tables 3.2 and 6.6). Further details of growth and the way in which we should guide it, and indeed of the breeding mare, are given in Chapters 7 and 8.

RATION FORMULATION USING THE DE AND NE SYSTEMS

Both the DE and NE systems require the accumulation of two sets of information:

(1) nutrient content of feeds (Appendix C); and
(2) nutrient requirements of horses described in the *same* terms as for feed (Table 6.5 for DE).

These two issues are covered in the text and tables.

Energy needs

The principle that has been adopted in calculating the *daily* requirements for many nutrients is to assume that the needs of various functions of the horse are additive, i.e. a factorial system has been used, the factors or functions being maintenance, work, growth, reproduction, etc. This approach is not precisely supported by biological evidence, but it is the simplest approach and the values (coefficients) can be modified and the factors augmented as new information comes to hand, or as the activities of horses are extended. The daily energy needs for:

- maintenance (m) are a function of BW;
- work (w), a function of BW, intensity (I) and time (T) spent;
- growth (g), a function of weight gain (G) and BW relative to mature size;
- pregnancy (p), a function of the mare's BW and stage of pregnancy (S_p); and
- lactation (l), a function of stage of lactation (S_l) and yield per day (Y).

Rationing (kg feed per/day) is based upon energy allowances, and in the case of a pregnant, lactating mare's energy requirements would be summated from:

$$m(BW) + p(BW \cdot S_p) + l(S_l \cdot Y) = \text{energy needs (MJ/day)} \qquad (4)$$

which when divided by the energy content of the feed (for that energy system), gives the kg feed per day. Alternatively, appetite (defined as kg feed per day) would need to be estimated and then the required energy density of the diet calculated from:

$$\text{needs(MJ)/appetite(kg)} = \text{MJ/kg diet}$$

The proportions of roughage and concentrates are then calculated:

$$\text{MJ/kg diet} = x(\text{MJ/kg roughage}) + (1-x)(\text{MJ/kg concentrates}) \qquad (5)$$

where x is the dietary fraction or proportion (typically 0.6–0.7) of roughage and the remainder of the diet (1 − x) is concentrate (therefore typically 0.4–0.3), ignoring water. This gives the proportions of roughage and concentrate making up the total appetite (kg/day).

The formulation of a ration requires estimates of: (1) total daily dry food intake, (2) the energy content of feeds, and (3) daily energy requirements (DE system, Tables 6.5 and 6.7). The two energy systems proposed for use, supplying information on (2) and (3), are the NRC DE system and the INRA NE system. The justification for each is that for:

- DE, digestibility is the most potent factor that segregates feeds that otherwise have similar GE values;
- NE, roughages and concentrates can be clearly segregated according to the efficiencies (k) by which the ME is utilized for maintenance and productive purposes.

(Note: the $q \times k$ values will differ somewhat according to the function of the horse, maintenance, fattening, milk secretion, etc.; however, for simplicity in application the INRA system assumes that efficiencies (k) of ME for maintenance apply to all functions. Moreover, the energy requirements for growth account for only 20% of the total energy requirements of the growing animal (Vermorel & Martin-Rosset, 1997); the remaining 80% is consumed in maintenance. In the final analysis the practical value of each system depends importantly on the reliability of its feed evaluation system (Appendix C). Precise nutritional definitions are elusive for roughages and succulents. Nevertheless, the relative value of the two systems in practice will depend on the development of this evaluation.

The DE system

The DE required in MJ/day divided by dry feed intake in kg/day gives MJ of DE per kg of dry feed needed. The DE contents of the roughages and of the concentrates are given in Appendix C and their required dietary proportions can then be roughly calculated using Eq. (5) above, examples of which are given in Appendix A. Equation (5) has been used in deriving the proportions in Table 6.8. The greater the intensity of physical activity the higher the proportion of cereals required. As the speed of the horse increases, the energy expended rises steeply on an hourly basis (Table 6.7). Hence, the types of problem encountered can be quite different in horses undertaking strenuous effort compared with those asked to respond in a leisurely fashion. Futhermore, compared with ponies, large horses tend to require a higher proportion of concentrates in the ration when both are subjected to hard work.

The recommendations in Tables 6.5 and 6.8 are likely to be somewhat in error where low energy densities are required by working horses. In those situations the feed energy requirements are likely to be underestimated and so the French NE system is discussed below.

Table 6.7   DE demands of maintenance and work on the flat (based on NRC 1989).

Bodyweight (kg)	200	400	600
Approx. feed capacity per day (MJ DE)	60	100	150
Maintenance requirement per day (MJ DE)	31	56	81
	Energy requirements for work above maintenance (MJ DE)*		
Walking (1 h)	0.4	0.8	1.3
Slow trotting, some cantering (1 h)	4.2	8.4	12.5
Fast trotting, cantering, some jumping (1 h)	10.5	20.9	31.4
Cantering, galloping, jumping (1 h)	25.0	50.0	75.0
Strenuous effort, racing, polo (1 h)	36.0	72.0	108.0
Slow trotting, some cantering (10.4 h, 100 km) calculated from above	43.5	87.0	130.5

*1 kg concentrate provides about 12 MJ DE.

Table 6.8 Effect of a range of required energy densities (MJ DE/kg air-dry feed) on the cereal content of the daily ration when hays of two energy contents are available.

Energy density of ration required	Oats (%)		Barley (%)	
	7.2[1]	7.8[1,2]	7.2[1]	7.8[1,2]
7.5	7	0	5	0
8.0	19	5	14	4
8.5	30	19	23	14
9.0	42	32	32	24
9.5	54	46	41	34
10.0	65	60	50	44
10.5	77	73	59	54
11.0	88	86	68	64

[1] Energy content of hay (MJ DE/kg): 7.2 MJ/kg, medium quality; 7.8 MJ/kg, good quality.
[2] Hay can be assumed to contain 86% DM and where haylage of 45% DM is to be used it may be substituted for the hay of 7.8 MJ DE in the proportions 1.8–1.9 kg haylage per 1 kg hay. Similarly, 1.6–1.7 kg haylage of 50% DM could be used.

The NE system

The NE system was introduced in France by INRA in 1984 and updated in 1990.

Energy

The NE system first provides the NE content of feedstuffs for maintenance. The feed values in this scheme are expressed in dimensionless horse feed units (UFCs), i.e. relative to a reference value of 1 for barley, where 1 kg standard barley has a NE value of 9.414 MJ (assuming barley contains 140 g moisture/kg). Thus:

$$1 \text{ UFC} = 9.414 \text{ MJ NE} \tag{6}$$

In the INRA tables, and Appendix C, UFCs are given per kg DM, so that UFC values are: barley 1.16, maize 1.33, etc. (i.e. each determined UFC value is divided by its fractional DM content. For barley with 86% DM the value is 1/0.86 = 1.16).

Energy value of feed

The NE contents of feeds for *maintenance* are calculated from their ME contents and the coefficients of their respective efficiencies of utilization for maintenance:

$$NE = GE \times dE \times ME/DE \times k_m, \tag{7}$$

where dE is the digestibility calculated from the digestibility of organic matter (OM) and k_m is NE/ME for maintenance.

The UFC of a feed is its NE value relative to that of standard barley:

$$UFC = (ME.k_m)/9.414 \text{ (as per Eq. 6)}$$

where NE of barley is 9.414 MJ/kg and ME is in units of MJ/kg (note: UFC is unitless).

Maintenance was chosen as it represents 50–90% of energy expenditure of horses and the NE feed values for maintenance are considered by INRA to be equivalent to those for physical activity, or work, a common equine function (i.e. $k_m \equiv k_w$). The NE for both maintenance and activity is expended mainly in ATP synthesis. In using the system one should apply both the INRA feed *and* requirement values.

UFC values were chosen as a basis of feed formulation for two main reasons. First, they are approximately additive, that is, different combinations of feedstuffs yielding the same total UFC should have the same productive effect, or the UFC value of a feed is not influenced by other feeds with which it may be combined. Martin-Rosset and Dulphy (1987) showed that in the horse digestibility of feed was also not influenced by feeding level, and that the digestibility of forage was not affected by the addition of concentrates to the ration, in contrast with the effects in sheep. (The variable that has the largest influence in the discrimination of feedstuffs is dE, which is accounted for in both DE and NE feeding systems.) The second and most important reason for selecting NE is that k_m draws a clear distinction between the productive values of forages and concentrates. The energy costs of mastication and propulsion of digesta through the GI tract and the heat of fermentation of forages in the hindgut are greater than the heats of ingestion and digestion of starch; and the efficiency of utilization of VFA derived from forage fermentation is less than the efficiency of glucose metabolism from starch. These costs and efficiencies both affect the value of k_m. In other words, the formulation of mixed feeds from the DE values of their constituent ingredients, with the objective of deriving a variety of mixtures with the same productive energy values, exaggerates the value of forages. The k_m value of standard barley is 0.79 and that of an average grass hay is 0.62. The following comparisons (Table 6.9) between concentrates and forages on a DM basis will exemplify the point. The ratios of the means indicate that relative to concentrate values the DE system overvalues these roughages by 15% for productive purposes. However, the large effects occur only with materials at the extremes, i.e.

Table 6.9 Comparison of the DE and NE values for concentrates (C) and forages (F) per unit DM.

	DE (MJ/kg)	UFC
Barley (C)	15.2	1.163
Maize (corn) (C)	16.1	1.35
Oats (C)	13.4	1.01
Maize silage (F)	11.2	0.88
Grass hay (F)	7.3	0.53
Barley straw (F)	6.8	0.28
Ratio F/C	0.566	0.480

maize and barley straw, otherwise the bias introduced by the DE system (ingestion only) is small (see Table 6.10). INRA have provided data to support the adoption of the UFC system, and the conclusions they draw are in line with proposals put forward by Frape and Tuck (1977).

Energy cost of eating

Chewing, ingestion and digestion of feed involve muscular activity. In addition, there is an increase, above the fasting rate, in general metabolic rate. Vernet *et al.* (1995), using indirect calorimetry in respiration chambers, measured the energy expended by sport horses during ingestion of feeds of different types. These feeds were measured at the level of 1.26 times the maintenance energy requirement, when maintenance energy demands were increased by an average of 38.8% during eating a mixed meal and heat production (maintenance) is increased by 75–95% during a meal of straw or hay (Vermorel & Martin-Rosset, 1997). For single feeds the energy costs of eating per kg DM intake are given in Table 6.10.

The energy cost of eating is mainly accounted for by the energy expended in chewing. An additional amount is contributed by cephalic activation of the sympathetic and parasympathetic nervous systems. In other words, the sight and smell of feed stimulates a physiological response of the whole GI tract and of some other organs. The measurements by Vernet and colleagues included only the increments in expenditure during actual eating and so excluded much of, for example, the HI of bacterial fermentation. The data do indicate that with poor-quality feeds, such as straw, over a quarter of the ME they contribute is lost during the process of taking the feed into the stomach. For mixed feeds the energy costs of eating are 6–10% of ME (Vermorel *et al.* 1997), and although the horse expends a similar amount of energy to that used by ruminants per minute it expends two or three times as much as the ruminant per kg DM, owing to a lower rate of eating (Vermorel & Martin-Rosset, 1997).

Protein

The scheme also accounts for the digestible crude protein (DCP) content of feeds, and in the case of forages these are reduced in proportion to their NPN contents.

Table 6.10   Energy cost of eating as a proportion of the ME value of the individual feed measured at 1.26 times maintenance (Vernet *et al.* 1995).

	Proportion of ME/kg feed DM expended in eating
Pelleted maize	0.010
Pelleted sugar-beet pulp	0.042
Long hay	0.102
Wheat straw	0.285

The useful protein is calculated from the estimated amounts of amino acids absorbed from the small intestine (plus amino acids absorbed from the large intestine) and this useful protein is termed MADC (matières azotées digestibles corrigées) (g/kg DM) (see forage K values in 'Protein and the evaluation of N absorbed from the large intestine', not to be confused with the coefficient k).

ENERGY AND PROTEIN REQUIREMENTS BASED ON INRA FEED UNITS, EXPRESSED AS UFC AND MADC

The recommended allowances in UFC for particular physiological functions were determined by a factorial method, or by feeding experiment. The scheme therefore partly allows for differing k values for maintenance, growth, lactation and work (note: INRA assume the same k value applies for maintenance and work – see 'Energy value of feed', above) for each feedstuff. Thus, k values partly depend on the particular mix of nutrients, that is, the substrate for metabolism, i.e. VFAs, fats, glucose and the proportions, or balance, of the amino acids (see Appendix C). In these respects, the variation in k is greater for growth than for maintenance. Products of poor-quality roughage have relatively less value for growth than for maintenance. Although k values differ for different functions UFC values will differ less as they are dimensionless ratios.

Maintenance

The maintenance requirement of an idle horse is:

$$0.038 \text{ UFC/kg BW}^{0.75}$$

or

$$4 \text{ UFC/day for a gelding weighing } 500 \text{ kg}$$

The maintenance requirement of working horses, cf horses at the maintenance level of activity, cf. 'easy list', is assumed to be increased by 5–15% and that of stallions is increased by 10–20%. The higher maintenance heat production assumed for working horses results from a higher metabolic rate, and may account for the difference in actual feeding rates of racehorses compared with their needs calculated in DE (see Chapter 9).

Protein for maintenance of adult horses

Dietary protein required for maintenance of adult horses may be of poorer quality than that needed for growth and so, within the range of dietary proteins given to horses, no adjustment for amino acid balance is needed in the requirement proposed by INRA. Daily protein requirement for maintenance of an adult horse is:

$$2.8\,\text{g MADC}/\text{kg BW}^{0.75} \qquad\qquad (8)$$

so that a 500-kg horse requires 295 g MADC daily.

Protein for maintenance related to DM intake and endogenous faecal N losses

The horse needs protein which is digested, yielding amino acids (and probably dipeptides) which are absorbed. The apparent digestibility of protein differs only to a small extent between sources. Apparent digestibility is expressed as:

$$\frac{(\text{N intake} - \text{faecal N})}{\text{N intake}}$$

Meyer (1983b) concluded that precaecal apparent N digestibility was 0.5–0.6. As this is 'apparent' it makes no allowance for endogenous N secreted into the lumen of the GI tract, whereas true digestibility does. Mean true values should be slightly greater than this range indicates and true protein digestibility in the small intestine ranges from 0.45 to 0.80. Endogenous faecal N loss is proportional to bodyweight:

$$\text{Daily endogenous N} = 52\,\text{mg} \times \text{kg BW}^{0.75} \ \text{(Slade } et\ al.\ 1970\text{)}$$

although endogenous loss also varies with the level of feeding. This may account for the much higher estimate of Meyer (1983b) of 180 mg N/kg BW$^{0.75}$, giving:

$$\text{Endogenous N} = 3\,\text{g N}/\text{kg DM intake} \qquad\qquad (9)$$

At constant DM intake, apparent N digestibility increases as the N content of the diet is increased, because the endogenous loss forms a smaller proportion of the total N; and likewise true digestibility, for which the endogenous N loss is subtracted from the faecal N, is similar for a range of dietary N concentrations. For most protein sources true digestibility falls in the range 0.7–0.9, and no allowance should be made for any change in dietary N concentration.

Protein sources that contain antinutritive factors can increase endogenous protein secretions. In the pig, toasted field bean protein, which may still contain some antinutritional factors, leads to a negative apparent N absorption in the small intestine (ileal N digestibility), although true protein digestibility is reasonable. If this were to apply to the horse, it would have two consequences:

(1) the measurement of endogenous N production using protein-free diets may seriously underestimate endogenous secretions; and
(2) if only nonamino acid-N is absorbed from the large intestine then amino acids contained in endogenous protein secretions passing the ileo–caecal valve would be lost to the horse and the stimulation to secretion by antinutritional factors would compromise feeding value.

A consequence of this is that field bean protein may have a much lower amino acid value than field pea protein, despite similar apparent N digestibilities and not too dissimilar true protein digestibilites. A conclusion concerning the horse may not yet be drawn. It should be noted that whereas apparent N digestibility reflects the net absorption of N to be metabolized by the horse, true digestibility does not, and requirements in that case must also be in terms of truly digested amino acids. On the other hand, apparently digested N will over-estimate usable N because a major proportion of that N absorbed from the hindgut will be NPN from which the horse gains no material benefit.

Endogenous urinary N

Horses given an N-free diet will continue to lose N in the urine, through the metabolism of tissue proteins. This is taken to be the minimal endogenous renal N loss, and was provisionally calculated (Meyer 1983b) as:

$$\text{Daily endogenous renal N} = 165\,\text{mg N} \times \text{kg BW}^{0.75} \tag{10}$$

There is renal conservation of N in that endogenous urinary loss is minimal in horses on low-N diets and the amount increases with an increase in dietary protein, whereas endogenous faecal-N loss is relatively constant with a change in dietary protein content.

Endogenous integumental N

Losses of N from skin and hair should be allowed for. An average value of:

$$\text{Daily integumental N} = 35\,\text{mg N} \times \text{kg BW}^{0.75} \tag{11}$$

is taken (Meyer 1983b). This slightly underestimates the rate of loss during seasonal hair shedding.

Total endogenous N

A reasonable total endogenous N value from the three sources of losses is:

$$\text{Daily basal endogenous N} = 380\,\text{mg N} \times \text{kg BW}^{0.75} \tag{12}$$

(i.e. 180 + 165 + 35), equivalent to 2.4 g crude protein.

N balance has been achieved during feeding trials with about 350 mg digestible N/kg BW$^{0.75}$. Evidence indicates that horses thrive better with some accumulation of protein reserve. Therefore, in crude protein (CP) terms the maintenance requirement (m) for DCP is taken to be:

$$\text{DCP}_{(m)}\ (\text{g/day}) = 3.3\,\text{g} \times \text{kg BW}^{0.75} \tag{13}$$

(i.e. CP = N × 6.25).

The energy requirement for maintenance, according to German workers (Meyer 1983b), is:

$$DE_{(m)} = 0.6 \text{ MJ} \times \text{kg BW}^{0.75}$$

(in accordance with the INRA value of 0.038 UFC or 0.36 MJ NE $\times$ kg BW$^{0.75}$).
So, in relation to energy, requirement for maintenance is:

$$5.5 \text{ g DCP/MJ DE}$$

or

$$9.2 \text{ g DCP/MJ NE} \tag{14}$$

Breeding mare

Pregnancy

The amounts of energy deposited in the products of conception (fetus + placenta + fetal membranes + udder) are approximately:

- 8th month, 0.636 MJ/100 kg BW/day; and
- 11th month, 1.954 MJ/100 kg BW/day.

In the 8th–9th, 10th and 11th months of pregnancy, respectively, 14, 41 and 45% of the total energy at birth is deposited in these products (Table 6.11; see also feed needs, Table 6.5). The INRA system assumes an efficiency of ME utilization (k) for pregnancy of 0.25. However, the daily allowance in UFC is adjusted according to the mare's condition, as undernutrition of a mare in good condition has no adverse influence on the foal's birth weight, or on its growth rate. The effect of this adjustment on fertility at foal heat, etc. has not been established.

Lactation

The mean characteristics of milk adopted in the INRA system are given in Table 6.12 (see also Table 7.2). The efficiency of ME use (k) assumed in estimating feed requirements is 0.65.

Protein requirements

Prior to the 8th month of pregnancy the mare's protein requirement does not materially exceed that for maintenance. Fetal growth, as a fraction of birth weight,

Table 6.11 Weight gain and composition of the fetus (Martin-Rosset *et al.* 1994).

Month	Weight gain (g/kg birth weight)	Fetus weight gain, fraction of birth weight	GE content (MJ/kg)*	Protein content (g/kg)
8	190	0.18	4.18	115
9	190	0.20	4.60	130
10	300	0.30	4.94	153
11	310	0.31	5.36	171

*Of conceptus (fetus + fetal membranes + uterus + udder).

increases each month from the 8th, when it is about 0.18–0.23 of that at birth (Table 6.11). In breeds weighing 500-kg mature weight:

$$\text{Fetal weight (kg)} = 0.00067x^2 - 20 \tag{15}$$

where x is time in days from fertilization (Meyer 1983b).

$$\text{Birth weight (kg)} = 0.45\, BW_m^{0.75} \tag{16}$$

where BW_m is mare's weight (kg) (Gotte 1972 in Meyer 1983b).

Equations 15 and 16 indicate that fetal growth rate accelerates towards the end of pregnancy and that birth weight is a function of mare's weight, i.e. small mares drop small foals. The mean birth weight of a foal out of a 500-kg mare is 47.6 kg.

The protein content of the fetus also increases from 115 g/kg in the 8th month to 171 g/kg at birth (Table 6.11) and the protein retention in fetal membranes, uterus and mammary gland is about 0.2 of that in the foal. The fraction of the total protein (a) retained in the foal at birth was estimated (Meyer 1983b), giving the additional daily protein requirement of a pregnant mare for foal's growth *in utero*:

$$\text{DCP (g/day)} = 6.15a(BW_m)^{0.75} \tag{17}$$

where BW_m is mare's weight (kg) (Table 6.13).

Thus, the mare's dietary protein requirement increases from the 8th month of pregnancy.

The milk production of mares increases up to about the 3rd month, whereas its protein content decreases from about 24 g/kg milk in the 1st month to 21 g/kg in the 4th month (Table 6.12). Meyer (1983b) assumes that the efficiency of utilization of DCP for the products of conception and milk protein synthesis is 0.5, leading to the recommendations given in Table 6.13.

INRA protein requirements for pregnancy
For the growth of the uterus, including its contents, and of the udder, the amount of protein retained daily is 5 and 21 g/100 kg BW in the 8th and 11th months, respectively. INRA assumes a metabolic efficiency for DCP of 0.50–0.55. Thus, the extra concentrate protein required by a 500-kg mare is approximately 45–50 g at 8 months

Table 6.12 Mare's milk yield and composition (Martin-Rosset *et al.* 1994).

Month	Milk yield (kg/100 kg BW)	Milk composition	
		Energy (MJ/kg)	Protein (g/kg)
1	3.0	2.41	24
2	2.5	2.09	24
3	2.5	2.09	24
4	2.0	1.99	21

Table 6.13 Protein requirements of breeding mares (g DCP/day) (based on Meyer 1983b; these estimates differ from INRA tables, Anon 1984).

Month	Fraction (a) of total protein retained in foal at birth	Body weight of mare (kg)		
		200	500	800
Pregnancy				
8th	0.14	227	451	641
9th	0.22	250	500	700
10th	0.23	260	528*	740
11th	0.31	282	561	798
Last days	0.10			
Lactation				
1st		585	1163	1655
3rd		611	1214	1727
5th		468	931	1324

*For example, this value is the sum of 150 g for fetus (Eq. 17) + 30 g for membranes, etc. (0.2 of fetus) + 348 g for mare (Eq. 13).

and 190–210 g at 11 months (Table 6.14 values less mare's maintenance requirements; the values in this table have been updated in accord with more 1990 INRA data). These values are considerably lower than Meyer's values (1983b).

INRA protein requirements for lactation

The mean composition of milk is given in Table 6.12. Assuming a metabolic efficiency for DCP of 0.55, an INRA requirement of 44, 38, and 36 g DCP/kg milk in months 1, 2–3 and 4 of lactation, respectively, is derived [also lower than Meyer's estimates (1983b)]. This gives a requirement, including maintenance of 950 g for a 500-kg mare in the 1st month, in agreement with the values quoted by Martin-Rosset *et al.* (1994), and approximately in accord with the revised values given in Table 6.14.

Growth

The efficiency of ME use for growth has not been established and will depend both on the energy substrates and on the proportions of fat and protein laid down (note: the index of the second term in Eq. 18 probably derives from the increase in fat deposition resulting from faster growth rates). Maintenance accounts for 60%, or more, of the energy expended, depending principally on the rate of growth allowed. The total requirements are assumed to fit the following relationship:

$$\text{UFC/kg BW}^{0.75}/\text{day} = a + bG^{1.4} \qquad (18)$$

where a = coefficient of maintenance requirement, b = coefficient of gain, and G = average weight gain (kg/day) (Table 6.15).

Table 6.14 Daily energy (UFC) and protein (MADC, g) requirements of horses as proposed by INRA (Anon 1984; updated by INRA, Ed. Martin-Rosset (1990)). All figures in brackets are assumed average daily gain (kg).

Adult weight (kg)	450		500		600	
	UFC	MADC	UFC	MADC	UFC	MADC
Maintenance	3.9	275	4.2	295	4.8	340
Light work	6.6	450	6.9	470	7.5	510
Medium work[1]	7.6	515	7.9	540	8.5	580
Intense work[2]	6.9	470	7.2	490	7.8	530
Mare, gestation						
month 8	3.8	315[5]	4.1	340[5]	4.7	395[5]
months 9–10	4.3	425	4.7	460	5.4	535
month 11	4.4	445	4.8	485	5.5	565
Mare, lactation						
month 1	8.2	865	8.9	950	10.5	1125
month 2	7.0	700	7.6	770	8.9	910
month 3	7.0	700	7.6	770	8.9	910
Growth[3]						
6 months	4.2 (0.70)	500	4.5 (0.75)	530	5.1 (0.85)	600
8–12 months	5.1 (0.70)	560	5.5 (0.75)	590	6.2 (0.85)	660
20–24 months	6.3 (0.40)	380	6.8 (0.45)	420	7.8 (0.55)	480
32–36 months	5.9 (0.15)	300	6.5 (0.20)	330	7.6 (0.30)	390
Stallion[4]						
resting	5.8	400	6.1	420	6.3	440
breeding	6.6–8.0	480–620	7.0–8.4	490–630	7.0–8.5	520–650

[1] 2 h daily.
[2] 1 h daily.
[3] Based on equations: daily UFC/kg $BW^{0.75}$ = a + $bG^{1.4}$, and daily MADC (g) = a $BW^{0.75}$ + bG, where a = coefficient of maintenance, b = coefficient of gain, and G = body weight gain (kg/day) (see Table 6.15).
[4] Including 0.5 h medium exercise daily.
[5] These values agree with Martin-Rosset *et al.* (1994) resulting from a deposition of 5 g/100 g BW and 0.5–0.55 metabolic efficiency for protein in products of conception.

Assuming 1 kg gain at 250 kg BW, requirement = 5.5 UFC
(maintenance 3.6 + gain 1.9)
Assuming 1 kg gain at 350 kg BW, requirement = 7.0 UFC
(maintenance 4.7 + gain 2.3)
Assuming 0.5 kg gain at 300 kg BW, requirement = 5.0 UFC
(maintenance 4.2 + gain 0.8)

Economy may be achieved by allowing the slower growth rate in winter on mixed feeds, followed by accelerated compensatory growth on pasture in the spring (see Chapter 8).

The protein requirements for growth are similarly calculated.

INRA energy requirements for growth: protein for growing foals

The rate of body weight gain, as a fraction of mature weight, declines progressively with increasing age. Concurrently, the protein content of that gain decreases and the

Table 6.15 INRA coefficients used in the estimates of requirements for growth of light breeds (daily UFC/kg BW$^{0.75}$ = a + bG$^{1.4}$, and daily MADC (g) = a BW$^{0.75}$ + bG).

Age (months)	UFC		MADC	
	a	b	a	b
6–12	0.0602	0.0183	3.5	450
18–24	0.0594	0.0252	2.8	270
30–36	0.0594	0.0252	2.8	270

a, Coefficient of maintenance requirement, b, coefficient of gain; G, average weight gain (kg/day).

Table 6.16 Digestible CP recommendations (DCP, g/day)* for growing foals at moderate and fast growth rates (after Meyer 1983b).

Age (months)	Weight of mature animal (kg) and growth rates									
	200		500		500		800		800	
			Moderate		Fast		Moderate		Fast	
	Average BW (kg)	DCP (g)	Average BW (kg)	DCP (g)	Average BW (kg)	DCP (g)	Average BW (kg)	DCP (g)	Average BW (kg)	DCP (g)
3–6	85	290	170	503	175	544	232	635	292	870
7–12	130	260	258	474	288	546	360	695	444	736
13–18	160	225	333	461	365	493	500	731	548	621
19–24	175	230	388	454	425	488	612	688	612	622

* Assumes requirements for maintenance of 4.5 g DCP/kg BW$^{0.75}$ at 3–6 months, decreasing to 4.1 g DCP/kg BW$^{0.75}$ at 19–24 months of age.

fat content increases as age increases. The protein content of each kilogram gained decreases from 197 g at 3–6 months to about 170 g at 2 years of age in horses of 500-kg mature weight. Breeds with lower mature weights grow faster, relative to their size, and reach near mature weights at a younger age than do larger breeds. With these considerations for growth in mind, the recommendations for DCP proposed by Meyer (1983b) are given in Table 6.16 (see Table 6.14 for INRA recommendations).

INRA protein requirements for maintenance of growing horses

The metabolic efficiency of DCP for growth is set at 0.45 and, as the amino acid turnover rate in growing horses is greater than in adults, so the maintenance requirement for MADC is higher during growth:

MADC maintenance requirement of growing horses
= 3.5 g/day/kg BW$^{0.75}$
(2.8 g MADC/day/kg BW$^{0.75}$ (19), for adults).

Table 6.17 Digestible CP requirements per kg BW$^{0.75}$ in working horses (Meyer 1983b).

	Light work	Moderate work	Heavy work
DM intake (g)	70	80	100
Faecal N losses (mg)	210	240	300
N losses in sweat (mg)	12	30	50–145
DCP requirement (g)	3.7	4.1	5.3
DE requirement (MJ)	0.68	0.83	1.0
DCP/MJ DE (g/MJ)	5.4	4.9	5.3

INRA lysine requirements for growing horses

Lysine is the limiting amino acid for growth in most conventional diets. The lysine requirement for growing horses is set at 6 g/kg dietary DM at 6 months, and 4 g/kg dietary DM at the yearling stage.

Work

INRA protein requirements related to energy requirement

(See also Chapter 9, where requirements are considered independently of N balance.)

For work the INRA recommendations are that protein needs are less than proportional to energy needs, whereas the NRC (1989) assumes that protein needs are directly in proportion to energy needs for maintenance and all amounts of work. Nevertheless, for simplicity INRA recommends that the work of an adult horse above maintenance requires 60–65 g MADC/UFC, i.e. protein needs proportional to those of energy (Table 6.14), but growing horses in work would require more protein.

Protein is not wasted when chemical energy in muscle is transformed into kinetic energy. Energy for work is derived mainly from glycogen and free fatty acids. However, in extended hard work blood urea concentration rises, as a consequence of protein catabolism. Therefore protein requirements (Table 6.17) are generally considered to rise above maintenance needs to achieve N balance in working horses, accommodating:

- muscle protein anabolism;
- N losses in sweat of 1.4 g N/l (although urinary losses decrease);
- increased feed intake that increases endogenous faecal N; and
- some muscle protein catabolism in hard work (adenine nucleotide cycling also increases in exercise, increasing uric acid excretion; see Chapter 9).

Thus, from Table 6.17 the protein requirements for work are taken to be 5.5 g DCP/MJ DE, in Eq. 14 for maintenance.

PROTEIN, MINERAL AND MICRONUTRIENT FEED VALUES AS DETERMINED BY THE INRA SYSTEM

Protein and evaluation of N absorbed from the large intestine

The protein value of feeds is expressed as g DCP/kg DM, that is, dietary CP multiplied by the apparent digestibility coefficient. A proportion of the CP is converted to bacterial protein and the horse benefits only from amino acids that are absorbed, but of both bacterial and dietary origins. The proportion of DCP N that is amino acid-N is less in green forages, and especially in silages, cf. soya and other concentrate proteins. This observation is reflected in the K coefficient by which DCP of forages is multiplied to give MADC (corrected digestible N matter), that is:

K = 0.90 for green forages
K = 0.85 for hays and dehydrated forages
K = 0.80 for straws and chaffs
K = 0.70 for good grass silages
in comparison with
K = 1.00 for soya, cereals, etc.

$$DCP \times K = MADC \ (g/kg \ DM) \qquad (20)$$

The absorption of amino acids takes place almost entirely from the small intestine. The absorption from the large intestine of amino acids released from forage and bacterial proteins may be of significance for horses receiving poor-quality forage diets, adequate in fermentable energy and supplemented with NPN. However, INRA proposals (Tisserand & Martin-Rosset, 1996) assert that amino acids synthesized within the large intestine do not materially contribute to the amino acid supply of the horse. Thus amino acid absorption from the hind gut is not likely to be a significant factor for most diets. These conclusions are consonant with evidence from other sources referred to in Chapters 1 and 2.

The resolution of the issue concerning the absorption of amino acids from the large intestine is of practical significance. Bacterial amino acid synthesis in the hind gut is considerable and 50–60% of faecal N in the horse is accounted for by microbial protein (Tisserand & Martin-Rosset, 1996). Cuddeford and colleagues (1992) reported that the apparent crude protein digestibility coefficient for high temperature-dried lucerne was 0.74 in comparison with a value of 0.36 for field-cured timothy hay, cut at full-bloom (lucerne had twice the N-digestibility of timothy). Gibbs *et al.* (1988) showed that the precaecal N-digestibility of high protein lucerne hay was three-fold that of coastal Bermuda grass (in contrast to the data with sun-cured lucerne of Klendshoj *et al.*(1979) in Table 6.18), and the lucerne led to a greater N-retention. However, the net uptakes of N from the large intestine did not differ and the overall N-digestibility of the lucerne was only 29% greater than that for the Bermuda grass hay, and according to Schmidt *et al.* (1982) an

Table 6.18 True N digestibility (N g/kg N intake) in the digestive tract of the pony of mixtures of two ingredients providing equal amounts of CP and of hays (Klendshoj *et al.* 1979) in Martin-Rosset *et al.* 1994.

	CP intake (g/kg)	Small intestine	Large intestine
Bermuda grass hay + crimped oats	122	690	130
Bermuda grass hay + micronized oats	122	580	320
Bermuda grass hay + soyabean meal	122	730	130
Bermuda grass hay	117	360	370
Lucerne hay	150	220	570
Lucerne hay	181	380	460

Table 6.19 CP (g) true digestibility coefficients of absorbable intestinal amino acids and MADC (g) (Jarridge & Tisserand 1984).

	Intake (g)		Small intestine	Large intestine		Digestion, total tract (g)	
	CP	Amino acids	True digestibility[1]	Entry[1] (g)	True digestibility[1]	DCP	MADC
Concentrates	180	171	0.85	26	0.90	148	147
Spring grass	180	162	0.70	49	0.80	128	117
Barley–maize	110	105	0.85	16	0.90	90	90
Grass hay	110	99	0.50	49	0.75	65	54[2]
Grass silage	110	82	0.50	41	0.75	65	44[2]

[1] Of amino acids.

[2] See K values in 'Protein and the evaluation of N absorbed from the large intestine', for grass hay 65 × 0.85 = 54, for silage 65 × 0.70 = 44, thus allowing for the NPN content of the large intestinal true digestibility of 0.75.

amount of N equivalent to 25–30% of dietary intake flows from the ileum into the caecum, irrespective of feeding level or type. The important point is that if most of the N that is absorbed from the large intestine is ultimately lost in the urine then the true differences in protein value between the lucerne and timothy, found by Cuddeford and colleagues, was much greater than two-fold. Some evidence indicates that a maximum of 10–12% of the plasma pool of amino acids is of microbial origin and derived from the hindgut. Thus, the range in coefficients for N retention is much wider than that of true digestibility. Stated differently, where the true digestibility coefficient is low, the potential retention coefficient is very much lower, whereas with high true digestibility coefficients there is a smaller decrement in the potential retention coefficient (note: the actual retention depends on the tissue needs).

Note that in Table 6.19 the true absorption coefficient of amino acids for concentrates and for barley–maize was the same, i.e. 0.85 (147 ÷ 171 vs. 90 ÷ 105); however, the amino acid balance of the cereals is likely to have been poorer, so in that case the potential N retention for growth would be lower for the cereals.

Calcium

(See also Chapter 3.) The bioavailability, or true digestibility, of dietary Ca varies considerably. The principal factors controlling this are:

- amount of dietary Ca (0.7 at requirement intake to 0.46 at several times requirement);
- amount of dietary P (10g P added/kg diet containing 4g Ca/kg reduced true digestibility from 0.68 to 0.43);
- vitamin D status (of less significance in the horse than in some other domestic species);
- dietary oxalate and phytates (Ca:oxalate <0.5 causes NSHP; phytates bind Ca. Implicated oxalate-rich grasses include: napier, guineagrass, buffel, pangola, green panic, paragrass, kikuyu, setaria and probably some species of millet grass. Lucerne contains oxalic acid, but has a high Ca availability).
- age of animal.

The major part of dietary Ca, given to growing horses, is directed to bone calcification. Work by Lawrence *et al.*(1994) indicated that although maximum bone mineralization was reached by 6 years of age, 76% of maximum was achieved by 1 year. As bone growth is most rapid in the first year of life the demand for dietary Ca is very pronounced during that time. There is some speculation that horses predisposed to exercise-associated myopathies (see Chapter 11) have a temporary failure in their ability to control intracellular Ca concentrations. However, there is no evidence that supplementary dietary Ca, above the dietary requirement, has a prophylactic influence on the risk of myopathies.

Phosphorus

The bioavailability of plant P varies with the proportion present as phytates. Phytate-P is digested by phytase (EC 3.1.3.8) present in the gut principally, or entirely, as microbial phytase. The digestibility varies between 0.25 and 0.35. The addition of yeast cultures to the diet has been shown to increase phytate-P use, presumably by stimulating microbial activity in the hindgut. Excess dietary P interferes with Ca utilization. Savage (1991) and Savage *et al.* (1993b) found that diets containing four times the NRC (1989) estimated requirement for P in growing foals, but the requirement for Ca, caused lesions of dyschondroplasia (OCD), without signs of NSHP (see Chapter 8). Cortical bone porosity was increased in the growing foals by the high P diet and the extent of osteoid-covered surfaces of cancellous bone decreased with time, despite a depression in growth rate. On the other hand, a diet containing over three times the NRC (1989) estimated requirement for Ca had no adverse effect.

Sodium, potassium, magnesium, trace elements and vitamins

Table 6.20 gives the requirements for vitamins, trace elements, sodium, potassium and magnesium, supplements of which should be unnecessary if a commercially prepared mixed feed is used at recommended rates. Where compounded feeds are not given, a proprietary mixture of trace elements and vitamins should be used, because in high concentrations such nutrients are toxic and the normal horse owner is unlikely to have facilities for their proper handling and weighing. Sometimes compounded feeds intended to form the entire concentrate portion of the ration are used as supplements to oats, so diluting their effects in so far as protein, minerals, vitamins and trace elements are concerned. Thus, a pellet providing 4000 iu vitamin A/kg, mixed 50:50 with cereals and in turn fed 50:50 with hard hay will provide approximately 1000 iu vitamin A/kg total diet. Dilution of this kind is frequently a cause of incorrect Ca:P ratios in rations.

Electrolytes

A discussion of the electrolytes – sodium, potassium and chloride – is given in Chapter 9, where problems of hard training are tackled. The potassium and magnesium needs of normal activity should be automatically met where good-quality roughage is available. The sodium needs (Table 6.20) can be met by providing NaCl (common salt), for simplicity ignoring the contribution made by natural ingredients. Sodium comprises 40% of NaCl so that the NaCl allowance should be two-and-a-half times the sodium allowance, that is, 8.75 g salt/kg provides 3.5 g sodium. In order that any excessive salt intake is satisfactorily counteracted, clean drinking water, free from contamination with salts, should always be available (Table 6.21), excepting the restrictions to its use discussed in Chapter 4.

Trace elements

The dietary allowances for trace elements, as distinct from those for the major minerals and electrolytes, do not change appreciably per unit of feed with a change in the function of the animal from growth to work or to various phases of reproduction. The secretion of iodide in mare's milk may slightly increase her basal needs for iodine, but the margin given in Table 6.20 should satisfy all needs. The difference between this allowance and the toxic level (the margin of safety) existing for selenium and iodine (the elements for which the difference in absolute quantities is probably least) is adequate in the hands of responsible individuals. However, it is unwise and potentially dangerous for the normal horse owner to handle pure forms of trace elements.

Vitamins

The dietary allowances for the fat-soluble vitamins A, D and E per unit of concentrate feed (Table 6.20) again are not varied much per unit with a change in the

Table 6.20 Minerals and vitamins per kg diet adequate for horses (based on NRC 1978).

	Adequate levels	
	Maintenance of mature horses	Mare, last 90 days of gestation, and lactating and growing horses
Sodium (g)	3.5	3.5
Potassium (g)	4.0	5.0
Magnesium (g)	0.9	1.0
Sulphur (g)	1.5	1.5
Iron (mg)	40	50
Zinc (mg)	60	80
Manganese (mg)	40	40
Copper (mg)	15	30
Iodine (mg)	0.1	0.2
Cobalt (mg)	0.1	0.1
Selenium (mg)	0.2	0.2
Cholecalciferol (vitamin D) (μg)[1]	10 (400 iu)	10 (400 iu)
Retinol (vitamin A) (mg)[2]	1.5 (5000 iu)	2.0 (6666 iu)
D-α-tocopherol (vitamin E) (mg)[3]	30	30
Thiamin (mg)	3.0	3.0
Riboflavin (mg)	2.2	2.2
Pantothenic acid (mg)	12	12
Available biotin (mg)	0.2	0.2
Folic acid (mg)	1.0	1.0

[1] 1 iu is equal to the biopotency of 0.025 μg cholecalciferol (vitamin D_3) or ergocalciferol (vitamin D_2).
[2] 1 iu is equal to the biopotency of 0.3 μg retinol (vitamin A alcohol). Grass carotene has 0.025 of value of vitamin A on a weight basis.
[3] 1 iu vitamin E is the biopotency of 1 mg DL-α-tocopheryl acetate. Where 50 g supplementary fat of average composition is added per kg feed, the requirement rises to 45–50 mg α-tocopherol/kg, equivalent to 79–88 mg DL-α-tocopheryl acetate. Also see p 299 for working horses.

Table 6.21 Characteristics of a good water source.

	mg/l
Ammonia (albuminoid)	<1.0
Permanganate value (15 min)	<2.0
Nitrite, N	<1.5
Nitrate, N	<1.0
Calcium	50–170
Lead	<0.05
Cadmium	<0.05
pH 6.8–7.8	
Total dissolved solids	<1000

function of the individual, bearing in mind also the capacity of the horse to store them. A reappraisal of the situation would be necessary if there was any radical change in the basic raw materials traditionally used for feeding horses, for example from dry cereal grains to root vegetables, or to high-moisture conserved cereals, the elimination of conserved pasture, and so on. An example of this is given for vitamin E and breeding mares (Chapter 4). Appendix B gives examples of dietary errors encountered by the author in practice. Such errors cover the whole range of dietary attributes.

SIMPLE RATION FORMULATION

The principal chemical components of a stabled horse's diet are depicted in Fig. 6.11. The amounts of protein, calcium and phosphorus needed per gram of ration are given in Table 6.6 and the levels in the concentrate – hay mixture can be calculated by multiplying their protein contents (g/kg), etc. (see Appendix C) by their proportions and summing. Thus, for hard hay and oats:

$$55\,g \times 0.6 + 96\,g \times 0.4 = 71.4\,g \text{ protein/kg mixture}$$

If the protein requirement is 100 g/kg, then the discrepancy of 28.6 g can be made good by substituting some soyabean meal (other suitable protein sources may be substituted in inverse proportion to their lysine contents) for some of the oats, but attributing the soya (or other protein) with a protein content that is the difference between its content and that of oats, that is:

$$440\,g - 96\,g = 345\,g \text{ protein (see Appendix C)}$$

Then the proportion of soya to include and of oats to remove per kilogram is:

$$28.6/345.0 = 0.083, \text{ i.e. or } 8.3\% \text{ or } 83\,g/kg \text{ mix}$$

For simplicity, the effects of this substitution on the energy, calcium and phosphorus contents of the mix can be ignored. A similar calculation using Appendix C, and detailed in Appendix A, will have been adopted for calcium and phosphorus, but

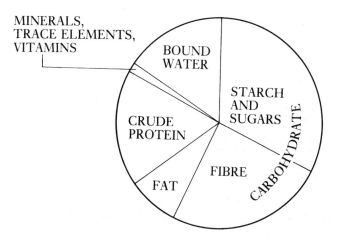

Fig. 6.11 Principal components by weight of a stabled horse's diet (excluding drinking water).

the deficits can be made good by adding limestone flour and/or dicalcium phosphate without making any substitution as for soya/oats. A calcium source of this type will normally be essential unless large amounts of lucerne or other legume forage are provided.

Comparison of the DE and NE systems for ration formulation

Exact comparisons between the DE and NE systems are not possible from tabulated data, for the reason that a definitive description is not provided for the feeds selected and so an assumed identity of feed composition between systems may not be assured. Moreover, the assumption made by the author that hay and barley had the same effect on capacity for feed is not justified, but was necessary for a simple comparison to be made in Table 6.22. Several general conclusions are reached.

- The expected selection against poor hay by the NE system did not materialize.
- The NRC (1989) DE system generally seems to assume higher requirements of both energy and protein for the functions examined, and this counterbalances a relatively higher value of the usable energy in hay assigned by DE.
- A change in the assumed feed intake of exercising horses had a large effect on the composition of the mixture selected by both systems.

The NE system should ultimately lead to economies in the selection of roughages for horses, as analytical work enlarges the fund of data on raw materials. These data should predict the extent of precaecal digestibility and the energy yield during microbial digestion. The relative complexity of the NE system indicates that adjustments for a variety of husbandry practices may be accommodated. However, direct comparison of the two systems under several practical environments is required to determine their comparative effects on work, lactation, pregnancy, growth, health etc.

The ME of a feed is its DE less the associated energy lost in urine and that lost from the intestines as combustible gases (principally hydrogen and methane). On average somewhat less than 20% of the energy apparently digested is lost to the horse by excretion in equine urine and by expulsion as methane. Vermorel *et al.* (1997) estimate that with normal diets horses lose about 4.2% of the GE of diets (8.0% of DE) in urine, but obviously greater amounts are lost with high protein diets metabolized as energy sources. Methane energy loss, owing to the activity of the cellulolytic and methanogenic bacterium *Archaea methanogenesis*, amounts to 1.95 SE 0.45% of GE in horses (Vermorel *et al.* 1997) and it may be assumed that a loss of energy amounting to about 30–65% of that lost as methane is lost as heat of fermentation (Webster *et al.* 1975). These losses differ amongst feeds and losses during intermediary metabolism in tissues differ as between the nutrients, glucose, VFA etc. The ratio ME:DE is approximately 0.85 for grass hays and 0.90 for mixed diets in horses (Vermorel *et al.* 1997) and km % values vary from about 60 to 80 amongst feeds in the horse (Vermorel & Martin-Rosset 1997). Thus the efficiency of DE use for maintenance varies over a range of approximately 50% to 70% in the horse. The practical value of a NE system to replace a DE system depends substantially on this variation being reliably estimated amongst feeds that are available at economic prices.

Computers and diet formulation

(This section has been written with the assistance of John C. Dickins.)

During the past 30 years, computers, with an array of programmes of increasing power, have been used in the formulation of commercial feed mixtures. Pro-

Table 6.22 Application of DE (NRC) and NE (INRA) systems (kg/day) to the formulation of simple daily rations based on moderately poor grass hay, barley grain and extracted soyabean meal. Mean requirement values of a horse, 500 kg at maturity, for energy and protein only are used with feed intakes (kg DM/day; NRC 1978). Some of the outcome values are impractical. However, the purpose of the table is to compare outcomes for similar functions. See Table 6.23 for assumed analytical values of feeds used.

Growth (intake/day, kg)		DE system	NE system	Exercise (intake/day, kg)		DE system	NE system
Foal							
weaning (4.2)	Hay	0.33	—	Light (12)	Hay	11.60	9.36
	Barley	3.18	—		Barley	0.40	2.64
	Soya	0.69	—		Soya	—	—
Foal							
6 months (5)	Hay	0.93	1.05	Moderate (12)	Hay	9.48	8.03
	Barley	3.35	3.59		Barley	2.52	3.97
	Soya	0.72	0.36		Soya	—	—
12 months (6)	Hay	0.71	1.92	Protracted (12)	Hay	5.25	8.96
	Barley	4.61	3.68		Barley	6.39	3.04
	Soya	0.68	0.40		Soya	0.36	—
18 months (6.5)	Hay	0.07	0.99	Light (10)	Hay	7.90	6.27
	Barley	5.58	5.51		Barley	2.10	3.73
	Soya	0.85	—		Soya	—	—
24 months (6.6)	Hay	0.56	1.14	Moderate (10)	Hay	5.79	4.93
	Barley	5.40	5.46		Barley	4.21	5.07
	Soya	0.64	—		Soya	—	—
Mare							
maintenance	Hay	5.30	5.92	Protracted (10)	Hay	1.55	5.87
(7.45)	Barley	2.15	1.53		Barley	7.79	4.13
	Soya	—	—		Soya	0.66	—
gestation	Hay	4.04	5.02				
last 90 days	Barley	3.01	2.16				
(7.35)	Soya	0.31	0.17				
lactation	Hay	4.06	4.85				
0–3 months	Barley	4.93	4.80				
(10.1)	Soya	1.11	0.45				
3–6 months	Hay	4.74	4.30				
(9.35)	Barley	4.22	4.66				
	Soya	0.39	0.39				
Stallion (10)	Hay	7.91	5.20				
	Barley	2.09	4.80				
	Soya	—	—				

grammes of lesser complexity are also available for use on the home personal computer. For this development to be truly successful it is requisite that several needs are met. These are that:

(1) the nutrient requirements of the animal are reasonably well established;
(2) the nutrient content and cost of a wide range of available feed materials are known and that the nutrient contents apply to the batch of each material available to the current user;
(3) the needs for, and nutrient and nonnutrient characteristics of, roughages to be used are accommodated in any formulation. [By subtracting the nutrient content of the forage portion of the ration from the total requirement (per kilogram of total feed) the remainder gives the required nutrient content of the concentrate portion. This procedure makes assumptions concerning the appetite of the horse (kg DM/day) discussed earlier (see 'Relationship of capacity for feed to bodyweight' and 'Energy needs') and below ('Appetite'), this chapter;
(4) physical, nonnutrient chemical and microbiological characteristics of all feeds concerned are known; and
(5) the effects of the characteristics in (3) and (4) on each class of horse are understood and allowed for during formulation.

Where this information is comprehensive it is possible to generate the formulae of mixtures that should be safe, nutritionally adequate, acceptable to the average horse and economical to use.

Principles of least-cost feed formulation

The number of chemical characteristics of each feed material that is accommodated in a formulation programme depends on its power. The simplest takes into account protein (g/kg) and energy (MJ DE or MJ NE) and the more comprehensive include ranges of essential amino acids and fatty acid residues, minerals, trace elements, vitamins, oil, fibre (crude, acid detergent, dietary, etc.) and ash. Certain of these characteristics provide limited nutritional information, but their declaration may be a statutory requirement on product labels within the EU (see Chapter 5). The computational technique is typically known as 'linear programming', as it is assumed that each additional amount of a nutrient contributed by one ingredient, substituting for another ingredient's nutrient, causes a similar quantitative response. More complex, nonlinear models of feed formulation (e.g. integer and quadratic programming) are also accommodated in some computer packages allowing, for example, for the calculation of a formula in whole units of the ingredient. The term 'least-cost' is used as the computation calculates a mix at minimum cost, consistent with formulation nutrient levels that are not less than the minimum requirements plus safety margins, that is, minimum nutrient constraints. The safety margins allow for variation in requirements of individual animals and deviations in ingredient composition from those assumed. Maximum and even minimum formulation levels for many ingredients are also set as ingredient constraints for reasons that are listed below.

The requirements of the formulation, expressed in terms of nutrient and ingredient constraints, take the form of a set of linear inequalities. For example, if x_1, $x_2 \ldots x_n$ represent the amounts (g/100 g) of different ingredients in the mix with protein contents $a_1, a_2, \ldots a_n$ (g/100 g), then the inequality:

$$a_1x_1 + a_2x_2 + \ldots a_nx_n \geq b \tag{21}$$

expresses the requirement that the protein content of the feed should not be less than b. In addition to such constraints, a similar linear combination of x_i, with cost coefficients in place of the analytical values, represents the total cost to be minimized. The solution provides a set of x_i (the formulation) which satisfies the constraints at minimum cost.

Feed manufacturers maintain, and regularly update, a database of analytical and cost information on a wide range of potential ingredients together with a set of nutrient and ingredient maxima and minima for each type of feed. Least-cost formulations are then produced automatically from this information periodically, as required. The computation also produces other values that give guidance on the sensitivity of the formula to small changes in the constraints, or costs.

Graphic example of linear programming

(An alternative simplified form of ration formulation is given in Appendix A.) The linear programming method of computing a least-cost formulation can be illustrated graphically for three ingredients, as shown in Fig. 6.12, whereas the computer may handle 30 or more potential ingredients. The two axes at right angles represent the percentages of the first two ingredients (x_1 and x_2) in the diet. The percentage of the third ingredient (x_3) in the diet is then determined, since all three inclusion rates must sum to 100.

Each point within triangle (a) ABC represents a particular formulation since the percentages of ingredients x_1 and x_2 determine the percentage of ingredient x_3. Negative inclusion rates are impossible, so admissible diets are restricted to the region defined by the triangle ABC (the hatched area), where the line AB represents mixtures of 100% ingredient x_1 and 0% ingredient x_2 to 100% of x_2 and 0% of x_1, excluding ingredient x_3. Point C represents 100% of ingredient x_3 and at point D the mix would contain 50% each of ingredients x_1 and x_2 and 0% of ingredient x_3. As indicated, a mix at point M_1 would contain 40% of ingredient x_3 and 50 and 10%, respectively, of each of the other two ingredients. Several of these mixes would be completely unacceptable; thus, ingredient and nutrient constraints are imposed. Reasons for such constraints include the following:

• A particular ingredient may contain an essential nutrient that is not accommodated in the programme and so the ingredient's inclusion at a predetermined minimum level may be necessary.

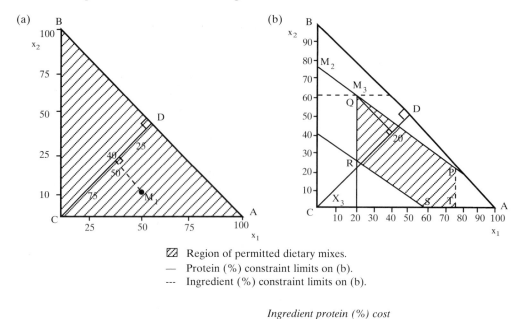

(a) and (b) diagrams

☒ Region of permitted dietary mixes.
—   Protein (%) constraint limits on (b).
--- Ingredient (%) constraint limits on (b).

Ingredient protein (%) cost

	Protein %	£/1000 kg
Bran, x_1	15	150
Barley, x_2	10	125
Beans, x_3	25	175

M_1: $x_1 = 50\%$
$x_2 = 10\%$
$x_3 = 40\%$

Ingredient (%) constraint limits

	Minimum	Maximum
Bran, x_1	20	75
Barley, x_2	—	60
Beans, x_3	—	—

Protein (%) of mix

	Minimum	Maximum
Mix	14	19

M_2: $x_1 = 0\%$ M_3: $x_1 = 20\%$
$x_2 = 73\%$ $x_2 = 60\%$
$x_3 = 27\%$ $x_3 = 20\%$

Direction of decreasing cost of mix. In this example, with beans costing £175/1000 kg, slope = −0.5.

Fig. 6.12 Derivation of a hypothetical least-cost mix of three ingredients with ingredient and protein constraints.

- Excessive amounts of some ingredients, e.g. molasses, may reduce horse acceptance of the product, or may cause a digestive upset. This upset can result from unacceptable concentrations of antinutritive substances, such as antiproteases and glucosinolates (Chapter 5), or from excessive concentrations of some nutrients.
- Manufacturing rate and quality can be facilitated, for example, by the inclusion

of minimum amounts of fat and pellet binders or exclusion of large amounts of abrasive ingredients.

- The provision of moderate amounts of each of several ingredients, rather than large quantities of one ingredient, is normally to be preferred. This policy should reduce the risk of problems arising when there are deviations in composition and quality of ingredients compared with the values attributed to the ingredient(s) in the programme.

A linear analytical, or ingredient, constraint on the diet is represented by a straight line in the graph (b) (Fig. 6.12) such that diets satisfying the constraint must lie on one side of the line. In this diagram PQ and RS are constraints restricting the protein in the diet to the range 14–19%, and TP and QR are constraints limiting the inclusion of bran to the range 20–75%. Feasible diets must therefore lie within the region PQRST. The restriction of barley to a maximum of 60% does not affect the outcome in this example.

Diets of equal cost lie along parallel lines. In Fig. 6.12 the costs assumed give a line with a negative slope of 0.5. By moving the line in the direction of the arrows, so reducing cost, we find the diet of least cost. This will lie on the boundary of the feasible region, normally at a vertex. If no ingredient constraints had been applied, the mix M_2, containing 73% of ingredient x_2 and 27% of ingredient x_3, would have been obtained. As a result of applying the four constraints the least-cost mix is that of M_3 at point Q, with the proportions 20, 60 and 20%, respectively, of ingredients x_1, x_2 and x_3.

Figure 6.12 also shows that, in this example, the critical constraints determining the position of the optimum are the lines PQ (minimum protein requirement) and RQ (minimum bran requirement), whereas the outcome would be unaffected by changing the position of the bran and protein maxima. Increasing the cost of beans to £250/1000 kg would reduce the slope of the cost line (make it more negative) to −0.8, and so change the optimum from M_3 to P. As the cost of beans increases past £200 the least-cost composition abruptly shifts from that at M_3 to that at point P. In practice, any such large change would be introduced in several small discrete steps, at intervals of time, to avoid possible implications regarding the health of the horse. The professional formulator will know the acceptable maximum size of each step for particular ingredient substitutions.

In conclusion, the practice of linear programming has improved the reliablility

Table 6.23 Assumed feed analytical values used to compare DE and NE systems in Table 6.22.

	DE (MJ/kg DM)	CP (g/kg)	UFC (per kg)	MADC (g/kg)
Hay	6.88	82	0.41	40
Barley	14.98	110	1.16	92
Soyabean meal	14.73	499	1.06	437

Table 6.24 Procedure for calculating ingredient make-up of a simple ration using NRC and INRA systems.

Calculation	NRC DE system	INRA NE system
(1) Assess daily feed capacity of horse	kg/day, Fig. 6.4	kg/day, Fig. 6.4
(2) Assess daily energy requirement of horse	DE/day, Table 6.5	UFC/day, Table 6.14
(3) Divide (2) by (1)	DE/kg	UFC/kg
(4) Calculate proportions of roughage: cereal	Appendices A and C	Appendices A and C
(5) Assess protein requirements per kg feed or per day	Table 6.6, CP	Table 6.14, MADC*
(6) Assess Ca and P needs per kg feed (90% DM)	Table 6.6	Table 6.6
(7) Calculate protein, Ca and P of mix	Appendices A and C	Appendices A and C
(8) Calculate soya, Ca and P additions necessary	Appendix A, Table 6.6	Appendix A, Tables 6.6 and 3.2
(9) Estimate other mineral and vitamin needs and add	Table 6.20	Table 6.20

*For relationship of DCP to MADC see Eq. 20, and to obtain MADC/kg feed divide daily MADC requirement by feed capacity per day (note: DM and 90% of DM feed estimates).

and cost-effectiveness of commercial diets. It allows formulations to reflect economically the most recent established scientific evidence with minimum risk.

Procedure for calculating diets by the DE and NE systems

The sequence of calculations necessary in calculating a simple diet using the NRC DE and INRA NE systems is given in Table 6.24.

APPETITE

There is much yet to be learnt about factors that influence the appetite of horses, but reluctance or eagerness to eat can be assessed either as the amount of dry feed consumed consistently per 24 h, or the amounts consumed in single meals. The ingesta derived from very coarse, poorly digested, long-fibrous feeds, if present in significant amounts, will be retained longer in the large intestine and depress the daily intake of DM. Although this is generally a disadvantage, it can occasionally help to contain the appetite of overfat animals. The proclivity of a horse to start eating energy-yielding feeds also depends upon the relative absence of products of digestion, including glucose, in the small intestine. To a lesser extent, concentrations of blood glucose may play a small part.

Under conditions of natural grazing, individuals will feed during perhaps 15 or 20 periods throughout the 24 h. A series of small meals reflects not only the low capacity of the stomach, but also probably more directly the switching-off mechanism of digestion products in the small intestine. High caecal concentrations of VFAs, especially of propionate, have an immediate but small depressing effect on appetite by extending the interval between meals in ponies fed *ad libitum* and by reducing meal size at the time, yet they have no sustained effect over 24h. Lower increases in caecal VFA may even stimulate appetite. Although many factors influence the capacity for feed, an estimate of the daily appetite of average healthy animals for leafy hay and oats can nevertheless be given (see Fig. 6.4). In the stable, horses are fed generally by the bowl, and overfeeding, or apparent loss of appetite, frequently reflects lack of recognition on the part of the groom of differences in the bulkiness and energy density of feeds (see Table 5.3). For example, taking both these characteristics into account, many of the better coarse feeds currently available provide 20% more energy per bowl than is provided by a similar volume of crushed oats. Failure on the part of the horse to eat up may then simply mean that it has been given 20% more energy in its hard feed than it is familiar with, or requires.

Feed in an unacceptable physical condition, and more importantly with an unfamiliar or unattractive aroma, will also be a deterrent to the initiation of feeding. Stale feed, which has incurred the oxidation of many of its less-stable organic components, is unlikely to encourage a horse to eat. Occasionally spices are used to cover, or camouflage, inadequacies of general acceptability, but they are generally to be discouraged as horses can become 'hooked' on these additives.

There is often glib talk about the appetite of horses for particular feeds, but apparently they have a true appetite only for water, salt and sources of energy, and if given the choice would not select a balanced diet. Hence, it is necessary to induce horses to consume mixtures most appropriate to their needs. Thirst will sometimes drive a horse to consume poisonous succulent plants, as a source of moisture, in an otherwise arid landscape. The appetites for water and for salt are interrelated. Thirst depends to a considerable extent on dehydration, or increased osmotic presssure of body fluids. When blood is hypertonic, horses will normally drink; if much salt has been lost through sweating the body may be dehydrated, but the fluids will be hypotonic and thirst is then engendered by giving salt. The appetite for salt varies among horses – deprivation causes a greater craving in some than in others, despite a more uniform essentiality. On the other hand, an extreme thirst probably takes precedence over appetites for both salt and energy.

RATE OF FEEDING

A too-rapid consumption of cereals and concentrates by stabled horses is sometimes a problem needing attention. Although the consumption of long hay is a relatively slow process, cereals and other concentrates are normally eaten first at a particular meal (there is a case to induce the consumption of some roughage first; see Chapter 2), but the form in which the concentrate is given can have some impact on the rate at which it is consumed. German workers (Meyer *et al.* 1975b) showed that 1 kg feed in the form of oat grain or a pelleted concentrate took horses weighing between 450 and 550kg about 10min to consume, but meal took longer. Feeding time and the number of chewing movements were, however, increased by 3–100% if 10–20% of chopped roughage was added to the oats or concentrate. Finely ground meal mixes took longer to eat than crushed grains or pelleted feed; the addition of chopped roughage to fine mixes speeded up intake. Poor quality retarded consumption rate, suggesting a place for good-quality barley straw in occupying the time of greedy horses. Many horses develop vices, such as wood chewing, faeces-eating (coprophagy) or less frequently cribbing (forced-swallowing gulps of air) (see Plate 11.1). Boredom in isolated boxes is a contributory factor in vice initiation and the provision of long hay or good-quality straw does help to circumvent this, but does not eliminate their occurrence.

Sound teeth enable horses to grind roughage to a small particle size, but decaying teeth, or molars with sharp points, abrading the cheek and tongue (see Fig. 1.1), encourage bolting and poor mastication with the consequences of rapid intake, choking and sometimes colic. Dental treatment is clearly indicated, together with the provision of coarsely and freshly ground cereals and roughage chopped in short lengths. These may be given mixed together in a wet mash, which in turn may reduce wastage. The dampening of feed is, however, generally unnecessary except when large quantities of bran and beet pulp are given, in which case time for the thorough absorption of water should be allowed so that their true bulkiness is realized. Linseed also swells considerably on soaking and boiling; the latter, of course, is to be recommended, as described in Chapter 5. Where feeds are dusty, damping down lays the dust, so decreasing the likelihood of irritation to the respiratory tract in susceptible individuals. The damping of feed may also inhibit the segregation of its components. If feeds are too dry, minerals, for example, may sift to the bottom and be left unconsumed in the manger.

PROCESSING OF FEED

Preferred methods for processing raw cereal grains are coarse grinding, cracking, rolling, crimping (passing between corrugated rollers), steam flaking and micronizing. The overall objective of each is to improve digestibility and acceptability, but the last two may accelerate the onset of satiety during a meal. The processing of oats and barley is generally difficult to justify if the costs exceed 5% of the cost of the raw material, but it is necessary shortly before feeding for the small grains, milo and wheat. Steam flaking or micronizing tends also to extend shelf life, destroys most heat-labile antinutritive substances and increases the proportion of starch digested in the small intestine when intake rates are high. Normal cooking is, however, inadequate for the destruction of some mould toxins found in poorly harvested crops (see also Ch. 5).

FEEDING FREQUENCY AND PUNCTUALITY

Horses, like most other animals, are creatures of habit and their reactions are in part affected by an inner clock. Thus, the wise groom sticks to regular feeding times, week in and week out, and with equally regular exercise metabolic upsets and accidents can be avoided, and damage to stable doors through chewing and kicking can be decreased. Designs for feed mangers and water bowls recommended by the French Ministry of Agriculture are shown in Figs 6.13 and 6.14.

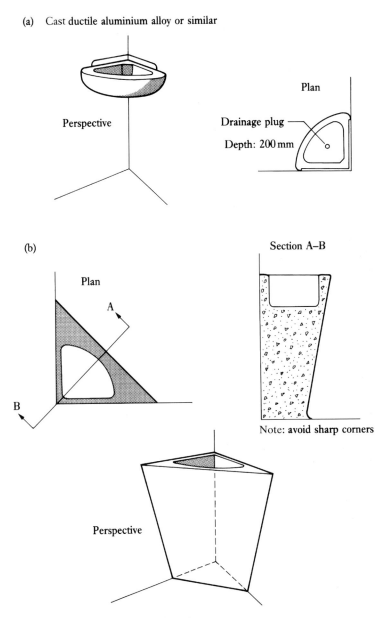

Fig. 6.13 Designs for feed mangers: (a) unbreakable manger; (b) manger embedded in concrete (Ministère de l'Agriculture 1980).

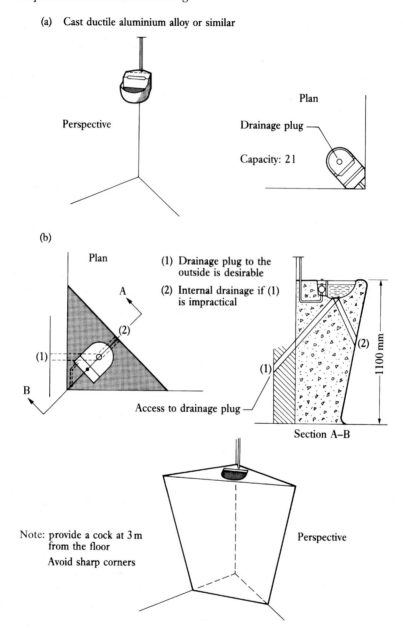

Fig. 6.14 Designs for water bowls: (a) unbreakable water bowl; (b) water bowl embedded in concrete (Ministère de l'Agriculture 1980).

CEREAL COSTS AND THEIR ENERGY CONTENTS

Cereals, and more particularly oats, of low bushel weight provide less DE per kilogram than do cereals of higher bushel weights. Digestive upsets are minimized, when changing from one feed or cereal type to another, by ensuring that the DE, or

NE, intake does not simultaneously rise. A bowl of an average sample of oats may provide only 56% of the DE of a similar bowl of average wheat (see Table 5.3). As a result of its bulkiness oats tend to be a safer feed than the other cereals although, through careful and expert feed management, no differences in this respect should arise. Therefore, in assessing the feeding value of the various cereals, account should be taken of the purchase price per unit of DE or NE (see Appendix C).

TRADITIONAL FEEDS

For some considerable time many stables in the UK have given their horses a diet based on grass hay, oats and a little cooked linseed. Before the advent of the tractor-drawn baler, hay was probably made for horses with greater care than it is today when the horse in many situations no longer reigns supreme in its claims to conserved forage and grazing privileges. In terms of nutrients, the diet now leaves a little to be desired, but the adult horse has some capacity to adapt to its culinary shortcomings, because it may take a considerable period of time for a large non productive animal to be depleted of essential nutrients. Many of these nutrients are made good where summer grazing is available; nutrients provided by productive pasture in summer will serve as a reservoir for 2 months of the ensuing winter, but not generally until the spring grass again refurbishes losses. Lactation of the mare and early growth of young stock rely traditionally almost entirely on the availability of grazing, any deficiencies of which are soon revealed.

A diet of oats, hay and linseed is alien to the natural inclinations of a browsing animal so that there is no good reason why other equally strange feeds should not be used if economies can be achieved and deficiencies rectified. There are now better protein sources than linseed and there are oils less prone to peroxidation than linseed oil as these other oils contain much lower concentrations of linolenic acid and contains added antioxidants (but see discussion of n-3 vs. n-6 fatty acids in 'Polyunsaturated acids', Chapter 5). Stabilized forms of groundnut, soya, sunflower and corn oils are widely available today. By understanding the necessity for protracted periods of adaptation and by recognizing the differences in bulk and energy density among the various cereals, the careful feeder has a wide choice of feeds. The principal advantage of oats is to the novice feeder, but this advantage could easily be supplanted (with a probable saving in costs per unit of energy) by the use of a variety of cereals and byproducts. However, the traditional feeder knows by long experience how many bowls of oats a performance horse will consume without refusal. Without careful weighing, the quantities of alternative mixtures required can be misjudged and overfeeding or underfeeding can result.

COMPLETE DIETS

Many horses and ponies receive diets based on compounded nuts and long hay, or haylage. In normal circumstances this hay should be preferred to a source of ground fibre if the quality is good and the price acceptable. Compounded feeds described as complete pellets are for consumption by adult horses required to carry out moderate work where no other feeds are to be given. These pellets have the advantage of providing a reasonably balanced diet relatively free from dust for horses subject to respiratory irritation and allergy. The absence of long fibre is, however, likely to cause a higher incidence of vices, such as wood chewing and coprophagy, but these annoyances are undoubtedly preferrable to the exacerbation of a respiratory problem. Rationing is simplified by the constant bulk and energy densities of the feed and storage space is minimized, but some problems of boredom require ingenuity to overcome.

SHELF LIFE OF FEEDS AND FEED CONTAMINANTS

The grinding of cereals increases their surface area exposed to the atmosphere and deterioration proceeds continuously, retarded only by the presence of natural antioxidants. Staleness detracts from feed acceptability. Legislation permits the addition of limited amounts of synthetic antioxidants to products for consumption by animals. The most commonly used of these are butylated hydroxyanisole (BHA), butylated hydroxytoluene (BHT) and ethoxyquin (EU legislation permits 15 chemicals as antioxidants to be included in animal feeds; Schedule 4, UK Feeding Stuffs Regulations 1995; Anon 1995c). These substances are safe, have the effect of extending shelf life and acceptability of feeds, and they cause no infringement of the Rules of Racing.

Some nutrients are particularly susceptible to destruction by light. Hence, concentrated feedstuffs should be covered during storage as the products of light destruction are also harmful to the stability of certain other nutrients. Atmospheric oxygen brings about the gradual destruction of fat-soluble vitamins and unsaturated fats. Among the more critical water-soluble vitamins, thiamin (vitamin B_1) and pantothenic acid are somewhat unstable. The loss of these nutrients and other unstable compounds detracts from the value of feedstuffs. However, whole grains in an uncrushed or unground form remain relatively stable for several years with only a slight decline in feeding value, although this, of course, assumes good storage, freedom from pests, moisture levels of less than 135 g/kg (the lower the better) and absence of cracked and moulded grains. Discoloration of cereal grains may indicate only superficial microbial damage during ripening as observed when one compares good-quality Scotch oats with their counterparts from Canadian or Australian sources. However, discoloration may be an indicator of more profound internal damage through fungal infection, which will seriously impair feeding value and stability in storage. Thus, bright and bold grains are to be preferred in the absence of other information. Further aspects of storage have already been discussed in Chapter 5.

Dressed seed corn or cereal grains that have been exposed to toxic pesticides should not be used for feeding to horses, although several pesticides leave quite harmless residues.

STUDY QUESTIONS

(1) What is the sequence of actions and decisions to be taken on introducing a rationing system and meeting the nutrient requirements of a stud and a riding stable?

(2) What are the merits and demerits of the two energy systems discussed: the NRC DE system and the INRA NE system?

(3) What should be the characteristics of a satisfactory feed store?

FURTHER READING

Martin, K.L., Hoffman, R.M., Kronfeld, D.S., Ley, W.B. and Warnick, L.D. (1996) Calcium decreases and parathyroid hormone increases in serum of periparturient mares. *Journal of Animal Science*, **74**, 834–9.

Martin-Rosset, W. (1990) Lalimentation des chevaux, techniques et pratiques. pp 1–232 [Ed. committee), INRA 75007 Paris.

Martin-Rosset, W., Vermorel, M., Doreau, M., Tisserand, J.L. and Andrieu, J. (1994) The French horse feed evaluation systems and recommended allowances for energy and protein. *Livestock Production Science*, **40**, 37–56.

Meyer, H. (1983) Protein metabolism and protein requirement in horses. *Fourth International Symposium on Protein Metabolism and Nutrition*, Clermont-Ferrand, France, 5–9 September 1983, No. 16. INRA, Paris.

National Research Council (1978) *Nutrient Requirements of Domestic Animals*. No. 6. *Nutrient Requirements of Horses*, 4th edn. revised. National Academy of Sciences, Washington DC.

Vermorel, M. and Martin-Rosset, W. (1997) Concepts, scientific bases, structures and validation of the French horse net energy system (UFC). *Livestock Production Science* **47**, 261–275.

Chapter 7
Feeding the Breeding Mare, Foal and Stallion

You should wean your foals at the beginning of winter, when it beginneth to grow cold, that is about Martinmas, or the middle of November, and wean them three days before full moon, and hang about their necks upon a piece of rope seven or eight inches of the end of a cow's horn, to catch hold of them upon occasion, after which bring them all into your stable, with racks and mangers pretty low set.

S. de Solleysel 1711

THE OESTROUS CYCLE AND FERTILITY

The natural season for maximum breeding activity in both the mare and the stallion in the UK is from April to November, but the breeding season can be shifted by artificially changing daylight length and by manipulating the diet; a geographical move to the southern hemisphere can, of course, have comparable effects. During the season the normal mare expresses consecutive oestrous cycles of approximately 22 days long; within each cycle there is a period of oestrus of varying intensities that lasts on average 6 days. The fertility of the oestrus is low at the start of the season, but the creation of large follicles that ovulate and generate corpora lutea can be stimulated by the extension of day length and dietary adjustment, as the reproductive activity of both mares and stallions is under the influence of daylight length.

Periods of darkness are associated with a rise in plasma concentrations of the hormone melatonin, much above that of day-time levels, so that during the winter melatonin is secreted for a greater length of time each 24h (Domingue *et al.* 1992). Ovarian activity and follicular growth occur during the late spring and summer months in response to increasing daylight length, reducing the nocturnal melatonin response, and acting on some endogenous biological rhythm. Sequential periods of 16h light followed by 8h dark are optimum for inducing ovarian activity. This seasonal pattern, modulated by light, can be advanced by about 21 days through stabling. Thus, an extension of daylight to 16h, an increase in the plane of nutrition and possibly a rise in ambient temperature during the cold months (December in the northern hemisphere) will stimulate the onset of normal cycling 2–3 weeks earlier in the first months of the year. High fertility tends to coincide naturally with the flush of grass in late spring. Successive oestrous periods will be of increasing fertility in healthy, barren or maiden mares and individuals that are increasing in bodyweight are more likely to conceive. Therefore, by starting with a lean individual in November and December, this objective is more likely to be achieved. It has, however, been suggested that forcing barren mares in December and January enhances the probability of twins; yet if early conception is desirable such a procedure is obligatory.

Freedom in the quantities of feed consumed seems possible where it is composed largely of hay. Doreau *et al.* (1990) offered pregnant mares of Anglo-Arab and Selle Français breeds ad libitum intakes of a 90:10 mixture of hay and concentrates from four weeks before foaling until five weeks after foaling. The hay was either of poor, or high, quality. The intakes for the poor and high quality mixtures before and after foaling averaged respectively 11.1 v 12.4 kg DM and 18.6 v 21.1 kg DM. The condition score of the mares receiving the lower quality diet was relatively poor at foaling and the consequential shortage of energy and inadequacy of protein (342 g, 426 g and 579 g MADC, see Chapter 6) during the last month of pregnancy, the first week of lactation and weeks 2–5 lactation, respectively led to poorer foal growth rates to 5 weeks of age without having influenced birth weight. Although the mares were initially in similar condition, the outcome may· have partly depended upon the extent of energy reserves of the pregnant mare. If slightly lower foal growth rates are acceptable adequate performance has been observed by Micol and Martin-Rosset (1995) under upland grazing conditions amongst French mares of heavy breeds (700–800 kg BW) such as Breton, Comtois and Ardennais, which may lose 17–25% of their body weight between foal weaning and the next foaling. Of this loss 12–14% is products of conception and 5–10% loss of body mass, indicating the mares' reserves had been liberally incorporated in foal tissues. Nevertheless, pregnant mares should be kept fit but not fat as this reduces foaling difficulties and provides greater freedom for controlling milk secretion by feeding during lactation.

The foal heat occurs within 14 days of foaling and subsequent heats occur at 22-day intervals in unbred mares. The recommended rate of feeding of lactating mares is given in Table 6.5, yet it has not been clearly established whether this is the optimum rate for maximum fertility of the foal heat and subsequent oestrous periods. Milky mares, nevertheless, have a greater tendency to resorb fertilized eggs at first oestrus. This could be the reason for a putative association between overfeeding during the last 3 months of pregnancy and a reduced subsequent fertility. Unfortunately the experimental evidence to support this assertion is conflicting. Jordan (1982) noted that no reduction occurred in conception rate among pony mares losing 20% of bodyweight during gestation, but allowed to gain weight during lactation. Heneke *et al.* (1981) reported that mares in this condition at foaling had reduced conception rates, longer postpartum intervals and more cycles per conception. Conception rates of mares in good condition at foaling, but who lost weight during lactation, were as good as those of mares in good or thin condition at foaling that maintained or gained weight in lactation. Evidence in the USA indicates that mares foaling in a fat condition should be allowed to hold their weight, rather than lose it, and that thin mares should gain weight during lactation in order to maximize the pregnancy rate at 90 days postfoaling. It is concluded that thin mares should be fed well in lactation to stimulate fertility.

Experience with dairy cows might suggest that if mares are fed too liberally through gestation and given inadequate feed during lactation they are more prone to a fatty liver condition, known to reduce fertility in the dairy cow. Observation of both horses and ponies shows that various stresses during late pregnancy and early lactation, accompanied by an inadequate and impoverished diet, predispose the mare to an extreme metabolic upset associated with loss of appetite, abnormal reactions, diarrhoea, hyperlipidaemia and eventual death. This represents a breakdown in energy metabolism with liver fat accumulation, as happens in the dairy cow. Any imposed weight reduction in obese pregnant mares should therefore take place before the last 3 months and, for preference, the fatness should be corrected before breeding is instigated. This may be achieved by providing the mare with good-quality hay, but no concentrated feed.

GESTATION

The gestation period of the mare commonly lasts for 335–345 days but may continue for 1 year. The period in part depends on the month of breeding. In the northern hemisphere early-bred mares (that is, those conceiving before the end of April) normally have a gestation period exceeding 350 days and up to 365 days. Those bred in May normally foal after 340–360 days, and those bred in June and July generally foal after 320–350 days. The critical factor may be day length during the last 3 months of pregnancy, as when the photoperiod was artificially extended to 16h in late gestation of quarter horse mares (Hodge *et al.* 1981), the gestation period was shortened by 11 days and the interval from parturition to first ovulation was decreased by 1.6 days in comparison with mares subjected to natural day length.

Where diets are grossly imbalanced in terms of protein and minerals, especially Ca and P, the foal will be adversely affected at birth and reduced milk yield and infertility will ensue.

Protein and energy requirements

(See Chapter 6 for feed requirements of the pregnant mare.) Definitive statements on protein and energy requirements for breeding mares are as yet not possible for two reasons:

(1)   A large well-fed mare has considerable reserves of energy and protein on which she can draw during pregnancy if daily intakes fall below recommended levels.
(2)   A reduction in intake generally seems to induce economies in metabolism so that deficiencies are partly offset.

Clearly the mare is capable of considerable adjustment to a variety of situations. However, in extremes, excesses or deficiencies of energy will lessen her reproductive efficiency. During winter or summer the ribs of a mare should not be seen, but should be detectable by touch with no appreciable layer of fat occurring between them and the skin. The condition of the over-fat mare can be improved by gradually reducing the cereal component of the ration while the protein and mineral mixture is maintained at the previously determined level of intake. Meyer (1983b), in Hanover, concluded that the mare should be 18% above normal weight before parturition to achieve a high fertility after parturition.

Pregnant mares are normally kept on pasture. Australian workers (Gallagher & McMeniman 1988) established that grass/legume pastures in south-east Queensland could support the nutritional requirements of TB brood mares by providing DE intakes of 68.0 and 91.7 MJ/day (the latter figure is approximately 10% higher than the NRC 1989 recommendation) and digestible N intakes of 91.2 and 138 g/day during mid and late pregnancy, respectively.

During periods of inadequate energy intake epinephrine secretion increases fat mobilization from body reserves leading to a rise in plasma NEFA concentrations in blood. In the normally fed human subject receiving three meals daily plasma NEFA concentration decreases after each meal and rises before the next meal. In the horse there is a decrease in plasma NEFA after a meal only where the ration is providing inadequate energy for long-term energy balance (Sticker *et al.* 1995). Thus, this measurement may provide a useful means of assessing energy status of the horse.

PARTURITION

In the 24 h before birth of the foal, the mare should be fed lightly with good-quality hay and a low-energy cereal mixture including bran, or proprietary horse and pony nuts (10–11% protein, 3% oil, 14–15% crude fibre; Table 7.1), with access to restricted quantities of warm water. The first feed after parturition can effectively be a bran mash and the second can include some bran with small quantities of good-quality proprietary stud nuts (16–17% protein, 3% oil, 8% crude fibre; Table 7.1) or a cereal protein mixture. Obese mares tend to be less active and so poorer muscle tone may lead to birth difficulties and delayed expulsion of the placenta, which should be passed during the first hour after birth. The rate of concentrate feeding up to day 10 should be restricted in order to avoid excessive milk secretion and digestive disturbance in the foal. However, inadequate amounts of energy may contribute to the metabolic abnormalities outlined in 'The oestrous cycle and fertility', this chapter. Recommended allowances are given in Table 6.5.

Perhaps 5% (Rossdale & Ricketts 1980) or 10% (Jeffcott *et al.* 1982c) of foals may be lost through perinatal mortality, including stillbirths and postnatal deaths. Of these, significantly more are male. Although nutrition is a vital factor, the significance of it in this statistic is entirely unknown. Birthweight is a crucial characteristic in determining the prospects of foals and, despite the influence nutrition can have on this, size of the dam is a major controlling influence. So, an acceptable minimum weight must depend on the breed and the purpose intended for the individual offspring. In TBs, an early rapid growth rate is normally expected and required for work at an early age. For this, foals of less than 35 kg probably should not be kept.

Table 7.1 Composition of foal milk replacer, stud concentrate mixture and horse and pony mix to be given with hay and water.

Foal milk replacer (see footnote for mixing[1])	(%)		Stud mixture[2] (%)	Horse and pony mix (%)
Glucose	20.0	Oatfeed	—	25.0
Fat filled powder (20% fat)	5.0	Oats	46.0	33.0
		Wheat bran	15.0	20.0
Spray-dried skimmed-milk powder	40.0	High-protein grassmeal (16% crude protein)	15.0	10.0
Spray-dried whey powder	32.7	Extracted soyabean meal	15.0	4.0
High-grade fat[3]	1.0	Molasses	5.0	5.0
Dicalcium phosphate	1.0	Fat[3]	1.0	1.0
Sodium chloride	0.2	Limestone	1.0	0.9
Vitamins/trace elements[4]	0.1	Dicalcium phosphate	1.1	0.5
		Salt	0.75	0.5
		Vitamins/trace elements[4]	0.1	0.1
Total	100.0		100.0	100.0

[1] Disperse in clean water at the rate of 175 g/l (for 2 days after colostrum at 250 g/l). Also can be pelleted and mixed with stud mixture as a weaning feed for orphan foals.
[2] This mixture is satisfactory as a creep feed and postweaning diet. However, a mix specifically for young weaned foals to be fed with grass hay could to advantage contain an extra 5% soyabean meal replacing 5% oats.
[3] High-quality tallow and lard, including dispersing agent. Stabilized vegetable oil could alternatively be added at time of mixing.
[4] To provide vitamins A, D_3, E, K_3, riboflavin, thiamin, nicotinic acid, pantothenic acid, folic acid, cyanocobalamin, iron, copper, cobalt, manganese, zinc, iodine and selenium.

Where twins are born, their total weight approximates that of large single births with a mean in the region of 55 kg for TBs, implying that it is practical to retain only the heavier of the two.

ACQUIRED IMMUNITY IN THE FOAL

The mare must pass adequate passive protection to her foal through the colostrum, and she therefore should have been situated in the foaling area for, at the very least, 2 and preferably 4 weeks before foaling. This means she will confer some immunity to the strains of microorganism peculiar to her environment, for example, those causing scours, joint-ill and septicaemia. Newborn foals will normally first suck within 30–180 min of birth. Colostrum is rich in protein (particularly immunoglobulins), dry matter and vitamin A. If foals are deprived of colostrum, an injection of about 300 000 iu vitamin A is in order. Immunoglobulins do not pass through the dam's placenta and can be absorbed efficiently through the intestinal wall of the foal only during the first 12–24 h of life. The major causes of colostrum deprivation in the foal are premature birth and delayed suckling, small intestinal malabsorption, premature leakage of milk through the teats or death of the mare.

The immunoglobulins are concentrated by the mare in her udder within the last 2 weeks of gestation, when their level in the mare's serum falls. There is, therefore, a selective concentration of this protein fraction in the mammary gland.

If the foal is suckled normally, the concentration of the immunoglobulin fraction in the colostrum, 12–15 h after birth, is only 10–20% of the initial value. It is known that the protein content of mare's colostrum is around 19% during the first 30 min after parturition, but by 12 h the level falls to about 3.8% and after 8 days it reaches a fairly constant level of 2.2% (Ullrey *et al.* 1966). The foal absorbs γ-globulin as intact undegraded molecules throughout the first 12 h of life. At birth the foal is agammaglobulinaemic but it responds to the colostrum, its serum γ-globulin rising for 12 h to 8 g/l serum (Jeffcott 1974b–d). Amounts of specific antibodies so acquired by the foal's blood decline from 24 h of age; by 3 weeks the values are halved and by 4 months the titre of antibodies provided by the mother is barely detectable.

The foal's own system for building active immunity in the form of autogenous γ-globulins first provides detectable products at 2 weeks of age in the blood of colostrum-deprived foals and at 4 weeks in those reared normally. By 4 months of age the γ-globulins have attained adult plasma concentrations. Up to this age, therefore, the foal is more susceptible to infection than is an adult in the same environment, particularly when it has received an inadequate quantity of colostrum, or colostrum at the wrong time.

If the mare has ejected much of her colostrum before foaling then it will be necessary to give the foal colostrum from another mare, preferably one accustomed to the same environment or, failing this, cow's colostrum rather than milk. Commercial sources of cow's colostrum are now available (see 'Bovine colostrum', this chapter). After 18 h the colostrum has little systemic immune value, although it does have some beneficial local effects within the intestinal tract.

A simple field test has been developed in which the turbidity of plasma is assessed following the addition of zinc sulphate; the results correlate well with concentrations of blood globulin in foals indicating whether sufficient antibodies have been absorbed in the neonatal period. This subject is discussed further under the section on orphan foals ('Blood plasma by parenteral, or oral, dosing'), this chapter.

NEONATAL PROBLEMS

Hygiene is generally outside the scope of this book, yet the importance of cleanliness in the foaling area cannot be overemphasized. It is essential that the foal receives colostrum to provide it with some protection (passive immunity) from potentially harmful organisms in the environment. Nevertheless, the consumption of excessive quantities of milk can overload the digestive capacity of the foal and the milk may then become a substrate for rapid bacterial growth in the intestines. This situation can precipitate diarrhoea despite the consumption by the foal of normal quantities of colostrum. There are two problems related to the feeding of the foal, yet unrelated to disease caused by microorganisms.

Haemolytic icterus

The foal's blood differs immunologically from that of its dam and on rare occasions the fetus may react with the dam's immune system, causing the production by the mare of isoantibodies to the foal's red cells. These antibodies are transmitted to the colostrum and the suckled foal may absorb sufficient to initiate a considerable destruction of red cells, precipitating an anaemia and jaundice, known as haemolytic icterus. In severe cases, the foal's urine will be discoloured with haemoglobin. If a mild attack is detected before icterus has occurred, the foal should not be allowed to nurse its dam for 36h. Where the mare has previously produced foals with the condition, she may still carry similar antibodies. In this case the foal should automatically be given colostrum from another mare at the rate of 500ml every 1–2h for three to four feeds, followed by milk replacer until 36h, when it may be returned to its dam. In the meantime the mare should be milked out by hand.

If the problem is anticipated, blood samples can be taken from the foal and the red cells allowed to settle. The abnormality causes red cell haemolysis (pink plasma rather than the normal straw colour). In these circumstances the foal may be severely anaemic and red cells from the mother may be slowly infused into a vein after syphoning off as much plasma as is conveniently possible. For preference, however, the source of red cells should be three or four geldings that have not previously received transfusions so the risk of immunological reactions is minimized. A simple procedure has been proposed as a means of precluding damage *before* colostrum is taken by foals that are considered to be at risk. One drop of umbilical blood is mixed with four drops of saline and five drops of mare's colostrum on a clean microscope slide, checking after several minutes for agglutination reactions.

Passing the meconium

At birth much of the large intestine, including the caecum and rectum, contains a substance, the meconium, which is normally completely voided within the first 2–3 days of life. Suckling usually sets up a reflex, promoting defaecation of this material. If this does not occur the normal passage of colostrum and milk may become blocked so that the gases formed during their fermentation cause distension and pain to the foal. It may then go off suck, act abnormally, crouch, lift its tail and flex its hocks in an effort to pass the offending material, or roll over in pain. Eventually a yellowish milk dung reaches the rectum, the meconium is cleared and the symptoms subside.

The problem is treated conservatively by administration of a lubricant through a stomach tube, plus one or two enemas of soap and water, or liquid paraffin, and the injection of pain-relieving drugs. If the foal goes off suck for an extended period it should be given a fluid feed by stomach tube or appropriate intravenous solutions. I.v. feeding of glucose and an isotonic electrolyte solution (see Table 9.2) is a life-saving procedure in cases of severe enteritis with consequent dehydration. In normal circumstances the foal will eat a quantity of its mare's faeces. In so doing it introduces beneficial microorganisms into the intestinal tract, which compete with pathogens present in the general environment.

LACTATION

At a given stage of lactation the composition of mare's milk is remarkably similar among the various breeds of horse. The composition changes rapidly during the first days of lactation and then more slowly (Figs 7.1–7.4 and Table 7.2). Milk contains about 2 MJ gross energy/kg. Eight TB and two standardbred mares receiving a diet of concentrates and hay showed that they achieved yields of 16, 15 and 18 kg daily (3.1, 2.9 and 3.4% of bodyweight daily, or 149, 139 and 163 g/kg $BW^{0.75}$) at 11, 25 and 39 days postpartum, respectively (Oftedal *et al.* 1983). Doreau *et al.* (1991) reported that, during weeks 0–5 of lactation, primiparous French Anglo-Arab mares, weighing 522 kg after foaling, produced less milk and of a lower fat content than was produced by multiparous mares (14.6 v 16.6 kg/day and 16.5 v 20.2 g/kg respectively). This difference was associated with a lower dry matter intake during pregnancy and lower plasma NEFA concentration during pregnancy and lactation for the primiparous mares. French draft (Breton, Comtois) mares, weighing 726 kg yielded 20 kg/day in the first week, rising to 27.5 kg/day by week 8 (Doreau *et al.* 1992) (Table 7.3). Milk yields are markedly influenced by the mare's innate ability, by feed consumption during the latter stages of pregnancy, and, more importantly, by water availability (Table 7.4) and intake of energy and nutrients during lactation. Experimental work with quarter horse and TB mares has shown that a reduction in the

Table 7.2 Nutrient content of milk in quarter horse, TB, Dutch warmblooded saddlebred horse mares and some other breeds (Bouwman & van der Schee 1978; Gibbs *et al.* 1982; Oftedal *et al.* 1983; Schryver *et al.* 1986; Doreau *et al.* 1988; Saastamoinen *et al.* 1990; Martin *et al.* 1991).

	Dutch and some other breeds			Quarter horse	TB
	Initial	28 days	196 days	2–150 days	24–54 days
Total solids (g/kg)	130	100–112	105	105	105
Gross energy (MJ/kg)	2.5	1.9–2.3	1.8–2	—	—
Fat (g/kg)	27	11–13	7	13–12	13
Protein (g/kg)	33	17–20	18	21–16	19
Ash (g/kg)	5.3	2	2.8	3	4
Calcium (g/kg)	1.2	0.7	0.8	0.8–0.6	0.9
Phosphorus (g/kg)	1	0.4	0.5	0.36–0.21	0.6
Lactose (g/kg)	58	66	66	—	69
Magnesium (mg/kg)	—	90	45	65–43	68
Potassium (mg/kg)	—	700	400	580–370	120
Sodium (mg/kg)	—	225	150	190–160	180
Copper (µg/kg)	—	450	200	—	200
Zinc (µg/kg)	—	2500	1800	—	2000

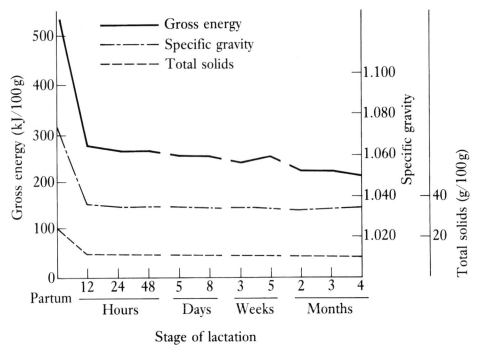

Fig. 7.1 Changes in specific gravity and concentration of gross energy and total solids in mare's milk at various stages of lactation (after Ullrey *et al.* 1966).

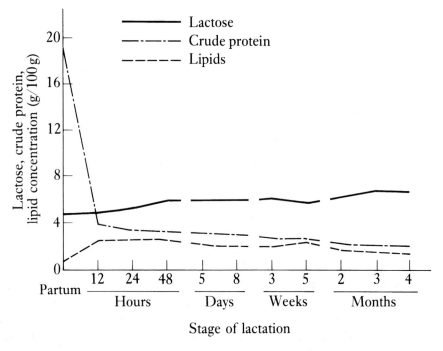

Fig. 7.2 Changes in concentration of lactose, crude protein and lipids in mare's milk at various stages of lactation (Ullrey *et al.* 1966).

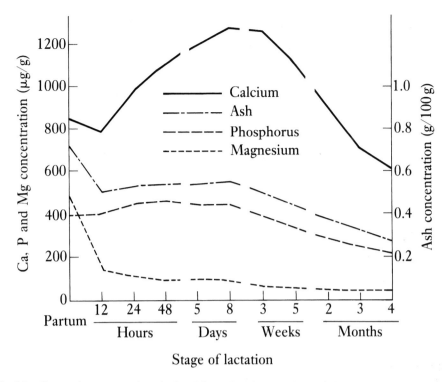

Fig. 7.3 Changes in concentration of ash, calcium, phosphorus and magnesium in mare's milk at various stages of lactation (Ullrey *et al.* 1966).

Table 7.3 Typical milk yields (kg/day) by mares of various bodyweights during the 1st to 25th week of lactation.

	1–2nd	4–5th	5–12th	20–25th
Quarter horse (500 kg)	10	14	10	—
TB, standardbred (494 kg)	12–16	14–16	18	—
Dutch saddlebred horse (600 kg)	14	16	19	11
French draft (726 kg)	20	25	27	—

energy intake to 75% of that recommended for lactation by the NRC (1978) does not lead to a parallel decrease in foal weight at 75 days (Banach & Evans 1981a,b). Undoubtedly, the thriving mare has considerable capacity to adapt within limits to a restricted diet.

Mares weighing between 400 and 550 kg may be expected to yield 5–15 kg milk daily during the first few weeks, 10–20 kg daily in the 2nd and 3rd month, falling to 5–10 kg by the 5th month; but normal yields assume an ample supply of water. Observations with Dutch saddlebred horses indicated that suckling frequency averaged 103 times per 24 h in the first week, falling to 35 times per 24 h by week 10, and on each occasion the foal suckled for 1.3–1.7 min. (Small amounts taken frequently

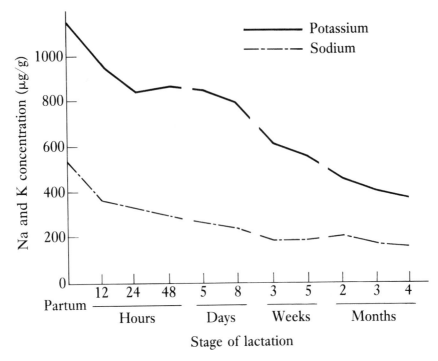

Fig. 7.4 Changes in concentration of potassium and sodium in mare's milk at various stages of lactation (Ullrey *et al.* 1966).

are unlikely to cause digestive disturbances.) At birth the foals weighed 57 kg on average and the weight doubled in the first 37 days of life.

Suckling TB and standardbred foals have been shown to take in daily milk dry matter equivalent to 3.1, 2.1 and 2% of bodyweight at 11, 25 and 39 days postpartum, when weight gain averaged 1.14 kg daily (Oftedal *et al.* 1983). The comparable daily intakes of gross energy were 39, 32 and 37 MJ, respectively.

Effects of dietary energy, protein and urea on milk yield

Energy intake can be voluntarily modified to affect milk yield and composition. The voluntary food intake of draft broodmares was less when equal quantities of tall fescue hay and concentrates were given, compared with 95% hay and 5% concentrates (Doreau *et al.* 1992), causing differences in the protein and fat contents and in the fatty acid profiles of the milk (Table 7.5). The milk from the high concentrate diet was richer in C18:2(n-6) (linoleic) and poorer in C18:3(n-3) (α-linolenic) acid. Thus, increased fibre and inadequate energy contents of feed may increase the fat and the protein contents of milk, but decrease yield. The effect on fat is similar to that in ruminants.

Dietary protein quality is known to affect the protein content of mare's milk. Glade and Luba (1987b) gave lactating mares daily 1.55 kg moderate quality pro-

Table 7.4 Water requirements in the stud.

(a) Water provided by pasture herbage (kg) per kg herbage DM and total (including herbage) minimum water needs (kg) per kg feed DM consumed.

Water content of herbage per unit DM		Water requirement of horses per unit DM consumed	
Spring growth	4	Last 90 days of gestation	3
Dry summer	2.5*	First 3 months of lactation	4
Mild winter	3	Breeding stallion	3
		Weaned foal	3
		Barren mare	2

*When 'burnt' and bleached the proportion of water may be as low as 0.15–0.3.

(b) Minimum supplementary water requirements (l) per mare daily, assuming average milk yields.

Liveweight of mare: (kg)	200	400	500
Last 90 days of gestation:[1]			
in stable	12.7	22.3	26.4
on pasture[2]	0	0	0
First 3 months of lactation:			
in stable	27.7	41.8	49.6
on pasture[2]	7.3	10.9	12.3

[1] The breeding stallion's needs are similar to those of the pregnant mare.
[2] These amounts will be insufficient for mares on parched pasture in environmental temperatures exceeding 30°C where shade should be provided.

Table 7.5 Effect of proportions of hay to concentrates in French draft mares (726 kg) on lactation (Doreau *et al.* 1992).

	Hay:concentrates 95:5	hay:concentrates 50:50
DE (MJ/kg diet)	9.3	12.9
DCP (g/kg diet)	129	142
NE (UFC)*	0.65	0.89
MADC (g/kg DM)	74	100
Feed intake (kg, DM/day)	22.9	21.4
Milk yield, week 4 (kg/day)	23.4	26.4
Fat, mean weeks 2 and 4 (g/kg milk)	14.7	11.6
CP, mean weeks 2 and 4 (g/kg milk)	26.3	24.4

*See Chapter 6, French INRA system.

tein/500-kg BW, or the same amount but with half provided as soya protein. This increased the protein content of the milk at 7 days from 25.3 to 33.2 g/l. A difference persisted until the 4th week of lactation. Plasma methionine and lysine were in higher concentrations in foals nursing the soya-supplemented mares and their withers' height at 7 weeks was greater. Poor pasture would therefore benefit from supplements of both good quality protein and starch. Urea supplementation of mares, given low protein diets, increases plasma and milk urea concentrations and reduces feed intake with adverse effects on milk intake and growth of their foals (Martin *et al.* 1991). Dietary energy source and amino acid intake within normal limits influence milk yield and composition and foal performance, but the evidence is insufficient for the recommendation of changes to existing requirement values.

Mineral composition of milk

The mineral composition of TB mare's milk was measured by Schryver *et al.* (1986) (Table 7.6) and other values are given in Table 7.2. The dry matter content was found to decline sharply from 12% in the 1st week to 10.8% in the 3rd, and then slowly to 10.2% by the 15–17th week of lactation in the study of Schryver and colleagues and as shown in Table 7.2.

Parturient paresis

Some high-yielding dairy cows suffer from a condition known as milk fever, or parturient paresis, which is probably caused by a sudden draining of blood calcium into the milk after parturition without an equivalent mobilization of bone calcium. A successful treatment entails giving cows a low-calcium diet from 2 to 4 weeks before parturition and then providing a diet relatively rich in calcium during lactation. A regulated but high dose of vitamin D given 8–10 days before parturition has also been beneficial in some instances, but the dose has to be carefully calculated to avoid a gross and toxic excess. Pony and horse mares with a history of tetany (cf. paresis in ruminants) associated with depressed blood calcium in lactation might well benefit from being given a diet containing 1.5–2 g Ca/kg and approximately 3000 iu vitamin D/kg throughout the last 2 weeks of gestation. It is difficult to predict parturition with sufficient accuracy 10 days before foaling and therefore single doses of large amounts of vitamin D at that time cannot be recommended. The diet should, of course, be adequate in other respects and during lactation the total diet of such mares should contain 5–6 g Ca/kg. Horses that are suffering from this form of tetany are slowly given calcium gluconate intravenously while cardiac action is continuously monitored.

Table 7.6 Mineral composition of mare's milk – electrolytes (mmol/l) and Cu and Zn (μmol/l) (Schryver *et al.* 1986).

Weeks postpartum	Ca	P	Mg	K	Na	Cu	Zn
1–2	33.3	29.2	4.64	17.0	9.4	12.1	44.4
3–4	27.9	22.6	3.66	13.0	7.5	7.5	35.2
5–6	23.1	19.5	2.96	10.0	8.0	6.8	30.6
7–8	21.6	19.0	2.47	11.5	6.5	5.0	29.1
9–14	20.7	18.2	2.18	9.8	6.4	2.7	27.5

Grazing and drinking behaviour in the postnatal period

Grazing, lactating Welsh pony mares were shown (Crowell-Davis *et al.* 1985) to spend about 70% of their time feeding, whereas, excluding nursing, their foals at 1 week of age spent 10%, rising to about 50%, of their time grazing at 21 weeks. Foals tended to eat when their mothers were feeding, which was of greater intensity early in the morning and the evening. Movement to water sources was generally carried out by the entire herd and the frequency of this increased with increasing environmental temperature to 35°C. However, foals under 3 weeks of age did not drink, and half the foals were not observed to drink before weaning (also see water requirements, Chapter 4).

Growth rate of the suckling foal reflects the rate of milk secretion of its dam. Among environmental influences on milk yield, the quality of pasture and free water availability are major influences. The supplementary feeding of lactating mares, which takes place in the stable at night or on the pasture, is another influence. Safe water sources are essential for lactating mares in hot weather.

Blood biochemical values of the foal

Normal blood characteristics of the foal are given in many texts and the subject is referred to in Chapter 12. The serum values given in Table 7.7 indicate that the neonatal foal is deficient in globulins which must be rectified by consumption of colostrum before any other feed. Serum alkaline phosphatase activity decreases with age, probably reflecting the relative decrease in the rate of bone growth with increasing age, or emphasizing that mishaps in proper bone development may be established before or shortly after birth. Blood glucose and fat are in low concentrations at birth, reflecting the low energy reserves of the neonatal foal and its inability to maintain body warmth in a cold environment unless colostrum and milk are provided.

Creep feeding of the foal

Foals will start to nibble hay and concentrates at 10–21 days of age. If the milk supply of the dam or the amount of grass is inadequate the provision of creep from

Table 7.7   Protein (g/l), neutral fat (g/l), and glucose (mmol/l) concentrations and alkaline phosphatase (ALP) activity (King Armstrong Unit) in the blood serum of TB foals (Sato *et al.* 1978).

Days from birth	Total protein	Albumin	Globulins	Neutral fat	Glucose	ALP
0	45.8	26.3	15.3	0.36	3.07	100
10	54.8	26.5	27.6	1.18	8.04	76
20	56.2	25.4	29.6	1.04	7.51	57
50	53.6	24.3	29.2	0.62	7.16	62
90	56.0	26.3	30.0	0.90	6.06	50
120	60.8	31.1	29.4	0.45	6.08	45

this time may well enable a normal growth rate to be achieved. However, the principal objective of creep feeding is to accelerate the anatomical and physiological maturation of the Gl tract of the foal so that ultimate weaning presents no particular stress or hazard and digestive disturbances caused by abnormal fermentation of ingesta are prevented.

Where the mare secretes minimal quantities of milk, a creep feed based on dried skimmed milk provided from about 2 weeks of age is recommended. The composition of this feed should be changed gradually to that of a stud nut (16–17% protein, 3% oil, 8% fibre) or a growing foal diet (17–18% protein, 3% oil, 7% fibre) (see Tables 7.1 and 8.6) from 10–14 weeks of age. The use of milk pellets as a creep feed before weaning is contraindicated as it defeats the prime objective.

Control of growth abnormalities

Foals that are growing well with mares on pasture have no particular need for additional dry feed until 2 months before weaning, say at 2–3 months of age. Here the functions are to compensate for a waning milk production in the dam, to redress the effects of a decline in pasture quality and probably more importantly to accustom the foal to the dietary regime it must expect after weaning. Thus, the feed should be a concentrate of the type given in Tables 7.1 and 8.6 on which the foal will be maintained throughout the forthcoming winter and spring. Supplementary creep feeds should be restricted in quantity to 0.5–0.75 kg/100 kg BW. Such a restriction will give a measure of control over the incidence of growth-associated ailments including epiphysitis and contracted tendons (flexural deformity) (Plate. 7.1). Where evidence of either of these is at hand a restriction of growth rate, imposed by cutting the supplementary feed and by reducing the mare's feed for a period of 3–4 weeks, should not prejudice ultimate mature size if carefully regulated. The extent of restriction must depend on how serious the problem is. Foals that stand high on their toes at birth should be exercised regularly and allowed to grow at a submaximum rate if the condition is to be contained. Overtly contracted tendons at birth are relatively untreatable and possibly result from intrauterine malpositioning. If therapy is possible, splints or extension shoes are used and the foals are exercised regularly with restricted access to feed until the abnormality is satisfactory (see also Chapter 8).

The vertical growth of normal foals is very rapid throughout the first 3–4 months of life. Access to creep feeders should be controlled by regulating the width of the entrance rather than by restricting its height. Foals of breeds with mature weights of 550 kg can be weaned easily when consuming nearly 1 kg of creep feed and 0.5 kg hay (or equivalent grass) daily by which time the foal should be in excess of 140 kg. These restrictions normally ensure that growth will not falter during the first postweaning week and by the end of the second week the rate of gain will be at least 1 kg daily in healthy stock. Several days before weaning it is appropriate to remove the mare's daily concentrate allowance and access to hay and pasture also may be restricted.

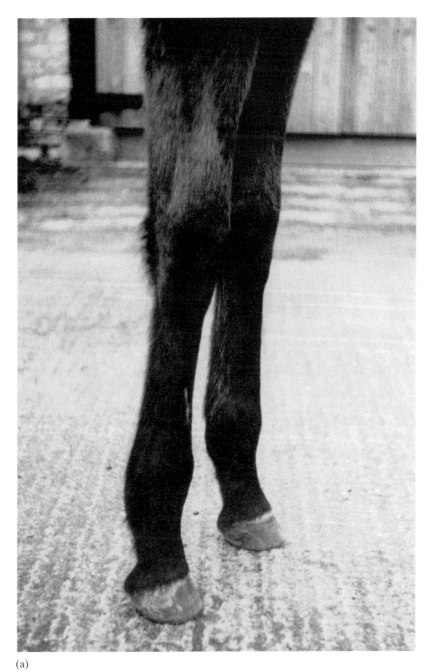

(a)

Plate 7.1 Chronic contracted tendons in a yearling showing enlarged fetlock joints and upright stance. Later improvement was achieved by desmotomy of the superior check ligaments. Before surgery (a) and after surgery (b).

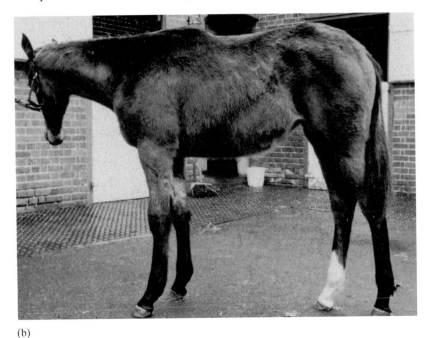

(b)

Plate 7.1 *Continued*

Epiphysitis (Plate 7.2), probably more correctly called metaphysitis, is not uncommon in faster growing, larger, fine-boned foals of TB, saddlebred or modern quarter horse breeding. It is encountered particularly in the fetlock joint at the end of the metacarpus and in the 'knee' joint at the distal end of the radius. Where it is slight, the foal will probably aright matters, but where severe, supplementary feeds should be restricted to good-quality roughage, the milk intake should be limited and the animal should be boxed until the worst of the 'bumps' subside. A restriction in the rate of weight increase allows joint maturation to continue without the stress of excessive pressure on the joints. Light exercise must then be undertaken daily and the normal feeding regime gradually reinstated. Exercise may, however, be damaging in severe epiphysitis, and in this case no analgesics should be used. Problems of this nature can arise in less than a week and may be complicated by angular deformities in one limb together with epiphysitis in the opposite limb. Successful treatment is contingent on immediate action. Attention by the farrier to the hooves should allow small misalignments of the limbs to be corrected during growth.

Glade *et al.* (1984) have advanced an intriguing explanation of the relationship between epiphysitis and the energy and protein consumption at each meal. Excessive intakes suppress normal postprandial hyperthyroxaemia (raised plasma concentration of the thyroid hormone T_4), because an intense insulin secretion stimulates T_3 formation from T_4, in turn inhibiting the TSH (thyrotrophin) and thus T_4 secretion. As T_4 is required for bone maturation, hypothyroidism is known to cause skeletal manifestations similar to epiphysitis and OCD, whereas insulin stimu-

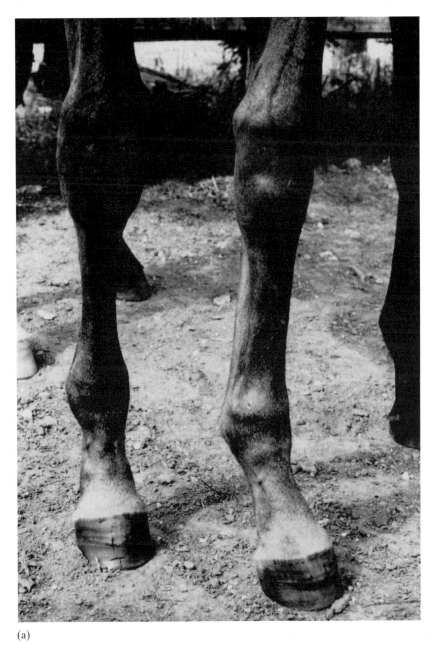

(a)

Plate 7.2 Epiphysitis in the lower (distal) end of the metacarpus and the upper (proximal) end of the proximal phalanx (fore fetlock) (a) and (b); and spavin of the hock (c) in foals. (Photograph 7.2 (c) conrtesy of Dr Peter Rossdale, FRCVS.)

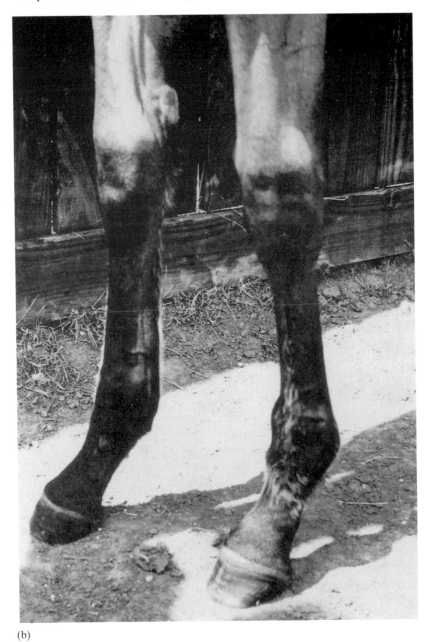

(b)

Plate 7.2 *Continued*

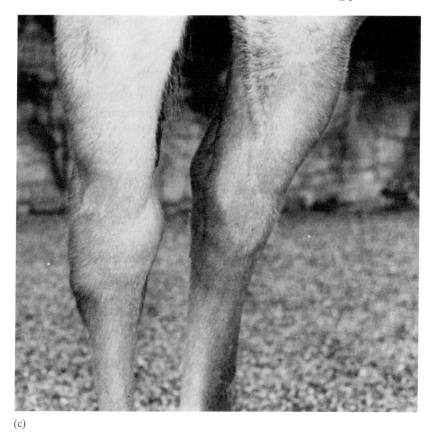

(c)

Plate 7.2 *Continued*

lates the formation of immature cartilage. Each of these diseases is characterized by enlarged growth centres, failure of bone formation from cartilage, occasional cartilage necrosis and cyst formation. The solution would seem to lie not only in the control of dietary energy and protein and the correction of errors in mineral and trace-element nutrition, but also in raising the number of daily feeds and decreasing their individual size. The logical extension of this may be to change to a system of earlier weaning and feeding foals *ad libitum* a complete mix described under 'Complete Mix', Chapter 8. This will encourage nibbling and avoid large postprandial surges in blood glucose and amino acids that stimulate insulin secretion.

Worming of the foal should not coincide with weaning, but the first dose of anthelmintic may be given at 2–3 months of age or ideally at 4 months of age and thereafter at 6-weekly intervals (see Chapter 11).

WEANING PROCEDURE

Restricted feeding of the mare limits milk secretion, but after weaning the udder should not be milked out. Some breeders rub camphorated oil into the udder.

The psychological attachment of the foal to its dam is greatest between the 2nd and 12th weeks of lactation with a peak around the 3rd week, at which time separation leads to the greatest agitation of both. Risk of injury to the foal at weaning as a result of excitement induced by separation is a major factor to be contained. On large stud farms, three alternative procedures are practised. These are:

(1)   All the mares are removed at the same time from a year's crop of foals.
(2)   One or two mares are removed at a time, starting with the first foaled, or most dominant mare, and allowing a few days to elapse before the next is removed.
(3)   The foals are separated for increasing time periods, such that the foals are allowed to nurse three times per day, twice and then only once on successive days, keeping mares and foals in sight of each other.

Methods (1) and (2) may require access to another farm to ensure that weaned stock are out of sight, sound and smell of their mothers. Method (2) may lead to some foals being kicked by more aggressive mares when searching for milk. The last method protracts the drying-up procedure, is more laborious, and not without risk.

It is, however, recommended that unless there are other mitigating circumstances mares should be abruptly removed from foals, starting with the mare having the largest and most independent foal, or with the most dominant mare, which is likely to cause problems to other foals. Several days should elapse before the next most dominant mare is removed, leaving the foals in familiar surroundings out of sight, sound and smell of their mothers. It is helpful to leave a gentle dry mare with the foals and any foal having a cold or other debilitating condition should not be weaned until it has regained health.

Foals at first may become frantic and it is important to ensure that all have company, that there is ample space for play, that the pasture has a clean water source and a shelter, and that the shelter and fences are free from protruding nails, splinters and loose wire. The pasture should also be of good quality, without a worm burden, and free from flints and rabbit, or other, holes that might cause leg injury.

Early weaning

The procedures outlined for both liquid and dry feeding of orphaned foals can be followed for early weaning of foals, but the procedure is both labour intensive and an interruption to normal activities on the stud farm. From both a practical and an economic point of view therefore a general recommendation for early weaning may not be given. Nevertheless, there will be circumstances where it has a practical value.

All early-weaned foals must first receive colostrum during the first day of life, before they receive other feed. Weaning at 3–5 days of age, when the mare is not imprinted on the foal and under hygienic conditions, is safe. A separation of 6h without feed normally eliminates fretting, but initially leads to a lower rate of

Table 7.8 Performance of TB foals weaned at 5 days of age (Pagan *et al.* 1993a).

Age (days)	Average BW (kg)	Milk DM intake (kg/day)	Pellet intake (kg/day)	Coarse feed intake (kg/day)	Total DM intake (kg/day)
5–14	72	1.28	0.26	0	1.54
15–28	85	1.34	0.44	0	1.78
29–42	100	1.29	0.63	0	1.92
43–56	115	1.32	0.89	0.10	2.31
57–70	130	1.21	1.00	0.16	2.37
71–84	144	1.25	1.40	0.19	2.84
85–98	158	0.70	1.87	0.51	3.08
99–112	171	0	2.09	0.64	2.73
113–126	182	0	2.03	0.46	2.49
127–140	194	0	2.15	0.47	2.62
141–154	206	0	2.21	0.74	2.95
155–168	220	0	2.27	0.83	3.10

growth than occurs in foals suckling their dam. The reasons for this are the shock of weaning and a lower milk dry matter intake. In traditional systems this lower intake is caused by a lower meal frequency than occurs with suckled foals. Compensation by larger meals precipitates diarrhoea. Early-weaned foals should receive a minimum of six meals of milk daily. The milk should contain about 120–130 g DM/l. Eight to ten meals daily would increase the early growth rate without harm; however, retardation of early growth is subject to partial compensation by 6 months of age. Dry pellets, containing dried skimmed milk and other high quality protein sources may be offered *ad libitum* during this time. The milk replacer should contain about 25% protein and 16% fat in the dry matter. Some evidence suggests that medium chain triacylglycerols may be a useful source of fat for foals with a compromised digestive function. Neutral fats, such as coconut oil, that contain medium chain fatty acids are hydrolysed more rapidly by pancreatic lipase. Any unhydrolysed medium chain fat is absorbed directly across the small intestinal mucosa. The total daily feed should ideally average 20–25% protein and 12–15% fat. Pagan *et al.* (1993a) reported on the performance of TB foals weaned at 5 days of age (Table 7.8).

FEEDING THE ORPHAN FOAL

The artificial rearing of relatively small foals should not be lightly attempted. TBs of less than 40 kg are normally destroyed. Orphan foals are deprived of the warmth of the dam and in cold weather should be covered with lightweight quilted material. Normally a nurse mare is desirable, but in the meantime artificial feeding is necessary. The initial concentration of milk provided should be 22% of dry matter for the first 1–2 days, dropping by 1% daily until a normal concentration of 14–15% of dry matter is reached and maintained until weaning. If there is diarrhoea the milk can be diluted, or preferably replaced, for a short period with a glucose–electrolyte solution, which provides sodium, potassium, chloride, organic base and glucose in particular (see Table 9.3). However, all neonatal foals must receive an adequate source of immunoglobulins following birth before they receive *any other* organic food.

COLOSTRUM

The need of the newborn for colostral immunoglobulins is crucial to its survival in normal environments. If the mare is lost after the first 24h the prospects of foal survival are greatly enhanced as sufficient colostrum should have been sucked. Where this is not the case the maintenance of a bank of frozen colostrum is an asset and its value is enhanced where it has been derived from mares in a similar microbiological environment to that experienced by the foal.

Hygiene in the collection of colostrum is vital and preservation should be carried out by experienced persons to ensure that no bacterial contamination has occurred, otherwise organisms will proliferate when the frozen colostrum is thawed. Thus, minimum quantities, sufficient for each feed, should be stored in each container and then thawed at one time. The colostrum should be consumed immediately on warming to preclude undesirable bacterial growth. The foal should receive about 500 ml colostrum by nipple or stomach tube every hour for three or four feeds before 12 h of age. If plasma immunoglobulin concentration can be measured and it is found to be less than 4 g/l, and the foal is less than 12–15 h old, then 2 l colostrum, or an amount to raise plasma levels to 8 g/l, should be administered by stomach tube in stages over several hours. Colostrum deprivation in a normal environment inevitably leads to very low serum IgG concentrations and septicaemia with a variety of bacterial species (Robinson *et al.* 1993).

Equine colostrum is the ideal, yet bovine colostrum has a value. Chong *et al.* (1991) raised in isolation 21 of 22 full-term Welsh Mountain pony foals, under conditions free from equid herpesvirus (EVH-1/4), but without mare's colostrum. The foals were given antibiotic prophylaxis and bovine colostrum during the first 24 h. This was followed by mare's milk replacer until weaning. The foals remained free from EHV-1/4 infection.

Evaluation of colostrum

The efficacy of colostrum depends on its content of IgG. In the stud the IgG content can be approximated using a hydrometer, designed for the purpose, that relates the specific gravity (SG) of the colostrum to its IgG content. Evidence indicates that the estimate is influenced by the fluid temperature, which therefore must be controlled to obtain reliable data. Colostrum with an SG of ≥1.085 (equivalent to >7000 mg/dl IgG) at 25°C is suitable for freezing as a source for foals deprived of adequate colostrum from their dams. Colostrum with an IgG concentration of SG > 1.03

(>3000 mg/dl) at 25°C is considered to be adequate for direct ingestion from the mare's udder within the first 12 h from birth, so long as no milk has been ingested previously. The IgG level of foal serum should exceed 800 mg/dl between 18 h and 24 h postfoaling.

Bovine colostrum

Sachets of certified bovine colostrum powder, rich in IgA, IgG and IgM are available commercially and they store well. This powder can be added preferably to water or the very first feeds of glucose and water given to newborn foals. These other feeds should *not* be given *before* the colostrum. A reserve supply of this product is a good insurance policy for most studs.

Blood plasma by parenteral, or oral, dosing

Where the concentration of immunoglobulins in the foal's plasma is less than 400 mg/dl and where colostrum is unavailable, one can use blood plasma, preferably from a donor gelding horse or unrelated mare which has never received a blood transfusion and which has been on the farm for some time. The dose is about 20 ml/kg BW, given i.v. over a period of 1–2 h, that is, an amount totalling approximately 1 l per foal. If the foal trembles, the rate of dosing should be reduced and rapid recovery will soon follow. It should raise the antibody titre of its blood to about 30% of the donor level. Oral dosing with plasma helps in cases of enteritis, but it should normally be given aseptically i.v.

Plasma therapy, given i.v., is sometimes indicated in horses of all ages. It may be necessary in foals exceeding 12–24 h of age which have received insufficient colostrum and immunoglobulins. It is also frequently indicated for septic foals, or for horses suffering considerable blood loss, where the blood pressure will be low. Therefore a supply of donor plasma held at −20°C is a very convenient source. Frozen plasma is normally thawed in a water bath at 37°C. This is a lengthy process and recent evidence indicates that careful thawing at a defrost setting in a microwave oven, alternating with short periods of agitation, until no ice particles remain, is a satisfactory and quick procedure. The process, carefully handled, causes no apparent damage to plasma proteins.

Taurine

Colostrum contains a high concentration of the nonprotein amino acid, taurine. The blood plasma of sick neonatal foals contains low concentrations of this amino acid, presumably because the undeveloped metabolic system is unable to synthesize taurine from cysteine and cystine, as occurs in the adult. There may be a case for including taurine in foal milk replacer diets, and certainly it is essential to include taurine in colostrum replacers for foals that do not receive colostrum naturally. A dietary concentration of 300 mg/kg dietary DM may be appropriate.

Enteral feeding

The requirement for frequent small feeds has been satisfactorily met by giving low residue solutions, containing sodium and calcium caseinate, glucose and fat, providing 4.2 kJ/ml through indwelling 12 French enteral feeding tubes. This fluid is given every 4 h by gravity flow at the rate of 0.35 ml/kg/min, building up to the total energy needs over 4 days. However, if gas colic occurs then an increase in the interval between feeds should help, and drinking water should be offered free choice. The tubes are placed with the tip in the distal oesophagus and the exterior portion secured by stented suture to the nostril (a stent is a mould for holding graft in place made of Stent's mass. In this case, it is a device for securing a tubular structure). An extension tube is placed in the free end and secured with a plastic guard to the halter. The tube is flushed with water and capped when not in use. This procedure avoids the risk of trauma to the pharynx caused by frequent intubation. Mineral oil is sometimes added at the rate of 1–4 ml to each feeding if constipation occurs, although its use should not persist as it will interrupt the absorption of fat-soluble nutrients.

FEEDING FREQUENCY AND METHOD

Bucket feeding of milk is the traditional method of feeding, although with compromised foals other approaches may be required. The dictum of a little and often applies forcefully to orphan foals. Frequent small amounts reduce the risk of digestive upset and of hypoglycaemia. Where the foal is to be trained to a bucket, the head can be drawn into it with a finger in the mouth – initially this may require the assistance of another person, but soon the foal will adapt to the procedure. All feeding utensils should be clean before each feed. The intake of liquid milk replacer, or a 50:50 mixture of skimmed milk and whole cow's milk, should be at the rate of 280 ml every 1.5 h so that the daily energy intake amounts to 9–10 MJ DE. The initial feeding can take place to advantage close by a horse acting as a decoy, but never at the stable door to avoid the association of it with feed. In order to minimize affinity with man, orphan foals should not be fondled.

Liquid milks are normally given at body temperature, but can equally be given cold. Within a few days the daily intake will attain 9–18 l and if the foal is permitted to drink freely it may reach 36 l. However, intake should be restricted to a maximum of 18 l in a large foal and with any evidence of diarrhoea the quantity should be reduced until the problem has subsided. Once the first few days are over the liquid can be provided in four then three feeds per day and any excess disposed of.

Automatic liquid-milk feeders of French design have proved very successful on large studs for groups of foals, and they avoid the problem of humanizing. They are electrically operated; the water is warmed and mixed with milk replacer powder (Table 7.1) at an adjustable rate. Fresh liquid is prepared to replace that used up as the foals drink. The appropriate concentrations of dry matter suggested above in

'Feeding the orphan foal' should be adhered to, as solutions that are either too weak or too concentrated can precipitate looseness or constipation. New foals are rapidly trained by experienced foals in the same yard.

Creep pellets in the form of stud nuts, or a concentrate mix, together with milk pellets and a little best-quality leafy hay from 7 days of age will encourage dry feeding. Access to *fresh* faeces from a healthy adult horse that has been wormed regularly will provide bacteria of the appropriate kind for seeding the intestinal tract. Any strongyle or ascarid eggs in the faeces should be immature and therefore of low infectivity, and so will passively traverse the GI tract of the foal; but the foal's faeces should be removed regularly. If progress is normal, liquid milk can be discontinued from 30 days of age and intake of dry feed will rise rapidly. At this time the foal may be consuming as much as 2–3 kg dry feed daily, although the consumption of hay will still be rather slight.

FOSTERING

The least-troublesome nurse is frequently an old coldblooded mare, especially of piebald breeding, or even a nanny goat; the worst type is a young flighty TB. Prospective individuals should be checked for disease, and their milk should be examined. The udder and the tail should be thoroughly washed and disinfected. The mare can be brought to the stable hooded and disorientated by walking around the area. A strong smelling substance such as camphorated oil can be placed on the muzzle and the same substance smeared on both the mare's own foal and the orphan. When these two are held together for a while the mare confuses their sounds.

A fostering gate is a boon for nurse mares, enabling foals to be suckled without being kicked. It will have facilities for feeding and watering but the mare should be allowed out for exercise at regular intervals. The crate should have a gate at each end; the critical dimensions are a length of about 250 cm, a width of 65 cm and a gap at both sides of one end 90 × 40 cm, the lower edge of which is 70 cm from the floor for access by the foals to the udder.

THE SICK NEONATAL FOAL

In addition to the tenets of good husbandry there are several important issues upon which action should be taken to further the prospects of the unhealthy neonatal foal.

Immunoglobin status should be measured. Plasma concentration of IgG should be raised to a minimum of 8 g/l, by administration of approximately 2 l colostrum, derived preferably from the dam. This should occur before 12 h of age. For foals approaching 24 h of age, or for those that are septicaemic, or hypothermic, 2–4 l equine plasma, containing at least 1.6 g globulins/l, from a suitable donor, should be administered aseptically and i.v.

Energy reserves are likely to be even lower than those of healthy foals. The risks of hypoglycaemia and hypothermia are considerable. The foal should be held in a warm draught-free box, but with clean air and neighbouring the dam.

Following administration of the colostrum (the injection of serum may take place subsequent to a meal of milk/dextrose, etc.), milk, preferably derived from the dam, should be given orally (bucket or bottle), or by indwelling nasogastric tube, every 1–2 h, so that the 24-h intake is approximately 20% of the foal's bodyweight. By remaining within smelling distance of the dam the foal is more likely to be accepted later on, and by milking the dam regularly her supply should not dry up. She is then more likely to suckle her foal at a future date. Fresh drinking water should be available. If the dam's milk is unavailable then a low fat (2%) cow's milk, fortified with 20 g dextrose/l, may be used.

The foal's urine should be monitored for the presence of glucose and the blood should be sampled before several meals to ascertain the status of glucose, TAG, ammonia, urea, potassium and haematocrit. Where TAG is elevated it may indicate hepatic malfunction and fat intake should remain low (with some still present), so long as there is no hyperglycaemia and urinary glucose. With hyperglycaemia the dextrose intake per meal should be reduced to lower urinary glucose level to less than 1%, reducing the risk of renal damage. With elevation in both plasma TAG and glucose a mixture of dextrose and fructose might replace a pure dextrose solution and the frequency of feeding should be increased, giving smaller amounts per meal. Elevated plasma ammonia should be countered with a reduction in milk protein intake. If milk replacer has been used the milk must have been spray-dried and not roller-dried. The detection of an elevated haematocrit and/or elevated plasma albumin concentration can indicate dehydration, compensation for which should be made with Ringer's solution.

If there is evidence of obstruction of the intestines and lack of bowel motility (ileus), impactions, or malabsorption, then total parenteral nutrition (TPN) will be necessary. This could be used as an adjunct (parenteral nutrition, PN) to part oral feeding, where digestive ability is poor. For a full account of the problems associated with PN and jugular catheterization the reader is referred to *The Equine Manual* (Higgins and Wright 1995). An i.v. solution for TPN over 4 days is given in Table 7.9. Alternatively the vitamin supply can be given by sterile intraperitoneal or subcutaneous injection. Monitoring should occur as indicated above, including that of electrolyte balance. Thrombophlebitis is not infrequent and there are several causes. The risk of it may be decreased by complete sterility and absence of septicaemia, the use of the most suitable catheters, the inclusion of heparin in the solution and by replacing approximately half the energy requirement with lipid emulsion (homogenized soya oil, e.g. Intralipid, Kabi Pharmacia Ltd, Milton Keynes). Lipid clearance should be monitored and fat supplementation should be avoided in cases of liver failure. Lipid use can reduce the risk of hyperglycaemia and it will reduce the total osmolality of the solution. If severe proteinuria is presented the amino acid (especially glycine) content of the PN solution should be reduced until the renal problem is resolved.

Table 7.9   Hypertonic i.v. solution for TPN of foals (2.7–3 l/day for a 45-kg foal).

5% amino acid solution	1000 ml
50% dextrose solution	500 ml
KCl	30 mEq
NaHCO$_3$	30 mEq
Injectable vitamin preparation	+*

*Commercial preparation of fat- and water-soluble vitamins.

Sugar tolerance

Glucose can be utilized by foals of all ages and maltose is successfully digested and absorbed by foals of 4 days of age, or more. Intestinal intolerance to specific foods, including lactose, occurs in some foals and older horses. Normal neonatal foals are tolerant of both lactose and glucose, as determined by a rise in blood glucose following oral administration of these sugars (note: galactose, a component of lactose, is an epimer of glucose and is rapidly converted to glucose by hepatic UDP (uridine diphosphate)-galactose-4-epimerase). Neonatal foals are, however, intolerant of the disaccharides maltose and sucrose, so these sugars are unsuitable for inclusion in milk replacers intended for feeding to foals of less than 5–7 days of age.

Infections and hygiene

Both respiratory and enteric pathogens are major causes of morbidity in young foals. Browning *et al.* (1991) surveyed 326 diarrhoeic foals in the UK and Southern Ireland from 1987 to 1989. They found that Group A rotaviruses were major pathogens in all age groups. Other pathogens included *Aeromonas hydrophila*, whereas coronavirus, parvovirus, *Campylobacter* spp., *Salmonella* spp. and *Bacteriodes fragilis* are currently likely to be minor putative pathogens in western countries. Rotavirus is a predominant cause of enteritis among foals in the USA. Whether vaccination for this virus may be an effective defence is yet to be established, but disinfection, hygiene and sound management practice will continue to be the primary control measures against the ravages of foal diarrhoea.

Gastric lesions in foals

Gastric lesions among foals showing no signs of gastric disease are of frequent occurrence under 10 days of age, but rare over 70 days of age (Murray *et al.* 1990). In common with horses under training, the lesions are situated predominately in the squamous mucosa immediately adjacent to the margo plicatus along the greater curvature. These squamous cells are less extensively protected by mucus than is the glandular mucosa. Lesions were found by Murray and colleagues to occur with lower frequency in the squamous fundus, the glandular fundus and in the lesser curvature. Lesions were more prevalent where there had been diarrhoea, and it has been suggested that a role of Gl pathogens, such as rotavirus, may be implicated. However, environmental stress with adrenal medullary hormone secretion, causing mucosal ischaemia, could also be involved.

THE STALLION

The stallion is subject to the same seasonal influences as affect the breeding cycles of the mare: his fertility is greatest in the summer and least in the winter. A large number of studies have been undertaken at Colorado State University to increase our understanding of the mechanisms that control the breeding behaviour of mares and stallions. Reference to their publications should be made where detailed information is sought.

The seasonal changes in the blood concentrations of luteinizing hormone, follicle stimulating hormone and testosterone, and consequential changes in testicular size, sperm production and libido are functions of changes in photoperiod. Recrudescence in reproductive activity, as in the mare, is a response to increasing daylight length. Thus, evidence suggests that improved fertility in the early months of the year may be obtained by following a regime similar to that proposed for mares, in which artificial light and richer feed are provided in late December and January. At no time should the stallion be allowed to fatten and higher fibre, but balanced feeds, are quite satisfactory out of the breeding season. Poorer quality hay supplemented with horse and pony nuts should then be satisfactory and they will facilitate the imposition of a rising plane of nutrition as the breeding season approaches when stud nuts, or an equivalent concentrate mixture with good-quality hay, should be introduced (Table 7.1). The energy requirements of the stallion rise during the breeding season as a consequence of increased physical activity, in particular that of pacing his run or stall, yet space for physical exercise is important to the stallion in all seasons.

There is little evidence to support the use of special supplements to enhance the fertility of stallions, but a diet of 0.75–1.5 kg cereal-based concentrates plus hay per 100 kg BW daily and clean water (Table 7.4) should suffice.

STUDY QUESTIONS

(1) What issues should be considered in deciding the optimum weaning age for a stud and for an individual foal?
(2) On making arrangements and provisions for colostrum-deprived foals what are the important issues?
(3) What action would you take in the event of diarrhoea occurring:
    (a) in a suckled foal,
    (b) in a weaned foal, and
    (c) when there is an outbreak among several foals of similar age?

FURTHER READING

Browning G.F., Chalmers, R.M., Snodgrass, D.R., *et al.* (1991) The prevalence of enteric pathogens in diarrhoeic Thoroughbred foals in Britain and Ireland. *Equine Veterinary Journal,* **23**, 405–9.

Chong, Y.C., Duffus, W.P.H., Field, H.J., *et al.* (1991) The raising of equine colostrum-deprived foals; maintenance and assessment of specific pathogen (EHV-1/4) free status. *Equine Veterinary Journal,* **23**, 111–15.

Doreau, M., Boulot, S., Bauchart, D., Barlet, J-P. and Martin-Rosset, W. (1992) Voluntary intake, milk production and plasma metabolites in nursing mares fed two different diets. *Journal of Nutrition,* **122**, 992–9.

Higgins, A.J. and Wright, I.M. (1995) *The Equine Manual.* W.B. Saunders, London.

Martin, R.G., McMeniman, N.P. and Dowsett, K.F. (1991) Effects of a protein deficient diet and urea supplementation on lactating mares. *Journal of Reproductive Fertility,* Suppl. **44**, 543–50.

Murray, M.J., Murray, C.M., Sweeney, H.J., Weld, J., Digby N.J.W. and Stoneham, S.J. (1990) Prevalence of gastric lesions in foals without signs of gastric disease: an endoscopic survey. *Equine Veterinary Journal,* **22**, 6–8.

Pagan, J.D., Jackson, S.G. and DeGregorio, R.M. (1993) The effect of early weaning on growth and development in Thoroughbred foals. *Proceedings of the 13th Equine Nutrition and Physiology Society,* University of Florida, Gainsville, 21–23 January 1993, pp. 76–9.

Rossdale P.D. and Ricketts S.W. (1980) *Equine Stud Farm Medicine*, 2nd edn. Cassell (Baillière Tindall), London.

Chapter 8
Growth

For it is easy to demonstrate that a horse may irrecoverably suffer in his shape and outward beauty as well as in strength by being underfed while he is young.

W. Gibson 1726

Normal growth patterns, growth quality and conformation in the horse are to a considerable extent beyond the control of the stud. Nevertheless, breeding and diet play increasingly measurable roles; consequently, their contribution will undoubtedly accelerate. A key to that input is an agreement of objectives that can be clearly characterized and measured.

IDEAL CONFORMATION

Selection of breeding partners for shape and size has undoubtedly concerned breeders since domestic horse husbandry began. However, the shape of a horse that is required to maximize its performance in the type of activity for which it has been bred has received scant objective attention. Variability in shape within breed is considerable. Mawdsley (1993) found a wide variation in linear traits within TBs, providing possible scope for progress by performance testing. (Variation in tissue type for work is discussed in Chapter 9.) In order that progress is achieved to improve the performance potential and working lifespan of foals it is necessary:

- for measurable objectives to be agreed within breeds at specified ages;
- for progeny performance testing to be conducted; and
- that allowance is made for corrections to malformations by the farrier and others.

How does normal growth unfold?

BIRTHWEIGHT AND EARLY GROWTH

Growth proceeds through a process of cell division and enlargement initiated by the fertilization of the egg. Cells differentiate, forming the various embryonic tissues. Soon after birth the number of cells in most tissues has reached a maximum and further growth is accomplished by hypertrophy (enlargement) of the individual cells; but in some tissues, for instance epithelial tissue, cell replication continues throughout life in order to replace cells that are sloughed off, or in hepatic tissue neoplasia occurs to compensate for malfunction. Not all tissues, organs and structures increase in size at the same rate so that during growth the shape of the animal changes. The potential for a maximum rate of growth, measured in kilograms of daily bodyweight gain, persists until about 9 months of age in the horse, at which time it gradually declines and ceases as the adult size and shape are attained. However, the overall rate of growth measured relative to the existing weight, i.e. say at 50 days of age (kg daily gain)$_{t50}$/(BW$_{t50}$), is initially slow (Fig. 8.1); it accelerates to a maximum before birth and then declines. By the 7th month of gestation, merely 17% of the birthweight and only 10–15% of the birth dry matter have been accumulated. Thus, the accretion of most of the energy and minerals present at birth occurs during the last months of gestation.

Partly because the mature number of cells in many tissues of the adult has been achieved by birth, or shortly afterwards, the maximum adult weight of horses and ponies is, to a large extent, determined by birthweight. As a rough guide, birthweight constitutes 10% of the adult weight; among TBs, individuals weighing less than about 35 kg at birth are very unlikely to reach 152 cm (15 hands) in adult life. One study revealed that the proportion of foals with a birthweight of less than 40 kg that actually raced was much smaller than the proportion of those weighing more than 40 kg (Platt 1978). Horses out of small mares by small stallions will be small as adults, but will achieve their mature size slightly sooner than will the products of large parents. Differences between breeds in rate of attainment of mature weight are greater than the differences in rate of attainment of mature height.

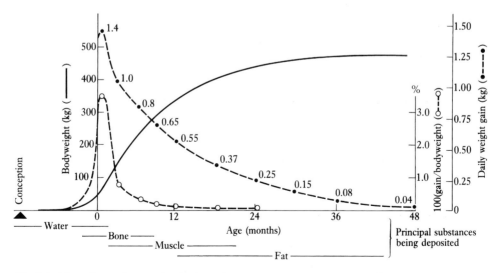

Fig. 8.1 Normal growth curve for the horse (mature weight of example, 500 kg).

Table 8.1 gives data for several breeds. As a general rule, foals attain 60% of their mature weight, 90% of their mature height and 95% of their eventual bone growth by 12 months of age. In the period from birth to maturity, the growth coefficients (the rate constant for the growth of a tissue, or structure, relative to that for the whole empty body) of the three major equine tissues in order of increasing magnitude are bone, muscle and fat, implying that bone is the earliest maturing tissue and fat the latest. This, and the earlier assertion that overall growth rate declines from birth (weight gain per unit empty liveweight) signify that ultimate height is determined in very early life, that early growth demands diets rich in bone-forming minerals, protein and vitamins, and that with increasing age an increasing proportion of dietary carbohydrate is required. The early rapid extension of the long bones of the legs, pronounced in tall breeds, render them subject to deformation through early malnutrition.

Early nutrition and weaning age

The nutrition of the foal at birth is affected not only by the feeding of the mare but also by the physiological efficiency of the uterine environment. This may to some extent be affected by the age of the mare, as indicated by the data in Table 8.2.

Table 8.1 Percentage of mature bodyweight and of withers' height attained at different ages in various horse breeds (Hintz 1980a).

Age	6 months		12 months		18 months	
	Weight	Height	Weight	Height	Weight	Height
Shetland pony	52	86	73	94	83	97
Quarter horse	44	84	66	91	80	95
Anglo-Arab	45	83	67	92	81	95
Arabian	46	84	66	91	80	95
TB	46	84	66	90	80	95
Percheron	40	79	59	89	74	92

Table 8.2 Effect of age of TB mare on bodyweight and height at withers of foals (Hintz 1980a; data based on 1992 foals).

Age of mare (years)	Age of foal			
	30 days		540 days	
	Foal weight (kg)	Height (cm)	Weight (kg)	Height (cm)
3–7	93.0	108.0	393.7	152.4
8–12	97.5	110.5	401.4	153.7
13–16	98.0	110.5	396.9	153.0
17–20	95.3	109.2	391.0	152.4

Differences in birthweight brought about by nutritional deviations in the pregnant mare can, however, be proportionately lessened by nutritional adjustments in early postnatal life. Although birthweight has a major impact on ultimate size for both genetic and environmental reasons, weaning age, in conditions of good management, has little influence. Foals weaned after receiving colostrum can achieve growth rates equal to that of those weaned between 2 and 4 months of age. These in turn may grow faster than foals weaned at 6 months of age. The appropriate weaning age, therefore, for any particular stud farm, turns on the most convenient and reliable management practice.

LATER GROWTH AND CONFORMATIONAL CHANGES

Initial and ultimate weights and heights differ as between colts and fillies (Table 8.3). The differences are small and limited studies in England (Green 1969) failed to detect sex differences in linear measurements, or any between early- and late-born foals. Nonetheless, some substantial evidence suggests that foals born late in the season are heavier and taller than early-born foals (Table 8.4) despite a somewhat shorter period of gestation. As previously indicated (Chapter 7), gestation length seems to be a function of daylight length in late gestation. Birth weight and subsequent rates of growth, as implied above in this chapter, also depend on the mature size of the breed. Micol and Martin-Rosset (1995) recorded that the rate of weight

Table 8.3 Effect of sex on growth of TB foals (Hintz 1980a; data based on 1992 foals).

Age (days)	Bodyweight (kg)		Withers' height (cm)	
	Colts	Fillies	Colts	Fillies
2	52.2	51.3	100.3	99.7
60	136.5	134.7	118.3	118.1
180	244.9	235.9	134.6	133.0
540	435.5	401.4	154.3	152.4

Table 8.4 Effect of month of birth on bodyweight and withers' height of TBs (Hintz 1980a; data based on 1992 foals).

Month of birth	Age			
	30 days		540 days	
	Weight (kg)	Height (cm)	Weight (kg)	Height (cm)
February–March	95.3	109.2	396.9	153.7
April	97.5	110.5	402.8	153.7
May	100.7	111.1	403.7	153.7

gain from birth to weaning on pasture of the heavy French breeds–Breton, Comtois and Ardennais (700–800 kg mature BW) is 1.3–1.7 kg/day. Even under harsh upland conditions the foals gain at the rate of 1.3–1.5 kg/day.

On the whole, differences in growth rate, after the neonatal period, have little influence on mature size. Although maximum height may be approached soon after 12 months of age, this may be delayed without a reduction in the ultimate measurement by reducing the rate of feeding slightly. Similarly, 90% of mature weight may be achieved at 18 months, but delayed until 24 months by the same restriction. In Fig. 8.2 the changes in height at the withers over the first 12 months of pony and TB foals are compared. Although the foals achieve very different heights the pattern of growth is similar. Height at the withers largely reflects linear growth of the long bones in the front legs. Long bones increase in diameter or thickness throughout their length, but no increase in length occurs within the shank, or diaphysis, after birth (Figs 8.3 and 8.4). They increase in length by growth in a metaphyseal plate at both the near and distant ends (proximal and distal) from the body. The rate of growth at each end is different; Table 8.5 gives values recorded for crossbred ponies (Campbell & Lee 1981). Correction of distortions in bone growth, owing to bad feeding practices and other causes, is possible during the phase of rapid growth of the end in question. Thus, for the distal radius or tibia, such correction could be imposed up to 60 weeks of age, whereas fetlock distortion, a fairly common condition, requires treatment by 3 months. In either case, temporarily restricted growth will either not or only slightly influence ultimate size.

The balanced adult horse of good conformation has a height at the withers that equals its length from the tip of the shoulder to the point of the buttocks (see Fig.

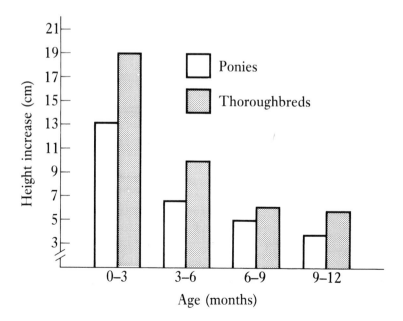

Fig. 8.2 Height increase (cm) at withers of pony and TB foals (Campbell & Lee 1981).

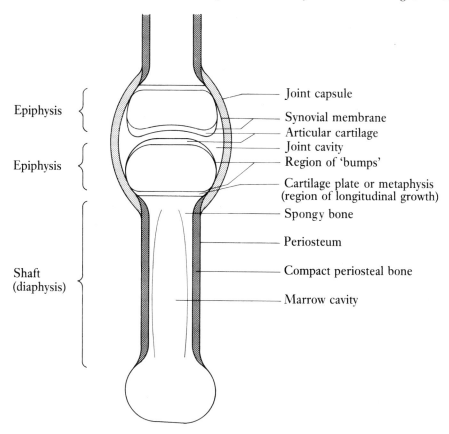

Fig. 8.3 Long bones and their articulation in late growth (joint cavity has been expanded for purposes of visualization). Note the regions of growth: cartilage at either end of diaphysis; cartilage of epiphysis; and periosteum of shank.

6.1). By contrast, a foal is taller than it is long (Plate 8.1) so that it inevitably trots wide behind. Fillies at birth tend to be fairly level across the top, but may be as much as 5 cm higher over the croup than over the withers at 1 year and then balanced again by 5 years. The length of their bodies tends to be greater than their height at the withers. Colts are generally higher over the hip at birth but level by 3 years. Some horses that have grown poorly in the front legs are lower over the withers than the croup at maturity and, as nearly 60% of the bodyweight is normally carried by the front legs, this unusual conformation can force additional weight and stress on the forequarter, increasing the risk of damage.

Similar height increases at the withers and at the hip are found in TB foals between 14 and 588 days of age when given feed at recommended rates. The hip height is about 1 cm greater than the withers' height throughout this period (Thompson 1995). Cymbaluk and colleagues (1990) found that quarter horses fed *ad libitum* cf. limit-fed between the ages of 6 and 24 months may have a slight tendency to gain relatively more in hind body mass than in fore body mass, but to

Bone

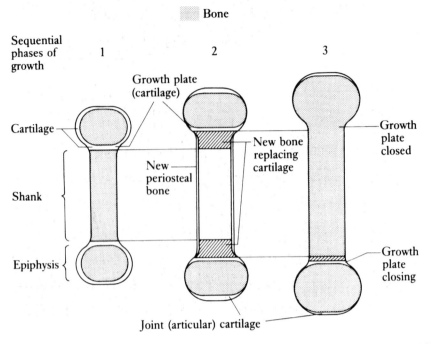

Fig. 8.4 Typical growth of long bone, for example radius or tibia (after Rossdale & Ricketts 1980).

Table 8.5 Growth of leg bones in three male and three female crossbred ponies (Campbell & Lee 1981).

Bone	Length of bone (cm)		Increase at each end (%)		Age at closure of growth plates (weeks)	
	0–7 days	2 years	Proximal	Distal	Proximal	Distal
Femur	21.6	30.9	24	20	55	55
Tibia	21.3	29.1	22	17	60	60
Radius	20.7	28.1	12	24	54	69
Humerus	16.0	22.6	31	11	80	70
Metatarsal	22.7	23.7	5	5	40–44	40–44
Metacarpal	18.4	19.4	5	5	40–44	40–44
Phalanges 1	5.9	6.6				
Phalanges 2	2.3	2.5				
Phalanges 3	5.6	6.2				

gain slightly less in croup height than in withers' height. Thus, adult conformation could be influenced by rate of feeding during growth, and very restricted feeding may restrict development of the hindquarter powerhouse. The extent to which compensatory growth occurs after 24 months is unclear.

When fed liberally, the growth of ponies up to 9 months of age may increase to nearly 1.5 kg/day, although by 12 months the rate may have fallen to half this amount. Under hill conditions, where young stock can receive little in the way of

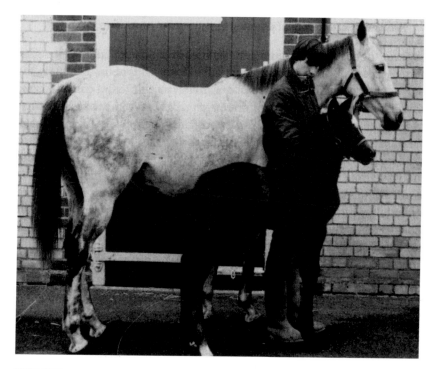

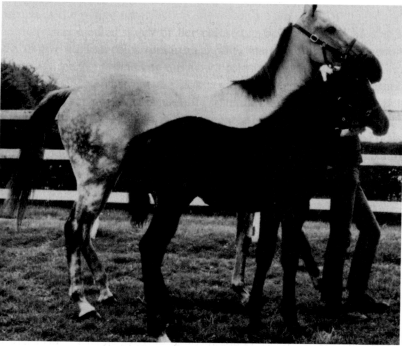

Plate 8.1 Grey colt (51 kg at birth) by Vitiges out of Castle Moon at Derisley Wood Stud, Newmarket: 5 days old (top) and 82 days old (below). Note that the height of the young foal exceeds its length, a characteristic that changes as adult proportions are attained. Over a period of as little as 77 days the length of the foal has increased considerably so the body proportions are already more like those of the adult. The height of the croup tends to increase faster than that of the withers during the first few months, but by 3 years of age in the TB the withers' height has caught up, through differential growth of the leg bones.

concentrates, growth can be strikingly depressed. Morgan horses on range were shown to be between 23 and 46 kg lighter and 2.5 cm shorter at 5 years of age than their contemporaries in the same environment receiving supplementary cereal-based concentrates (Dawson *et al.* 1945). With only moderate restrictions, such horses may attain normal mature height at 5 years, whereas with more liberal feeding they can reach it by 3 years (Kownacki 1983). New Forest filly ponies, fed from 6 months of age throughout the winter at a rate that only maintained bodyweight, grew faster during the following summer on rough pasture than their counterparts permitted to gain 0.4 kg/day in the winter (Ellis & Lawrence 1978a,b, 1979, 1980). Although bodyweight may remain constant in such deprived circumstances, parts of the skeleton grow differentially so that the restricted fillies in this example were tall, thin and shallow-bodied. With a delay in the closure of growth plates these differences are partly corrected during later growth. Catch-up growth of this kind is not without risk among potentially tall breeds, and although the evidence against interrupted growth curves is slight, a smooth growth curve should be sought for TBs, saddlebreds and the like (Fig. 8.1).

It is concluded that compensatory growth occurs in horses with some small influence on ultimate conformation, but the influence of this on the extent of active working life is yet far from clear. The subject of growth abnormalities is discussed in 'Developmental orthopaedic disease', this chapter.

Supplementary feeding in the first winter

With full access to good summer pastures, yearlings require no supplementary feed (that is, assuming the land has no specific trace-element deficiencies). The amounts of supplementary feed necessary during the previous winter will depend, first, on the quality and, second, on the amount of grazing available. Winter pastures have little value as a stimulus to growth in young stock, but will provide reasonable roughage to complement cereal-based concentrates. Until the spring flush is available, stock at this age should receive from 1.25 to 1.5 kg concentrates containing 16% protein and appropriate vitamins and minerals per 100 kg BW daily (see Figs 6.2–6.4 for height and girth equivalents).

When exercise is limited by in-wintering, nuts or other concentrate feeds should be given to young stock in two meals daily to avoid intestinal upsets that might cause oedema in the legs. Stabled, weaned horses may be given free access to good-quality hay, and when they are receiving concentrates at the rate of between 1.25 and 1.5% of bodyweight, they will consume hay or other forage dry matter at a rate of about 0.5–1.5% of bodyweight. This should allow for a normal growth rate during the winter and early spring. In the study of New Forest and Welsh ponies, the maintenance requirement of the filly foals amounted to between 14 and 14.6 MJ DE/100 kg BW, equivalent to 1.75 kg average hay or 1.1–1.2 kg average nuts per 100 kg BW (Ellis & Lawrence 1980). At the end of the first winter, that is, when yearlings are 11 or 12 months old, colts and fillies should be separated. The first oestrus should occur before May, unless winter conditions and feed have been poor.

Long yearlings

From 90% of mature weight long yearlings (during their second year of life when 80–90% of mature weight has been attained by light breeds of horses) can be fed most economically on maintenance rations as for mature horses. This should include good-quality hay and a concentrate offered at a rate of 0.75–1% of bodyweight and providing cereals, a concentrate protein source, minerals and vitamins. However, when these more slowly growing stock are out on good-quality pasture continuously they should need no other source of feed, apart from a trace-element mixture should a specific deficiency exist.

Meat production

The heavy breeds (Breton, Comtois and Ardennais (700–800 kg mature BW) are grown in France for meat in high forage systems. The systems include: (a) colts castrated at 18 months and surplus fillies slaughtered at weights exceeding 700 kg off grassland in mixed systems with cattle; (b) following grazing and finishing indoors at 10–15 months of age (450–500 kg) BW with continuous gains of 1.0–1.4 kg/day; (c) slaughtered supplemented with cereals on grassland at the end of the grazing season at 18 months (550–580 kg BW); or (d) at 22–24 months (620–670 kg BW) after finishing indoors during the second winter. The interesting obsevation was made by Micol and Martin-Rosset (1995) that as concentrate allowance to finishing horses is increased the ad libitum intake of hay decreases by 1.26 kg DM for each 1.0 kg increase in concentrate. On the other hand, where maize silage is given (>30% DM) the silage DM intake decreases by 0.81 kg for each 1.0 kg increase in concentrate, so overall intake increases. The concentrate improves silage acceptability as the horses consume 24 g DM/kg BW daily of a 100% hay diet, but only 16 g DM/kg BW daily of the 100% silage diet.

EFFECTS OF DIETARY COMPOSITION

Nutrient requirements are dealt with in Chapter 6. Certain aspects of those requirements which have engaged the interests of research workers over the last 10 years are considered here, especially where it may affect the quality of the skeleton and mobility of the horse.

Dietary protein

The NRC (1989) has published dietary crude protein requirements for growing horses. The needs of the horse are for protein that is digested, yielding products that are absorbed. The digestibility of proteins differs to a small extent between sources. Digestibility is normally expressed as apparent digestibility, which is:

$$\frac{(\text{N intake} - \text{faecal N})}{\text{N intake}}$$

As endogenous faecal N loss is proportional to bodyweight and not to N intake

$$52\,\text{mg N} \times \text{kg BW}^{0.75}$$

(Slade *et al.* 1970), apparent N digestibility increases as the N content of the diet is increased. By comparison, true digestibility, for which the endogenous N loss is subtracted from the faecal N, is similar for a range of dietary N concentrations. For most protein sources true digestibility falls in the range 0.7–0.9 and for most common sources it is in the range 0.75–0.85. One should strictly make some allowance for digestibility in calculating protein requirements, although dietary concentration of a selected protein does not affect its true digestibility.

Three diets, containing all nutrients at concentrations recommended by the NRC (1978), except protein, have been compared in Arabian, TB and standardbred foals from 4 months of age (Schryver *et al.* 1987). The dietary protein and lysine concentrations were: 90, 140 and 200 g/kg and 2.8, 7 and 12.6 g/kg diet DM, respectively. The lowest protein intake depressed the rate of daily weight and height gain and gain in forecannon circumference, whereas there was no significant difference in the responses to the other two diets. At 9 months of age the foals that had received the 90 g protein/kg diet were changed to 200 g. After a further 140 days there was no significant difference in bodyweight, height or cannon circumference among the groups. The NRC (1989) recommends 131 g protein/kg diet (90% DM, i.e. 146 g/kg DM) from 4 to 6 months of age. Breeders have frequently limited the protein content of the weaned foal's diet to avoid the development of skeletal defects. Nevertheless, there is little evidence that high intakes of protein are a cause of this frequent problem in several breeds (see 'Developmental orthopaedic disease', this chapter). In the experiment of Schryver and colleagues, rate of feed intake, bodyweight gain, height gain and the increase in forecannon bone girth were all greatest with the highest protein intake, although differing not significantly from the 140-g diet. Higher protein and lysine allowances increase cannon girth, whereas inadequate protein intake tends to produce more spindly bones that are only slightly shorter. Some evidence, however, does indicate that bone density is slightly lower in foals that are allowed to grow at their maximum rate, and therefore bone strength for a given body size may be slightly compromised by very rapid growth rates.

With growing horses of all ages it is important that the energy allowance is not excessive with respect to the protein allowance. Commencing at 120 days of age, poorer body condition has been observed in foals given 125% of the recommended NRC (1978) energy allowance but only 100% of the protein allowance, compared with foals given either 100 or 125% of both NRC energy and protein allowances. Possibly as a result of this and other evidence, the NRC (1989) recommendation for the protein allowance of yearlings is 20% greater than the earlier recommendation, although for weanlings a similar recommendation is given. Most evidence also indicates that increased growth rates can be achieved by providing more dietary protein, of average quality, than the amount recommended by the NRC (1978). On the other hand, the INRA (French) recommendation, given as MADC (see Chapter 6), is 66% of the NRC (1989) crude protein value at 6 months and only 54% at 12 months. Very approximately the MADC value should be 70% of the crude protein value to be equivalent. The crucial factor is which proposal leads to the healthier ultimate development of the frame.

In experiments by Thompson *et al.* (1988) energy, protein, Ca and P intakes, relative to NRC (1978) recommendations, were varied in the diet of growing foals. The results showed that inadequate dietary Ca caused lower increases in bone mineral content and in long bone length. Raised intakes of energy, or of protein, increased the rate of growth in length of several bones; but inadequate intakes of protein relative to that of energy compromised the development of cortical bone and led to smaller increases in withers' height. Thus, fast rates of bodyweight increase caused by excessive energy intakes introduce an increased risk of inadequate bone development, where the balance of nutrients is incorrect, and increase the risk of conformational and musculoskeletal abnormalities (Thompson *et al.* 1988; Cymbaluk *et al.* 1990).

In the experiment of Schryver and colleagues (1987) Ca metabolism was not adversely affected by high protein intakes, even though the 200-g protein diet contained nearly twice as much S as was contained in the 90-g protein diet. Other evidence indicates that a protein intake that is 130% of the NRC (1978) recommendation for foals causes an increase in the urinary excretion of Ca and P. This is primarily the result of the increased S-amino acid consumption causing a decrease in renal tubular filtrate pH and a consequential reduction in the renal tubular reabsorption of Ca and P (see cation anion balance, Chapters 9 and 11 and Appendix D). The difference between studies in the risk of urinary Ca loss may be due to:

(1)   differences between studies in the actual acid:base balance of the diets; and
(2)   the age at measurement. There is a more rapid turnover of the amorphous calcium phosphate salts, found in greater abundance in the bones of younger animals, compared with the turnover of the more stable apatitic crystals that predominate in the bones of older animals.

There is therefore a theoretical risk to bone calcification and dietary supplementation with sodium bicarbonate may be indicated with high protein diets. Optimum amino acid balance of these diets is also indicated, as excessive imbalanced protein could exacerbate urinary Ca loss.

Limiting amino acids

The first and second dietary limiting amino acids for the growth of foals are lysine and threonine, when they are given a diet of maize and oats, in equal amounts, plus soya, dried lucerne and coastal Bermuda grass hay (Graham *et al.* 1994). An excessive consumption of these amino acids does not materially compromise the tubular reabsorption of Ca. The lysine requirement, as a proportion of the diet, decreases rapidly with increasing age, but in the Finnhorse breed it exceeds 31 g/day up to 10 months and it is important for normal blood haemoglobin and red blood cell formation (Saastamoinen & Koskinen 1993). An imbalance in dietary amino acids is likely to increase urinary losses of S, for a given rate of N retention, and therefore, as indicated in 'Dietary proteins' above, excessive poor quality dietary protein may compromise Ca balance.

Summer pasture generally contains more protein, and of a higher quality, than protein given to in-wintered horses. If inadequate dietary protein is given to foals during their first winter Scandinavian evidence indicates there is a fall in the concentrations of plasma albumin (Mäenpää *et al.* 1988a,b). According to this evidence, total plasma protein fell from 68 g/l in November to 56 g/l in April and reductions occurred in the plasma concentration of several of the free essential amino acids: isoleucine, leucine, lysine, phenylalanine, threonine and valine. Perhaps more consistently under these conditions decreases in plasma lysine, methionine and valine have been reported (Saastamoinen & Koskinen 1993). These decreases can be associated with a marked interruption to the normal pattern of growth to 12 months of age. On the other hand, other reports indicate that the effect of inadequate dietary protein on serum total protein has been slight, or absent, in adult ponies (Reitnour & Salsbury 1976) and among weanlings, when growth rate has been influenced (Godbee & Slade 1981), although Saastamoinen and Koskinen (1993) also found a reduction in the concentration of blood haemoglobin of their in-wintered foals.

Complete feed mixtures for growing horses

Complete mixtures of concentrates and chopped roughage are given *ad libitum* routinely to growing cattle, but to suggest that similar mixtures should be given to groups of growing TBs and horses of other breeds may seem an affront, or at the very least inappropriate. When a number of such stock are reared together through both the first and even the second winter months, there is much to recommend such a procedure. It does require sheltered *ad libitum* feed hoppers, as by ensuring that the feed is always present, risk of colic or of laminitis (founder) is nonexistent in foals. Complete feed mixtures should initially contain a cereal-based concentrate and good-quality chopped hay in a ratio of 2:1, gradually falling, as the growth rate declines, to a ratio of 1:1. Stock will normally consume a total amount of dry feed daily equivalent to 3–3.5% of their bodyweight. The physical make-up of the mix and particle size should be so regulated that no segregation between the hay and the concentrate arises. Molasses is often a good material to include in the mix as an aid in preventing segregation and some ingenuity is required to ensure that no bridging occurs in hoppers. The author's own evidence (Frape 1989) with such mixtures has been very encouraging, as assessed by conformation and control of DOD.

Preparation for sale

TB yearlings are normally prepared for the autumn sales by excluding them from pasture. A diet based on 1.5 kg good-quality stud nuts or other concentrate, plus 1–1.5 kg hay/100 kg BW is a typical practice. Traditional concentrate mixtures contain 75–90% of oats, plus bran, soyabean meal, dried skimmed milk and sometimes a vitamin/mineral premix. There is no particular reason, however, why these horses should not be provided with a complete mixture of concentrates and chopped hay as advocated for younger stock, but in a ratio of 0.5:0.5 concentrates to hay. Table 8.6 gives proposed feed mixtures for growing horses and Table 8.7 rates of feeding (on the basis of 90% DM). The composition of these diets conforms with the principles of equine growth previously propounded in this chapter.

Table 8.6 Feed mixture for growing horses for providing with chopped forage or grass hay or as a supplement to poor pasture.

	Concentrate mixtures (%)		
	Weaned foal	Yearling	Presale of long yearling
Oats	41.1	44.7	59.0
Wheatbran	15.0	15.0	10.0
High protein grassmeal	15.0	15.0	10.0
Extracted soyabean meal	18.0	15.0	10.0
Molasses	7.5	7.5	7.5
Feed-grade fat	1.0	1.0	2.0
Limestone	1.2	0.7	0.5
Dicalcium phosphate	0.5	0.5	0.4
Salt	0.5	0.5	0.5
Vitamins/trace elements*	0.2	0.1	0.1
Total	100.0	100.0	100.0

*See Table 7.1.

Table 8.7 Daily feed allowances (kg) for growing horses (500-kg mature weight).

	Concentrates	Hay (clover/grass)
Foals (kg BW)		
100 (4–5 weeks)	0.5	—
130–180	2.2–3.2	1.3–1.9
180–230	2.9–3.9	1.8–2.3
230–270	3.6–4.8	2.2–2.8
270–320	4.0–4.6	2.7–3.2
Yearlings (kg BW)		
310–360	3.5–4.5	3.0–3.7
360–410	3.0–4.2	3.6–4.1
410–460	3.0–3.8	4.0–4.5

DEVELOPMENTAL ORTHOPAEDIC DISEASE

Disturbances of skeletal growth and development in the horse have been included within the umbrella term developmental orthopaedic disease (DOD) (Jeffcott 1991). The major disorders included are:

- dyschondroplasia (DCP) (or osteochondrosis [OC]; osteochondritis dissecans [OCD] is a similar condition, but occurs where inflammation is present and a separation of a piece of articular cartilage and of the underlying bone within the joint has occurred);
- physitis (physeal dysplasia or epiphysitis);
- angular limb deformities and flexural deformity contracted tendons; and
- vertebral abnormalities (wobbler disease; also see vitamin E, Chapter 4).

The likely causes of DOD include:

- congenital malpositioning and/or malnutrition;
- high rates of postnatal growth with unbalanced diets;
- probably biomechanical trauma;
- endocrinological dysfunction;
- toxicity, including iodine toxicosis;
- heredity. Many investigators have demonstrated this in several breeds (Sandgren 1993); note: whereas most colts with signs are destroyed, most fillies are retained. Sire index is related to DOD, yet stallions having a propensity to sire affected offspring may not show signs themselves.

Dyschondroplasia is a generalized disease of young animals resulting from a disturbance of the endochondral ossification of the growth cartilage. It therefore involves the articular/epiphyseal and metaphyseal growth plate regions of long bones. When it affects the articular-epiphyseal cartilage it may result in synovitis. The differentiation and maturation of cartilage cells, and the matrix they secrete in these regions, are affected. The proliferating chondrocytes produce and secrete the extracellular cartilage matrix of proteoglycans and collagen type II. This cartilage may be faulty, especially in a dietary Cu deficiency, and the capillary endothelial cells fail to penetrate the distal region of the hypertrophic zone. Hypertrophic chondrocytes secrete: (a) metalloproteinase enzymes (including lysyl oxidase, necessary for linking adjacent protein chains), essential for cartilage development, and (b) a basic fibroblast growth factor (bFGF), which stimulates proliferation and migration of endothelial cells and so the angiogenesis of this region in the fetus. The capillary invasion appears to be inhibited by large carbohydrate meals and the hypertrophic zone then fails to mature. The immature region extends, osteoid does not form normally, so calcification is faulty and the bone is weakened. Subchondral fracture lines may be apparent, loose pieces of tissue appear in the joint (OCD) and synovitis develops. Bone fragments occur particularly at the distal extremity of the tibia (Jeffcott 1991). Secondary osteoarthrosis, occurring in the articulations of the cervical spine, causes stenosis of the vertebral canal and signs of ataxia (wobbler syndrome).

There are particular predilection sites of dyschondroplasia. These are the shoulder and fetlock in the foreleg, the stifle and hock in the hindleg and the cervical spine (Jeffcott 1991). During the suckling phase the incidence of DOD is slightly greater in colts than in fillies, and certain breeds may be more prone. These include standardbreds, TBs and quarter horses; however, it may be that a closer scrutiny has been paid to these breeds. The pony was thought to be exempt from risk, but the author's own studies in Newmarket (D. Frape unpublished observations) show this not to be so.

Clinical signs may, or may not, be present, but biomechanical stress can precipitate them. Dyschondroplasia may lead to joint stiffness, joint distension (e.g. 'bog spavin'), angular limb deformities and flexural deformity. The first signs occur at various ages: in very young foals, shortly after weaning, or in yearlings and older animals, particularly with the onset of training. However, the peak incidence seems to occur between weaning and the end of December in northern latitudes.

Intake of dietary protein, soluble carbohydrate and energy and effects of exercise

Diets providing excessive amounts of soluble carbohydrate (digestible starch) and total energy [129% of the recommendations of the NRC (1989)], given to foals from 130 days of age, caused widespread lesions of dyschondroplasia (Savage *et al.* 1993a,b). In both of these studies part of the energy increase was achieved with extra fat and so it may not be concluded that the adverse effects were solely, or necessarily, the result of a raised postprandial insulin response (see below). A dietary protein level of 126% of NRC (1989) recommendations had no significant adverse effect, compared with 100% of those recommendations (Savage *et al.* 1993a). High protein diets do not seem to predispose foals to dyschondroplasia.

The author increased the protein allowance and quality of a foal weaning diet bringing about an increase in growth rate and a reduction in the clinical evidence of physitis, and flexural and angular deformities among successive crops at three TB studs (Frape 1989). Foals weaned at 3–4 months of age were given this diet *ad libitum* as a coarse mix diluted with 100 g molassed chaff/kg. Glade (1986) postulated that nutritionally induced effects on cartilage growth are mediated by the endocrine system. A single meal initiates insulin secretion and a T_4 response. High plasma concentrations of insulin, resulting from the consumption of large amounts of glucose-yielding carbohydrate, can inhibit growth hormone. The insulin apparently stimulates an early postprandial clearance of T_4 and its conversion to T_3. Glade suggested that dyschondroplasia has similarities to hypothyroidism (see Chapter 3, iodine) and the episodic transient postprandial hypothyroidaemia, produced by a high carbohydrate meal, could cause dyschondroplasia. The *ad libitum* feeding in the author's study (Frape 1989) was intended to exclude large intakes of glucose-yielding carbohydrates during any one hour of the day, so that hyperinsulinaemia would be avoided.

The effects of high energy diets on the abnormal development of joints seem to be independent of their effects on growth rate, but the frequency and severity of lesions are less where foals are exercised. Such exercise is generally to be recommended, except where severe lesions already exist. With adequate dietary energy Raub *et al.* (1989) demonstrated that as little as 20 min of medium trot 5 days per week between 147 and 255 days of age in TB and quarter horses was sufficient to increase the radiographic density of the medial side of the third metacarpal and to increase its circumference (the increase in density may simply reflect an increase in circumference). An increase in the breaking strength of the bones should then have been achieved.

Restricted dietary protein

There is a tendency for TB and many other breeders to provide weaned foals and yearlings with less dietary protein than is recommended by the NRC (1989). O'Donohue (1991) found that of the nutrients he measured, dietary protein was the only one for which the daily allowance of short yearlings (towards the end of the first winter housing period), on 46 Irish stud farms, was on average less than that recommended by the NRC. The effect of this will have been to lower the rate of growth, with the objective of reducing the risk of DOD. There is some evidence that DOD is more prevalent in overweight horses, although from evidence we and others have produced (reviewed by Frape 1989), this may be caused by excessive dietary energy and otherwise badly balanced diets. O'Donohue found that vitamin and mineral supplementation in studs was a 'hit and miss' affair. Supplementary allowances of vitamins A and D ranged from 0 to 18 times the NRC (1989) recommendations. Nevertheless, faster growing horses may still be more prone to osteochondrosis (Sandgren 1993). Yet in the latter Scandinavian evidence the faster growing horses were proportionately heavier at birth.

Normal feeding practice and meal frequency

One consequence of rising labour costs has been to decrease the number of feeds per day. If energy-rich feeds are fed in large amounts, say once daily, rather than in smaller amounts in three meals, then there may be an increasing risk of DOD from very large insulin responses. O'Donohue (1991) found that once-daily feeding of TB short yearlings was gaining in popularity, that there was a tendency for DOD to be more prevalent where once daily feeding was practised, and that 67% of the 1711 foals he examined in Ireland showed some signs of DOD, although only 11.3% were deemed to need treatment (cf. 10–16% in Scandinavian standardbreds; Sandgren 1993). Of the treated cases, angular limb deformities and physeal dysplasia together constituted 72.9% of the total. O'Donohue indicated, in agreement with the author's experience in England (D. Frape unpublished observations), that the peak clinical expression of DOD occurred between weaning and the end of December, during the introduction of concentrate feeds. Lesions of OC have been detected between birth and 3 months of age (Sandgren 1993), so that the stress of weaning may simply exacerbate an existing condition. With increasing labour costs automated or *ad libitum* feeding systems, to provide more frequent feeds to yarded stock, could be a welcome development, as the author has found (Frape 1989).

Minerals, trace elements and DOD

In Chapter 3 in discussing the nutrition of Ca, P and trace elements it was pointed out that excessive dietary P consistently produces lesions of dyschondroplasia, although clinical signs of NSHP normally do not occur when adequate dietary Ca is provided. Savage *et al.* (1993b) gave foals from 130 days of age a diet providing 388% of NRC (1989) P recommendations, but 100% of the Ca recommendations, and found severe lesions of dyschondroplasia without clinical signs of NSHP. The incidence of dyschondroplasia was much greater than in those given 342% of the Ca with 100% of the P recommendations, or in those given 100% of both recommendations. The high P diet would have a lower cation:anion balance (see Chapter 9 and Appendix D) and may therefore have caused some acidosis which would induce bone Ca mobilization.

Pregnant mare and foal diets containing less than 10 mg Cu/kg may cause DOD and flexural deformity. Lysyl oxidase is a Cu-containing enzyme required for the cross-linking of protein chains in elastin and collagen of cartilage through the lysine residues, by oxidizing the ε-amino group. Failure of this function disrupts normal bone cartilage development. Dietary Cu supplements for the pregnant mare and concentrations of up to 30–45 mg Cu/kg in the weanling's feed may decrease the incidence of dyschondroplasia and physitis in growing horses. These levels of Cu are absolutely safe and well below any toxic threshold, as the horse has a considerable resistance to chronic Cu toxicity. Cu supplementation of the lactating mare has no measurable effect on the Cu content of the milk, which is quite low.

Summary of control of DOD

In the rearing of foals:

- Avoid large meals rich in glucose-yielding carbohydrate, but ensure the protein requirements are met with balanced amino acids.
- Provide adequate dietary Cu, but with a trace element balance.
- Provide adequate dietary Ca with a correct Ca:P ratio.
- Provide adequate exercise daily.
- Avoid breeding from stallions that are genetically predisposed to produce dyschondroplasia-affected foals.
- In studs where there are risks of dyschondroplasia, the foals should be given no more than the recommended NRC rates of DE intake.

- Where clinical DOD occurs, the energy intake of foals should be decreased and the joint trauma minimized by restricting the foal to hand-walking exercise until signs disappear.
- In studs where good labour is scarce consider the introduction of automated feeding to avoid large meals, so long as careful daily observation, by experienced persons, of each foal is not compromised and so long as obesity is avoided.

STUDY QUESTIONS

(1)   What factors are important in selecting a suitable pasture for weaned foals?
(2)   What action should be taken with a foal presenting mild signs of orthopaedic disease, or with moderate signs, at an early age, or postweaning?

FURTHER READING

Cymbaluk, N.F., Christison, G.I. and Leach, D.H. (1990) Longitudinal growth analysis of horses following limited and *ad libitum* feeding. *Equine Veterinary Journal*, **22**, 198–204.

Mawdsley, A. (1993) *Linear assessment of the Thoroughbred horse*. MEqS thesis, Faculties of Agriculture and Veterinary Medicine, National University of Ireland, Dublin.

O'Donohue, D.D. (1991) *A study of the feeding, management and some skeletal problems of growing Thoroughbred horses in Ireland*. MVM thesis, Faculty of Veterinary Medicine, University College, Dublin.

Raub, R.H., Jackson, S.G. and Baker, J.P. (1989) The effect of exercise on bone growth and development in weanling horses. *Journal of Animal Science*, **67**, 2508–14.

Sandgren, B. (1993) *Osteochondrosis in the tarsocrural joint and osteochondral fragments in the metacarpo/metatarsophalangeal joints in young standardbreds*. PhD thesis, Swedish University of Agricultural Sciences, Uppsala.

Thompson, K.N. (1995) Skeletal growth rates of weanling and yearling Thoroughbred horses. *Journal of Animal Science*, **73**, 2513–17.

Thompson, K.N., Jackson, S.G. and Baker, J.P. (1988) The influence of high planes of nutrition on skeletal growth and development of weanling horses. *Journal of Animal Science*, **66**, 2459–67.

Chapter 9
Feeding for Performance and the Metabolism of Nutrients During Exercise

But if you intend the next day to give him an heat (to which I now bend mine aim) you shall then only give him a quart of sweet oats and as soon as they are eaten, put on his bridle, and tie up his head, not forgetting all by ceremonies before declared.

T. De Gray 1639

Feed costs are major outgoings in the breeding, training and use of sport horses. Corbally (1995) found that in the Irish sport horse breeding industry feed was the second largest cost after labour, representing 16% of the total cost when the costs attributable to amortization of buildings, and brood mares and the current expenditure on labour, stud, veterinary surgeons, schooling, bedding, farrier, showing, tack and equipment, and registration were also included. At equestrian sports centres feed was the second largest cost representing 19.4% of the total. It is therefore important to the economy of the industry that feed has the optimum composition and is used judiciously. This chapter, however, considers the principles underlying the metabolism and feeding of nutrients, so that composition of feed and methods of its provision may be manipulated to optimize performance and minimize the risks to the horse and rider. To achieve brevity the discussion is restricted to the relationship of performance to the functioning of tissue cells and their nutrition.

WORK AND ENERGY EXPENDITURE

A major role of feed for working horses is the conversion of the chemical energy of feed into locomotion at speeds varying from 160 to over 900 m/min (6–35 mph) for distances varying from 1 to 150 km or more. This enormous range may superficially lead to equal fatigue in fit horses, but quite different processes of nutrition physiology are involved at the two extremes of distance and speed. At the one extreme a flat race of 6–8 furlongs (1.2–1.6 km) would *theoretically* increase the day's energy needs by a mere 4% (Plate 9.1), a barely perceptible effect, whereas at the other they would be increased by five- to sixfold. Training regimes recognize both this and the different responses of breeds to contrasting forms of work. These distinctive training procedures induce profound and dissimilar physiological changes in the attainment of fitness. Diets must be formulated to conform with these changing needs, but appetite may flag in the process. First, an adequate and optimum nutri-

Plate 9.1 Rathgorman, ridden by Chris Bell, cantering 8 furlongs (1.6 km) on the all-weather track at Dunkeswick, West Yorkshire, watched by trainer Michael W. Dickinson. As part of the training for National Hunt racing the horses canter 5 days per week. If a horse races every 7–21 days it has an easy canter on the day before and for several days after a race. A regular work programme is maintained during the racing season but with some variety in location, scenery and type of work.

tion for a particular purpose implies an optimum supply of nutrients to each tissue and cell and the efficient disposal of waste products.

In sprint races, horses obtain much of their muscular energy from anaerobic pathways of respiration, whereas in extended work, such as endurance competitions, energy is derived almost exclusively through aerobic pathways. (Anaerobic respiration is the breakdown of organic nutrients in the absence of oxygen but with the release of energy captured by ATP.) A day's hunting, with episodes of hill climbing, cantering and jumping, and periods of waiting and walking, combine both processes and rates of energy expenditure. The expenditure during a long day by a hunter carrying a heavy huntsman exceeds the average daily consumption of feed energy by several-fold. The processes of energy metabolism are summarized here, as a greater understanding of them should allow more rational feeding and it should foster comprehension of each future development in the feeding of working horses.

What is work?

Before proceeding further it may be appropriate here to review the meaning of 'force', 'power', 'work' and related terms and see how they have been applied to measurable and practical work of working horses. First, if a horse was a smooth ball on a frictionless horizontal surface no energy would be required for it to continue moving at a constant speed in a straight line. Fortunately for both the horse and feed manufacturer all movement of the horse involves friction, both in the horse's motion (an assumed value of 2 J to move 1 kg BW 1 m horizontally is made) and in anything the horse may be pulling. This requires a force to overcome the friction and the force acting over a given distance equals the work output. If hill climbs intervene, the force of gravity has to be overcome as well. The extra energy required above maintenance is

$$2J[BW \text{ (kg)} \times \text{horizontal distance (m)}] + \frac{\text{work done}}{0.298}$$

where 0.298 represents the ratio of work done pulling/net energy used (cf. gross efficiency; Table 9.1). Force is measured in newtons (N), joules per metre, J/m^{-1}, or kilograms per metre per second squared, $kg/m/s^2$. Work output is measured in newtons per metre, N/m, or the distance over which the force is applied; note: N/m = J.

The working horse

One of the most traditional forms of work for a horse must be that of pulling a plough. Several groups of workers have accurately measured the work of ploughing; most recently Pérez *et al.* (1996) reported on five Chilean crossbred draught horses weighing 547 kg, pulling a mould board plough, cutting a furrow of depth and width 12.6 and 22.3 cm, respectively. Of course the measured performance is influenced by the soil characteristics, but the values they obtained (Table 9.1) agree well with those obtained by others and are informative to the nutritionist.

The riding horse

Energy expenditure and work type

Endurance work is at the rate of 154–224 m/min (5.75–8.35 mph) over variable slopes, whereas flat racing is at about 940 m/min (35 mph). For comparison, the stage-coach horse, pulling a heavy load, covered about 40 km (25 miles) at speeds similar to those required of endurance horses – 200–214 m/min (7.5–8 mph). Strenuous effort tends to depress appetite so recovery from extreme effort requires several days for reserves to be refurbished. The data in Fig. 6.4 and Table 6.7 show that the capacity of larger horses for feed is proportionally less than that of smaller horses, so that after undertaking strenuous work they may require a longer period for recovery, especially if it is assumed that the weight of the rider is proportional to the weight of the horse. The information in Table 6.2 is based on experimental results and although there has been some extrapolation it does indicate greater demands for feed energy than the theoretical estimates for endurance work given in Table 6.7

Table 9.1 Mean draught load, distance travelled, work done and net energy expended, in addition to maintenance, during a 6-h day by five Chilean draught horses ploughing a field (Pérez *et al.* 1996). The horses were exercised under submaximal conditions, although there was a significant increase in blood cortisol concentration.

Liveweight (kg)	547
Force applied, or draught load[1] (N)	905
Distance travelled (km)	13.6
Speed (m/s)[2]	0.93
Power developed (kJ/s or kW, i.e. kilowatts)	0.83
Work done in pulling above maintenance (kJ/horse)	12 276
Work done over 6-h day (kJ/kg BW$^{0.75}$)[3]	108
Estimated net energy for work, including horse moving itself (kJ/day)[4]	130 584
Estimated net energy used for work, including horse movement (kJ/kg BW$^{0.75}$)	1 153
Estimated gross efficiency of work, including horse movement, but excluding maintenance (%)	9.4
Net energy for work expressed as multiple of maintenance	2.24

[1] Mean force exerted by the five horses was equivalent to 17.7% of their bodyweight, or a draught power of 780 watts, J/s^{-1}.

[2] 18.5 h would be required to plough 1 ha.

[3] This is greater than that achieved by oxen, as the horse requires about two-thirds of the time to achieve a similar task.

[4] Energy requirement for this work of 6 h is about 125% of that required for 2 h medium trot daily including rider and maintenance.

(Yet the NRC estimates in Table 6.7 seem to exceed the INRA estimates; see Table 6.22.)

Blaxter (1962) calculated that the energy expenditure for vertical effort in horses, in addition to any horizontal effort, amounted to 17 times that expended in horizontal movement above the costs of energy metabolism at rest. A 400-kg horse is calculated to expend 0.67 kJ/m in horizontal work and 11.4 kJ/m vertically. Exercise over uneven and hilly ground can therefore be much more arduous than that on the flat. Sprint work on the flat must be considered a special case as it is almost entirely confined to young horses which may be still growing. Extended work, for which there is more published experimental evidence, is generally undertaken by older horses.

Work, especially, entails an increase in nutrient supply to the muscles, and in converting glucose or free fatty acids to high-energy phosphate compounds – ATP and CP or phosphocreatine (PCr) – which the muscles use as an immediate source of energy, there is an increase in the rate of production of wasteproducts, more particularly carbon dioxide and heat. The supply of nutrients and the effective disposal of wastes entail large physiological adjustments, which training seeks to encourage. The most critical feeds for work are those that provide energy, water and electrolytes. Electrolytes are those elements that in solution carry an electrical charge; they include sodium, potassium, magnesium, calcium, chloride and phosphate.

ENERGY SUBSTRATES AND THEIR EXPENDITURE

Most potential energy for muscular work is absorbed from the intestinal tract as glucose, VFAs (acetic, propionic, butyric and smaller quantities of related acids), longer chain fatty acids, neutral fats and amino acids. Absorbed glucose, propionate and glucogenic amino acids are potential sources of blood glucose and of liver glycogen, a storage starch, whereas absorbed long-chain fatty acids, neutral fats, ketogenic amino acids and particularly acetate and butyrate are potential sources of blood fats and fatty acids, storage fat and acetyl-CoA. Blood glucose and its precursors are also, of course, potential sources of body fat through the key substance acetyl-CoA.

ATP and CP (PCr) formation and use

The liver, while storing energy in the form of glycogen and fat (and also as protein), serves the vital role of maintaining normal levels of blood glucose through the breakdown and restorage of the glycogen. Muscle cells also store glycogen and form high-energy phosphate compounds – CP and ATP – necessary for muscular relaxation and contraction by drawing on blood glucose and fatty acids as fuels. The complete release of chemical energy from them requires a supply of oxygen (O_2) reaching the muscle cells through the arteries. However, an immediate and rapid release of energy, particularly important in short sprint races, can be achieved through the process of glycolysis, in which glucose is broken down to pyruvate in the muscle cell without the consumption of O_2, and also by the release of energy from previously stored ATP and CP. The further and complete breakdown of pyruvate and of fatty acids demands the presence of O_2 and this process takes place exclusively in the mitochondria, through the agency of what are known as β-oxidation of fatty acids and the TCA cycle.

ATP resynthesis

Energy for muscle contraction is derived from CP (PCr) and ATP which are formed from the energy derived during the combustion of glucose and fatty acids. TBs possess a high proportion of PCr-rich fast twitch fibres in skeletal muscles (Harris & Hultman 1992a), so that loss of adenine nucleotides (AN) during intense exercise is greater than in man (see 'Ammonia and the alanine vehicle', this chapter). After repeated gallops a 50% *loss of ATP* was recorded (Harris *et al.* 1991c; Sewell & Harris 1991; Sewell *et al.* 1992a), associated with a decrease in running speed and fatigue (Harris *et al.* 1991b), and with lower glycolytic rates and muscle (ATP)/(ADP). The accumulation of ADP stimulates AN degradation to inosine monophosphate (IMP) with an increase in plasma NH_3 concentration. There is a critical pH below which ADP rephosphorylation declines (see 'Ammonia and the alanine vehicle', this chapter), PCr acting both as an intracellular buffer and a reservoir for this rephosphorylation. Supplementation of human subjects with *creatine monohydrate* has increased muscle total creatine and PCr contents (Harris *et al.* 1992). The energy reserves for this synthesis depend upon training, diet and the inherent characteristics of the horse. The slow rate of ATP resynthesis, when supported by free fatty acid (FFA) oxidation, means that inadequate glycogen storage in active muscle fibres causes early fatigue despite an abundance of fatty acids.

The TCA cycle

The operation of the TCA cycle in the mitochondria requires the diffusion of O_2 to these organelles and the removal of CO_2 to the blood. During light work this process occurs quite smoothly, and with each turn of the cycle one unit of oxaloacetate is produced which is required for the subsequent metabolism of acetyl-CoA in the presence of O_2. However, when larger quantities of oxaloacetate are present, the metabolism of acetyl-CoA is likely to proceed more rapidly. When larger quantities of fatty acids are dissimilated to acetyl-CoA, the cycle must turn at an accelerating rate. This is also achieved by the provision of extra quantities of oxaloacetate from outside the mitochondrion. There must, therefore, be adequate quantities of pyruvate requiring an ample supply of glucose, lactate, glucogenic amino acids or even glycerol. During extended work, the trained healthy horse finds no difficulty in breaking down fatty acids to CO_2 as sufficient O_2 is taken in by normal respiration. In fact, such work leads to an accumulation of glycerol, signifying that it is not called on to form pyruvate units in any great quantities. If the utilization of fatty acids was interrupted, this would probably be expressed as a retarded metabolism of acetyl-CoA and there would be a buildup of blood ketones (acetoacetate and 3-hydroxybutyrate).

Work in Newmarket (Frape *et al.* 1979) has shown that these ketones accumulate in the plasma only after work stops (Fig. 9.1), possibly implying that no bottleneck occurs to the complete combustion of fats. This postexercise rise in plasma ketones probably reflects the redistribution of blood from muscle to adipose tissue from which NEFAs will then be flushed, rapidly increasing the plasma concentration of their metabolites, owing to a reduced need for their cellular consumption. Hyperlipidaemia, recognized in starving horses, reflects a blocking to fat metabolism in animals relying overwhelmingly on residual fat stores in the relative absence of carbohydrate substrates. The activity of enzymes required in this fat breakdown may also be depleted in an associated deficiency of dietary protein. Figure 9.2 gives a brief account of the pathways by which energy sources are metabolized in horse muscle cells and liver.

Glycogen, glucose, FFAs and VFAs

The amounts of muscle and liver glycogen stored at any one time is very variable, as reserves are considerably depleted during extended exercise and are repleted only over several meals. Nevertheless, the author calculates that on average liver reserves of glycogen are 5–10% of those in skeletal muscles, making assumptions concerning muscle mass and that the liver represents 1.5% of bodyweight.

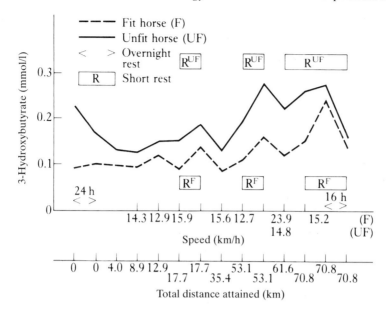

Fig. 9.1 Effect of rest periods during an endurance ride on plasma 3-hydroxybutyrate in a fit and an unfit horse (Frape *et al.* 1979).

Glycogenolysis, or the breakdown of glycogen, continues following the completion of intensive exercise. Recovery of these glycogen stores is promoted by giving a meal of 1–2 kg cereal 1.5 h following exercise, or by the i.v. infusion of 6 g glucose/kg BW, 30 min postexercise. This may have implications for horses competing on successive days. On the other hand, in trained horses glycogen stores do not limit sprint performance, as training can increase glycogen capacity of limb muscles by 39% (Guy & Snow 1977a).

With increasing rates of energy expenditure the preference for glucose as a substrate increases, although the proportions of glucose and FFA used change rapidly with distance. Glycogen utilization rates of 2.68 and 1.06 mmol glucosyl units/kg dry muscle/s with total consumptions of 27.3 and 32.5% of the initial store were caused, respectively, by 800- and 2000-m gallops (Harris *et al.* 1987). This suggests that over the greater distance at a somewat lower velocity (aerobic) combustion of FFA increases considerably as a proportion of total energy expenditure.

In addition to FFA derived from the hydrolysis of fat stored in adipocytes, VFAs absorbed from the hindgut are an important energy source. The contribution of acetate to oxidation in the hindlimb was reduced from 32% in horses given roughage to 21% in those given a diet with an oats:roughage ratio of 0.52:0.48 (Pethick *et al.* 1993). This may be one reason why higher roughage intakes are recommended for endurance horses (also see 'Sweat and dehydration', this chapter).

Pyruvate–lactate: the oxygen debt

The anaerobic production of energy by glycolysis would soon be halted by an excessive accumulation of pyruvate. The cunning mechanism has therefore evolved

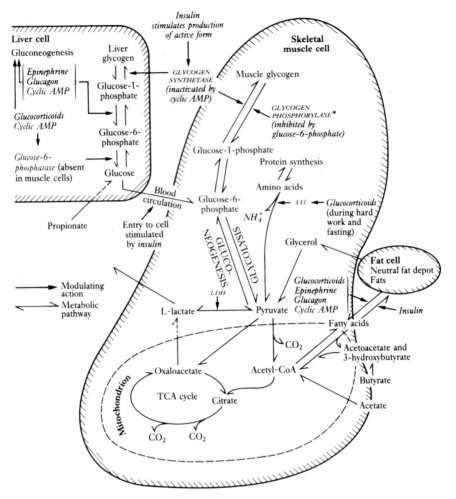

Fig. 9.2 Major pathways of energy metabolism in muscle and liver cells and their modulation by the endocrine system (intermediate steps have been excluded). Glycolysis: the oxidation of glycogen or glucose to pyruvate and lactate. Gluconeogenesis: The formation of glucose or glycogen from non-carbohydrate sources. The principal substrates for this formation are glucogenic amino acids, lactate, pyruvate, glycerol and to a lesser extent propionate.

*Bursts of muscular action with the secretion of glucagon and epinephrine and nervous stimulation of the release of Ca^{2+} initiate the formation of cyclic adenosine monophosphate (AMP) and then the phosphorylation and activation of glycogen phosphorylase.

in which pyruvate is reversibly converted to lactate as an intermediate 'waste' product. This reduction even more importantly oxidizes reduced NAD^+, essential for triggering an important glycolytic step. Therefore, in sprint racing, lactate diffuses into the blood from the muscle cell and there accumulates until sufficient O_2 is available for its hepatic conversion to pyruvate. This mechanism allows more energy to be obtained by glycolysis than would otherwise be possible. As the O_2 debt has eventually to be repaid, the yield of energy per unit volume of O_2 consumption, owing to anaerobic processes, is only half that achieved by aerobic processes. However, during recovery from sprint races O_2 is abundant so that the advantage of the rapid availability of chemical energy outweighs the minor disadvantage of interest payment.

Lactic acid threshold

Whether sufficient O_2 is present in muscle cells depends on the speed and the extent of training of the horse. The anaerobic lactic acid system in unconditioned TBs is probably untaxed until work rates exceed around 600 m/min (22 mph) on the flat (Williamson 1974). The anaerobic threshold of standardbred trotters is said to be 300–400 m/min (11–15 mph) on the flat (Milne *et al.* 1976), although a striking increase in blood lactate of standardbred trotters was not observed until speeds exceeded 600 m/min (Lindholm 1974), or 684–750 m/min (25–28 mph) (Lindholm & Saltin 1974), and Williamson (1974) recorded similar blood responses in these breeds undertaking the same types of exercise. The threshold of exercise-induced lactate accumulation may be taken to be 4 mmol/l lactate in the plasma, and the velocity for a particular horse at which this occurs is termed V_{LA4}.

In a study by workers at the University of Glasgow (Snow & MacKenzie 1977b) cantering for up to 22 km caused no stress on nutrient reserves in the blood in that, immediately after the exercise, blood glucose, glycerol and FFAs, as well as pH, rose slightly, which indicated an adequate rate of their mobilization and a sufficient irrigation of muscle cells with O_2. Snow and MacKenzie (1977a) found that at a maximum work rate of 864 m/min (32 mph) over 3×600 m, with two 5-min intervals between the gallops, anaerobic respiration was necessary in that blood glucose, glycerol, FFA and lactate were all elevated and there was a fall in pH. The increased fund of blood glucose was associated with a considerable rise in adrenocorticosteroid secretion; training for 10 weeks before the gallops caused a further increase in blood glycerol and lactate with smaller changes in FFA and pH than was the case with the untrained horses. As corticosteroid secretion was even greater after training, an explanation may be proferred that training confers a more efficient mobilization of reserve fat, and both it and glucose are used to a greater extent by trained horses, resulting in faster times. The lesser fall in blood pH indicates that training also leads to better ventilation and oxygenation of the muscles, as this effect of training coincided with a greater, rather than a smaller, increase in blood lactate.

TRAINING METHODS

Muscle type and breed

Long-distance riding requires a preponderance of muscle fibres capable of slow contraction but resistant to fatigue, whereas sprinting ability demands the presence of a higher proportion of fast-contracting fibres that happen to be readily fatigued. These fast-contracting fibres are categorized as having myosin with high ATPase activity at pH 9.4 and a high glycolytic activity (type II fibres). They are subdivided into fast twitch, low oxidative (FT) and fast twitch, high oxidative (FTH) fibres. Fibres with low myosin-ATPase activity (type I fibres) are present in relatively higher proportions in the skeletal muscles of horses more suited to long-distance riding. These are known as slow twitch, high oxidative fibres (ST). Standardbred trotters older than 4 years have 54% FTH, 22% FT and 24% ST fibres in the gluteus medius muscle (Lindholm & Piehl 1974). All muscles have about the same ratio of low oxidative to high oxidative fibres, but a mixture of anaerobic and aerobic training increases the proportion of FTH fibres and decreases the proportion of ST and possibly also of FT fibres.

The purpose of training is to modify muscular action and indeed the whole metabolism of the horse so that it functions at maximum efficiency with minimum fatigue at the speed and over the distance at which the sights are set. In this process feed energy is converted into work and Table 6.7 gives the approximate amounts of feed energy required by horses of different weights undertaking work of a variety of intensities. In all but one instance the work covers a period of 1 h, although it is appreciated that the most strenuous effort could not be sustained continuously for such a period. Nevertheless, the method allows comparisons of different degrees of effort. On the one hand, walking creates practically no further demand on the requirements for maintenance, whereas at the other extreme, racing – and more particularly lengthy endurance work – create an energy demand that exceeds the capacity of the horse for the immediate replenishment of the losses through feeding. Glycogen and fat reserves are therefore heavily drawn upon.

Muscle hypertrophy

Adaptation to hard muscular work entails changes not only in the blood vascular system but also in the skeletal muscles. Hypertrophy of muscle occurs during training, the extent of which varies with the breed of horse and with the type of work. This change is reflected in an increase in apparent N retention in quarter horses (Freeman *et al.* 1988). Losses of N in sweat were not measured and part of the apparent retention may be accounted for by this loss. The increased retention of N appeared to continue for a month of rest following the exercise period, indicating a higher maintenance requirement for protein in horses that have been in training.

Training effects

As the intensity of training increases, the demand for energy release by way of a particular metabolic pathway rises, so increasing the need for the appropriate enzymes to catalyse the reactions. Sprint racing imposes greater demands on anaerobic metabolism and longer races call predominantly on aerobic processes. The latter implies a more intense use of the TCA cycle in the mitochondria. Training for sustained work is therefore seen to bring about an increase in the number of these cellular organelles and their associated enzymes. Training for sprint racing, however, increases glycolytic activity and a twofold increase in the activity of the enzymes aldolase (a key enzyme in the glycolytic pathway) and ALT has been observed. The latter promotes the formation of the amino acid alanine (see 'Ammonia and the alanine vehicle', this chapter, for a function of this amino acid) from pyruvate so lessening the conversion to lactate and therefore probably lessening fatigue by moderating a fall in pH.

Interval training

The object of interval training is to increase the volume of work accomplished in a single training session by providing exercise in bouts, separated by recovery periods. This system allows some recovery of the muscle glycogen reserves and a reduction in heart rate and plasma lactate concentration during rest periods. The reduction in plasma lactate will be accelerated by trotting the horse between bouts, rather than by walking or cantering. Repeated high intensity interval training has caused loss of both performance capability and body condition and weight and so should be carefully monitored.

MUSCLE ENERGY RESERVES AND FEEDING BEFORE EXERCISE

Hormonal effects

A normal meal of grain leads to an elevation in plasma glucose and insulin concentrations. Insulin elevation can last for a period of up to 8 h. It is an anabolic hormone promoting glucose and fat storage, whereas exercise requires the mobilization of energy reserves for combustion. An intense exercise bout 2–6 h after a meal is associated with decreased FFA or NEFA availability, leading to a rapid fall in blood glucose concentration, as the contribution of fat to energy expenditure is decreased. For example, Lawrence *et al.* (1993) fasted standardbred horses overnight and then gave them no grain (controls) or 1, 2 or 3 kg maize grain 2.5–3 h before a warm-up and intense exercise over 1600 m at a heart rate of 206 bpm. The controls maintained steady plasma glucose concentrations, whereas plasma glucose declined in those given any grain, a response similar to that observed by others, where horses were fed 3 h, or 4 h, before exercise. Plasma FFAs were initially higher in the controls, but FFA declined in these horses during the intense exercise. In the grain-fed horses plasma FFA remained steady during exercise, but the concentration was always less than in the controls. The effect of a meal on repeated bouts of intense exercise during a day probably depends on the size of the meal, timing, exercise intensity, etc. Lawrence *et al.* (1995) found that a meal neither improved nor impaired the performance of repeat exercise bouts. The reason for this may be that intense exercise stimulates epinephrine and norepinephrine (adrenaline and noradrenaline) secretion which overrides the effect of insulin on glucose and fat metabolism, so that fat mobilization and release of NEFA into the blood would not be impaired during second, third, etc. bouts of intense exercise. (Also see 'Feeding before endurance rides', this chapter.)

Blood distribution effects

(Also see 'Exercise warm-up', 'The vascular and respiratory systems', this chapter, and Fig. 9.7.) Experiments with ponies (Duren *et al.* 1992) indicated that exercise of 7.8 m/s on a 6.3° incline at 75% of heart rate maximum for 30 min, 1.4 h after feeding, led to higher GI tract and skeletal muscle blood flows compared with those in fasted ponies. There was an increase in heart rate, cardiac output, stroke volume and arterial blood pressure, whereas accommodation may not have been possible with more intense work. This work is therefore normally delayed for 5–8 h after feeding, although the optimum time interval is influenced by the dietary proportion subject to fermentation and type of activity.

Glycogen use and muscle type

The accelerated use of glucose is accommodated by a stimulation to glycogen deposition in the muscles of adequately fed trained horses. This training can increase the glycogen capacity of TB muscles by a third. Blood glucose cannot, however, be maintained indefinitely during work, and endurance rides of up to 150 km indicate a gradual decline and an exhaustion of muscle glycogen. In one 80-km (50-mile) ride, blood glucose on average fell by 40%, whereas FFA rose eight-fold (Hall *et al.* 1982). In trained horses, blood-glucose concentration is not affected by endurance rides of 50 km (31 miles), but is decreased 23% during a ride of 100 km (62 miles) (Essén-Gustavsson *et al.* 1984).

The net loss of muscle glycogen is extremely small in light work. Lindholm (1974) observed that standardbred trotters, for instance, lost 0.3 mmol glycogen/(kg × min) when trotting at 300 m/min (11 mph), whereas trotting at a maximum rate (750 m/min or 28 mph) led to a loss of 14 mmol glycogen/(kg × min). Three aggregate minutes of maximal trotting were proved to cause a 48% decrease in muscle glycogen, but the decrement was not equally distributed among the fibre types. Maximal work causes a striking depletion of ST fibres in addition to a loss from the other two types. On the other hand, slow trotting leads to a gradual depletion of ST fibres after which the FTH fibres become active and depleted. Thus, there seems to be a preferential fibre recruitment with increasing speed and duration.

Glycogen 'loading'

The stimulation of glycogen accretion has been considered from both safety and efficacy aspects. Starchy diets and regular exercise can increase muscle glycogen, but this increase is not maintained if the starch content of the diet is subsequently reduced. Harris and Hultman (1992a) found no difference between diets in this loading, yet it probably occurs during rest more effectively on a high carbohydrate than a high fat diet, following intense exercise that depletes type II muscle fibres (fast twitch) of glycogen (Pagan *et al.* 1987a). Aerobic work is ineffective and glycogen loading may cause poorer performance (Topliff *et al.* 1985, 1987; Pagan *et al.* 1987a) and an increased risk of exertional rhabdomyolysis ('tying-up'), according to a widely held view, although this may not be associated with either lactic acidosis, or excessive dietary carbohydrate. Hodgson (1993) speculates that horses predisposed to it exhibit a temporary failure in the control of intracellular $[Ca^{2+}]$.

Tissue fat as an energy source

Horses described as 'moderately fat' (condition score 7.5) require more energy for maintenance than do those in 'moderate condition'. The reserves of fat in the latter state are adequate as a source of energy for exercise, indicating that a relatively lean condition is satisfactory.

Exercise warm-up

Horses have a high aerobic capacity and rapid kinetics of gas exchange compared with other mammalian species that have been measured. The maximum rate of oxygen uptake per minute (VO_{2max}; ml/kg/min) in racehorses is double that of human athletes and VO_2 increases very rapidly at the onset of high intensity exercise. The kinetics of gas exchange are affected by a warm-up prior to high intensity exercise which results in horses reaching steady state VO_2 faster. Tyler *et al.* (1996) found that a warm-up for 5 min at 50% VO_{2max} in standardbred racehorses increased the aerobic contribution to total energy requirement from 72.4 to 79.3% when they were run to fatigue (1–2 min) at 115% of VO_{2max}. The maximal accumulated O_2 deficit was lower in the warm-up horse, i.e. 34.7 vs. 47.3 ml O_2 eq/kgBW. A warm-up is very desirable for maximum performance during high intensity exercise, as it should allow greater use of aerobic energy sources and it may reduce the risk of injury. The effects are likely to be conferred by providing the time necessary for the redistribution of the blood supply and an increase in circulation rate.

Recovery

Recovery from intensive exercise is promoted by postrace aerobic trotting, compared with no exercise, as clearance of lactic acid is accelerated.

Bleeders

Slight pulmonary bleeding seems to be a normal consequence of strenuous work and does not reflect any dietary abnormality, but is simply the physiological stress of a massive increase in nutrient and gaseous irrigation of muscle tissues. Pulmonary capillaries necessarily have very thin walls to allow rapid exchange of respiratory gases across them and stress failure is greatly increased at high lung volumes. The capillary walls of TBs are not strong enough to withstand the stresses that develop as a result of the high capillary pressures accompanying extremely high cardiac outputs during strenuous exercise. There is some evidence that low environmental temperatures (range for most observations −10 to +17°C) increase the risk (Lapointe *et al.* 1994).

Bone remodelling and training

(See also 'Acid–base balance', this chapter.) The commencement of training causes a remodelling process to occur in the long bones to accommodate the stresses on the skeleton. This process involves the mobilization of bone salts and their redeposition. There is a risk of microfractures occurring if training is intensified too rapidly. Ca balance data indicate that bone density is low during the first 2 months of training and that Ca balance and bone density are still increasing after 3–4 months. Diet must be adequate in Ca to allow for this physiological adaptation.

Causes of withdrawal from training

During the period 1990–1992 causes of illness or fatal injury among 496 racehorses in California were analysed (Johnson *et al.* 1994). Musculoskeletal injuries accounted for about 80% of TB and quarter horse submissions, among which forelimb fractures were prominent. Poor mineral nutrition is likely to have played some role in this statistic.

THE ENDOCRINE SYSTEM

An examination of Fig. 9.2 will reveal that the changing demands of the horse for energy are monitored by a number of endocrine secretions, or hormones. Where rapid changes are necessitated, the signal for secretion by the appropriate glands is provided by the involuntary action of the autonomic nervous system, reacting to environmental stimuli, which in turn brings about other essential changes in cardiac muscle and the smooth muscles of arteries, intestines, etc. Endocrine secretions to a considerable extent mediate their effects by switching on and switching off some of the enzymes that regulate the chemical reactions in energy metabolism.

Insulin

The insulin molecule attaches to receptors on cell membranes to stimulate cellular uptake of glucose and glycogen synthesis and TAG 'clearance' (different cellular receptor), so that their blood concentrations are decreased after a meal. Insulin retards the breakdown of glucose. Horses accustomed to a high starch diet possess a high insulin sensitivity (so long as the β-cell function of the pancreas is normal) and therefore they are more inclined to hypoglycaemic shock when subjected to a fast than are horses normally fed hay diets and accustomed to deriving blood glucose from other sources. The insulin sensitivity of a horse can be determined by measuring its glucose tolerance and insulin response to a given dose of starch or glucose. Low sensitivity and poor tolerance increase the areas under the plasma response curve of insulin and of glucose. Large doses of glucose in insulin resistant, diabetic or fasted animals cause glycosuria.

In TBs, blood glucose normally peaks about 2 h from the start of feeding. The peak value is 6.5–8.5 mmol/l (117–153 mg/dl) at 1.5–3 h; then there is a gradual decline to a postabsorptive level of about 4.6–4.8 mmol/l (83–86 mg/dl). Ponies on roughage diets may maintain normal levels 1.4 mmol/l (25 mg/dl) lower than this. Blood insulin activity tends to fall slightly during work because of the catabolism of glucose, but some insulin is still probably required to ensure that glucose is available to the working muscle cell.

Glucagon

In order to sustain concentrations of blood glucose within normal limits the influence of insulin is counteracted by that of glucagon. Whereas the former promotes uptake of glucose by all cells of the body (with the exception of brain cells), glucagon appears to focus its effects primarily on the liver and adipose tissue. It achieves an increase in blood glucose by stimulating those enzymes that cause a breakdown of liver glycogen (Fig. 9.2) and by encouraging gluconeogenesis. In this latter function it works in concert with other hormones discussed below, a particularly important task in roughage-fed animals. These other hormones are the glucocorticosteroids produced by the adrenal cortex and epinephrine and norepinephrine secreted by the adrenal medulla.

Adrenal hormones (catecholamines, glucocorticoids)

The rapid initiation of intense work necessitates an immediate response in terms of energy mobilization. This is brought about by sympathetic nervous activity, which not only causes splenic release of red cells but also stimulates the adrenal medulla to secrete epinephrine and norepinephrine (adrenaline and noradrenaline). The extent of this reaction depends on the intensity of the work load – that is, the faster the horse is running the greater is the secretion. The medullary hormones affect several tissues, increasing the mobilization of fatty acids from adipose tissue and stimulating the production of glucose with a rapid rise in the blood concentration by the breakdown of liver glycogen and by amino acid metabolism (Fig. 9.2).

The glucocorticoids secreted by the adrenal cortex are somewhat slower in responding to work demand and their secretion depends on a hormonal signal from the anterior pituitary. Moreover, they stimulate a rise in blood glucose and the accumulation of liver glycogen by promoting gluconeogenesis, through the inhibition of protein synthesis, and they provoke the breakdown of depot fats to FFA and glycerol. Synthetic analogues of these secretions, when given repeatedly, have a comparable effect and cause muscular wasting. They also initiate bone problems through an inhibition of Ca absorption from the gut. The stimulation from glucocorticoids of amino acid mobilization is expressed by excitation of transferase enzymes, raising serum (Codazza *et al.* 1974; Sommer & Felbinger 1983; Essén-Gustavsson *et al.* 1984) and muscle (Guy & Snow 1977a) activities of ALT and AAT (see Fig. 9.2), observed after exercise.

In an analogous fashion to the response of glucagon, a reduction in blood glucose triggers the secretion of the glucocorticoids and their circulating level increases with agitation, trauma and psychological stress. Training leads to a greater adrenocorticoid response under such stress. This applies to sprint, endurance and other training so that normal concentrations of blood glucose are maintained more effectively in all circumstances. In contrast to the response of medullary hormones, amounts of plasma cortisol seem to be uncorrelated with the intensity and speed of work. (Note: acute stress in the horse leads to an increase in circulating total cortisol, whereas chronic stress generally seems to decrease plasma total cortisol concentration. This may be due to the production of an adrenocorticotropic hormone (ACTH)-release inhibiting factor, which could be an endogenous opioid.)

THE VASCULAR AND RESPIRATORY SYSTEMS

Blood volume of horses is about 9.7% of bodyweight so that a horse weighing 560 kg would contain around 51.2 l (SG 1.06). Blood volume is important for both O_2 and heat transport so that a large plasma volume is accompanied by a considerable skin blood flow. This volume can increase by 30% over 2 weeks' training (Erickson *et al.* 1987). The volume of blood discharged per beat (stroke volume) in a 560-kg horse at rest would be 1.2 l and as blood volume is proportional to bodyweight, cardiac output per ventricle ranges between 56 and 75 ml/(kg × min) at rest. The need for such a flexible system can be appreciated when it is realized that O_2 consumption of skeletal muscles can increase 100-fold in strenuous exercise.

The blood of the horse is a fluid containing, by volume, about 45% of red cells (erythrocytes), 1% of white cells (leucocytes) plus platelets, and 54% of plasma. The red cells have, as a major function, the transport of O_2 from the lungs to the muscles and other tissues. To accommodate an increased O_2 demand a reserve of red cells is held in the spleen. This splenic reserve is very large in TBs so that they can increase the numbers of red cells in circulation by between 30 and 60%. Thus, in heavy work, the O_2-carrying capacity of the blood may increase from rest by as much as 8.8 volumes %. Between rest and galloping at 700 m/min (26 mph), the O_2-carrying capacity changed from 15.9 to 21.4 volumes % in one study and from 16.35 to 25.19 volumes %, a 54% increase, in another study (Milne 1974) using TB, quarter horse, Arabian and standardbred horses. The extent of change is increased by training and the effect of a change in capacity is augmented by a redistribution of the blood supply to skeletal and cardiac muscle. The degree of splenic release tends to be proportional to speed; in one set of measurements the PCV rose 32% at 350 m/min (13 mph) and 55% at 700 m/min (26 mph) (Williamson 1974; for other observations see Figs 9.3 and 9.4). To assess the red-cell count and haemoglobin content of blood

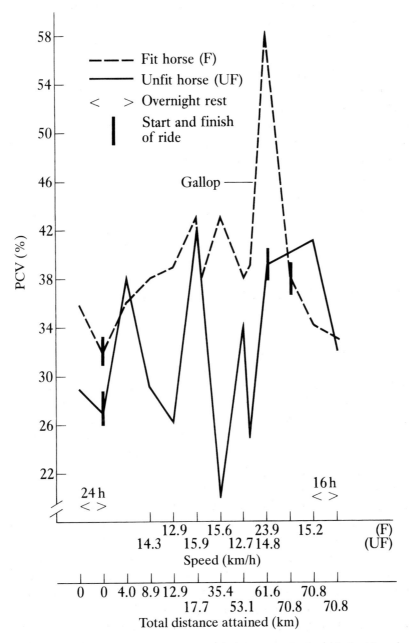

Fig. 9.3 PCV of the blood during an endurance ride by a fit Arab stallion and a less-fit pony gelding (Frape *et al.* 1979).

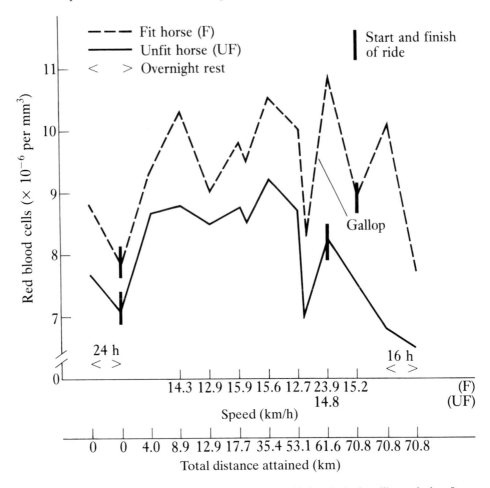

Fig. 9.4 Changes in erthyrocyte count during an endurance ride in a fit Arab stallion and a less-fit pony gelding (Frape *et al.* 1979).

as affected by dietary inadequacies and other factors, the most consistent results are obtained after splenic release has been accomplished.

Blood and waste disposal

The supply of blood to muscles, increased during work, is the principal vehicle by which the wasteproducts of energy metabolism, water, carbon dioxide and heat, are removed from muscle cells. In the absence of their efficient disposal, pathological cellular changes would result. To facilitate this objective the capillaries of the skeletal and cardiac muscles become dilated during increased work, while those in the visceral region contract. This in turn diminishes the digestion and absorption of nutrients and gives credibility to the tenet that horses should not be fed before hard work (see 'Muscle energy reserves and feeding before exercise', this chapter).

Body temperature

At high work rates heat production is 40–60 times basal levels, and body temperature can rise appreciably (Fig. 9.5). The mean muscle temperature of five standardbred trotters increased from a normal of 37°C to 41.5°C during a race of 2100 m at a mean rate of 708 m/min or about 26 mph (Lindholm & Saltin 1974). Carlson (1983b) calculated that if a horse could work at a moderate intensity for 1 h with an O_2 consumption of 30–40 l/min, the total waste heat would amount to 38 MJ. The principal mechanism for exhausting this waste heat from the body, when atmospheric humidity is not excessive, is the evaporation of sweat and of moisture from the surface of the lungs. The majority of heat is dissipated in this way through the skin, rather than the lungs, accompanied by electrolyte losses, accelerating the onset of fatigue. Dilatation of subcutaneous blood vessels diverts blood from skeletal muscles, contributing to a decrease in work capacity, at a cost influenced by the depth of subcutaneous fat (Webb *et al.* 1987b). If circulatory adjustments fail to maintain heat balance, ventilation rate is increased (Fig. 9.6), which induces respiratory alkalosis. Training is important, as the rise in core temperature during exercise is proportional not to exercise intensity, but to the percentage VO_{2max}, which is increased by training, so ameliorating a temperature rise.

Following feed ingestion the amount of blood distributed to the GI tract increases, competing with the redistribution necessitated by a rise in heat production. Accommodation is achieved by augmenting cardiac output, redistributing regional blood flow, or by combining these mechanisms (Fig. 9.7).

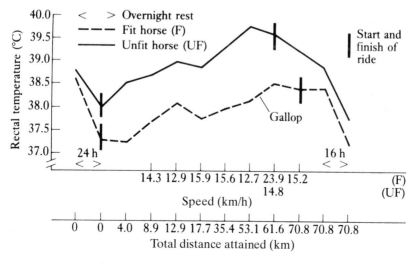

Fig. 9.5 Effect on body temperature of an endurance ride by a fit Arab stallion and a less-fit pony gelding in cool weather (Frape *et al.* 1979).

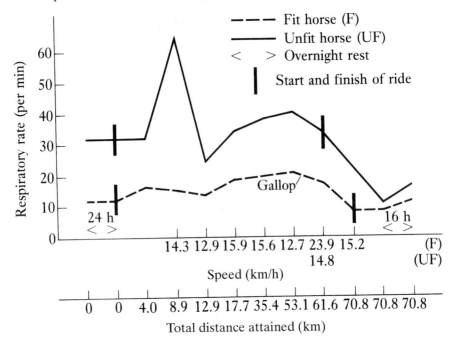

Fig. 9.6   Effect on respiration rate of an endurance ride by a fit Arab stallion and a less-fit pony gelding (Frape *et al.* 1979).

Feed consumption leads to:

- ↑ coeliac blood flow
- ↓ skeletal muscular and blood flow, or
- ↑ cardiac work

Exercise leads to:

- ↑ skin blood flow
- ↑ water-electrolyte loss
- ↑respiratory and heart rates
- ↑pulmonary blood flow

Fig. 9.7   Heat flow and blood redistribution as a consequence of eating and exercise (Frape 1994). (↑, increase; ↓, decrease.)

Blood as a buffering system

The red cells contain haemoglobin which not only carries O_2 but also acts as a buffer to the lactic acid produced in the contracting muscles. This is probably why the pH of blood falls only during galloping, whereas the haemoglobin content rises appreciably during only moderate exercise. As haemoglobin is a major buffer, blood pH is changed very little for a considerable change in bicarbonate content. However, in performing this function haemoglobin carries less O_2. In extreme cases of acidosis the horse becomes cyanotic, or O_2-starved.

Blood glucose concentration

Blood glucose concentration is the expression of a dynamic balance of glycogen breakdown and glycogen synthesis and the production of glucose from other sources

– amino acids, lactic acid and propionate (gluconeogenesis). The resting level is somewhat higher in horses intensely trained for sprint races. This state is brought about by the stimulation of the two systems leading to the formation of glucose and by increasing the efficiency of fatty acid utilization, sparing glucose. Thus, blood glucose fluctuates throughout the 24h, rising in TBs from postabsorptive values of 4.7 mmol/l to a peak of approximately 6.5–7.5 mmol/l, 2–3h after a large feed of oats, following moderate work (those fed after strenuous exercise may show no plasma glucose response) (Frape 1989). In contrast, blood glucose concentrations reached values of between 4.7 and 6.4 mmol/l in ponies after receiving a pelleted meal to appetite, whereas after 3h of fast the levels fell to between 2.8 and 5 mmol/l (Ralston *et al.* 1979). Plasma pyruvate also tends to increase with glucose, whereas fatty acids and glycerol are in lower concentrations. The fluctuation between peak and resting levels of glucose varies with the type of diet, in that foods containing more grain and less roughage lead to higher peaks and lower troughs. Horses subjected to a rougher diet develop a greater faculty for gluconeogenesis and can therefore resist a depression in blood glucose more readily during a fast, but are unable to meet the demands of an excessive rate of exertion. Furthermore, breeds such as TBs, with higher insulin sensitivity, experience a greater fluctuation in blood glucose than do ponies given the same diet at similar times of the day.

Haematocrit and blood viscosity

In order for blood to travel freely through the small capillary bed of muscles, it is essential that it retain its fluidity, and where the PCV exceeds 55% there is an exponential increase in blood viscosity. The interest shown in providing horses with additional red blood cells before races can therefore be a misguided activity. The importance of a low viscosity is recognized in horses subject to dehydration in hot climates and during long-distance rides (Figs 9.8 and 9.9). For this reason it may be no chance that the PCV of Arab horses (Fig. 9.3) is lower than other hotblooded horses and that Arab horses are highly adapted to both hot climates and extended work. A voluminous splenic pool may therefore be a disadvantage for some purposes. Plasma albumin, a reserve protein and a major contributor to blood viscosity, displays a decreased concentration during training for reasons that may be understandable in this context. The dilutions achieved still provide adequate osmotic pressure and do not necessarily impute a dietary protein deficiency. In fact the increase in dietary carbohydrate, normal at this time, may contribute to albumin catabolism.

Pulse, respiratory rate and fitness

Heart rate is linearly related to speed of the horse and varies between the approximate limits of 30 and 240 contractions/min. After a gallop at 700–800 m/min (26–30 mph) the rate may be achieving 240 beats/min with an output from each ventricle

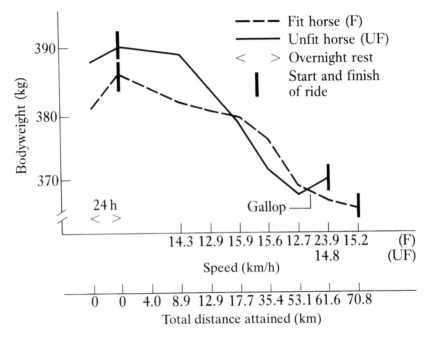

Fig. 9.8   Effect on bodyweight of an endurance ride by a fit Arab stallion and a less-fit pony gelding in cool weather (Frape *et al.* 1979).

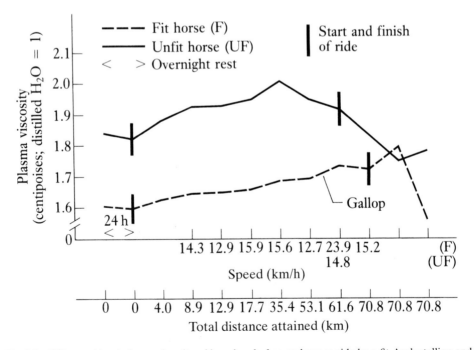

Fig. 9.9   Effect on blood plasma viscosity of breed and of an endurance ride by a fit Arab stallion and a less-fit pony gelding (Frape *et al.* 1979).

of 3–4 l/s. Heart rate, particularly after work, is a good indicator of fitness. In the preride checks of endurance horses it is agreed that pulse rate should fall within the limits of 36–42 beats/min (Fig. 9.10) and respiration rate between 8 and 14/min (Fig. 9.6). Both are higher in unconditioned horses. After an endurance ride and 20-min rest the pulse rate should have fallen to less than 55 and the respiration rate to 20–25/min. In exhausted horses, the rate of both these is greater (tachycardia and hyperpnea) and the occurrence of muscular spasms more likely. The ratio of pulse to respiratory rate should fall within the limits of 2:1–5:1. During heavy exertion and heat stress the pulse and respiration rates have been known to rise to 85 and 170, respectively, a ratio of 1:2, that is, an inversion of the pulse:respiration rates. Poorer horses and those suffering from adrenal exhaustion tend to exhibit lower heart:respiratory rates both before and after exercise. After a 20-min rest during an endurance ride horses exhibiting heart rates exceeding 70 or heart:respiratory rate ratios of less than 2:1 should be eliminated. Hyperventilation of the lungs may simply reflect a shortage of oxygen and normal respiratory acidosis (not commonly seen in endurance horses), or it may indicate an increase in body temperature (easily measured per rectum) brought on by hot weather or inadequate training, or both (Fig. 9.5). Alkalaemia normally follows a raised body temperature (see 'Body temperature', this chapter).

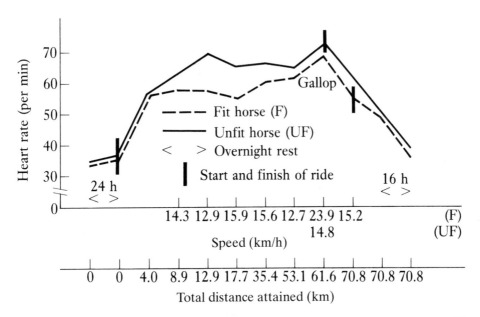

Fig. 9.10 Effect on heart rate of an endurance ride by a fit Arab stallion and a less-fit pony gelding (Frape *et al.* 1979). (Note: raised heart rate of unfit horse between 53 and 61 km may have resulted from anxiety when partner galloped off.)

RESULTS OF EXERCISE

Sweat and dehydration

A horse working at a moderate intensity for 1 h with an O_2 consumption of 30–40 l/min would need to dissipate 38 MJ waste heat (Carlson 1983b). The disposal of this quantity by evaporative processes alone would entail the loss of just over 15 l water. Although this is an oversimplification, it is a reasonable estimate of events in high environmental temperatures and low relative humidities. Sweat losses are quite modest in horses racing for distances of up to 3 km, but it seems that body water losses in sweat (and urine) and from the lungs during a prolonged exercise can approach 10–12 l/h and total as much as 40 l. Typically, bodyweight falls by 5–9%, principally from evaporative losses (Fig. 9.8), but the extent of loss depends on the level of fitness and the availability of water and electrolytes during the exercise. The evidence in Fig. 9.8 and Table 9.2 clearly indicates a greater rate of bodyweight loss and of sweat production (carrying the fixed ions of Na, K, Cl, Ca, P and Mg) by an unfit horse than by a fit one, although in both horses the molar sum of cations approximately equals that of the anions in the sweat.

Water losses are more aptly compared with total body water, which, in a horse of 450–500 kg bodyweight, would amount to 8–14% of the total of about 300 l in extended exercise. Of this 300 l about 200 l is in intracellular fluid (ICF) and 100 l is in extracellular fluid (ECF), made up of the water in blood plasma, interstitial fluid, lymph and contents of the GI tract.

Electrolytes and their losses

The 100 l of ECF contain 14 000–15 000 mmol readily exchangeable Na, representing nearly all the exchangeable Na of the body, the total Na content of which is about 1 kg. Of this total, 40% is located in the skeleton and during extended Na depletion this depot appears to be partially mobilized. The 200 l of ICF contain 20 000–

Table 9.2 Electrolytes (mmol/l)[1] in evaporated sweat of horses participating in an endurance ride[2] and other data obtained from exercising horses.

Horse	Cl	Na	K	Ca	P	Mg	pH of venous blood
Fit at mid-ride[2]	910	710	215	—	—	—	—
Fit at finish[2]	1180	880	270	—	—	—	7.29
Unfit at finish[2]	3060	2120	780	—	—	—	7.36
Harris[3]	155	135	41	3	<0.3	2	—
Mean of other data[4]	231	173	49	—	—	—	—
Hoyt et al. 1995[1]	26	19	36	0.051	0.083	—	—

[1] Hoyt et al. 1995a; data given as mmol/MJ DE consumed for work above maintenance.
[2] Frape et al. 1979.
[3] P.A. Harris (1996) personal communication.
[4] Meyer et al. 1978; Carlson & Ocen 1979; Rose et al. 1980a; Snow et al. 1982.

30 000 mmol readily exchangeable K, most of the body's reserve. The bulk of the exchangeable Cl is present in the ECF, where it is the major fixed anion, but its concentration is substantially lower than that of Na – it is probably of the order of 10 000–12 000 mmol/100 l.

Sweat

The primary route of fixed-ion loss by working horses is sweat produced for the purpose of preventing an excessive rise in body temperature, so that dehydration through sweating entails a loss of both water and electrolytes with a contraction of the volume of body fluid. The electrolytes of horse sweat consist principally of Na, Cl and K, with lesser quantities of Ca and P (Table 9.2) and a small proportion of Mg, amounting to about 10 mmol/l. Note that the values attributed to Hoyt *et al.* (1995a) for miniature horses are defined per MJ DE consumed for work above maintenance. They found relatively greater losses of K than of Na and concluded that dietary needs during exercise were increased over the basal dietary supply by threefold, sevenfold and sixfold, respectively for Na, K and Cl. It would be important to establish the validity, or reason, for the differences in relative loss of Na and K. Changes in the composition of the blood plasma depend on the proportions lost of each of these constituents and of water and on the movement of ions into and out of ICF space. Extended exercise, with minimal water consumption leading to dehydration, will normally precipitate a substantial reduction in the concentration of plasma Cl; little change is frequently detected in the amounts of K and Na, although hypokalaemia is not rare. The explanation for the hypochloraemia is revealed by a comparison of the Na and Cl contents of sweat (Table 9.2) with their respective concentrations in blood serum (Table 3.1), demonstrating that a much greater proportion of Cl than of Na in body fluid is lost. Extended work, in hot dry weather, by a 450–500-kg horse expressing the above plasma changes may yield losses of as much as 35 l of water, 3500 mmol Na, 1500 mmol K and 4200 mmol Cl (equivalent to 80 g Na, 59 g K and 149 g Cl). The loss of Na, for example, represents over 200 g of sodium chloride, that is, much more than a horse would eat in a day. These values are at variance with the relatively higher values for K reported by Hoyt *et al.* (1995a), which are equivalent to 0.43 g Na, 0.93 g Cl, 1.41 g K, 2.03 mg Ca and 2.56 mg P/MJ DE consumed for work above maintenance.

The total concentration of electrolytes in sweat is higher than that in blood plasma, so a decline in the plasma concentration, despite considerable dehydration, is readily understandable. Most studies have revealed a decline in plasma electrolytes during endurance rides taking place in hot weather. For example, Carlson *et al.* (1976) reported reductions of 4.2 mmol Na/l, 0.9 mmol K/l, 10.3 mmol Cl/l and no change in Ca. On the other hand, the changes can be very variable depending on such factors as relative losses of water to electrolytes, plasma pH changes and time of collection after the ride. In horses losing a considerable amount of water, Rose *et al.* (1977) detected increases of 6 mmol Na/l and 0.3 mmol K/l plasma, no change in Ca and a decrease of 6.8 mmol Cl/l.

Thirst, drinking and osmotic pressure of blood

Thirst is in part controlled by the osmotic pressure of the blood and therefore it is frequently necessary to rectify electrolyte loss (Tables 9.3 and 9.4) before dehydrated horses will drink adequate quantities of water. Electrolyte mixtures are widely available commercially. If the horse is at the same time acidotic then plasma K may rise, despite a K deficit, as intracellular K is exchanged for H^+ ions. Subsequently, renal losses of K may increase, ultimately causing severe K depletion. A further discussion of the means of assessing K status and the major causes of depletion and their therapy are given in Chapter 11.

Horses should be allowed to drink frequently during extended work and at least 2 min should be conceded on each occasion. If the weather is very hot then a drink at least every 2 h is desirable. Hard work diverts much of the blood supply to the skeletal muscles away from the splanchnic bed of blood vessels serving the GI tract. It is thought that this inhibits the efficient absorption of water so that the amounts consumed on any one occasion should be relatively small. The consumption of large amounts is also to be avoided because of the large difference between it and blood in osmotic pressure. After work, very dehydrated animals should be given about 4.5 l (1 gallon) every 15 min, preferably containing 30 g electrolytes. If the horse will not drink, administration by stomach tube is sometimes necessary. Dehydration is frequently accompanied by coldness and fatigue, muscular tremors, colic, thumps, lack of appetite and a low pulse:respiration rate ratio. Severely dehydrated animals are sometimes given a 5% glucose–electrolyte solution i.v., while heart rate is monitored, when their ability to absorb fluid from the gut is in doubt (Table 9.3).

Where there has been a contraction of blood volume the administration of electrolytes has only a transient effect on increasing that volume. Nevertheless, rectification of losses and of the acid–base balance must be considered. The total osmotic pressure of the blood depends to a large extent on the colloids it contains, the principal ones being proteins, more especially albumin. However, protein loss during work will be minimal and reflect only its metabolism as an energy source apart from very slight losses through pulmonary haemorrhages. The rebuilding of energy reserves, particularly in respect of muscle glycogen, will take several days after very extended hard work.

Fatigue

Fatigue during extended exercise partly results from a decline in blood glucose concentration, and the rate of this decline is affected by velocity, by the extent of glycogen stores, by dietary manipulation that spares glycogen mobilization and by training that promotes a greater use of both fat and glycogen during maximal exercise (Snow & Mackenzie 1977a). Fatigue is first exhibited by a decrease in speed during the course of running. Contributory factors to fatigue therefore also include a loss of muscle ATP, depletion of muscle glycogen, metabolic acidosis, and blood lactate and ammonia accumulation. Fatigued horses may still possess a mean of 400–

Table 9.3 Composition (mmol/l) of various electrolyte solutions for i.v. use[1] (Rose 1981).

	Na	K	Ca	Mg	Cl	Glucose	Bicarbonate	Lactate	Acetate	Gluconate	Propionate
0.9% NaCl	154	—	—	—	154	—	—	—	—	—	—
5% Dextrose	—	—	—	—	—	278	—	—	—	—	—
Hartmann's solution	131	5	2	—	112	—	—	28	—	—	—
Ringer's lactate	130	4	3	—	109	—	—	28	—	—	—
Normosol R[2]	140	5	—	1.5	98	—	—	—	27	23	—
Dilusol R[3]	140	5	—	1.5	98	—	—	—	27	23	—
Normosol M[2]	40	13	—	3	40	278	—	—	16	—	—
Balanced electrolyte solution[4]	137	5	3	3	95	—	—	—	27	—	23
Solution to treat acidosis[5]	137	20	—	—	97	—	60	—	—	—	—
5% NaHCO₃ (hypertonic)	600	—	—	—	—	—	600	—	—	—	—
5% Dextrose saline (hypertonic)	154	—	—	—	154	278	—	—	—	—	—

Note: i.v. solutions should contain a bicarbonate precursor (lactate, acetate, gluconate or propionate), as the sole use of fixed bases such as chloride can cause metabolic acidosis, hypokalaemia and hyperchloraemia.

[1] Give only 1–2l/h of solutions containing 5% dextrose, but up to 3–5l/h for other solutions, all of which should be at 37°C.
[2] Abbot Laboratories, Illinois, USA.
[3] Diamond Laboratories, California, USA.
[4] Merritt 1975 in Rose 1981.
[5] Rose 1979 in Rose 1981.

Table 9.4 Composition (g/kg DM) of mixtures for administration by stomach tube and as a daily supplement.

Mix*	Glycine	Sodium chloride	Monopotassium phosphate	Magnesium sulphate	Potassium chloride	Calcium carbonate	Calcium gluconate	Sucrose or glucose*
(1)	470	270	190	13			57	
(2)		325			325	175		175
(3)		170			70			100

Note: Mixes (1) and (2) may be squirted as a thick slurry to the back of the mouth *after* water has been given, to rectify a water deficit and thirst. Mix (3) is a suggested simple mixture to be given in the amounts shown daily as a dry supplement mixed with bran or other palatable dry material to a 500-kg horse during periods of extreme effort in hot weather. Fresh water must be freely owailable. (N.B. It is usually more 'palatable' dry than as a solution.)
In order to provide approximately isotonic solutions add 230 g mix (1) to 6 l water, 120 g mix (2) to 6 l water every 2–3 h for a 500-kg horse.
* May be replaced by molasses.

500 mmol glycogen/kg muscle tissue, but the muscle fibres used predominately in submaximal extended exercise, that is, the type I fibres (slow twitch oxidative fibre) may be almost deplete and the type IIa fibres (fast-twitch oxidative) show a reduction in glycogen content. Although glycogen loss is pronounced in sprint exercise and fat is a major energy substrate in submaximal extended exercise, glycogen loss inevitably occurs during extended exercise.

Measurement of fitness and exhaustion

A medium-paced canter is normally accomplished with aerobic respiration so that the ratio of lactate to pyruvate in the blood is unchanged. However, during galloping the ratio rises sharply and an effect of training is to induce smaller changes in blood lactate and in the lactate to pyruvate ratio. Blood lactate concentration may nevertheless be unrelated to racing speed, but as it is a relatively strong acid it tends to be correlated with blood pH. Blood lactate concentration is partly an expression of adaptation to training and Persson (1983) proposes that the estimation of blood lactate at a work-induced heart rate of 200 beats/min could be used to gauge fitness in training, where lower lactate concentrations down to 2 mmol/l would reflect greater fitness. Heart rate would be measured by telemetry on the track and the results would be unaffected by track conditions that affect speed. In addition to dehydration and electrolyte depletion, blood acidity is a dominating factor in determining exhaustion (Table 9.5) and serum levels of the enzyme CK provide an indication of the severity of exercise and metabolic acidosis. In one study where horses raced at the rate of 700 m/min (26 mph) serum AAT rose 50%, serum Ca rose 13%, but serum CK rose 227% (Williamson 1974). It is thought that the increase in serum concentration of muscle enzymes after exercise is explained by an increase in the permeability of muscle cell membranes, owing to hypoxia (lack of oxygen). Thus, inadequate training or severe work loads will induce a greater increase in the serum concentration of CK.

Table 9.5 Effect of racing 1900–2500 m on concentractions of blood lactate and acidity and their relationship to exhaustion (Krzywanek 1974).

	Blood lactate (mequiv/litre)			Blood pH		
		After race			After race	
	Before race	Immediately post	15 min post	Before race	Immediately post	15 min post
Exhausted	0.58	23.14	24.37	7.379	7.086	7.105
Not exhausted	0.63	17.99	16.92	7.379	7.164	7.213

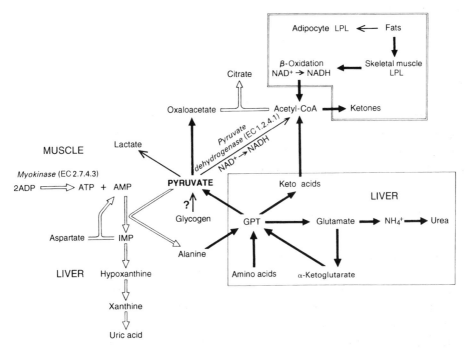

Fig. 9.11 Proposal for reactions stimulated (➡), not measurably influenced (⇨) or suppressed (→) by high protein or high fat diets during equine exercise of moderate intensity in the postabsorptive state. Reactions proposed to be stimulated by dietary fat (▣) and by dietary protein (□). GPT, glutamic-pyruvic-transaminase (EC 2.6.1.2); LPL, lipoprotein lipase (EC 3.1.1.34); IMP, inosine monophosphate. (Frape 1994.)

Ammonia and the alanine vehicle

The accumulation of ammonia (NH_3) in the blood of exercised horses probably contributes to fatigue, although the experimental evidence is weak. Ammonia causes pyruvate to accumulate, which in O_2 limited states is converted to lactate, resulting in a decrease in muscle pH. Blood ammonia concentrations have been decreased by dietary monosodium glutamate supplementation. *Plasma ammonia* and uric acid concentrations rise and attain maxima during recovery (Harris *et al.* 1987; Miller-Graber *et al.* 1987; Miller-Graber & Lawrence 1988), signifying an increased rate of AN cycling. A greater demand for ATP is partly met by the myokinase reaction ($2ADP \rightarrow 1ATP + 1AMP$) (Fig. 9.11; Frape 1994). Deamination of AMP, which supplementary $NaHCO_3$ may ameliorate, produces NH_3 and IMP, yielding uric acid (Harris *et al.* 1987) and alanine as the ammonia vehicle. Additional dietary protein may, or may not, aggravate this (Fig. 9.11).

In summary of fatigue

The first requirement of exhausted dehydrated horses is water, followed closely by electrolytes. Bodily energy resources must then be rejuvenated and if the weather is cold the horse must be kept warm but not hot. Metabolically some ATP may be reformed by the myokinase reaction in the presence of pyruvate, but generally the fuels for ATP synthesis are depleted. After normal hard work a horse should be cooled by gentle exercise of the muscles through walking, ridding the muscles of wasteproducts, but access to light grazing or hay should not be ruled out. After this relaxation of 1–1.5 h, tepid water should be given before a light meal of concentrates. It can be concluded that exhaustion during long-distance work is an expression of nutrient depletion whereas during sprint work it is principally the result of raised blood lactic acid and a consequential fall in blood pH.

ACID–BASE BALANCE OF THE BLOOD

What is an acid?

An acid is a substance, such as lactic acid, that yields hydrogen ions in solution. The acidity of blood or other solutions is expressed as the pH (the negative logarithm of the H^+ ion concentration). Acids and bases are produced during the metabolism of nutrients and abnormalities in the acid–base balance result from dysfunction, or overloading, of general metabolism and respiration. The normal pH of arterial blood is 7.5 and that of venous blood 7.4. The blood carries CO_2 to the lungs, partly in the form of weak carbonic acid (H_2CO_3), one of the principal acids of fizzy drinks. It and haemoglobin act as the principal buffers in blood; that is, they prevent the pH from shifting appreciably and so prevent death from this cause. In the plasma, CO_2 reluctantly and slowly forms H_2CO_3, but on diffusing into the red cells this reaction is accelerated 13 000-fold by the presence there of the enzyme carbonic anhydrase. Despite the accelerated change only 1 part in 800 of CO_2 forms H_2CO_3. This is described in Fig. 9.12, which shows that nearly all this H_2CO_3 dissociates into H^+ and bicarbonate (HCO_3^-) ions. The former are partly buffered by haemoglobin and the latter to a large extent diffuse back into the plasma so that around 20 times as much CO_2 is carried as HCO_3 as remains in the form of dissolved gas. Now, the dissociated form of H_2CO_3 in blood solution is in a constant (K) proportion to the undissociated form, as shown below:

(a) CELLULAR GAS EXCHANGES

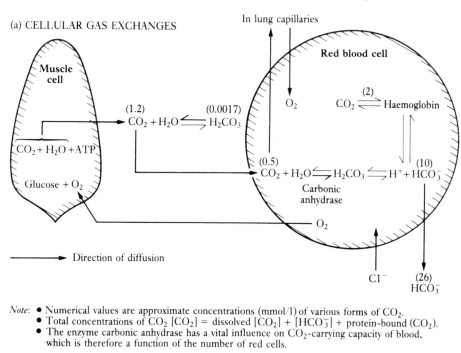

Note: • Numerical values are approximate concentrations (mmol/1) of various forms of CO_2.
• Total concentrations of CO_2 [CO_2] = dissolved [CO_2] + [HCO_3^-] + protein-bound (CO_2).
• The enzyme carbonic anhydrase has a vital influence on CO_2-carrying capacity of blood, which is therefore a function of the number of red cells.

(b) PLASMA REACTIONS

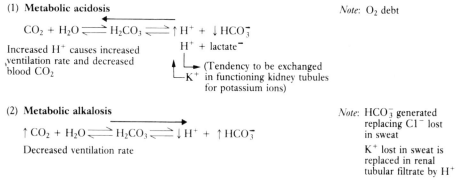

(1) Metabolic acidosis

$$CO_2 + H_2O \rightleftharpoons H_2CO_3 \rightleftharpoons \uparrow H^+ + \downarrow HCO_3^-$$

Increased H^+ causes increased ventilation rate and decreased blood CO_2

H^+ + lactate$^-$

(Tendency to be exchanged K^+ in functioning kidney tubules for potassium ions)

Note: O_2 debt

(2) Metabolic alkalosis

$$\uparrow CO_2 + H_2O \rightleftharpoons H_2CO_3 \rightleftharpoons \downarrow H^+ + \uparrow HCO_3^-$$

Decreased ventilation rate

Note: HCO_3^- generated replacing Cl^- lost in sweat

K^+ lost in sweat is replaced in renal tubular filtrate by H^+

(3) Respiratory acidosis

$$\uparrow CO_2 + H_2O \rightleftharpoons H_2CO_3 \rightleftharpoons H^+ + HCO_3^-$$

Respiratory insufficiency causes increased ventilation rate

Note: Plasma [H^+] tends to rise

(4) Respiratory alkalosis

$$\downarrow CO_2 + H_2O \rightleftharpoons H_2CO_3 \rightleftharpoons H^+ + HCO_3^-$$

Overheating or pain causes increased respiratory minute volume

Note: A decrease in [H^+] and possibly [HCO_3^-] results from respiratory changes

Increased direction of reaction

$\uparrow\downarrow$ Initial changes in plasma concentration of metabolite

Fig. 9.12 Effect of metabolic state on carbonic acid in blood plasma and associated relationships in muscle cells and red blood cells.

$$\frac{[H^+] \times [HCO_3^-]}{H_2CO_3} = K$$

If acid is produced during muscular activity, or during intestinal colic (see Ch. 11), this increases the H^+ ions in the numerator and these react with HCO_3 forming CO_2 and water (Fig. 9.12). In this process the HCO_3 concentration falls, but the H^+ ion concentration does not rise as much as it would in the absence of the HCO_3 buffer, and so the ratio $HCO_3:CO_2$ governs the pH of the blood.

Acidosis and alkalosis

Fatigue during exercise is associated with a deviation of blood pH from the ideal range, causing metabolic acidosis or, with overheating, respiratory alkalosis. In lactic acidosis there is increased production of H^+ ions owing to an O_2 debt. With overheating there is increased loss of CO_2, resulting from a high respiratory minute volume (Frape 1994). Metabolism for all important functions occurs most efficiently when arterial pH approaches the normal value of 7.5. Deviations are associated with losses of important ions through the kidneys and intestinal tract, causing a burden on metabolism which has eventually to be rectified. A lowering of pH results from an excessive rate of acid production during exceptional work and from disease states of the intestinal tract and associated organs, kidneys and lungs in particular. Organic acids produced in metabolism have only short-term consequences as they should be ultimately disposed of by metabolism and respiratory compensation. Fixed (that cannot be metabolized to CO_2 and water, and exhaled) acids and bases absorbed from digesta have a longer term influence and attention has been focused on the fixed cations and anions present in diet and their deducible effects on acid–base balance. Bone acts as a buffer to prolonged dietary imbalances of this kind. An acid diet, containing an excess of fixed anions, leads to bone resorption and osteoporosis so that renal excretion of those anions may proceed.

Base excess

A quantitative measure of the acid–base status of a horse, as affected by metabolism, health and diet, is known as the base excess (BE) (Fig. 9.13). This is the base content of the venous blood measured by titration with a strong acid to a pH of 7.4 at normal CO_2 tension. Base deficit is the same as negative BE and its measurement requires titration with a strong base, again to a pH of 7.4. At that pH, BE is zero and plasma bicarbonate equals approximately 22–25 mEq/l. In the normal horse, bicarbonate should range from 24 to 27 mEq/l and the BE from 2 to 5 mEq/l.

Function of the lungs and kidneys

The lungs have a vital short-term role in the acid–base balance by providing a route for the excretion of carbonic acid (Fig. 9.12). Although the kidneys also function in

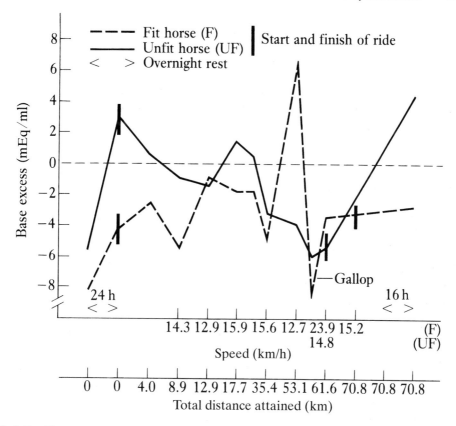

Fig. 9.13 Changes in plasma base excess (BE) during an endurance ride by a fit Arab stallion and a less-fit pony gelding (Frape *et al.* 1979). BE of unfit horse declined from start. Its respiration rate was consistently the higher, which could imply that a higher ventilation rate was removing more CO_2 from blood and carbonic acid reaction was moving to left (Fig. 9.12b '(4) Respiratory alkalosis' equation). Lap 8 was run at lowest speed (Fig. 9.10) and, although respiration rate of unfit horse did not fall (Fig. 9.6), the low rate of other horse immediately allowed the BE to rise. This lap was followed by a rest period before the fit horse was galloped.

this role, quantitatively they are insignificant as the lungs dispose of 200 times more of the acid. However, the kidneys fulfil a longterm role in disposing of nonvolatile acids and bases from the diet [see 'Diet BE and "fixed" dietary cation–anion balance (DCAB)', this chapter]. In this balance, and in that of acid–base, an important principle is disclosed. This is the principle of electroneutrality in which all urinary and lung excretion and in all movement across normal cell membranes in the horse no electrical charge can accumulate, as in a battery. The number of anions has to correspond with the number of cations moving in the same direction. This may lead to pH changes of various bodily fluids and the concept forms a basis for an understanding of the issues involved in acid:base and cation:anion balance.

Like most laws, this one is only approximately true as it may conflict with that of iso-osmolarity (see Glossary), in which all bodily fluid compartments (among which water is exchangeable through semipermeable membranes) approach isotonicity. In achieving this, a small voltage gradient occurs as, for example, between intracellular fluid in the muscles and extracellular fluid. These principles play a central part in muscular ailments, such as 'tying-up' (see Ch. 11). In so far as their acid–base status is concerned, horses are subject to four types of abnormal metabolism. Only those conditions relating to physical work will be discussed here.

Metabolic acidosis

In strenuous exercise an O_2 debt with a tissue accumulation of lactic acid occurs to varying degrees. It is also possible that this form of acidosis may exist in combination with a B-vitamin deficiency, causing an incomplete metabolism of pyruvic acid, but this proposition has not yet been substantiated. The build-up of H^+ ions causes hyperventilation and thus an increase in respiratory minute volume with a fall in blood partial pressure of carbon dioxide (PCO_2). This shifts the carbonic acid equation to the left (Fig. 9.12), which brings about some respiratory compensation of the acidosis and prevents an excessive fall in pH. In an analogous situation, the raised H^+ ion concentration and pain of laminitis will have a similar respiratory effect so that in its advanced state respiratory alkalosis supervenes. Sodium bicarbonate at a rate of between 250 and 300 g (3000–3600 mEq) over 24 h is sometimes given as a therapy for metabolic acidosis, but its unconsidered use can produce untoward effects.

Metabolic alkalosis

Alkalosis can occur in chronic laminitis as well as from exhaustion after long-distance cross-country work in hot weather; however, a reduction in ventilation rate in the worst-affected horses may induce slight acidosis. During extended work a depletion of body potassium and chloride develops in varying degrees. H^+ ions are excreted as a substitute for the depleted potassium, and bicarbonate fills the anion gap after loss of chloride. Tetanic spasms in extremis may sometimes result from a decreased availability of ionized calcium brought on by alkalosis. Unfit horses may present signs of adrenal exhaustion. The consequent failure to secrete aldosterone precipitates an excessive urinary loss of sodium with potassium retention and a further decline in wellbeing. Sodium bicarbonate therapy would severely aggravate alkalosis and the provision of a balanced electrolyte and glucose solution containing sodium, potassium and chloride (see also Ch. 11) is to be recommended. The want of potassium is not truly attested by its plasma level as a consequence of a shift from intracellular to extracellular space (see page 401). Electrolyte losses and hypocalcaemia, related to a raised pH, in exhausted endurance horses, are associated with a condition known as synchronous diaphragmatic flutter, in which heart and respiratory contractions coincide. In addition to electrolytes, calcium gluconate solutions are frequently given i.v. (Table 9.3).

Respiratory acidosis

This is the build-up of blood CO_2 (hypercapnia) which arises with the respiratory insufficiency of sprint work. It is obvious that the condition may also obtain with several diseased states of the lungs, including respiratory allergy, and the resulting decline in oxygen tension (hypoxia) in the blood may cause metabolic acidosis.

Respiratory alkalosis

Hard work in hot weather brings about overheating which directly affects the respiratory centre of the brain, causing rapid breathing. Rapid ventilation rate flushes CO_2 from the blood. The pain of colic and laminitis also accelerates respiration rate with similar consequences. The effect is to induce a slight rise of blood plasma pH (Fig. 9.14), although an absence of change in endurance horses may reflect a concomitant fall in both CO_2 and HCO_3.

Factors controlling plasma pH

Plasma $[H^+]$ is regulated by several independent variables, especially:

- partial pressure of CO_2, i.e. PCO_2, controlling $[H^+]$ and $[HCO_3^-]$;
- strong ion difference (SID), i.e, $[Na^+] + [K^+] - [Cl^-] - [lactate^-]$;
- weak electrolytes $[A_{total}]$, represented by [albumin].

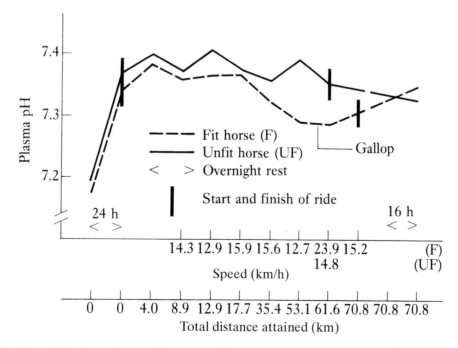

Fig. 9.14 Effect of an endurance ride on venous blood plasma pH in a fit Arab stallion and a less-fit pony gelding (Frape *et al.* 1979).

A decrease in plasma SID is normally associated with a decrease in plasma pH. This points to an increase in plasma [H^+], implying there is insufficient [HCO_3^-] available with which it can combine to maintain the constant K in the bicarbonate equation (see 'What is an acid?', this chapter). (Note: the function of dosing with $NaHCO_3$ is to provide that bicarbonate ion. Also, the Na^+ ion balances the lactate$^-$ ion, the concentration of which in plasma does not necessarily decline).

DIET BE AND 'FIXED' DIETARY CATION–ANION BALANCE

Diet BE

Mineral nutrition has a role in acid–base balance. This follows from the premise that excess base, or acid, in feed may be estimated as the difference between the sums of mineral cations and anions, approximated by:

$$\text{Dietary BE (mEq/g)} = (Na + K + Ca + Mg) - (Cl + P + S)$$

The effect that diet and metabolic acid production have on plasma BE is shown in Appendix D. Typical horse diets have a dietary BE of 200–300 mEq/kg estimated from the fixed-ion content. Thus, for a horse consuming 10 kg feed/day, the BE would amount to approximately 2500 mEq, which is similar to that provided by 200 g sodium bicarbonate – a quantity frequently given in therapy over a period of 24 h. Horses with a serious deficit of bases may require, over that period, double the amount contained in a normal ration. The dietary optima assume that the protein content is probably higher than that provided for the average adult horse. For a diet containing 10% crude protein, the dietary BE might be of the order of 30 mEq/kg less.

The absorption of Ca, Mg and P is limited and variable so that dietary cation anion balance (DCAB) is frequently measured in a simplified form as mEq/kg dietary DM:

$$\text{Dietary BE (mEq/kg)} = (Na + K) - (Cl + S)$$

(Note: in some studies the dietary S, normally providing 70–80 mEq/kg, is ignored.) On this basis, by including 10 g $CaCl_2$/kg in the 'medium' diet DM (Table 9.6) to give the 'low' diet and 13 g $NaHCO_3$/kg to give the 'high' diet, the urinary and blood pH were affected.

In the GI tract H^+ ions are exchanged for Ca^+ and Mg^{2+} ions by consumption of the 'low' diet. Urine is excreted in an electrically neutral state and during exercise immature horses given this diet could experience a considerable loss of urinary HCO_3^- and Ca^{2+} (hypercalcuria), accounting for metabolic acidosis and a negative Ca balance, demineralization and a weakening of the skeleton. Cooper *et al.* (1995) found raised urinary Ca^{2+} and Cl^- values in horses given a diet with a DCAB of −25.7

Table 9.6 The effect of dietary BE on urinary and blood pH.

	Low	Medium	High
mEq/kg diet	22	202	357
Urinary pH	5.38	7.69	8.34
Blood pH	7.368	7.400	7.402

and relatively greater urinary losses of P, Na^+ and K^+ in horses given a diet with a DCAB of 370.4. Diets with a DACB >200, as indicated above, reduce the risk of metabolic acidosis. Lower blood pH, PCO_2 and $[HCO_3]$ values are found at rest (Baker *et al.* 1992) and subsequent to anaerobic exercise (Stutz *et al.* 1992) with a dietary (Na + K – Cl) balance of 5–21 cf. 107–125 or more mEq/kg. Horses receiving 107 mEq/kg, or less, recovered normal blood glucose postexercise more slowly than did those receiving 201 mEq/kg, or more. To maximize the transitory buffering effect of diet the optimum time for anaerobic exercise is 3.5–4.5 h after feeding (Stutz *et al.* 1992), despite an increased fluid load (Meyer 1992). Popplewell *et al.* (1993) recorded faster times over 1.64 km and lower heart rates postrace, 2–4 h after a meal with a balance of 295 cf. 165 mEq/kg DM, despite higher blood lactates. Moreover, a balance of 354 mEq/kg DM achieved a greater Ca balance cf. 223 mEq/kg (Wall *et al.* 1993). The optima for balance and total mineral ion load is, however, likely to differ between types of work – sprint vs. extended.

Dietary protein

The effect of dietary protein level on performance and metabolism during exercise is discussed in this chapter under 'Dietary protein requirements and exercise'. The effect of excess dietary protein, or that inevitably oxidized, on acid–base balance has not been addressed. The effects may be small, but a basis for drawing conclusions should be summarized. Firstly, it must be remembered that only absorbed products of protein digestion can influence cation–anion balance.

Many native proteins are rich sources of P, as phosphorylated amino acids, e.g. phosphoserine, and so contribute to the fixed anions. The oxidation of neutral amino acids has no effect on the acid load, but three classes of these nutrients do. Specifically, the basic (cationic) amino acids (lysine, arginine and histidine) yield neutral end products plus a proton (H^+) and thus are strangely acidogenic. The sulphur-containing amino acids (methionine and cysteine) are also acidogenic, as they generate sulphuric acid when oxidized. The dicarboxylic amino acids (aspartic and glutamic acids) are anionic, but metabolizable, and consume protons when oxidized, and so reduce the diet's acid load. Protons can be excreted in the urine but primarily as the ammonium ion ($H^+ + NH_3 \rightarrow NH_4^+$). Glutamine is the principal amino acid involved in renal ammonium ion genesis, and although not a dietary essential nutrient it is used up in this process (see 'Protein assimilation', this chapter, glutamine drinks as an aid to recovery from prolonged exercise). Horse diets frequently have lysine and threonine as first and second dietary essential limiting amino acids and consequently there is interest in adding lysine-HCl to make good any limitation. When supplied as their Cl^- salts they are a fixed anionic, or acid, source. However, if 0.1%, or 1g lysine-HCl/kg diet is added, it provides only 5.5 mEq acid/kg.

The effect of the rapid metabolism of amino acids during intense exercise on acid load may be important at the margin, but the amounts of acid produced are dwarfed by that produced through carbohydrate metabolism. Moreover, it is not established whether the rate of acid production from protein is markedly influenced by the dietary level of protein.

Effect of DCAB on digestibility and Ca and Mg retention

The cation–anion balance of diets has been artificially adjusted by additions of calcium chloride, ammonium chloride, potassium citrate and sodium bicarbonate, giving ranges in balance from 20 to 400 mEq/kg diet. There is an increase in dry matter digestibility with an increase in positive balance. Hypocalcuria, promoting an increased retention of Ca and Mg, can also result from an increase in the balance, reducing the chronic risk of osteoporosis. A low dietary cation–anion balance induces metabolic acidosis, causes hypercalcuria, decreased Ca and Mg balances, and when horses are exercised within 4h of feeding, it leads to poorer work output and a slower recovery, owing to a poorer dietary buffering effect (Popplewell *et al.* 1993).

Supplements

Cofactors

The activity of enzymes depends on the presence of the necessary cofactors and an increasing demand for these enzymes implies an increasing need for the cofactors. These cofactors include magnesium and zinc together with forms of the vitamins thiamin, riboflavin, niacin, pantothenic acid and vitamin B_{12}, all of which play major parts in carbohydrate and/or fat metabolism. The horse derives these B vitamins from its diet and by microbial synthesis in its intestine. As the intensity of work increases, the composition of the diet and the amount of feed consumed change as a consequence of the increased consumption of starchy cereal grains. This will alter not only the dietary supply of B vitamins, but also the intestinal synthesis of the vitamins, and it is an open question whether the rate of their absorption is exceeded by tissue demand when horses are in intensive training.

The microbial fermentation of starch yields a higher proportion of propionate in the VFA. Metabolism of this acid requires adenosylcobalamin (vitamin B_{12}), as the coenzyme of methylmalonyl-CoA mutase, and work with ruminants has revealed that such diets may create a dietary requirement for this vitamin (Agricultural Research Council 1980), the lack of which causes an accumulation of propionate, depressing appetite. Uncontrolled observations by the author (D. Frape unpublished observations) of horses in training have shown that their blood concentrations of vitamin B_{12} are lower than in many other horses and that the palates of those with flagging appetites may be whetted by supplements of the vitamin. A reasonable inference is that an increased consumption of cereal grains by horses increases propionate production and hence the dietary requirement for vitamin B_{12}. An analogous argument may be put for thiamin, functioning as cocarboxylase in the cleavage of pyruvate. Studies by Topliff *et al.* (1981) suggested that the exercising horse may have a thiamin requirement double that of nonworking horses. TBs in training have low serum folate concentrations, but whether this simply reflects a lower potency of training diets has not been established. Thus, there is no conclusive evidence concerning the effect of work on the requirement for B vitamins functioning as enzyme cofactors.

Water and electrolyte loading

Despite normal variation in the levels of each of the dietary electrolytes, adaptation serves to maintain the pH of body fluids in the normal physiological range. Outside this normal range tissue pH may be altered through an overload of these compensatory mechanisms. Excretion of excess fixed ions requires water as a solvent, increasing water demand. Complete water restriction for 20 h before exercise has been shown to reduce the intestinal water content by 10% at the start of protracted exercise, and that exercise reduces it by a further 15–20%, through sweat losses, regardless of whether drinking had been allowed. Horses that are severely dehydrated are exceptionally exhausted (Carlson *et al.* 1976), and are reluctant to drink. Even during a ride, spontaneous drinking may not occur when water is offered if there has been an iso-osmotic loss in sweat, unless the fall in plasma volume exceeds 6% (Sufit *et al.* 1985) and especially if electrolytes are not given. Water and balanced electrolytes are therefore frequently given during a ride.

Sodium chloride feeding before a ride could encourage water intake *then* and so should improve water balance during a subsequent extended ride. Loading the horse with electrolytes tends to increase their temporary accumulation, and that of water, in the large intestine (Slade 1987). This could act as a reserve of water, Na^+ (Meyer 1992) and of Cl^- (Coenen 1992a) for extended work. The daily ileo–caecal flow of water and electrolytes per kilogram of bodyweight is in the range of 100–140 ml water, 300–420 mg Na^+, 50–70 mg K^+ and 100–140 mg Cl^-. Absorption along with water from the large intestine, during ingesta fermentation, is estimated to be 75–95% for Na^+, over 90% for Cl^- and 30–55% for K^+ (Meyer 1992). These nutrients can revive tissue depleted of water, Cl^-, Na^+, K^+ and Ca^{2+} through sweating (Rose *et al.* 1977). The benefit of such an electrolyte reserve is promoted by the presence of fermentable material in the hindgut, as this allows the continuous absorption of the reserve ions during a ride.

The amounts of dietary electrolytes necessary to keep a horse in electrolyte balance have been assessed to be 1.3–1.8 g Na, 3.1–3.9 g Cl and 4.5–5.9 g K, 8.5 mg Ca and 10.7 mg P/Mcal DE (0.3–0.4 g Na, 0.7–0.9 g Cl and 1.1–1.4 g K, 2 mg Ca and 2.6 mg P/MJ DE). Thus, the requirements for Na, K and Cl during exercise are increased over maintenance needs by three-, seven- and sixfold, respectively, according to Potter's group in Texas (Hoyt *et al.* 1995a). Natural feed, given subsequently to rides, is likely to contain much more K than Na. There should therefore be about twice as much Na as K and 1.2 times as much Cl as Na in supplements given with these feeds (Table 9.4). Organic anions can make up the residue. Small amounts of Ca and Mg may also be included.

Sodium bicarbonate

Anaerobic exercise causes a rise in plasma K^+, released from the contracting muscle fibre. A failure of its reuptake may result from an inhibition of the Na–K pump of the fibre membrane, owing to decreased ATP availability (Harris & Snow 1988) through inadequate buffering of H^+ ions within active fibres (Harris & Snow 1992). The loss of intracellular K^+ leads to an altered transmembrane potential that may contribute to fatigue during exercise.

To counter this, oral sodium bicarbonate ($NaHCO_3$), which increases plasma SID, results in a smaller rise in plasma NH_3 through a lower ATP loss and IMP formation (Greenhaff *et al.* 1991b), HCO_3^- accelerating H^+ removal (the bicarbonate system is the major proton acceptor in the body). A positive effect seems to occur only where the duration of exercise is 2–3 min, accounting for its particular use with standardbreds engaged in races over 1.6–2.4 km. However, the use of alkalizing agents prerace is discouraged, or leads to disqualification, in many racing jurisdictions. No effect of $NaHCO_3$ was observed over 1 km (Greenhaff *et al.* 1991b). Lawrence *et al.* (1987a, 1990) reduced race times by 1.1 s over 1.61 km with standardbreds treated orally with 0.3 g/kg BW mixed with 20 ml corn syrup and 10 ml water, cf. powdered confectioner's dextrose and salt mixed with corn syrup and water, 2.5 h before exercise ($P < 0.1$). Treatment increased both blood pH and lactate disappearance rate postrace. Harkins and Kamerling (1992) treated TBs with 0.4 g $NaHCO_3$/kg BW in 1 l water, cf. 1 l water only, 3 h before a 1.61-km race, increasing venous HCO_3^- and pH. Postrace there was an increase in venous blood pH and lactate in the $NaHCO_3$ group, with no change in race times, or in venous partial pressure of carbon dioxide (venous $vPCO_2$). (Note that the lactate$^-$ ion will be neutralized by Na^+ and it does not determine the pH. Alternatively, it can be argued that the increase in $[Na^+]$ reduces plasma $[H^+]$ through the maintenance of electroneutrality, despite an increase in plasma $[Lac^-]$. This increase in $[Lac^-]$ is probably the result of increased efflux of Lac^- from muscle cells caused by the extracellular alkalosis, so reducing fatigue.) The optimum dose and time are 0.4 g $NaHCO_3$/kg BW (in 1 l water) 2–4 h prior to work, as assessed by blood pH and HCO_3^- (Greenhaff *et al.* 1990b; Corn *et al.* 1993), although doses of 1 g/kg BW have achieved higher values with a peak 4 h postadministration.

Analysis of venous blood from standardbred pacers before racing has revealed HCO_3^- values in excess of 40 mmol/l, indicating higher doses than 0.4 g/kg. A dose of

Na equivalent to 20% of the body's total exchangeable Na should increase plasma volume, which could have an opposite effect on sprint performance to that of a buffer, and may account for variable results following $NaHCO_3$ administration. Moreover, the value of the large intestine as a source of Na may be modulated by acetate production (Argenzio *et al.* 1977), which varies with the interval from, and nature of, the last meal. Lloyd *et al.* (1993) administered 1 g $NaHCO_3$/kg BW, cf. a similar molar dose of NaCl or water only. The $NaHCO_3$ extended exercise to exhaustion on a treadmill and increased blood lactate, but a comparison with the two untreated groups, where all horses had access to water, indicated poorer endurance with $NaHCO_3$, possibly from a higher fluid load. Yet Hanson *et al.* (1993) gave horses, with and without free access to water, 1 g $NaHCO_3$/kg BW in 4 l water and found no significant difference in plasma volume. The response is, however, more complicated as alkalosis, caused by the 1 g dose/kg, led in both studies to hypercapnia and some hypoxaemia through respiratory compensation. This ventilatory depression was not thought to have affected performance, and was probably associated with reduced tidal volume because of the tendency for a 1:1 linkage of respiratory frequency with stride. Moreover, intracellular $[H^+]$ may exchange for extracellular K^+, causing hypokalaemia after administration of 'milk shakes' of sodium bicarbonate. Both plasma K^+ and Ca^{2+} concentrations decline with alkalosis so that cardiac and skeletal muscle contractions could be disrupted, which may contribute after exercise to the distress of synchronous diaphragmatic flutter, and other signs.

Thus, the optimum dose, method of administration and overall effects of $NaHCO_3$ and water have yet to be determined. Nevertheless, some value may result from treatment with 0.4 g $NaHCO_3$/kg BW (in 1 l water) 2–4 h prior to gallops lasting 2–3 min.

Carnosine

A more enlightened approach to combating the rise in intracellular $[H^+]$ may be to alter the intracellular concentration of the imidazole dipeptide buffers carnosine (β-alanylhistidine) and its N^2-methyl derivative, anserine. Carnosine contributes 30% of the buffering in equine skeletal muscle (Harris *et al.* 1991a), and in type IIb fibres (prominent in equine muscle) it may account for up to 50%, with a concentration of 188 mmol/kg DM muscle (Sewell *et al.* 1991a,b; Sewell 1992b). However, dietary supplements of histidine have not given convincing responses in delaying fatigue.

FAT SUPPLEMENTS AND EXERCISE

Horses participating in competitive long-distance rides are required to make effective use of body fat reserves as a source of energy to conserve glucose sources, as a severe depression in blood glucose is a measure of fatigue. Dietary oils and fats are well utilized (McCann *et al.* 1987; Hollands & Cuddeford 1992; Potter *et al.* 1992b) to fulfil this function, although vegetable fats are generally more readily digested than animal sources. Fats given to man and seemingly to the horse dely gastric emptying of carbohydrate, and so improve glucose tolerance, by lowering the postmeal peak plasma glucose response. Fats are not subject to microbial fermentation and their greater use decreases colic and laminitis risks and may promote intramuscular and hepatic fat metabolism, increasing performance at submaximal and intense rates. Work in Texas with quarter horses used for cutting shows that the benefits of a 10% fat supplement are adaptive and take 3–4 weeks to materialize (Julen *et al.* 1995). Fat supplements can delay the decline in blood glucose during endurance rides, accelerate the rate of recovery of resting pulse and respiration rates (Hintz *et al.* 1978a,b; White *et al.* 1978) and promote the recovery of resting blood glucose, reducing the risk of fatigue-related injuries. The practical problems of adding large amounts of fat to the diet, however, have to be addressed.

Fat yields less CO_2/mole ATP generated, decreasing plasma PCO_2. Therefore relative to the effects of a carbohydrate diet there is a small increase in plasma pH brought about by decreases in both $[H^+]$ and $[HCO_3^-]$. Hard training (Hambleton *et al.* 1980) and fat supplementation with anaerobic (Pagan *et al.* 1993b) and extended aerobic (Pagan *et al.* 1987c) exercise are followed by an elevation in plasma FFA, whereas resting FFA are lowered by supplementation (Harkins *et al.* 1992). Thus, there may be a stimulation to β-oxidation, or to both fat mobilization and metabolism (Figs 9.15 and 9.16), sparing glycogen. This probably results from an increase in citrate production which inhibits phosphofructokinase, one of the rate limiting enzymes of glycolysis. This in turn results in an accumulation of glucose-6-phosphate, inhibiting glucose phosphorylation and sparing glucose oxidation. Citrate synthase (EC 4.1.3.7) activity of muscles, of types I and IIa, is increased by fat supplementation, supporting this analysis and indicating that dietary fat increases oxidative metabolism of skeletal muscles. The addition of 100 g corn oil to the diet of sprint-trained Arabian horses (Taylor *et al.* 1995) caused, at fatigue, greater plasma glucose and lactate⁻ (11 vs. 8 mEq/l) concentrations, offset by increases in plasma $[Na^+]$ and $[K^+]$ and decreases in plasma $[Cl^-]$. Thus, the fat also raised SID and minimized the decrease in pH of plasma.

Three points to note concerning fat supplementation are that:

(1) increased plasma lactate may possibly cause fatigue independently of an effect on plasma pH;
(2) the increase in lactate in Arabians is less than that observed in TBs. This may be associated with the higher proportion of slow-twitch, oxidative muscle fibres (type I), and greater oxidative enzyme activity in Arabians;
(3) The greater rise, postexercise, in plasma lactate (and alanine, a precursor of pyruvate and of glucose) with fat supplementation possibly results from the accelerated rate of glycogenolysis coupled with a reduction in the activity of pyruvate dehydrogenase (PDH) complex and decreased oxidation of pyruvate. PDH is a key regulator of fat and carbohydrate metabolism (Fig. 9.11). There can in fact be a synergistic effect of the combined treatment with

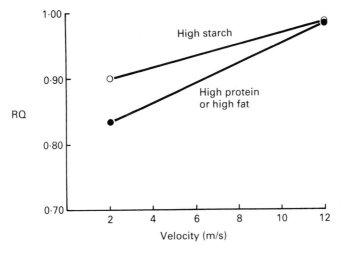

Fig. 9.15   Relationship between RQ and velocity of horses given high starch, high protein or high fat diets. At high velocity there is no difference in RQ because high rate of energy expenditure demands glycolysis. At low to moderate velocity, fat and protein may be used. (Frape 1994.)

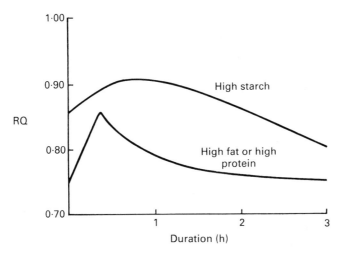

Fig. 9.16   Relationship between RQ and time during aerobic work (low to moderate velocity, 4–6 m/s) in horses given high starch, high fat or high protein diets. On high fat or high protein diets fat mobilization becomes predominant earlier, thereby conserving muscle glycogen. (Frape 1994.)

Table 9.7   Blood [Lac$^-$] during sprints in horses adapted to either a control or high-fat diet and administered either water or NaHCO$_3$ before exercise (Ferrante *et al.* 1994b).

	Blood [Lac$^-$] (mmol/l)
Control/water	5.73 ± 0.12
Control/NaHCO$_3$	6.26 ± 0.21
Fat/water	7.01 ± 0.20
Fat/NaHCO$_3$	9.47 ± 0.32

fat and $NaHCO_3$, leading to even higher plasma lactate levels (Table 9.7). There is a lower rate of pyruvate conversion to acetyl-CoA, as β-oxidation increases production of acetyl-CoA. However, much evidence in horses run at a constant velocity indicates a lower plasma lactate accumulation with dietary fat supplements (Table 9.10). This points to a substitution of β-oxidation for glycogenolysis, conserving glycogen stores.

During sprinting glycogenolysis is required as an anaerobic high power source to complement the greater oxidation of fatty acids, a low power source, associated with fat supplementation. Although plasma lactate may be increased there may be no significant decrease in plasma pH, or SID, if there are increases in plasma Na^+ and K^+ concentrations and a decrease in plasma Cl^- concentration.

Heat increment

The increased energy density and decreased intestinal residue achievable with dietary fat supplements could be the essential characteristic of fat (Hiney & Potter 1996). When 10% fat replaced starch, heat production fell from 77% of available DE to 66% and available NE rose from 16% of DE to 36% during work (Scott *et al.* 1993), reducing thermal stress (McCann *et al.* 1987), regardless of body fatness and in both temperate and hot weather (Potter *et al.* 1990). Waste heat production during exercise is very large. Approximately 80% of stored energy utilized for movement is lost as heat. The decrease in heat production achieved by fat supplementation in the main reflects diminished microbial fermentation in the hindgut.

Muscle glycogen

Several reports indicate no difference in resting muscle glycogen concentration between high starch and high fat diets (Hintz *et al.* 1978a; Pagan *et al.* 1987a; Topliff *et al.* 1987). Pagan *et al.* (1987b) equalizing DE intake between treatments, and Greiwe-Crandell *et al.* (1989) providing higher energy intakes with fat supplements, both found lower concentrations. Most other reports describe increased resting (postexercise muscle glycogen concentration may be no higher as glycogen utilization may be promoted by its higher muscle level) muscle glycogen following vegetable, or animal, fat additions at approximately 10% of the diet, to provide equal DE or ME intakes (Hambleton *et al.* 1980; Meyers *et al.* 1987, 1989; Oldham *et al.* 1990; Harkins *et al.* 1992; Jones *et al.* 1992; Scott *et al.* 1992; Julen *et al.* 1995) (Table 9.8). Effects of fat on hepatic glycogen capacity, which is 10% of that in skeletal muscle, are equivocal, as marginal increases (Hambleton *et al.* 1980) and decreases (Pagan *et al.* 1987b) are reported. Although fat concentrations of up to 20% of the dietary DM have been used without digestive upset, or any reduction in utilization, dietary fat levels of 15% have decreased glycogen storage compared with controls. A diet clearly needs to contain enough starch from which the storage is derived. Meyer and

Table 9.8 Resting muscle glycogen [1] as affected by added dietary fat in diets of differing energy densities but generally given to horses to equalize DE intake (Frape 1994).

Added dietary fat (g/kg diet)	0	20–30	50–60	80–100	140–150	SE	Reference
	Muscle glycogen (mmol/kg wet tissue)						
Animal fat	81	—	—	78	—	—	Hintz *et al.* 1978a
	94	—	109	143	—	10.5	Meyers *et al.* 1989
	88	—	—	127	—	2.6	Oldham *et al.* 1990
	93	—	—	145	—	2.1	Scott *et al.* 1992
	Muscle glycogen (mmol/kg DM)						
Vegetable oil[2]	—	200	255	292	240	—	Hambleton *et al.* 1980
	680	—	—	—	580	100	Pagan *et al.* 1987a
[3]	198	229	—	—	—	12	Harkins *et al.* 1992

[1] Gluteus medius, biceps femoris or quadriceps femoris.
[2] Fat added replaced maize grain giving diets of different energy densities but constant daily energy and protein intakes.
[3] Fat-added diet contained less roughage but more fat, starch and protein and was fed to provide equal DE intakes.

Sallmann (1996) found that when 2 g fat/kg BW daily was given as much as 0.4 g fat/kg BW was transferred to the hindgut, with a potential risk for disturbance of caecal microbial metabolism.

Respiratory quotient

Respiratory quotient (RQ) is the ratio at standard temperature and pressure (STP) of the volume, or moles, of CO_2 eliminated to the volume, or moles, of O_2 utilized in oxidation, i.e. CO_2/O_2 for carbohydrate = 1, fat = approx. 0.7 and amino acids = approx. 0.85. RQ rises with increasing speed (Pagan *et al.* 1987b), is lowered by training (Meyers *et al.* 1987) and is either not affected by dietary fat (Meyers *et al.* 1989) or is lowered by additional protein, or fat, during submaximal exercise (Pagan *et al.* 1987b) (Table 9.9; Figs 9.15 and 9.16). A lower RQ implies a lower rate of CO_2 production. A decreased PCO_2 may moderate a decrease in blood pH (through maintenance of the dissociation equilibrium of carbonic acid), so offsetting fatigue. RQ is positively correlated with muscle glycogen reserves during mild aerobic exercise and it declines as submaximal exercise progresses (Pagan *et al.* 1987b) (Fig. 9.16), indicating a sparing of glycogen. Higher stores of glycogen, with fat supplementation, accelerate their mobilization rate during anaerobic exercise (Oldham *et al.* 1990; Jones *et al.* 1992; Scott *et al.* 1992; Julen *et al.* 1995), yet a lower rate (Greiwe-Crandell *et al.* 1989), no difference (Hintz *et al.* 1978a) and a marginally greater loss of glycogen, with lower initial reserves (Pagan *et al.* 1987b), have all been reported during aerobic exercise with fat supplements (Figs 9.17 and 9.18). No clear picture emerges that fat would particularly benefit exercise in which extended aerobic metabolism dominated. Metabolic adaptation to a fatty diet may take as

Table 9.9   Relationship of added dietary fat to RQ in horses exercised on treadmills (Frape 1994).

Speed (m/min)	Slope (°)	Time (min)	Added dietary fat (g/kg diet) (SE)				Reference
			0	50	100	150	
					R	Q	
180	9	20	0.910	0.860	0.870	—	Meyers *et al.* 1989
300	0	90	0.830 (0.012)	—	—	0.750 (0.022)	Pagan *et al.* 1987b
360*	0	2	0.887 (0.019)	—	—	0.827 (0.038)	Pagan *et al.* 1987b
480*	0	2	0.890 (0.031)	—	—	0.873 (0.030)	Pagan *et al.* 1987b
600*	0	2	0.977 (0.027)	—	—	0.957 (0.043)	Pagan *et al.* 1987b

*Step-wise work test.

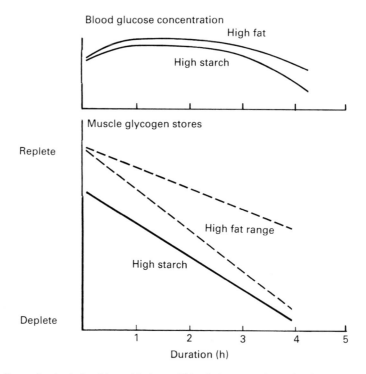

Fig. 9.17   Generalized relationships, with time, of blood glucose and muscle glycogen concentration in horses of moderate fatness during extended aerobic work (Frape 1994).

much as 6–11 weeks (Custalow *et al.* 1993) and some studies may not have allowed sufficient time for this, so that their outcome could have depended on design details and horse temperament. The interpretation is complicated by the expression of harder anaerobic work (Webb *et al.* 1987a) and faster speeds, with no greater

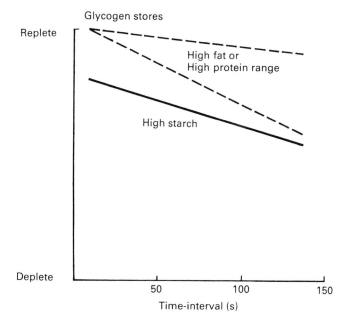

Fig. 9.18   Generalized relationship between glycogen stores and time-interval during intense anaerobic work (>600 m/min, >190 beats/min) (Frape 1994).

glycogen loss, at a constant heart rate (Oldham *et al.* 1990) by fat supplemented horses. As fat yields energy only by oxidation, mimimal glycogen sparing would be expected during maximal exertion, when its value is other than for conservation of glycogen.

Blood glucose

Some workers report similar (Worth *et al.* 1987) or lower (Meyers *et al.* 1989) blood glucose concentrations with added fat during aerobic exercise, but the majority (Hintz *et al.* 1978a; Hambleton *et al.* 1980; Webb *et al.* 1987a; Oldham *et al.* 1990; Harkins *et al.* 1992; Scott *et al.* 1992; Custalow *et al.* 1993) observed higher values during and after exercise of all types (Fig. 9.17; and Table 9.10), even with increased work effort (Webb *et al.* 1987a; Harkins *et al.* 1992).

Lower heart rates and a more rapid recovery of resting rates (Meyers *et al.* 1987), lower blood lactic acid concentrations during and subsequent to submaximal and strenuous standardized exercise tests (SETs) in fat supplemented horses (Pagan *et al.* 1987a,c; Webb *et al.* 1987a; Meyers *et al.* 1989; Pagan *et al.* 1993b) (Table 9.10), a higher velocity (m/s) at which venous blood lactate attains 4 mmol/l (V_{LA4}) (Pagan *et al.* 1993b) and lactate speed threshold (Custalow *et al.* 1993) possibly reflect a slightly lower RQ (Table 9.9 and Fig. 9.15), but an absence of blood pH measurements complicates fatigue assessment. Higher lactates during some higher speed SETs (Ferrante *et al.* 1993; Taylor *et al.* 1993) with fat cf. starch may represent a

Table 9.10 Effect of fat supplementation on blood plasma lactate and glucose concentrations during exercise at constant velocity (C) or at uncontrolled velocity (UC) and after postexercise rest. Data averaged over the sources used for each comparison of carbohydrate control and added fat (Frape 1994).

Added dietary fat (g/kg)		Plasma lactate (mmol/l)		Plasma glucose (mmol/l)		Reference
		Work	Rest	Work	Rest	
0	C	2.30	1.98	5.42	5.32	Hambleton *et al.* 1980; Meyers *et al.*
100	C	1.89	1.47	4.93	5.23	1987, 1989; Webb *et al.* 1987a;
						Worth *et al.* 1987
0	C	9.90	—	2.52	3.80	Pagan *et al.* 1987c
100	C	6.70	—	2.82	3.70	
0	UC	2.31	2.25	5.32	6.02	Hintz *et al.* 1978a; Harkins *et al.* 1992
100	UC	2.12	2.09	6.11	6.25	
0	UC	15.79	2.76*	6.57	5.29*	Webb *et al.* 1987a; Oldham *et al.* 1990
100	UC	18.61	1.87*	7.05	6.02*	Scott *et al.* 1992

*Data from Webb *et al.* 1987a only.

tendency for the response curves to crossover at these speeds (Custalow *et al.* 1993). The effects of high protein on blood lactic acid may be more prominent than those of high fat (Pagan *et al.* 1987a,c). Observations in human subjects indicate that high protein, high fat diets increase the activity of skeletal muscle lipoprotein lipase (LPL), whereas high carbohydrate diets reduce that activity (Jacobs 1981). As the insulin response to a carbohydrate diet exceeds that to high protein, high fat diets, and as insulin depresses muscle LPL activity, the observation of Jacobs is understandable. Increased energy generation from fat oxidation with high fat and protein diets may owe to the combined effects of increased muscle LPL hydrolysis of plasma TAG and increased use of plasma FFA (Figs 9.16 and 9.17), accounting for lower plasma lipids during aerobic SETs (Meyers *et al.* 1987). No certain conclusions can, however, be drawn from changes in venous blood FFA values (Frape 1993), but it is suggested that high fat diets may increase activity of muscle LPL (and possibly of TAG lipase), reciprocally reducing adipose tissue LPL activity, in contrast to the effects of starch (Fig. 9.11). As plasma TAG concentration tends to decline with fat supplementation, the increase in muscle LPL activity would seem to exceed the reciprocal decrease in adipose tissue LPL activity.

A dietary increase in either protein or fat normally results in decreased starch, reducing 'heating', anxiety, heart rate and excitability. Mixed fats contain lecithins, a component of which is choline. Holland *et al.* (1996) reported that vegetable oil, or especially oil enriched with additional soya lecithin, reduced the spontaneous activity and excitability of horses when the diet contained 100 g of this supplementary oil/kg. Choline is used in the synthesis of acetylcholine, a neurotransmitter found in parasympathetic nerve synapses and in voluntary nerves to skeletal muscles. The

Table 9.11 Provisional conclusions on effects of high fat diets given to exercising horses compared with diets of normal fat concentration that provide similar amounts of dietary fibre, protein and DE but more starch (Frape 1994).

Advantages
(1) Lower RQ during submaximal exercise (promoting fat catabolism) potentially extending endurance.
(2) Possible decrease in adipose tissue LPL activity (EC 3.1.1.34) and increase in muscle LPL activity.
(3) Increase in muscle glycogen stores, more glycolytic energy and possible delay in glycogen exhaustion during extended aerobic exercise.
(4) Increased, or sustained, blood glucose concentrations during extended exercise.
(5) Possibly delayed lactic acid accumulation during anaerobic exercise (lactic acid accumulation is proportional to the rate of glycogen expenditure, when other conditions are constant).
(6) Reduction in gut fill, which may benefit work at >200 beats/min, but which may compromise endurance.
(7) Reduced excitability of hotblooded horses and a possible reduction in risks of colic and laminitis in all types of horse.
(8) Fats contain lecithins, a component of which is choline which is a precursor of acetylcholine, the neurotransmitter. Lecithin feeding is associated with reduced excitability.

Disadvantages
(1) High cost of high-grade fat.
(2) Wide availability of poor-quality feed-grade fat and difficulty of assessing quality.
(3) Lack of stability of large fat supplements in mixed feed and practical problems of administration to the horse.
(4) Refusal of high fat diets, or delay in acceptance of equivalent intakes, i.e. lower palatability.
(5) Lower large intestinal fluid reserves for endurance events.

potential advantages and disadvantages of fat are proposed in Table 9.11 (also see Chapter 5, n-3 fatty acids).

Polyunsaturated fatty acids and T-bars (3-thiobarbituric acid reactive substances)

Linoleic acid is a dietary essential polyunsaturated fatty acid (PUFA). Oils especially rich in PUFA have no notable benefit for ponies given a diet deficient in PUFA for 7 months (Sallmann *et al.* 1992). However, chain length, or degree of unsaturation, may influence the exercise response (Pagan *et al.* 1993b). PUFAs in cell membranes are susceptible to attack with the removal of an H atom with its electron, leaving a radical subject to attack by O_2, yielding peroxyl radicals. Chain reactions, the breakdown of the cell membranes and several products that include malonyldialdehyde (MDA), n-pentane and ethane are a consequence. MDA can be measured colorimetrically with thiobarbituric acid (TBA). Strenuous exercise causes increased plasma thiobarbituric acid reactive substances (TBARs) and breath n-pentane per kilogram of bodyweight (McMeniman & Hintz 1992). However, the peroxidative stress of 3% corn oil was accommodated in exercising ponies given 42 iu vitamin E/kg dietary DM, through increased plasma glutathione peroxidase and superoxide dismutase activities and increased ascorbic acid concentration, despite higher muscle TBARs regardless of vitamin E concentration.

DIETARY PROTEIN REQUIREMENTS AND EXERCISE

A number of investigations have demonstrated that dietary protein concentration can influence running ability in horses. However, it is estimated that protein catabolism accounts for no more than 5–15% of the energy consumed during exercise, yet in both extended work (Rose *et al.* 1980) and exercise in the postabsorptive state following high protein meals (Miller-Graber & Lawrence 1988) plasma urea is elevated. The NRC (1989) has concluded that dietary protein requirements of horses are proportional to those for DE.

Protein assimilation

Protein assimilation of tissue occurs during work following rest (Meyer 1987), but its scale is unclear. Johnson *et al.* (1988) detected no increase in N balance of working ponies. Patterson *et al.* (1985) found that 1.9 g digestible protein/kg $BW^{0.75}$, equivalent to 5.5% of dietary protein (maize soya protein), was adequate for intense exercise (cf. 7% and 8.5% dietary protein), as measured by plasma total protein, albumin and urea N. Orton *et al.* (1985a) trotted growing horses for 12 km daily at 12 km/h, on either a 12–14 or a 6–8% protein diet. Exercise increased feed and protein intakes, and consequently growth rate, of horses on the low protein diet to equal that of those receiving more protein. The greater appetite provided protein, surplus to exercise requirements, utilized for growth. In a survey of racing TBs, Glade (1983a) found that protein intake (confounded with DE intake) was positively correlated with time to finish, implying that excess protein depressed speed. It is concluded that there is little justification for greatly increasing the daily protein intake of exercising horses to meet some putative increase in chronic requirement.

The relationship between dietary protein and extreme performance is, however, far from clear. Despite seemingly high NRC (1989) estimates, actual intakes are still higher. Yet an excess of 56% over the estimates among racing standardbreds (Gallagher *et al.* 1992a), and of 21% for racing TBs (Gallagher *et al.* 1992b) may reflect the natural protein content of palatable high energy feeds. RQ was lower in horses given a high protein, cf. control, diet and exercised at high speed (Pagan *et al.* 1987b), implying a stimulation to protein or fat metabolism in the postabsorptive state (Fig. 9.15), increasing urea yield (Frank *et al.* 1987) and water needs. Apart from this increased need, Hintz *et al.* (1980) observed no detrimental protein effect in horses during distance riding, where dehydration causes fatigue. Miller-Graber and Lawrence (1988) recorded a higher plasma urea N, 16–19 h after an 18.5% cf. 12.9% protein meal during 15 min work at 170–180 beats/min; but plasma NH_3 rose to the same extent in both groups, jugular lactate concentration increased less, the increase in plasma glutamine was marginally less and that of plasma alanine significantly less in the high protein group ($P < 0.05$). Others have also observed lower plasma lactic acid concentrations during intense (Pagan *et al.* 1987b,c) and less intense (Frank *et al.* 1987) exercise with high protein diets (24.6 and 20%, respectively, vs. 14.6 and 10% protein, respectively), decreasing heart rate and glycogen catabolism at high speed (Pagan *et al.* 1987b). High protein reduced the postexercise rise in plasma NH_3 only in untrained (Frank *et al.* 1987) and not trained (Miller-Graber & Lawrence 1988) horses. Observations in the fasting state may have excluded an adverse excess protein effect. Thus, Miller-Graber *et al.* (1991) performed the test 3–4h after a meal, when 9% cf. 18.5% dietary protein was inconsequential for hepatic, or intramuscular, glycogen use, or for venous blood lactate concentration. Nevertheless, 5min postexercise venous blood lactate:pyruvate ratio was higher with the 9% diet, possibly indicating a higher pyruvate dehydrogenase (EC 1.2.4.1) activity, or decreased NADH:NAD ratio, with that diet.

It is concluded that high protein diets, above the need for N balance, may confer some metabolic advantages to working horses (Table 9.12; Figs 9.11, 9.15, 9.16 and 9.18). However, outcomes such as increased urea production in the stable could increase environmental ammonia, contributing to respiratory stress. (It is of current interest that glutamine drinks taken postexercise by long-distance human athletes reduce the frequency of respiratory infections. The reason seems to be that during extreme exertion glutamine sources are depleted and glutamine is an essential fuel for the functioning of the immune system – the equine possibilities have not been examined to the author's knowledge.)

Specific amino acids

There is an increase in plasma free lysine and phenylalanine after exercise. As these amino acids are not normally catabolized for energy, their increase indicates an increase in net protein catabolism during exercise. Valine and isoleucine are glycogenic and readily oxidized to provide energy. Supplements, including these branched chain amino acids, have resulted in a lower plasma lactate accumulation, and lower heart rates, when given 30min before exercise.

Table 9.12 Some metabolic responses of untrained, exercised horses to dietary protein intakes well in excess of N balance, compared with responses at approximately N balance with diets providing similar amounts of starch.

- Increase in blood urea concentration.
- Decrease in venous blood ammonia concentration postexercise.
- Decrease in RQ during aerobic exercise.
- In the range 240–600m/min
    decrease in venous blood lactate concentration.
    decrease in hepatic lactate concentration.
    decrease in venous blood, lactate:pyruvate, ratio.

FEEDING METHODS

Feed sequence, protein utilization and plasma amino acids

The true digestibility of protein in the small intestine of horses ranges from 45 to 80%. At high rates of protein intake more will be degraded to NH_3 in the large intestine. Utilization of this by gut bacteria is between 80 and 100% (Potter *et al.* 1992c). Excessive protein intakes must inevitably increase the burden of unusable N either in the form of inorganic N or as relatively unusable bacterial protein. This burden is influenced by feeding sequence. The provision of a concentrate feed 2 h later than roughage, cf. simultaneous feeding, caused higher levels of plasma free, and particularly essential, amino acids, 6 and 9 h later, respectively (Cabrera *et al.* 1992), indicating improved nutritional value derived from delaying the concentrate. Plasma urea did not rise with this dissociated feeding, but it rose continuously for the 9 h after the mixed feeding, implying there was a large caecal flow of digesta.

Protein and heat production

Belko *et al.* (1986) found that the thermic effect of food in exercising men increased with the protein content, 150–270 min postprandially. No similar measurements of heat production in horses are available, although Frank *et al.* (1987) measured no difference in heart rate, or body temperature, between horses trained on diets containing 10 and 20% protein. However, the feeding sequence referred to above may influence heat production as deamination and urea synthesis are associated with additional waste heat. The optimal amount and preferred feeding practice for dietary protein are thus not established, but optima for both may exist for both intense and extended exercise. The French evidence (Cabrera *et al.* 1992) may suggest that the feeding sequence should be the reverse of normal practice. Improved husbandry may allow the achievement of maximal effects without a burden of amino acid degradation products. Deamination of AMP is clearly prominent during brief maximal exercise (Miller-Graber *et al.* 1987) (Fig. 9.11). Alanine is the principal vehicle for shuttling NH_3 from muscles to liver, but whether this is stimulated by extra protein given during rest is not definitely established.

Processing of cereals and precaecal digestion

The extent to which cereal starch provides glucose, or VFA, depends upon its precaecal and even its preileal digestibility. Kienzle *et al.* (1992) reported that the preileal digestibility of oat starch was higher than that of maize starch, with similar degrees of processing. Grinding of whole grain led to high preileal digestibility (%), for oats amounting to 98.1% and for maize 70.6%, while rolling, or breaking, had little effect (whole oats 83.5%, rolled oats 85.2%, whole maize 28.95%, broken maize 29.9%). Starch gelatinization enhances its small intestinal digestion, but at moderate, or high, rates of intake only. At low rates (<0.4% of bodyweight per meal) most sources of starch are digested in the small intestine (Potter *et al.* 1992a). Thus, end products, gut fill and possibly the optimum time interval between feeding and exercise are all influenced. A change in the proportions of fatty acids to glucose in the end products, which can be influenced by processing, may modulate exercise performance (Frape 1994).

'Hotting-up' and the heat increment of feed

In Chapter 6 the phenomenon of waste heat generated during the digestion and metabolism of feeds was outlined and a mechanism for the 'hotting-up' effects of certain feeds was adduced earlier in this chapter ('Fat supplements and exercise'). Many trainers and horsemen and women are reluctant to use energy-rich cereals such as maize and barley because of the alleged risks in this connection. However, the explanation here makes clear that where alternative feeds are rationed at rates that provide the same amounts of net energy, then energy-rich cereals will generate less rather than more *total* heat over a period of 12 h, or more. Measurements in polo ponies maintained at a constant bodyweight support the conclusion that energy-rich feeds do not necessarily exacerbate a heating effect of feed, but body temperature and metabolic rate increase after feeding (Fig. 9.19). The ponies received approximately equal amounts of net energy from either maize and lucerne hay, or oats and timothy hay. No significant differences in response, either before or after exercise, owing to diet, were noted (Wiltsie & Hintz in Hintz 1983). If anything, the maize-fed animals were less 'hotted-up'. Energy-rich feeds may possess other advantages for sprint horses – for example, causing less gut-fill, or nonfunctional weight. Where unnecessary problems of 'hotting-up' have arisen, it is partly the consequence of a lack of appreciation of the differences between cereal grains in their energy content and bulk density described in Chapter 5 and quantified in Table 5.3. The cooking of cereal starch may reduce extended 'hotting-up' by promoting digestion and thereby decreasing microbial fermentation.

In contrast to *total* heat production, the *rate* of fermentation, and of heat evolved, from indigestible and 'spill-over' starch is more rapid than that from fibre, and an increase in metabolic rate is caused by the greater peak blood glucose level achieved by digestible cereal starch cf. roughages. This also increases the rate of heat production. This increase in metabolic heat production rate is the principal cause of the heat increment of food in humans and it is greater with foods yielding large glucose and insulin responses and an enhanced glucose storage as glycogen. These rates can be controlled to a considerable extent by the amounts of feed and the way this feed is presented (see Chapters 5 and 6). This elevated 'hotting-up' is of shorter duration than that of the total heat increment attributable to a feed and its effects should have passed before horses are subjected to strenuous effort say 5 h later, which would follow small feeds.

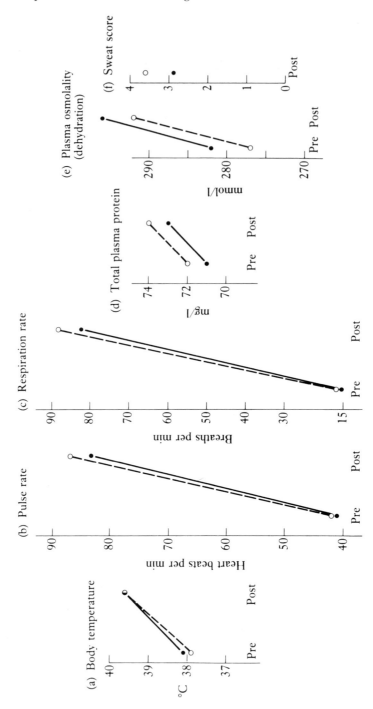

Fig. 9.19 Pre-exercise and postexercise values for polo ponies given 6.8 kg lucerne hay plus 3.2 kg maize (●) or 8.2 kg timothy hay plus 4.0 kg oats (○) continuously over 4-week periods of a reversal experiment to maintain constant bodyweight (Wiltsie & Hintz in Hintz 1983).

Gut fill and speed

Processing of roughage can accelerate rate of passage, reduce gut fill, but decrease hindgut utilization (Wolter *et al.* 1975, 1977, 1978). The extent to which feed is fermented will influence the weight of ingesta. Slade (1987) measured the speed of quarter horses galloping at up to 19.6 m/s (44 mph) over 137 or 229 m from a running start. Speeds differed according to digestibility of feeds, as reflected by differences in gut fill.

Feeding before endurance rides

Meyer (1987) concluded that endurance horses (cf. sprint horses) should be fed larger amounts of roughage (6–8 kg/day) to dilate the large intestinal volume, increasing water and electrolyte reserves. Poor-quality roughage should be avoided, as it causes a higher postexercise plasma lactate concentration and more extreme hypoglycaemia than found with good-quality roughage.

The maximum postprandial increase in caecal volume is 8–14 kg/kg DM ingested, depending on the fibre content, with 130–135 mmol Na/l flow from the ileum. The average Na content of the large intestine is 40 mmol/l. Meyer (1987) concluded from previous evidence (Meyer *et al.* 1982a) that the horse should be fed at least 5 h before an endurance race, depending on the feeding sequence (see 'Feed sequence, protein utilization and plasma amino acids'), as most of the residue will then have passed the ileo–caecal orifice. Meyer (1987) compared two rations: one contained 2 kg concentrates and 3 kg hay, providing 11.5 g Na and 80 g K, the other 2 kg oats, providing 1 g Na and 10 g K. Four hours postprandial the retention of water, Na and K was, respectively, 5.8 kg, 9.3 g and 48 g for the first ration and 0.6 kg, 0.2 g and 1 g for the second. In support of Meyer, Ralston (1988) found that horses failing to complete 160-km races had received mixtures containing less hay and more grain, and had been trained for greater distances each week (83 cf. 61 km). The energy intake per kilometre trained was less in those that failed. Some samples of roughage contain excessive K which can stimulate diuresis, and so loss of water.

Glucose solutions given before prolonged exercise could increase glycogen degradation rate, through decreased fatty acid mobilization, induced by the antilipolytic effect of elevated plasma insulin. Glucose given during hard exercise does not stimulate insulin secretion, but hyperinsulinaemia is protracted in TBs following a meal (Stull *et al.* 1987; Frape 1989), the timing of which may be critical before extended exercise. Ponies given a fluid providing 5.4 MJ DE following exercise daily exhibited lower heart rates and blood lactate concentrations during and following subsequent exercise (Lindner *et al.* 1991).

Feeding before sprints

In contrast, for short races there is some justification for a *light* concentrate ration between 4–5 h before the start, although timing is rather critical as the race should not coincide with elevated plasma insulin. A large meal would extend the time over which plasma insulin is elevated. The factors influencing the optimum interval between a meal and exercise are summarized in Table 9.13.

Meeting the increase in energy needs

The build-up of feed and of training before races may take 8–12 weeks, and in the case of event horses a typical build-up may start with 5 kg concentrated feed daily and finish with 8–8.5 kg 2 months later. The ration should be distributed among four daily feeds, and the hay, one-third of which should be given in the morning feed and two-thirds in the evening, should be reduced to 3.5–4 kg/day for a horse of average size during the last part of training (the optimum sequence of concentrate and hay may be otherwise in the last days before an event, as indicated in 'Feed sequence, protein utilization and plasma amino acids, this chapter). The protracted and intensive training for dressage (Plate 9.2) places emphasis on mental attitude and alertness, but it is equally important to achieve the right level of energy intake at each meal and overall. It is speculated that lecithin supplements may assist the mental composure of dressage horses, as may be indicated by the observations of Holland *et al.* (1996).

Feed allowances and work intensity

In all cases, feed intake should be increased as the work rate increases and the concentrated feeds should be severely restricted if work rate is reduced for any reason, whether this be of short or longer duration. On rest days, a horse that would normally receive 8 kg concentrates should then receive a maximum of 4 kg distrib-

Table 9.13 Factors influencing optimum interval between last meal and subsequent exercise.

- Weight of ingesta
    Size of meal
    Rate of passage
        proportion of roughage
        degree of grinding of roughage
- Dietary cation: anion balance
    ≈250 mEq/kg 3–4 h before sprint
- Potential for yielding glucose or VFA
    Digestibility of starch
        insulinaemia
        hindgut volume and fluid/electrolyte retention
    Fermentability of fibre
        Hindgut volume and fluid/electrolyte retention

Conclusions

- Type of meal: more roughage for aerobic exercise than for anaerobic.
- Optimum interval: sprint 3–7 h, extended aerobic exercise 5–8 h.

Plate 9.2 Wily Trout, a 16-year-old gelding, ridden by Chris Bartle and Pinocchio, a 15-year-old, ridden by Jane Bartle in 'Passage' during dressage training for the 1984 Olympic Games.

uted in three feeds, but with a greater allowance of hay of up to 5–5.5 kg for the average-sized horse.

Either underfeeding or overfeeding leads to inferior performance. A horse should not be fed in order to fortify it with reserves for future events, but rather the rates of feeding should be consistent with immediate needs. Overfeeding will create fatness, which causes a greater burden on the heart and the horse generally, and it interferes with the dissipation of heat. The synthesis of fat from excess carbohydrate, protein and dietary fat does not encourage those enzymes that participate in the breakdown of fat so necessary during work. Furthermore, overfeeding can lead to stocking up (oedema) of the legs, hives (bumps under the skin), forms of colic, founder, exertion myopathy and general overheating.

Concentrated feed should be given in a minimum of three meals per day with some hay, ample water and salt supplements available, in the morning, at noon and with ample hay in the evening. Many horses in training do not take sufficient salt from licks, hampering progress, and a feed source providing 60 g daily is recommended.

Feed intakes in practice

Observations both in the USA and the UK show that horses racing on the flat from 2 years of age, and weighing 470–530 kg, consume daily between 13 and 18.5 kg total feed, which amounts to between 2.7 and 3.7% of bodyweight (Mullen *et al.* 1979;

Hintz & Meakim 1981; Glade 1983a; Frape unpublished data). Of this, concentrates with, for example, cereals, nuts, bran and linseed, amount to between 30 and 60% of the ration. In one American study of 171 horses (Glade 1983a) the concentrates provided between 43 and 59% of the DE and between 39 and 64% of the crude protein of the total ration. Moreover, the total ration provided 163 MJ DE/500-kg BW and 1686 g crude protein [270% of the NRC (1978) estimated minimum requirement]. In the UK, daily rates of protein intake among both flat and National Hunt horses amount to between 1000 and 1400 g/day (the author's measurements). These figures are well below the average of the American horses, largely as a result of the lower protein content of horse hays produced in the UK.

Appendix B gives examples of dietary compositional errors encountered by the author in practice.

STUDY QUESTIONS

(1) How would you feed a horse from 96 h before (a) a sprint event, (b) a major dressage event, or (c) a long-distance event?
(2) How should you manage a fatigued horse?
(3) How would you introduce fat supplement feeding to a stable of working horses?

FURTHER READING

Baker, L.A., Topliff, D.R., Freeman, D.W., Telter, R.G. and Breazile, J.W. (1992) Effect of dietary cation–anion balance on acid–base status in horses. *Journal of Equine Veterinary Science*, **12**, 160–3.

Corbally, A.F. (1995) *The contribution of the sport horse industry to the Irish economy*. MEqS thesis, Faculty of Agriculture, National University of Ireland, Dublin.

Custalow, S.E., Ferrante, P.L., Taylor, L.E., *et al.* (1993) Lactate and glucose responses to exercise in the horse are affected by training and dietary fat. *Proceedings of the 13th Equine Nutrition and Physiology Society*, University of Florida, Gainesville, 21–23 January 1993, No. **504**, 179–84.

Duren, S.E., Manohar, M., Sikkes, B., Jackson, S. and Baker, J. (1992) Influence of feeding and exercise on the distribution of intestinal and muscle blood flow in ponies. *First Europäische Konferenz über die Ernährung des Pferdes*, Institut für Tierernährung, Tierärzliche Hochschule, Hannover, 3–4 September 1992, pp. 24–8.

Ferrante, P.L., Taylor, L.E., Meacham, T.N., Kronfeld, D.S. and Tiegs, W. (1993) Evaluation of acid–base status and strong ion difference (SID) in exercising horses. *Proceedings of the 13th Equine Nutrition and Physiology Society*, University of Florida, Gainesville, 21–23 January 1993, No. **504**, 123–4.

Foster, C.V.L. and Harris, R.C. (1992) Total carnitine content of the middle gluteal muscle of Thoroughbred horses: normal values, variability and effect of acute exercise. *Equine Veterinary Journal*, **24**, 52–7.

Frape, D.L. (1989) Nutrition and the growth and racing performance of thoroughbred horses. *Proceedings of the Nutrition Society*, **48**, 141–52.

Frape, D.L. (1994) Diet and exercise performance in the horse. *Proceedings of the Nutrition Society*, **53**, 189–206.

Gallagher, K., Leech, J. and Stowe, H. (1992a) Protein energy and dry matter consumption by racing standardbreds: a field survey. *Journal of Equine Veterinary Science*, **12**, 382–8.

Gallagher, K., Leech, J. and Stowe, H. (1992b) Protein energy and dry matter consumption by racing Thoroughbreds: a field survey. *Journal of Equine Veterinary Science*, **12**, 43–8.

Greenhaff, P.L., Hanak, J., Harris, R.C., *et al.* (1991) Metabolic alkalosis and exercise performance in the thoroughbred horse. *Equine Exercise Physiology*, **3**, 353–60.

Greenhaff, P.L., Harris, R.C. and Snow, D.H. (1990) The effect of sodium bicarbonate (NaHCO₃) administration upon exercise metabolism in the thoroughbred horse. *Journal of Physiology*, **420**, 69P.

Greenhaff, P.L., Harris, R.C., Snow, D.H., Sewell, D.A. and Dunnett, M. (1991) The influence of metabolic alkalosis upon exercise metabolism in the thoroughbred horse. *European Journal of Applied Physiology*, **63**, 129–34.

Greenhaff, P.L., Snow, D.H., Harris, R.C. and Roberts, C.A. (1990) Bicarbonate loading in the Thoroughbred: dose, method of administration and acid–base changes. *Equine Veterinary Journal*, Suppl. **9**, 83–5.

Hambleton, P.L., Slade, L.M., Hamar, D.W., Kienholz, E.W. and Lewis, L.D. (1980) Dietary fat and exercise conditioning effect on metabolic parameters in the horse. *Journal of Animal Science*, **51**, 1330–39.

Harkins, J.D. and Kamerling, S.G. (1992) Effects of induced alkalosis on performance in thoroughbreds during a 1600-m race. *Equine Veterinary Journal*, **24**, 94–8.

Harkins, J.D., Morris, G.S., Tulley, R.T., Nelson, A.G. and Kamerling, S.G. (1992) Effect of added dietary fat on racing performance in thoroughbred horses. *Journal of Equine Veterinary Science*, **12**, 123–9.

Harris, P. and Snow, D.H. (1988) The effects of high intensity exercise on the plasma concentration of lactate, potassium and other electrolytes. *Equine Veterinary Journal*, **20**, 109–13.

Harris, P. and Snow, D.H. (1992) Plasma potassium and lactate concentrations in thoroughbred horses during exercise of varying intensity. *Equine Veterinary Journal*, **24**, 220–25.

Harris, R.C. and Hultman, E. (1992) Muscle phosphagen status studied by needle biopsy. In: *Energy Metabolism: Tissue Determinants and Cellular Corollaries* (eds J.M. Kinney and H.N. Tucker), pp. 367–79. Raven Press, New York.

Harris, R.C., Marlin, D.J. and Snow, D.H. (1991) Lactate kinetics, plasma ammonia and performance following repeated bouts of maximal exercise. *Equine Exercise Physiology*, **3**, 173–8.

Harris, R.C., Marlin, D.J., Snow, D.H. and Harkness, R.A. (1991) Muscle ATP loss and lactate accumulation at different work intensities in the exercising Thoroughbred horse. *European Journal of Applied Physiology*, **62**, 235–44.

Hintz, H.F., Ross, M.W., Lesser, F.R., *et al.* (1978) The value of dietary fat for working horses. I. Biochemical and hematological evaluations. *Journal of Equine Medicine and Surgery*, **2**, 483–8.

INRA (1990) *L'Alimentation des Chevaux* (ed. W. Martin-Rosset). INRA Publications, Versailles.

Johnson, K.A., Sigler, D.H. and Gibbs, P.G. (1988) Nitrogen utilization and metabolic responses of ponies to intense anaerobic exercise. *Journal of Equine Veterinary Science*, **8**, 249–54.

Jones, D.L., Potter, G.D., Greene, L.W. and Odom, T.W. (1992) Muscle glycogen in exercised miniature horses at various body conditions and fed a control or fat supplemented diet. *Journal of Equine Veterinary Science*, **12**, 287–91.

Lawrence, L., Kline, K., Miller-Graber, P., *et al.* (1990) Effect of sodium bicarbonate on racing standardbreds. *Journal of Animal Science*, **68**, 673–7.

Lloyd, D.R., Evans, D.L., Hodgson, D.R., Suann, C.J. and Rose, R.J. (1993) Effects of sodium bicarbonate on cardiorespiratory measurements and exercise capacity in Thoroughbred horses. *Equine Veterinary Journal*, **25**, 125–9.

McCann, J.S., Meacham T.N. and Fontenot J.P. (1987) Energy utilization and blood traits of ponies fed fat-supplemented diets. *Journal of Animal Science*, **65**, 1019–26.

Meyer, H. (1987) Nutrition of the equine athlete. In: *Equine Exercise Physiology 2*, pp. 644–73. ICEEP Publications, Davis, California.

Meyer, H. (1992) Intestinaler Wasser- und Elektrolytstoffwechsel Pferdes. *First Europäische Konferenz über die Ernährung des Pferdes*, Institut für Tierernährung, Tierärztliche Hochschule, Hannover, 3–4 September 1992, pp. 67–72.

Meyer, H., Lindemann, G. and Schmidt, M. (1982) Einfluss unterschiedlicher Mischfuttergaben pro Mahlzeit auf praecaecale- und postileale Verdauungsvorgänge beim Pferd. In: *Contributions to Digestive Physiology of the Horse. Advances in Animal Physiology and Animal Nutrition.* Supplement to *Journal of Animal Physiology and Animal Nutrition*, **13**, 32–9. Paul Parey, Berlin and Hamburg.

Meyers, M.C., Potter, G.D., Evans, J.W., Greene, L.W. and Crouse, S.F. (1989) Physiologic and meta-bolic response of exercising horses to added dietary fat. *Journal of Equine Veterinary Science*, **9**, 218–23.

Miller-Graber, P.A., Lawrence, L.M., Foreman, J.H., *et al.* (1991) Dietary protein level and energy metabolism during treadmill exercise in horses. *Journal of Nutrition*, **121**, 1462–9.

National Research Council (1989) *Nutrient Requirements of Horses*, 5th revised ed. National Academy of Sciences, Washington DC.

Nielsen, B.D., Potter, G.D., Morris, E.L., *et al.* (1993) Training distance to failure in young racing quarter horses fed sodium zeolite A. *Proceedings of the 13th Equine Nutrition and Physiology Society*, University of Florida, Gainesville, 21–23 January 1993, No. 504, pp. 5–10.

Oldham, S.L., Potter, G.D., Evans, J.W., Smith, S.B., Taylor, T.S. and Barnes, W. (1990) Storage and mobilization of muscle glycogen in exercising horses fed a fat-supplemented diet. *Journal of Equine Veterinary Science*, **10**, 353–9.

Pérez, R., Valenzuela, S., Merino, V., *et al.* (1996) Energetic requirements and physiological adaptation of draught horses to ploughing work. *Animal Science*, **63**, 343–51.

Plummer, C., Knight, P.K., Ray, S.P. and Rose, R.J. (1991) Cardiorespiratory and metabolic effects of propranolol during maximal exercise. In: *Equine Exercise Physiology 3*, (eds S.G.B. Persson, A. Lindholm and L. Jeffcott) pp. 465–74. ICEEP Publications, Davis, California.

Potter, G.D., Arnold, F.F., Householder, D.D., Hansen. D.H. and Brown, K.M. (1992) Digestion of starch in the small or large intestine of the equine. *First Europäische Konferenz über die Ernährung des Pferdes*, Institut für Tierernährung, Tierärzliche Hochschule, Hannover, 3–4 September 1992, pp. 107–11.

Potter, G.D., Webb, S.P., Evans, J.W. and Webb, G.W. (1990) Digestible energy requirements for work and maintenance of horses fed conventional and fat supplemented diets. *Journal of Equine Veterinary Science,* **10**, 214–18.

Rose, R.J., Arnold, K.S., Church, S. and Paris, R. (1980) Plasma and sweat electroyte concentrations in the horse during long distance exercise. *Equine Veterinary Journal*, **12**, 19–22.

Scott, B.D., Potter, G.D., Greene, L.W., Hargis, P.S. and Anderson, J.G. (1992) Efficacy of a fat-supplemented diet on muscle glycogen concentrations in exercising thoroughbred horses maintained in varying body conditions. *Journal of Equine Veterinary Science*, **12**, 109–113.

Scott, B.D., Potter, G.D., Greene, L.W., Vogelsang, M.M. and Anderson, J.G. (1993) Efficacy of a fat-supplemented diet to reduce thermal stress in exercising Thoroughbred horses. *Proceedings of the 13th Equine Nutrition and Physiology Society*, University of Florida, Gainesville, 21–23 January 1993, No. 504, 66–71.

Sewell, D.A. and Harris, R.C. (1992) Adenine nucleotide degradation in the thoroughbred horse with increasing exercise duration. *European Journal of Applied Physiology*, **65**, 271–7.

Sewell, D.A., Harris, R.C., Hanak, J. and Jahn, P. (1992) Muscle adenine nucleotide degradation in the thoroughbred horse as a consequence of racing. *Comparative Biochemistry and Physiology*, **101B**, 375–81.

Sewell, D.A., Harris, R.C., Marlin, D.J. and Dunnett, M. (1992) Estimation of the carnosine content of different fibre types in the middle gluteal muscle of the thoroughbred horse. *Journal of Physiology*, **455**, 447–53.

Stutz, W.A., Topliff, D.R., Freeman, D.W., Tucker, W.B., Breazile, J.W. and Wall, D.L. (1992) Effect of dietary cation–anion balance on blood parameters in exercising horses. *Journal of Equine Veterinary Science*, **12**, 164–7.

Sufit, E., Houpt, K.A. and Sweeting, M. (1985) Physiological stimuli of thirst and drinking patterns in ponies. *Equine Veterinary Journal*, **17**, 12–16.

Taylor, L.E., Ferrante, P.L., Kronfeld, D.S. and Meacham, T.N. (1995) Acid–base variables during incremental exercise in sprint-trained horses fed a high-fat diet. *Journal of Animal Science*, **73**, 2009–2018.

Tyler, C.M., Hodgson, D.R. and Rose, R.J. (1996) Effect of a warm-up on energy supply during high intensity exercise in horses. *Equine Veterinary Journal*, **28**, 117–20.

Chapter 10
Grassland and Pasture Management

... but that grass which grows on wet grounds, or the winter-grass, abounds with little or no spirit, wherein a great deal of the true nourishment consists, and therefore it must needs beget a viscid and indigested chyle, which must also render those horses that are fed with it sluggish, dull and unactive.

W. Gibson 1726

GRASSLAND TYPES

In humid temperate climates natural succession favours the replacement of grassland by scrub and then woodland and forest. To sustain high-quality 'permanent' pasture requires perseverance in land management, through the grazing of domesticated animals and through the treatment of grassland as a crop to be cultivated. These pastures constitute the greater proportion of grazing and they contrast with the uncultivated areas of mountain, moorland, heath and downland, where wild grazing and browsing animals contribute to the distribution of plant species that evolves.

Fertility and grass species

The most fertile temperate pastures can theoretically support annually five, 500-kg barren or pregnant mares per hectare from grazing and conserved forage. The most productive swards contain more than 30% perennial ryegrass (*Lolium perenne*), a proportion of rough meadow grass (*Poa trivialis*), and the remainder of grasses consisting mainly of cocksfoot (*Dactylus glomerata*), timothy (*Phleum pratense*), other meadow grasses, Yorkshire fog (*Holcus lanatus*), species of bent grass (*Agrostis*) and fescue (*Festuca*). The proportion of white clover (*Trifolium repens*) depends very much on the use of nitrogenous fertilizers and the seasonal grazing pattern, but can amount to 25% of the cover. Other broad-leaved plants vary in abundance according to management. An extensive survey of grasslands in England and Wales (Hopkins 1986) (Table 10.1) indicated that *Lolium perenne*, *Agrostis* spp. and *Holcus lanatus* were numerically the most important species, contributing 35, 21 and 10% of the cover, respectively (of those swards exceeding 20 years of age the proportions were 22, 27 and 14%, respectively). The palatable *L. perenne* decreased in proportion with time and the unpalatable *H. lanatus* increased. The fescues (*F. arundinacea*, *F. rubra*) are also highly palatable to horses and generally their persistence requires a lower fertility than does perennial ryegrass. In England potential stocking density of all grazing animals is generally correlated positively with the contribution perennial ryegrass makes to the sward.

Poor drainage, low fertility and plant species

Where poor drainage has not been rectified, creeping bent (*Agrostis stolonifera*), Yorkshire fog, rough meadow grass and creeping buttercup (*Ranunculus repens*) thrive better than ryegrass. Other less-productive grasses which invade swards of this class in significant numbers include meadow foxtail (*Alopecurus pratensis*), couch (*Agropyron repens*), crested dog's-tail (*Cynosurus cristatus*) and barley grass (*Hordeum murinum* and *H. pratense*), together with red clover (*Trifolium pratense*) and bird's-foot trefoils (*Lotus* spp.). However, the decline in ground cover by sown species over the years occurs on all but the most fertile land regardless of the excellence of drainage. About 20% of the cover by sown species is lost after 5–8 years and a further 10% is lost during the next 4–12 years. Poorly drained soils provide a less suitable initial habitat for sown species and the proportion of them is less throughout the pasture's life. Improvement by heavy treatment with fertilizers, drainage and intensive management yields swards of open texture, subject to poaching in wet weather. Lush pastures of this description with little bottom are unsuited for grazing by horses without great care and experience.

In many river valleys, rushes and sedges appear in *Agrostis* pastures where drainage is impeded, or the land is otherwise neglected. The fertility may be potentially quite high, but in the more degenerate and derelict land, even in lowland areas, purple moorgrass or flying bent (*Molinia caerulea*), bracken (*Pteridium aquilinum*) and gorse (*Ulex europaeus*), quite useless for horses, may sometimes appear. On better-drained slopes of acid soils, between altitudes of 100 and 350 m (350–1100 ft) under annual rainfalls of 90–120 cm (35–45 in), fine-leaved fescues and bent grasses dominate pastures with a scarcity of clovers in the latitudes of 50–57°N of maritime

Table 10.1 Factors favouring species distribution in pasture* (after Hopkins 1986).

	Drainage	Soil nutrient status	Fertilizer N inputs	Grazing intensity	Hay cutting	Sward age	Elevation	Optimum pH
L. perenne	Good	good	high	hard	—	young	low	6–7
T. repens	Moderate to good	good	low	hard	No	younger	—	7–7.5
H. lanatus	Poor	low	low	moderate	yes	older	NS	5–6
F. rubra	Moderate	low	low	moderate	—	older	high	5–6
P. trivialis	Moderate	high	high	low	yes	—	—	6–6.5
Rumex spp.	Poor to moderate	fair	—	—	yes	—	—	—
Agrostis	Moderate	low	low	low	—	older	NS	5–6

* It should be recognized that these assessments relate to mixed swards, as pastures of single species only exist as such immediately after sowing.

NS, no significant correlation between elevation and frequency of the species.

climates. The specific distribution within these ranges depends on soil pH, latitude, aspect, soil drainage and grazing. These areas merge into the uncultivated rough and hill grazings in which there is the invasion of bracken fern and gorse at lower altitudes along with the fine-leaved fescues. On moorland, matgrass (*Nardus stricta*) and purple moorgrass, rushes (*Juncus* spp.), heather (*Calluna vulgaris*) and bell heather (*Erica cinerea*) may occupy a larger proportion of the area. Fertilizers generally will encourage nutritious grass species, whereas excessive grazing by ponies of the better fine-leaved fescue and *Agrostis* areas may lead to their suppression and the encroachment of economically useless *Nardus*, bracken, etc., and where the pH is low poor bone development occurs in young horses. The spread of shrubs and useless weeds may largely depend on drainage, the extent of cutting and the presence or absence of cattle.

Poor upland grassland is generally considered to be of marginal value for horse production. Nevertheless, in France upland pastures composed of *Nardus stricta, Festuca ovina* and including *Vaccinium myrtillus* (bilberry, blaeberry) are grazed successfully in summer by heavy breed mares (1.5 ha per mare and foal), during which time the mares gain in condition (Micol and Martin-Rosset 1995).

Herb strips

Herbs may be defined as broad-leaved plants with nonwoody aerial parts and so could, of course, include clovers. Like clovers, many other herbs are rich in protein, minerals and trace elements relative to the common grasses; some are relished by horses and espoused by enthusiasts. The dry matter of nettle, for instance, contains nearly 6% of lime, 5% of potash and 2% of phosphoric acid, but the fresh plant is usually not sought by horses and ponies. A few relevant chemical values of herbs are given in Table 10.2. Herbs may be especially useful on marginal land, in which the upper layers are frequently leached of nutrients by excessive rainfall, and many tend to stay green in winter, so they provide a succulent winter bite although their regrowth is protracted. When herbs are present in abundance they depress total yield per hectare of major nutrients, but this is less likely to be a critical issue in horse paddocks. In any case, their establishment in pasture as part of a normal grass and clover seed mixture is uncertain. Herb seeds are rather expensive, but many of them inevitably become established in permanent pastures through natural agencies. Herb strips are frequently sown on the headlands of fields. Table 10.3 gives suggested seed mixtures, which include some relatively noncompetitive grass species although these are not essential. Chicory (*Cichorium intybus*) is a successful herb for strip-seeding in temperate pasture areas. It can reduce nutrient losses and yields DM rich in K, P, Ca, Mg and Na, although poor in N. However, the economic worth of herb mixtures is unproven for horses in any general way.

Grass breeding

The IGER, Aberystwyth in Wales has crossed early- with late-flowering perennial ryegrasses to produce a more even DM production through the season. Ryegrass

Table 10.2 Mineral contents (g/kg DM) of perennial ryegrass and red clover at early maturity (Thomas *et al.* 1952; Worden *et al.* 1963) and of herb species (Hopkins *et al.* 1994) as means of four harvest dates from pasture sward.

	N	P	K	Ca	Na	Mg
Perennial ryegrass						
Head	22.0	4.2	17.0	2.3	—	1.3
Leaf	21.0	3.2	23.0	8.7	—	1.7
Stem	8.0	2.7	17.0	3.0	—	0.9
Red clover						
Head	37.0	4.1	21.0	11.0	—	2.8
Leaf plus petioles	45.0	2.9	17.0	21.0	—	3.4
Stem	16.0	1.5	17.0	11.0	—	2.4
Chicory (*Cichorium intybus*)	26.2	5.4	26.1	19.0	12.2	3.5
Yarrow (*Achillea millefolium*)	26.4	5.8	33.2	12.1	13.0	2.5
Dandelion (*Taraxacum officinale*)	21.3	3.3	29.3	7.8	10.4	2.1
Ribwort plantain (*Plantago lanceolata*)	23.6	4.5	17.8	20.5	10.0	2.2
Grass/clover	24.2	3.5	19.3	9.9	3.9	1.6

Table 10.3 Suggested herb mixtures (kg/ha), including some grasses, for sowing as a strip 8–10 m wide in horse paddocks.

	Based on	
	Davies 1952	Archer 1978a*
Chicory (*Cichorium intybus*)	3	2.2
Ribwort plantain (*Plantago lanceolata*)	3	1.1
Burnet (*Sangiusorba minor*)	4	2.2
Yarrow (*Achillea millefolium*)	1	0.6
Cat's-ear (*Hypochoeris radicata*)	2	—
Dandelion (*Taraxacum officinale*)	—	0.3
Sheep's parsley, wild parsley (*Petroselinum crispum*)	1	0.6
Meadow fescue (*Festuca elatior*) or creeping red fescue (*F. rubra*).	3	(13)
Timothy (*Phleum pratense*) or smooth meadowgrass (*Poa pratensis*)	3	—
		(7)
Crested dog's-tail (*Cynosurus cristatus*)	—	(7)
White clover (S100) (*Trifolium repens*)	3	—
Total	23	7 or 34

*Without the inclusion of grass seed the mixture should be introduced into an existing paddock by direct seeding if ground is well harrowed and the sward cut short.

× fescue (*Lolium multiflorum* × *Festuca gigantea*) hybrids have been bred to increase pasture persistance and to resist drought. Italian ryegrass containing 44% more Mg than standard Italian ryegrass indicates it may be useful for hay production in studs. All these ventures could be of particular value in equine husbandry. Where grass is grown specifically for silage a higher yield potential exists in some more exotic grass species, such as brome grasses. (For endophytes in seed production see 'plant disease control', this chapter.)

PASTURE AS AN EXERCISE AREA

The production by pasture of digestible nutrients for horses and ponies is clearly of economic importance. Its critical role is starkly revealed by the historical evidence of the delay of European military campaigns until the spring flush in May by armies dependent upon horses. However, other issues sometimes play even a dominant part in the selection of pastures, or their management, for horse husbandry. The thick cushion found in old pastures is better for exercise than is the open texture of heavily fertilized leys, in which numerous stones, upturned during ploughing, contribute to leg injuries. Many horses needing rest and gentle exercise between periods of hard work, barren and pregnant mares and 1–3-year-old growing stock are turned out to subsist on pasture. In each of these cases, leys or highly fertilized permanent pastures would initiate rapid and unwanted fat deposition. This creates unnecessary problems in the early stages of subsequent work, in late pregnancy and early lactation, or it contributes to leg abnormalities in growing horses and to the incidence of laminitis and colic. Therefore, a high degree of skill is needed in evolving pastures for horses and ponies that provide useful grazing and also yield saleable and reliable stock. The thick, matted turf of well-drained old pasture resists poaching in wet weather and can, therefore, provide exercise and maintenance areas for out-wintered stock. However, not only is the total annual yield of digestible feed generally lower in these pastures, but, especially where drainage is poor, the season of herbage growth is shorter, a fact probably of much greater economic significance. The length of the grazing season is generally greater the higher the fertility of the land.

NUTRITIONAL PRODUCTIVITY OF PASTURE

Within any pasture the nutritional quality varies from area to area. Therefore the feeding value of the whole pasture will depend on the stocking density and the amount of the most attractive herbage at any one time. In temperate grassland areas – excluding acid land with very high rainfall – the protein content of pasture is directly correlated with rainfall and inversely with soil temperature during the growth period. Pastures grazed by horses in northern temperate latitudes tend to produce the greatest yield of DE and protein during May and June, after which there is a precipitous decline in productivity from July to August when grasses flower. Clovers and other legumes, if encouraged, prolong pasture growth and extend the summer grazing season. Where persistent leafy strains of grasses have been established on fertile soils, this mid-summer fall in productivity is much less noticeable. More fertile deep soils are less inclined to dry out and the leafy strains of grasses continue vegetative growth much later into the summer. By grazing these pastures, the formation of seed heads is delayed, or avoided, and tillering encouraged, so that their productivity is further enhanced. Regrowth of succulent leafy material occurs in early autumn, but work with sheep indicates that the ME of autumn grass is utilized 40% less efficiently than that of spring grass of the same crude chemical composition. This poorer value should be recognized when foals are weaned in the late summer on pasture without supplementary feeding (see Ch. 7).

Minerals

Excepting horses confined to tropical grasses, a Ca deficiency is unlikely among grazing horses, even when the grazing and browsing are desiccated. In a parched terrain horses and ponies are deprived, first, of water (Table 7.4), energy and protein and, second, of P. However, stock can become deficient in Ca, P and Mg if they are confined to wet acid soils covered by poor-quality, fine-leaved grasses. Many ponies coming off such hill land present signs of 'big-head' and other consequences of bone demineralization. Horses seem to be less prone to grass tetany caused by Mg deficiency than are cattle, but a fall in serum Mg is possible when lactating mares are grazing on low-Mg soils and it has been suggested that part of the effect is through excessive amounts of K in lush herbage. Leafy material contains far more K than the horse requires in normal circumstances. The needs for Na and Cl are likely to be met in horses dependent upon pasture in temperate latitudes.

Vitamins

Green leafy material is a rich source of folic acid, and comparisons made by the author (unpublished observations) between horses in training for flat racing, given a cereal-based diet supplemented with folic acid and vitamin B_{12}, and grazing in-foal and barren mares, foals and yearlings indicated a 23% lower concentration of serum folate and a 33% lower concentration of serum vitamin B_{12} in the horses in training. Several other water-soluble vitamins are equally adequate in cereal-based and grazing diets.

There are normally large stores of vitamin A in the liver resulting from the consumption of green herbage rich in β-carotene, but after a very extended drought there can be a clinical vitamin A deficiency as a result of protein and Zn deprivation coupled with the scarcity of green herbage. It is unlikely that a deficiency of any of the other fat-soluble vitamins D, E and K would occur among horses confined entirely to pasture. However, a few isolated pasture species not found in the UK (see vitamin D, Ch. 4) can cause vitamin D toxicity, bone demineralization and soft tissue calcification.

Trace elements

Australian evidence (Langlands & Cohen 1978) suggests that general pasture improvement increases the uptake by grazing animals of Cu, Zn, Mn, P, Ca and Mg. Improvement in the drainage of waterlogged soils tends to increase Se and Zn availability, but it may reduce the availability of Fe, Mn, Co and Mo, and an excessive use of N fertilizers may decrease the concentration of several trace elements in the sward. However, the relationships are complex (Burridge *et al.* 1983).

The effect of drainage on Mo availability may be advantageous as peaty, poorly drained soils found in parts of Somerset and Southern Ireland precipitate Cu-deficiency problems through low availability of Cu and high availability of Mo in the soil, particularly where the soil pH is also high (see Chapter 3). These soils (pH in excess of 7.6–7.7) also tend to be deficient in available Mn and Co. Some of the soils contain more than 20 mg Mo/kg and an increase of Mo by 4 mg/kg depresses Cu availability to grazing ruminants by 50%. The effect may be seasonal and an excessive uptake by plants of Mo and sulphate in the absence of generous amounts of available Cu leads to deficiency signs in cattle and sheep. Hypocupraemia in horses occurs less widely, but it exists in several parts of the UK and particularly in Southern Ireland. On the other hand, the horse is much less susceptible to the effects of Mo and sulphate, as in ruminants ruminal microorgansims synthesize thiomolybdate that reacts with Cu, decreasing its availability.

Soils subject to a high rainfall, waterlogging and a low soil pH are prone to Se-deficient herbage, as may occur in hill areas and on sands and gravels, in, for example, Newmarket, associated with low blood concentrations of Se in horses. By contrast, seleniferous soils containing very high levels of Se are a cause of toxic signs in grazing animals, for example on glacial lake deposits in Southern Ireland. Shale, mudstone and clay soils contain higher concentrations of Se than chalk, limestone and sandstone soils (Thornton 1983) and many mountainous areas. Seleniferous soils are notorious in various regions of the world where accumulator plants store toxic amounts of soil Se. These accumulators leave Se residues which are apparently more readily absorbed by the roots of other plants, leading to alkali disease in grazing stock.

Some inland continental areas, and even alkaline soils in central England, can induce signs of I deficiency in the young stock of grazing mares. When seaweed is used in excessive quantities as a source, signs of I toxicity, similar to those of deficiency, have been observed. Deficiencies of Fe, Mn, Co and some other more exotic trace elements have not been recorded and are unlikely among grazing horses and ponies.

The correction of trace-element deficiencies by applying minerals to the soil is unsatisfactory for some elements as the uptake is scant and repeated treatment is necessary. Better absorption is generally achieved with foliar sprays, but these are expensive and translocation is slight so that frequent treatment is unavoidable. Injections of Se have proved successful in grazing horses, but these are relatively expensive and repeated treatment at intervals is again necessary. When horses are held for extended periods on grazing lands, supplementary feeding with relatively concentrated sources of trace elements seems at present to be the most practical solution.

(See also mineral blocks, Chapter 5 and below).

NUTRIENTS REQUIRED FOR PASTURE GROWTH AND DEVELOPMENT

Ultimately all life on this planet depends on sunlight and the fixing of atmospheric carbon by the action of chlorophyll present in bodies called chromoplasts in seaweeds and in chloroplasts in higher plants (see also 'Photosensitization' this chapter). Chlorophyll is green, but the colour is masked by other pigments in some species. Chlorophyll is somewhat similar to haemoglobin, but contains Mg in place of Fe. It absorbs red, orange and blue parts of the spectrum and uses this radiant energy to combine water with carbon dioxide in a reduction reaction, producing hexose sugar and oxygen, summarized as:

$$12H_2O + 6CO_2 \rightarrow C_6H_{12}O_6 \,(\text{hexose}) + 6O_2 + 6H_2O$$

It is clear that not only light and water are required, but also the process needs warmth, so pasture plant growth accelerates to a maximum in mid-summer, given adequate rainfall.

In addition to C, H, and O present as carbohydrates and fats, plant tissues contain a range of elements used in the synthetic process and present as components of proteins and many other tissue compounds. All these elements are present in many

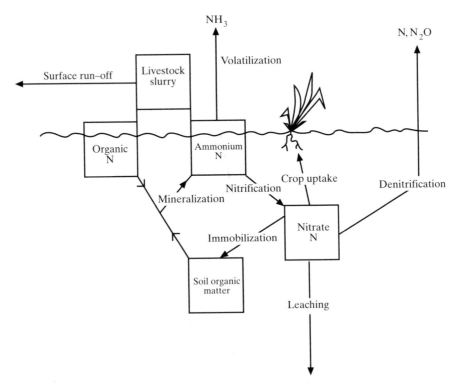

Fig. 10.1 Fate of slurry N in the soil. N can be lost as a result of surface run-off, volatilization of ammonia, denitrification and leaching (Anon 1986).

soils, but in most situations some are unavailable in optimum quantities for maximum plant growth, although in the height of summer water supply is frequently the limiting factor to this growth. The critical elements are:

- typically N, P and K, as provided in chemical fertilizers;
- the other major elements – Ca, S and Mg; and
- essential minor elements including Fe, Mn, Cu, Co, B, Mo, Zn, but also Na, Cl, Al and Si.

Leaching

Where fertilizers are applied to many soils, P and K are retained to a greater extent than is N. When N is not absorbed by plant roots, or used in microbial growth, much passes into drainage water. Concern exists over the pollution of streams, rivers and drinking water sources with nitrates leached from the soil (Fig. 10.1). It has been estimated that only 8–16% of N entering pastures leaves farms in the UK as meat or milk (ryegrass swards may recover 65–90% of N applied, but much is then recycled in dung and urine and ultimately lost) (Fig. 10.2). Apart from nitrates leached out, gaseous N is lost as ammonia (NH_3), nitrogen (N_2), and nitrous oxide (N_2O). Proportionately less N is lost when fertilizer is applied as ammonium sulphate than when it is applied as ammonium nitrate. Less still is lost when atmospheric N is fixed in root nodules of legumes by the bacterium *Rhizobium trifolii.*

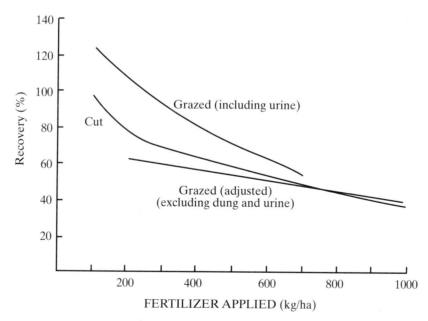

Fig. 10.2 Apparent recovery of N from cut and grazed swards in National Grassland Manuring trial GM24 during 1982–84 and apparent recovery from grazed swards when N inputs are adjusted for returns in dung and urine.

Chemical composition of herbage

Amounts of crude protein, soluble sugars and nitrogen-free extractives (NFE) in the dry matter of herbage are highest during the period of rapid leaf growth in the spring, next highest during regrowth in the early autumn, lower during the period of flowering in mid-summer and normally poorest during the winter when there is extensive dying back of the aerial parts of herbaceous plants. The months of the year when these phases occur in northern latitudes depend on the latitude, the lateness of the spring, rainfall, soil type and temperature. After the grazing of herbage, the first regrowth contains per unit of dry matter the highest protein, lowest crude fibre and highest NFE, or soluble carbohydrate and starch. These values change progressively as growth proceeds. For example, a study over 50 years ago (Fagan 1928) showed that in Italian ryegrass (*Lolium multiflorum*) from the 2nd to the 10th week of growth, the crude protein composition of the aerial parts declines from 19 to 7%, the crude fibre increases from 20 to 25% and the NFE increases from 44 to 60%. The changes can largely be explained by a rapid shift in the proportions of leaf to stem and leaf to flowering head (Tables 10.2 and 10.4). The horse digests fibre less easily than can domesticated ruminants so that shorter grass containing a higher proportion of leaf is a more valuable feed than herbage approaching maturity. Table 10.5 gives some mineral values for herbage found by the author in stud paddocks at Newmarket. Although these values change with stage of growth their digestibility is less affected by maturation than is that of energy.

White clover: its use and control and its relation to grazing intensity

Clovers tend to have deeper roots than do most grasses, especially the less-productive grass species. Clovers draw moisture and minerals from lower horizons in well-drained soils, so that they remain green during summer drought when grasses have gone to seed, they cause less N-pollution of drainage water and they may rectify an imbalance of trace elements between the upper and lower horizons of the soil profile.

Recent research interest has therefore refocused on making better use of white

Table 10.4 Effect of growth of timothy (*Phleum pratense*) on the ratio of leaf to stem and chemical composition (% DM) (Waite & Sastry 1949).

Sampling date	Leaf/ stem	Crude protein		Ether extract		Ash		Crude fibre		NFE	
		Leaf	Stem	Leaf	Stem	Leaf	Stem	Leaf	Stem	Leaf	Stem
20 May	2.57	21.7	14.1	3.8	2.9	7.1	9.9	19.1	23.5	48.3	49.6
2 June	1.30	17.2	11.4	4.7	2.5	6.5	7.9	23.8	29.7	47.8	48.5
16 June	0.39	18.5	7.6	4.1	2.6	8.0	6.6	26.1	32.6	43.3	50.6
30 June	0.35	12.3	4.4	3.3	1.7	8.8	5.0	26.9	31.7	48.7	57.2
14 July	0.20	11.1	3.4	3.2	1.3	9.0	5.0	30.6	32.4	46.1	57.9

clover (or other legumes in nontemperate climates) in permanent pastures and leys. A white clover sward can fix up to 200 kg N/ha annually, and use of this source can reduce N losses, mostly in winter, to a quarter of those resulting from equivalent fertilizer applications to a ryegrass sward. Although growth rate is slower where reliance is placed on clover N, this can benefit management of equine pastures. Pastures on acid upland soils may carry no clover. White clover should then be sown as seed inoculated with *Rhizobia* of the appropriate strain, otherwise it will fail. In the UK these strains are available from the Institute of Grassland and Environmental Research (IGER), Trawsgoed, Aberystwyth, Dyfed, SY23 4LL. Clover may also fail through lack of available phosphate, or excess ionic aluminium, caused by soil acidity. Liming at 3–5 t/ha is therefore to be recommended (Table 10.6). Also slugs, weevils or eelworms (soil nematodes) may cause damage.

It is advised that the clover content of pasture be held in balance with that of grasses. Clover growth in the UK reaches a maximum in July, when it may represent the dominant weight of leaf in the pasture. The growing points of clover plants are farther above the soil surface than are those of grasses. Grass shoots, or branch stems, grow as 'tillers' (stolons on the surface) and as rhizomes (sub-surface) which

Table 10.5 Mineral contents of grazed sward in stud paddocks at Newmarket, Suffolk (composition of DM) (data collected by author).

		Found	Normal range in UK
Calcium	(%)	0.34–1.6	0.3–1
Phosphorus	(%)	0.2–0.54	0.15–0.45
Potassium	(%)	1.5–2.5	1.6–2.6
Magnesium	(%)	0.12–0.2	0.11–0.27
Sodium	(%)	0.03–0.34	0.1–0.6
Sulphur	(%)	0.22–0.43	0.15–0.45
Molybdenum	(mg/kg)	0.9–2.8	0.1–5
Copper	(mg/kg)	4.5–12.3	2–15
Selenium	(mg/kg)	0.025–0.049	0.02–0.15
Zinc	(mg/kg)	21–34	12–40
Manganese	(mg/kg)	44–220	30–115
Cobalt	(mg/kg)	2.9–7.6	0.03–2

Table 10.6 Effect of lime in seed bed with an initial pH of 4.7 and superphosphate in seed bed annually or in alternate years on annual DM yield (t/ha) over 3 years* (Sheldrick *et al.* 1990).

Lime (t/ha)	Total* superphosphate application (kg P/ha)				
	0	3 × 15	2 × 30	3 × 30	1 × 120
3	5.19	6.22	5.54	7.35	7.50
6	6.82	7.38	7.09	7.45	6.38

*15 kg annually, 30 kg in years 1 and 3, 30 kg annually, or 120 kg in seedbed.

originate at the nodes of the original stem, from which nodes leaves also originate. In perennial grasses only a few of the stems are flowering stems, the remainder are vegetative stolons that in various grass species may be above ground, at ground level, or, in fact, are rhizomes below the soil surface. Grazing, especially by sheep, will remove relatively more leaf and growing points from clovers than from grasses. Therefore repeated grazing can reduce the proportion of clover in the sward should this be required. In contrast, cutting for silage, liming and phosphate fertilizer application will tend to promote clover and 100–120 kg N/ha, applied per silage cut, will create a predominance of grasses.

White clover may be introduced by strip seeding with a Hunter rotary strip seeder at the rate of 4 kg Huia or Menna seed/ha in early summer, following paraquat spraying at 0.4 kg active ingredient (a.i.)/ha. The clover should be allowed to become well established before grazing and that should be on a rotational basis so that the cover is not reduced below 4.5 cm height.

Fertilizers, soil fertility and pH

A soil pH of approximately 6.5–6.8 should be maintained, as below this range there will not be enough free Ca and much of the P will be fixed as ferric phosphate or aluminium phosphate. Above the range, several minerals, including Fe, will be less available.

Phosphate fertilizer use

Clover, in particular, benefits from P fertilizer application, with the adjustment of soil pH if the soil is definitely acid. Basic slag, as a slow release alkaline P source applied every 4–5 years, is now a rarity, owing to changes in the steel industry. Superphosphate is composed mainly of Ca $(H_2PO_4)_2$ + $CaSO_4$. $2H_2O$ (18% P_2O_5, or 8% P) and is a readily available source of P not requiring to be as finely ground as the less-soluble slag and rock phosphate. It therefore has an immediate effect on P-deficient pastures and is valuable for application to seed beds. Where superphosphate and lime are shown to stimulate clover growth the soil is likely to have been acid and deficient in available P, and where pastures are rich in legumes then in all probability the Ca and P contents of the soil are satisfactory. A visual survey of the frequency of clovers and scarcity of fine-leaved grasses can be used as a 'rule of thumb' in predicting the wellbeing of the soil in these respects. Triple superphosphate (45% P_2O_5, or 20% P) in small amounts, i.e. 30 kg P/ha annually, or 120 kg/ha every 4 years (Table 10.5), with liming, should sustain clover presence on many mildly acid soils (pH 4.5–5.5). The amounts of pit chalk or limestone required on acid soils in order to raise the pH to within the desirable range and to ensure P availability will depend on the pH value determined and the texture of the soil, but amounts between 1.25 and 7.5 t/ha can be used as a top dressing. In using limestone and rock phosphates it is critical that only finely ground material is purchased, as this characteristic will influence the availability of the Ca and P to the roots. Rock phosphates are not generally recommended for studs and are really suitable only on soils of a pH below 5. Limestone contains variable amounts of Mg and Mn that may be useful. Although limestone soils are rarely deficient in Mg, acid soils frequently are and dolomitic limestone would provide a useful source of this element.

Assessing fertility and fertilizer requirement

When first embarking on grassland management for horses and at intervals of, say, every 10 years thereafter, it is desirable to carry out chemical determinations on the soil in order to assess, at the very least, its pH, P, K and Mg status. Soil sampling must be carried out in a representative and sensible fashion, even within a field, so that distinctions can be drawn between clearly different soil types. Furthermore, the soil profile in old pastures can be such that the status is quite dissimilar in upper and lower layers reached by plant roots. Many soils may show surface deficiencies of available P and Ca, accompanied by a lower pH, whereas lighter soils – especially where hay crops have been taken – are frequently K-deficient in the upper layers. A full response to N fertilizers should not be anticipated if these primary deficiencies have not first been rectified.

Table 10.7 gives the indices used by the Agriculture Development and Advisory Service (ADAS) in the UK to classify grasslands in terms of their major nutrients. Where the value is 2 or over no fertilizer treatment for that particular nutrient is required at the time of measurement. Recommended rates of P, K and N treatment of grassland for grazing and for haymaking or silage are given in Table 10.8. Where fields are set aside for hay or silage a generous use of fertilizers is worthwhile. K fertilizer may be required only when a hay crop is taken, and it can be in the form of sulphate or 'muriate of potash' (potassium chloride), or as part of a compound fertilizer. Soluble Mg fertilizers are rarely used, but can be applied to increase the mobile Mg content of pasture herbage. Where K is liberally used this generally lowers the mobile Mg content of herbage. As horse pastures are frequently depleted of N, with yellowing of the grasses, the rates can be fairly high. On the other hand, the excessive application of N may cause laminitis and the pollution of drainage water. Even so, in order to maintain the quality of the herbage and to get an economic output from the pasture, moderate applications of N fertilizers should always be considered desirable unless there is a reasonable proportion of

Table 10.7   ADAS soil fertility scale (mg/l soil).

Index number	P	K	Mg
0	0–9	0–60	0–25
1	10–15	61–120	26–50
2	16–25	121–240	51–100
3	26–45	241–400	101–175
4	46–70	401–600	176–250
5	71–100	601–900	251–350

Table 10.8   Recommended rates of P, K and N treatment of grassland.

N, P or K index	For grazing (kg/ha) per year			For haymaking or silage – nutrients per cut (kg/ha)			
	P_2O_5	K_2O	N^1	P_2O_5	K_2O	N High rainfall	Moderate[2] rainfall
0	60	60	20–50	60	80	60	80
1	30	30	20–50	40	60	60	80
2	20	20	20–50	30	40	60	80
>2	0	0	20–50	30	30	60	80

[1] The darker green the grass the less required. The larger quantities should be divided among three applications.
[2] 45–65 cm (19–25 in) per year.

Table 10.9   Approximate composition (%) of straight fertilizers and farmyard manure.

Straight fertilizers		
Ammonium nitrate	N	34
Ammonium sulphate	N	20.6
Nitrochalk	N	15.5
Calcium cyanamide	N	20.6
Triple superphosphate	P_2O_5	45
Superphosphate	P_2O_5	18
Steamed bone flour	P_2O_5	29
Bone meal	P_2O_5	22
Guano	P_2O_5	13–27
Basic slag	P_2O_5	18–20 (45–50 lime)
'Muriate of potash' (KCl)	K_2O	60
Sulphate of potash with magnesia	K_2O	26 (5–6 Mg)
Kainit (Na, K, Mg)	K_2O	14
Kieserite ($MgSO_4$)	Mg	16
Calcined magnesite	Mg	60
Farmyard manure (kg/10 t)	N	15
	P_2O_5	20
	K_2O	40
	Mg	8

well-distributed clover. A sward content of 25–35% of red or white clover may yield the equivalent of 150–200 kg slowly released N/ha annually through N fixation. N application rates of 20–25 kg/ha, when required, 3–6 weeks before the pasture is grazed are recommended.

The composition of some straight fertilizers is given in Table 10.9. Compound fertilizers contain two or more nutrients in a reliable form and the weight of a nutrient in a 50-kg bag of any fertilizer is given by dividing the percentage of the nutrient by 2. Thus, a 50-kg bag of an N:P:K compound 20:10:10 will contain 10 kg

Table 10.10 Recommended rates of nutrient application to seedbeds in establishing a pasture (kg/ha)

P or K index	N						P_2O_5 K_2O (All mixtures)	
	Grass Spring sown	Autumn sown	Grass/ clover		Grass/ lucerne	Lucerne		
			(a)	(b)				
0	125	50	40	75	50	25	100	125
1	100	50	0	50	25	0	75	75
2	75	0	0	0	0	0	50	30
>2	0	0	0	0	0	0	25	30

(a) If autumn sown or if it is intended to rely on clover as the main source of N for the sward.
(b) If little reliance is to be placed on clover as a source of N for the sward.

N and 5 kg each of P_2O_5 and K_2O. In all situations, granulated, rather than powdered, inorganic fertilizers should be used and at least a week should elapse after treatment before horses and ponies are allowed onto established pastures. This will give sufficient time for the granules to percolate down to soil level, avoiding the consumption of any significant amounts by grazing animals.

The principal organic fertilizer is farmyard manure, preferably excluding horse manure, and its very approximate composition is given in Table 10.9. Its advantage over inorganic N sources is the slow release of N, but set against this is the expense of transport and distribution. It is probably more practical to distribute 50 t/ha on a few paddocks than to use half as much on a much larger area and there is some justification, particularly on light soils, for applications every 5–7 years. Fishmeal, now scarce, containing 10–11% N, is sometimes used in organic-based fertilizers. It has advantages, in terms of slow release, similar to those of farmyard manure.

Where land has been ploughed up and new permanent pastures, or leys, are being sown, readily available nutrients should be provided for the seedlings; recommended rates of seedbed application are given in Table 10.10.

SWARD HEIGHT

A considerable number of studies have investigated the effects of the intensity of grazing, or cutting, on the productivity of pastures. Horses and sheep graze more closely than do cattle. Ryegrass and white clover swards maintained at an average height of 3.6–3.8 cm with sheep have been shown to produce more DM annually than those maintained at 2.5 cm. With grazing cattle similar unfertilized sward heights of 6 cm were shown at IGER, Aberystwyth to be more productive and to retain a greater presence of clover than swards maintained at 4.5 cm. The DM yields differed by 17 and 15% and clover presence differed by 4 and 23% in years 1 and 2, respectively.

Table 10.11 Effect of maintained sward height on tillering and proportion of tillers that produce flowering heads mid-June in a Melle perennial ryegrass pasture in the UK (Treacher *et al.* 1986).

	Cutting date		Grazing height (cm)		
	Mid-April	Late summer	3	5	7
Tillers (1000s/m^2)	24	11	35	29	24
Flowering tillers (%)			20	27	47

Reference was made to tillering (see 'White clover: its use and control and its relation to grazing intensity', above) and therefore the spreading of grasses in a sward, as influenced by grazing. In contrast to the effect on clovers, grasses are, within limits, encouraged by grazing if adequate leaf remains, and consequently they produce thousands more tillers per square metre of sward (Table 10.11). (Clovers also spread by producing thousands of stolons per square metre, but grazing causes a proportionately greater loss of leaf than in grasses.)

It is clear (1) that repeated cutting for silage/hay leads to pastures that are less suitable for horse grazing, as there is a less firm mat of herbage for exercise, (2) that close cutting in spring reduces subsequent grass seed production, and (3) that repeated close grazing reduces the proportion of clover in the sward and the total annual yield of DM.

INTENSITY OF STOCKING WITH HORSES AND RUMINANTS

Horses should be removed from a pasture as soon as they have eaten the available herbage, if alternative land exists. They are more active than ruminants, and they can damage both the soil's structure in wet weather and the growing plants through prolonged trampling. In contrast to cattle and sheep, which spend periods ruminating, horses may spend up to 60–70% of the 24 h searching for the most delectable foliage. Overgrazing, poaching of the soil, or stocking grassland during a heavy frost damages plants and depresses the rate of regrowth, encouraging the spread of prostrate and opportunist annual weeds seen particularly in an arc around gateways, drinkers and feed troughs. One of the few advantages of old, matted, sod-bound pastures is that they may be less prone to damage in this way. However, the undergrazing of pastures tends to promote the nutritionally poorer grass species, partly through seeding, and parasite survival, as distinct from transmission, is extended.

The ideal stocking rate is no greater than that which will feed the horses in the growing season. It is preferable to stock few horses with the balance made up of cattle to clear the excess growth at the season's height. Mixed stocking, either on a rotational basis or together with horses, initially decreases the number of horses that can be maintained, but these few will receive a better diet and will imbibe fewer intestinal parasitic worm larvae. Moreover, the quality of the pasture can be maintained at a high level for many more years. By breaking up an area of land into paddocks, rotational grazing is facilitated and better parasite control is achieved, particularly where the grazing species are also rotated in each paddock. Occasionally lactating mares, or young stallions, will bully cattle, so careful judgment should be exercised and rotation rather than mixing of animal species thus carries certain advantages.

Herbage yield and horse productivity

How does horse grazing affect pasture productivity and animal productivity? What degree of defoliation is optimum for horse productivity? Close cutting, or grazing, reduces yield, but as herbage leaves mature and die they obstruct light to the newly emerging, actively growing leaves. Moreover, the digestibility of the pasture crop changes radically with stage of growth, and differs as between stem and leaf. Maximum rate of leaf growth and yield of digestible nutrients for horses over the year requires that leaf is harvested at some intermediate stage and that dead herbage is removed. Sound pasture management becomes increasingly important as the area available to each horse is reduced, when conditions become increasingly favourable to parasite transmission and bullying, and when excessive stocking depresses herbage regrowth (through removal of leaf and clover growing points; see 'White clover: its use and control and its relation to grazing intensity', this chapter). There is a limit to the amount of trampling that grass can survive, particularly when soil is wet. The resulting bare ground becomes infested with docks, nettles and thistles. Thus, satisfactory husbandry is based upon:

- grazing paddocks just before grass leaf growth declines, but well before senescence supervenes;
- allowing time for the grass to recover;
- replacing depleted nutrients;
- determining optimum interval between grazings (which varies with season, rainfall, ambient temperature and soil fertility);
- applying an understanding of the life cycles of the critical pasture worm parasites of horses (Chapter 11);
- understanding the temperament of each horse; and
- applying an understanding of each horse's environmental needs.

The horse digests forage less efficiently than do cattle. The horse therefore prefers, and obtains greater sustenance from, the younger leaves, but ample good pasture on its own can provide sufficient energy and protein for growth of horses over 5–6 months of age. Where the horse is forced by circumstances to graze less digestible parts of the herbage some evidence indicates that it compensates by increasing its intake. This observation is not acceptable generally. Experiments in the Netherlands (Smolders and Houbiers, personal communication) showed that grass cut at the haymaking stage, yielding 3700 kg DM/ha, was consumed daily by adult horses at the rate of 2.1 compared with 2.4 kg DM/100 kg BW of grass cut at the grazing stage, yielding 1900 kg DM/ha (these are equivalent to 100 and 113 g/kg metabolic weight, respectively). The younger grass had higher energy and protein contents and a lower crude fibre content. An optimum was, however, found. The DM consumption rate of grass with a DM content below 14% was less than that of grass above 14% DM. Thus, apart from extremely young grass there is a decrease in grass DM intake with increasing maturity of that grass. These Dutch horses obtained, as a percentage of their maintenance requirements, 170–200% energy and 460–500% digestible protein.

Any guidelines on pasture use must be very approximate as all pastures differ in productivity. The husbandry of growing horses has the objective of producing not only efficient weight gain, but also sound skeletal growth (see Chapter 8). This twin objective has a bearing on recommendations for stocking density. Aiken *et al.* (1989) measured the weight gain of quarter horse yearlings of 347 kg BW at stocking rates of 6.7–12.4/ha Bermuda grass (*Cynodon dactylon*) pasture. Weight gain per horse decreased in a complex way with increasing stocking rate. This probably related to the effect of grazing intensity on pasture yield of highly digestible herbage. Forage samples were vertically divided into top, middle and bottom canopy layer thirds for analysis. The top third was that over 8 cm. Forage availability decreased in each of the three canopy layers as grazing pressure increased, but particularly in the least dense, palatable top third. This third is the most photosynthetically active leaf tissue, the most productive both for horse growth and, of course, for pasture growth. With increased stocking rates grazing of the lower two thirds became unavoidable. As these layers contain stems and senescing plant tissue, animal performance was limited by the small proportion of leaf present. The data are summarized in Tables 10.12 and 10.13.

Stock carrying capacity

Grazing pressures on pastures with continuous grazing should be set low enough to ensure sufficient regrowth of vegetation, but high enough to prevent accumulation

Table 10.12 Productivity of Bermuda grass pastures over 56 days grazed by yearling quarter horses (Aiken *et al.* 1989).

Yearlings/ ha	Herbage mass (kg DM/ha over 56 days)	Herbage allowance (kg DM/100 kg BW)	*In vitro* digestible DM (g/kg DM)	Weight gain per horse (kg/day)	Withers' height increase (cm/56 days)
6.7	334	69	391	0.37	2.2
8.0	130	32	406	0.13	1.6
9.5	93	27	409	0.31	4.2
12.4	54	20	394	−0.31	2.0

Table 10.13 Productivity of Bermuda grass pastures in the top, middle and bottom layers of the canopy over 56 days when grazed by yearling quarter horses (Aiken *et al.* 1989).

Yearlings/ ha	Forage availability (kg DM/100 kg BW)			Digestible DM* (g/kg DM)		
	Top	Middle	Bottom	Top	Middle	Bottom
6.7	10	21	38	421	403	383
8.0	4	11	17	442	425	404
9.5	2	11	14	413	405	392
12.4	1	8	11	414	400	391

*Measured *in vitro* by the method of Goering & Van Soest (1975).

of mature, poorly digested forage. As horses select the tender, top layers of forage, this situation argues for complementary use of ruminants which clear and utilize mature and senescing plant tissue.

When an allowance is made for loss of much of the edible foliage as a source of horse feed, but ignoring extremes of climatic fluctuations, the carrying capacities have been estimated: 1 ha high-quality grassland should provide pasture and hay for three to four light horses of about 400 kg, or four to six smaller ponies. Low-quality permanent pasture, however, may support only one horse per hectare and, in the extreme, 25 ha dry range may be required to supply the needs of a single horse throughout the year. Average-quality grassland can produce sufficient growth for two horses, and with adequate moisture when fertilized for three horses per hectare, or as summer pasture only for double the number. On good temperate summer grassland in France heavy breed mares (700–800 kg BW) require 0.7–1.0 ha per mare and foal, or 1.5–2.0 ha when following cattle. With a rotational grazing system, the cutting of surplus herbage and N fertilization (80–150 kg N/ha), 2.0–2.5 growing horses are grazed per hectare. In contrast under harsh upland conditions in France only 0.5–0.7 growing horses per hectare are possible (Micol and Martin-Rosset 1995). Many TB studs produce barely enough grass for one mare plus followers up to yearlings sales per 1–1.5 ha and then provide none of their own hay. The latter should not in any event be produced from paddocks that have been grazed by horses during the previous year at the least, if worm control is to be practised assiduously.

Archer (1978a) found that only 10% of the area of long-established horse pastures was grazed. After ploughing and reseeding following an arable rotation, the grazing area was extended to 20–30%, most of which included the previously grazed areas. Horses will not graze near horse droppings and these areas are rejected if removal of the droppings is delayed for more than 25 h after they have been voided. Horse urine does not engender a similar instinctive reaction. The inborn habit reduces the transmission of parasitic worm larvae and leads to the establishment of both grazing and camping areas, with a consequent effect on productivity. Horses will, however, graze right up to cattle dung pats and graze evenly over areas after well-rotted cattle manure has been spread. Similarly, cattle will graze the longer grass around the horse dung pats. Ideally, therefore, horse droppings should be removed from pastures on a daily basis and not spread by harrowing. Although the latter practice will destroy more parasitic worms, it will enlarge the rejected area. The advantages of integrating ruminants with horse pasture management is obvious. Steers are better than milking cows as they remove less Ca, P and N from the soil and both are probably better than sheep for which, nevertheless, there are some staunch advocates. Pastures heavily fertilized for dairy cows are generally unsuitable for horses.

Even with the best will and resolute adoption of mixed grazing, some areas of rank growth of low feeding value will remain, swamping out any young basal growth. Such areas should be topped at least six times per season, with the toppings removed to avoid mounds of mouldy grass suppressing the underlying grass. The pursuance of this practice may prevent weeds from seeding, it will promote tillering and regrowth of young grass and will destroy some of the infective larvae.

All grazing animals thrive better if they have the companionship of other stock and this may be particularly true with highly strung hotblooded horses. Even goats, sheep, chickens, ducks or ponies can fulfil a useful role. The old adage 'to get his goat' implied that a favourite racehorse could be nobbled by stealing his mascot before a race. Gilbert White in his *Natural History and Antiquities of Selborne* (1789) relates in a letter dated 15 August 1775 to Daines Barrington how a lone horse made an abiding companion of a solitary domestic hen, consoling and protecting it, in so far as it was possible, from the trials of avian life.

In addition to companionship, all horses should have access to shelter from the sun at high noon and from cold, windy, wet weather. This shelter may be naturally formed by trees or be a simple three-sided covered structure.

Extending the grazing season and rejuvenation of pasture

A continuous supply of inorganic nutrients, including water, is required for the growth of pasture plants. In addition, roots require oxygen, so soil moisture must be present without waterlogging. For this reason, the structure, type and humus content of soil are critical in ensuring a continuous supply of moisture at all levels through which the roots of pasture plants permeate. Cultivation to assist aeration and to extend the grazing season is brought about by relieving compaction and poaching of well-established pastures. This is of most benefit in wet areas and paraplowing, moling or subsoiling all improve water table levels and increase early season yields.

It is not only economically important to extend the grazing season, but also ecologically desirable to reduce leaching of N from the soil, which occurs outside the normal season of growth. Mixed pasture species, including improved grass strains, assist in the achievement of this goal. Rye (*Secale cereale*) strip-seeded in autumn into permanent pasture increases early season growth and it may be useful on light-textured free-draining soils in dry areas where N fertilizer use is restricted.

GRAZING BEHAVIOUR

The opinion is held that some environmental factors contribute to bad behaviour, even cribbing, among horses at pasture. What can be considered a poor-quality environment as seen through a horse's eyes, is learnt by experience. The proximity of other animals as company, ample palatable forage and some natural shelter from extreme conditions probably stand high in the estimation of most horses.

A crude comparison with the ruminant indicates that the horse has a much smaller stomach, necessitating short grazing sessions at relatively frequent intervals throughout the 24 h. A study with mares allowed to graze during 12 h light and 12 h dark in North Carolina indicated that grazing took up 17.2 h daily, in which 89.7% of the daylight and 76.4% of the dark was occupied in the activity. Horses tend to graze at similar times, but generally without interference once a hierarchy has been established. Dominance hierarchy or patterns of agonistic and affiliative behaviour are apparent during feeding from a single source, when horses are in close proximity, one to another.

The rate of intake of forage during grazing depends on the quality of herbage, its density and the appetite and size of the horse. Recent work (Cross *et al.* 1995) confirms that grazing endophyte (*Acremonium coenophialum*)-infected tall fescue (*Festuca elatior*) (see 'Poisonous plants', this chapter) depresses grass DM intake and bodyweight maintenance of horses. On the other hand, healthy tall fescue and creeping red fescue (*Festuca rubra*) are two of the most palatable grass species for horses. Experiments in Kentucky (J.D.M. Cantillon & S.G. Jackson personal communication) with quarter horse geldings showed that the consumption of Johnson tall fescue (*F. arundinacea*) or lucerne *(Medicago sativa)* over 7 h of grazing averaged 5.5 kg organic DM, regardless of the time of day. This amount maintained bodyweight.

Soil ingestion

The ingestion of soil by animals while grazing can occur in significant amounts; it depends on the height of the herbage, the openness of the sward, the contamination of leaf by earth and on the species of animal. On rough terrain the soil intake by sheep is said to approach 20% of their daily intake of DM and measurements have shown that a 500-kg horse may ingest as much as 1–2 kg soil daily while grazing. Apart from K, the mineral and trace-element contents of soil are generally higher than they are in herbage DM, although their availability, or digestibility, varies according to the element and soil type. Most common mineral elements (possibly apart from I and Co) are required by plants in their growth. However, the proportions taken up by roots differ considerably from one element to another. Plants do not absorb in significant quantities heavy metal poisons such as Pb [cadmium (Cd) may be absorbed in greater quantities], although the leaves can become contaminated by industrial fallout and the soil and subsoil can be polluted by industrial seepage. Soil consumption may then be a cause of, for example, Pb or F toxicity.

High concentrations of toxic minerals in the soil may solely depend on their geological origin. The significance of seleniferous soils is discussed elsewhere (in this chapter: 'Trace elements', 'Toxic legumes and other species' and Table 10.20; and also in Chapter 3).

SUPPLEMENTS ON PASTURE

The nutrient value of pasture varies particularly with:

- season
- climate
- plant species
- soil type
- fertilizers and
- geological formation.

The prime effect of soil type is on the trace-element and mineral contents of the plant. The Ca content can be influenced to some extent by liming, and trace elements by foliar sprays. Possibly the most economical way of overcoming pasture deficiencies of trace elements, in particular, is by directly supplementing the horses. Supplementation carries with it the risk of harmful excesses of certain of the trace elements, unless horses are dosed individually. For many years mineralized salt licks have been available. These licks are unsatisfactory as they can be washed away and some individuals 'hog' them, whilst others neglect them. Feed blocks are another form of supplementation.

Feed blocks

Hardened and molassed mineral feed blocks containing all the important major and minor mineral elements on a cereal base have now been available for many years (Table 10.14). Blocks are hardened by pressure and binding agents, or by the use of chemical setting reactions. These reactions include the formation of calcium sucrosate (molasses combines with calcium oxide or hydroxide, but the product is said to restrict intake, owing to a bitter taste), or calcium sulphate (gypsum). The author's own experience with these blocks is that those that are well formed withstand quite wet weather, when they are held in appropriate containers, and that most horses at pasture feed from them. Nevertheless, there is considerable individual variability in consumption and with time and age of block. Feed blocks should be protected from contamination in a container, but that container should allow consumption by prehension as well as by licking; hence the horse should be able to bite the edges. Murray (1993) placed blocks in cut-down plastic barrels, stabilized in the lower half with concrete. The barrels measured approximately 0.7 m high by 0.4 m wide. The height had the advantage of allowing sheep to graze with the horses without the risk of Cu intoxication.

Table 10.14 Declared composition of molassed feed block used by Murray (1993).

Ca (g/kg)	90
P (g/kg)	50
Mg (g/kg)	40
Na (g/kg)	80
Cu (g/kg)	5
Zn (g/kg)	4
I (g/kg)	0.1
Se (g/kg)	0.015
Mn (g/kg)	1.5
Co (g/kg)	0.1
Vitamin A (iu/kg)	333 000
Vitamin D_3 (iu/kg)	66 000
Vitamin E (iu/kg)	2 500

A detailed examination of the consumption of molassed feed blocks was undertaken by Murray (1993) in Ireland. The following were the main conclusions:

(1) Weather in Ireland did not influence intake.
(2) Weight of block remaining influenced intake. Below a residual weight of 4 kg, intake decreased markedly.
(3) Intakes of Ca, Mg and P from herbage and the block together were higher than that recommended by the NRC (1989), and the calculated intakes of the various trace elements were at all times below the maximum tolerance levels given by the NRC (1989).
(4) Plasma Ca and Mg concentrations did not reflect supplementation, whereas plasma and hair Cu concentrations increased with supplementation.
(5) Peak attendances at the block occurred early in the morning and at midday; the horse spending the longest there ate the most. Dominance hierarchy within a group influenced access to the block and consequently affected intake.
(6) Daily intake per horse ranged from 23 to 283 g, but averaged over a week the range was from 32 to 215 g per day.

SAFETY OF GRAZING AREAS

Wooden fencing

The paddock principle of rotation allows a more complete recovery of the grass and prevents the excessive spread of flat weeds, such as plantain, daisy, buttercup and dandelion. All horse paddocks, structures and fences must be free from protruding nails, barbed wire, loose wire, tins, broken bottles, large stones and anything that can ensnare horses, particularly where young stock are reared as they tend to be more curious and flighty.

Treatment of posts with creosote takes a lot of beating when this preservative is properly applied by allowing the posts to soak for several hours in creosote heated to less than boiling temperature. This treatment should be carried out in the open, so avoiding inhalation of the fumes. The creosote should be allowed to dry thoroughly before horses have access to the posts. Horses should have no access to containers of creosote, tar and other chemicals containing phenolic substances. Phenol is very poisonous to horses, and is rapidly absorbed through intact skin. Wooden railway sleepers cut lengthways make good posts. Wooden fencing is readily protected from chewing by running electric fencing wire along the top (Plate

Plate 10.1   Effective use of electric fencing installed on post and rail fencing to discourage wood chewing by horses.

10.1). The tops of fence posts should be flat with no sharp edges and level with the horizontal poles nailed to their inner sides.

Electric fencing

Stud managers have been reluctant to use electric fencing for controlling the grazing of horses. However, research has shown that single strand reflective tape is effective, as long as the current is maintained continuously. One study in North Carolina with TBs, quarter horses and yearlings showed that horses, in groups of two or four, could be successfully rotated through 0.06-ha sections, back grazing the previous section. Occasional breakages of the fence occurred. The area was surrounded by three-wire electric fencing.

WATER SUPPLIES

A piped water supply should be clean and adequate (see Table 7.3) and where natural watercourses are used, freedom from contamination for at least a few miles upstream should be confirmed (see Chapter 4 for effects of drought on the safety of natural water sources).

SILAGE AND HAYLAGE AND THEIR SAFETY

(See also Chapter 5.) Silage quality is normally determined by the date, or time, of cutting. As grass grows during the spring and early summer yield increases but digestibility falls. In terms of the ruminant D-value a decrease of 1 unit of D-value between 71 and 51 is equivalent to an increase in dry matter yield of perennial ryegrass of 500 kg/ha, through delay in harvesting from the end of April to mid-June in the UK. The quality of forage legumes depends on the retention of leaf during harvesting. This is particularly true for lucerne, owing to the relatively high proportion of fibrous stem.

Silage should be made with safe raw materials free from weeds and contaminating pollutants. It should have undergone an adequate acetic acid fermentation and possess a high content of dry matter without any evidence of moulding. This can occur in the space at the tops of silos where the conditions are hot and humid. Good-quality haylage is quite satisfactory and material compressed in plastic sacks with a mild fermentation and with about 50% of dry matter is normally reasonably safe, but is usually on the expensive side. To prepare haylage, stands of pure grass are cut at the bud stage and wilted for 18–24 h before baling tightly. Careful harvesting should minimize contamination of the herbage with earth. The baled product has a relatively high pH (about 5.5) and should not be fed to horses for at least a week after baling. Haylage from an unfamiliar source should be given in small quantities initially, even where the stock have previously received silage or haylage. Owing to its moisture content its feeding value is about half that of high-quality nuts. However, its high content of soluble dry matter and ease of mastication facilitate a rapid intake of available energy; a few horses seem prone to gas colic if given inordinate amounts.

Badly made silage may be contaminated with pathogenic bacteria – *Listeria, Salmonella* spp. or *Clostridium botulinum* – any of which may cause problems. Some clostridial spores will inevitably be present on the ensiled crop. Horses and ponies, possibly through the absence of a rumen, are more susceptible than are ruminants to pathogenic clostridial bacteria. The toxins of *Clostridium perfringens* invoke severe enteritis and enterotoxaemia; the neurotoxins of *C. botulinum* are particularly dangerous because they cause a descending paralysis commencing with tongue and pharyngeal paresis, with dysphagia, progressive paralysis, recumbency and eventual death. These botulinum bacteria multiply in the silage clamp or bale during abnormal fermentation. It is particularly important, therefore, that clostridial growth is entirely prevented in ensiled forage intended for horses. The fermentation of sufficient water-soluble carbohydrate in this material brings about a fall in pH, which will inhibit the growth of clostridia if the content of dry matter is at least 25%. Where the dry matter content is only 15%, a pH of 4.5 will not inhibit the proliferation of clostridia and secondary fermentation of the silage occurs; lactic acid is degraded to acetic and butyric acids, carbon dioxide is evolved with extensive deamination and decarboxylation of amino acids, yielding amines and ammonia. The resulting rise in pH activates the less acid-tolerant proteolytic clostridia. The ammonia they release brings about a further increase in pH. These changes are accompanied by a characteristic smell and loss of the attractive vinegary aroma.

The likelihood of clostridial proliferation is hampered by wilting the fresh material for 36–48h before ensiling. In big-bale silage and haylage, dry matter contents of 30–60% are thus achieved. Satisfactory fermentation is facilitated in normal silage, with an adequate decline in pH, by harvesting forage containing not less than 2.4–2.5% of water-soluble carbohydrate in the fresh material [the National Institute of Agricultural Botany (NIAB), Cambridge suggest 10% in the dry matter]. In NIAB trials where 100 kg N/ha was given in March, mean values in the first conservation cut at 67D are given in Table 10.15.

A proportion of ryegrasses in any mixture obviously assists good fermentation and acid production during silage making. With less reliable material, i.e. with either one or more of the following qualities:

Table 10.15 Soluble carbohydrate content of grass with a ruminant D-value of 67 (based on NIAB 1983–4a).

	Soluble carbohydrate (g/kg DM)
Italian ryegrass	31
Hybrid ryegrass	26
Perennial ryegrass (intermediate and late)	24
Perennial ryegrass (early)	23
Timothy	14
Cocksfoot	13

- of poor fermentation potential;
- when the moisture content is higher than desirable;
- when soil may contaminate the herbage,

or to avoid the growth of undesirable microorganisms and aerobic deterioration during 'feed-out', it may be necessary to add molasses to ensure the presence of adequate soluble carbohydrate. However, viscid liquids of this type are difficult to apply uniformly so that additives are often used to achieve a pH of 4.2–4.6. Reliable additives are: (1) formic acid (2.3–3.0l/1000 kg), but corrosive, and (2) cultures of *Lactobacillus plantarum* (4×10^6 cells/g fresh grass), but more expensive. Both additives properly used can lead to a more rapid fall in pH, lower contents of ammonia-N and acetic acid, and higher contents of water-soluble carbohydrate and lactic acid than in untreated silage. Subsequent fermentation will reduce the pH further to a level that makes clostridial fermentation unlikely. If, however, a stable pH is not reached in silage with low DM content, then saccharolytic clostridia, which are present in the original crop as spores, will multiply and will initiate the secondary fermentation referred to above.

Big-bale silage, widely used in farming, and other plastic-wrapped silages have high DM contents, minimal fermentation, no additives and a pH between 5 and 6. They are subject to aerobic deterioration, owing to lower density than bunker silage and to the potential for punctures in plastic wrappings. They depend entirely on high osmotic pressure (low water activity) to inhibit clostridial proliferation and on the absence of tears in the plastic to prevent mould growth (see Chapter 5, plate 5.2 a–c). Deaths have occurred among horses consuming big-bale silage associated with the presence of clostridial organisms and/or toxins, without any contemporary effects on cattle. Great care should thus be exercised in selecting silage with the appropriate aroma, a high DM content and a pH of 4–4.5. These are essential prerequisites of silage, or haylage, used in horse feeding if health risks are to be minimized.

Dairy cattle given very extensively fermented ryegrass silage lost greater amounts of N in the urine compared with those given silage of restricted fermentation, or fresh grass. Extensive fermentation cf. restricted fermentation provided a higher proportion of propionate and a lower proportion of acetate in the VFA of the silage. Similar comparisons have not been undertaken with horses, although it is apparent that very extensively fermented herbage, or that containing significant amounts of ammonia, or n-butyric acid, is not very palatable to horses. Practice and experience are necessary adjuncts to good recipes in the achievement of the optimum product. High DM (45–60% DM), fermented, plastic-wrapped herbage is gaining in popularity and is generally more palatable to horses than is silage with a lower DM. Adult horses tend to consume more DM from a high than a low DM silage.

Metallic contamination of silage

Acidity causes the solution of heavy metals that may contaminate ensiled herbage. This is particularly important where pastures may have been used for game or clay pigeon shooting. Extremely high concentations of lead, causing death, have been detected by the author in silage from fields over which clay pigeon shooting has occurred. Dissolved lead is more toxic than metallic lead and horses are seriously and permanently injured, or killed, by exposure of this kind. Other heavy metals may be similarly dissolved.

Maize silage (also see Chapter 8)

In various areas of France and for many years, maize silage of 30–35% DM has been given to pregnant and lactating mares and to growing horses, as the main dietary constituent. Milk production is excellent and foal growth rate good. Supplemented with concentrates the silage is fairly well consumed by growing horses (Micol and Martin-Rosset 1995) and long experience has enabled breeders to use this material safely.

GRASSLAND IMPROVEMENT

Permanent pasture should be well drained and where it is low lying, or the subsoil is heavy, consideration should be given to piped drainage and, in the appropriate areas, pumped low-level ditches. Improvement in drainage reduces poaching and allows the stocking density to be increased. By lowering the water table, plant roots are encouraged to permeate lower horizons of the soil so that they gain access to a greater variety of minerals and to moisture for longer rather than shorter periods during a drought. This extends the grazing season and encourages ryegrass, cocksfoot, timothy and clover, discouraging rushes, tufted hair-grass or tussock grass (*Deschampsia caepitosa*), *Agrostis*, couch, Yorkshire fog and water-tolerant meadow grasses. Fertilizers, therefore, can be more effectively used, and the increased potential growth of the resulting sward decreases the concentration of parasitic worms. It also decreases the risk of liver fluke infection. Although horses have a pronounced resistance to liver fluke, they can become infested and their economic worth is thus prejudiced. Where mixed grazing is practised, the sheep and cattle can, of course, be treated before they are introduced. The snail burden of the land may also be reduced by copper sulphate treatment or, for example, by grazing with ducks.

If piped drainage is impractical, the subsoiling of impervious heavy land will improve its structure and for a shorter period bring about the advantages of more permanent drainage. Sometimes such marginal land is infested with quite inedible species, including cottongrass (*Eriophorum* spp.), rushes and, in better drained acid areas, by thick mats of *Nardus* grass. Here, initial improvement can frequently be effected through burning and hard grazing with steers. Most poor-quality permanent pastures can be improved by occasionally discing or heavy harrowing to break up a surface mat. This allows aeration, encouragement of better grasses, and some control of moss (*Lycopodium* spp.), mouse-ear chickweed and buttercup. In dry areas, however, little disc penetration will occur unless there has been recent rain.

Chemical control of weeds

Although the cutting of pernicious weeds such as docks (*Rumex* spp.) and thistles before they seed is helpful, it must be done on a regular basis five or six times a season, and even then control of creeping thistle is doubtful. Thus, the use of selective weedkillers based upon methozone (MCPA) and/or 2,4-D to control broad-leaved weeds without damage to mature grass is recommended. Creeping thistles, docks and in some upland studs bracken encroach progressively on grazing areas. The Long Ashton Research Station, Bristol has achieved effective control of these species with low dose sulphonyl-urea herbicides, although these are not entirely clover-safe.

Where small prostrate weeds, such as members of the buttercup family (Ranunculaceae), are widespread, spraying by tractor is most convenient but where there are patches of creeping thistles, docks and nettles, a knapsack sprayer can be used to advantage; this causes less damage to clovers and other valuable broad-leaved herbs. Spraying should always be carried out according to the manufacturer's directions (1) when there is little wind; (2) at a time of rapid weed growth, but before flowering; (3) refraining in periods of drought but avoiding rain for a few hours after spraying; and (4) not before newly sown pasture grasses have at least three leaves.

MCPA is effective against common nettle, daisy, dandelion, spear thistle (*Cirsium vulgare*), annual scentless mayweed (*Tripleurospermum inodorum*), black bindweed (*Polygonum convolvulus*), redshank (*P. persicaria*), cleavers (*Galium aparine*), common chickweed (*Stellaria media*), charlock (*Sinapis arvensis*), fumitory (*Fumaria officinalis*), poppy (*Papaver rhoeas*), fat hen (*Chenopodium album*) and small nettle (*Urtica urens*). It will check broadleaved dock (*Rumex obtusifolius*), bulbous buttercup (*Ranunculus bulbosus*), creeping buttercup (*R. repens*), colt's-foot (*Tussilago farfara*), creeping thistle (*Carduus arvensis*), curled dock (*Rumex crispus*), field horsetail (*Equisetum arvense*), perennial sowthistle (*Sonchus arvensis*) and soft rush (*Juncus effusus*). The most satisfactory control is achieved by spraying bulbous buttercup and chickweed in autumn and ragwort in early May. Bracken fern is very resistant to spraying, but the most effective time is August. Pastures should be left for 15 days before they are grazed in order to allow adequate time for the penetration of the herbicide, and the dead residues of poisonous weeds should be removed from the pasture. MCPA and 2,4-D are of low toxicity to animals, and pastures properly treated with them have not caused fatalities, regardless of how soon they have been grazed.

Poisonous plants

Horses and ponies do not normally graze growing poisonous weeds, but young stock or other stock freshly introduced to a pasture, following a period in stable, may eat

Table 10.16 Plants found poisonous to horses in Europe.

Species	Common name	Location	Toxic parts and effects reported
Aconitum napellus	Monk's-hood	Shady stream sides	Whole plant, espec. root, very potent
Agrostemma githago	Corn-cockle	Cultivated fields	Espec. seeds, saponins
Anagallis spp.	Pimpernel	Sandy coastal areas, cultivated land	Whole plant
Aquilegia vulgaris	Columbine	Damp places, lime-rich woodlands, grassland	Probably whole plant, mild hydrogen cyanide
Arenaria spp.	Sandwort	Dry places	Whole plant, mild
Atropa bella-donna	Deadly nightshade	Lowland, hedge banks, edges of woods	Whole plant, espec. berries
Bryonia dioica	White bryony	Lowland hedges	Red berries, vegetative parts, diarrhoea
Cannabis sativa	Hemp	Wasteland	Flowers, fruits, low toxicity, but horses at greater risk
Chaerophyllum temulum	Rough chervil	Road sides	Seeds and shoots
Chelidonium majus	Greater celandine	Shade, banks, walls	Stems
Cicuta virosa	Cowbane, water hemlock	Damp places	Whole plant, espec. roots, very poisonous, incl. hay
Conium maculatum	Hemlock (smooth spotted stems)	Damp places	Whole plant, mature hay unlikely to be harmful
Cyclamen hederifolium, C. europaeum	Sowbread	Woods, banks, especially Southern Europe	Whole plant
Daphne mezereum	Mezereon	Woodlands	Berries and bark
Datura stramonium	Thorn-apple	Waste lowland	Whole plant
Delphinium spp.	Larkspur	Cultivated land	Whole plant, potent
Digitalis purpurea	Foxglove	Hedgerows, wood margins	Whole plant
Equisetum spp.	Horsetail	Stream banks, damp woods	Whole plant, incl. hay, espec. horses
Euonymus europaeus	Common spindle-tree	Hedgerows	Whole plant
Frangula alnus	Alder buckthorn	Woods, hedges	Leaves, bark, fruit, diarrhoea
Helleborus spp.	Hellebore	Lime-rich woodlands, scrub	Whole plant, potent
Hyoscyamus niger	Henbane	Clearings, edges of woods	Whole plant
Hypericum spp.	St John's wort	Grasslands	Whole plant, photosensitivity
Iridaceae	Iris, narcissus, daffodil	Gardens	Bulb
Laburnum anagyroides	Laburnum	Gardens, self-sown	Whole tree
Lactuca virosa	Acrid lettuce	Road sides	Whole plant

Table 10.16 *Continued*

Species	Common name	Location	Toxic parts and effects reported
Liliaceae *(Allium oleraceum, Endymion non-scriptus, Fritillaria meleagris, Convallaria majalis, Colchicum autumnale)*	Field garlic, bluebell, fritillary, lily of the valley, meadow saffron	Woods, meadows	Bulb
Linum catharticum	Purging (fairy) flax	Pastures, meadows, uplands	Seeds
Linum usitatissimum	Common flax, linseed	Escaped from cultivation, common	Seeds
Lolium temulentum	Darnel	Waste	Possibly fungal infection
Lonicera xylosteum	Honeysuckle	Woodlands	Berries and leaves
Lupinus spp.	Lupin	Light dry soils	Whole plant, incl. hay, espec. seeds
Mercurialis annua	Annual mercury	Clearings, edges of woods	Whole plant
Mercurialis perennis	Dog's mercury	Poor land	Whole plant
Oenanthe crocata	Hemlock, water-dropwort	Damp places	Whole plant, espec. roots
Papaver spp.	Poppy	Cultivated land, wasteland	Whole plant
Paris quadrifolia (Liliaceae)	Herb Paris	Damp alkaline woodland	Low toxicity
Pteridium aquilinum	Bracken fern	Heaths, wasteland, woods	Whole plant, incl. rhizome and hay
Ranunculus spp., *R. sceleratus*	Buttercup, celery-leaved buttercup	Damp places, pastures	Whole plant, irritant, colic, inflammation
Rhamnus catharticus	Common (purging) buckthorn	Chalky thickets	Fruit, diarrhoea
Rhododendron spp., *Azalea, Kalmia*	Rhododendron, azalea, kalmia	Woodland	Whole plant
Saponaria officinalis	Soapwort	Hedges, banks, damp places	Whole plant, mild, saponins as for *Stellaria*
Senecio spp.	Espec. ragwort	Old pasture	Whole plant
Solanum dulcamara	Woody nightshade, bittersweet	Road sides	Whole plant
Solanum nigrum	Black nightshade	Cultivated lowland	All parts, espec. berries
Solanum tuberosum	Potato	Cultivated land	All parts incl. tubers when damaged
Stellaria spp.	Chickweeds	Ubiquitous annual	Whole plant, mild
Tamus communis	Black bryony	Woods, hedgerows	Possibly berries, mild toxicity
Taxus baccata	Yew	Woods, gardens	Whole tree, very potent

avidly a variety of such plants. Furthermore, during times of drought some that are deep-rooted or near watercourses and remain green entice the unwary horse. Weeds found in Europe that are poisonous to horses in both the growing and dried states are listed in Table 10.16 and those reported worldwide to be toxic to horses are listed in Table 10.17. After cutting and drying, or after being destroyed by sprays, weeds that retain their toxicity are more palatable than before and so should be removed and burnt.

Reseeding

If there is little pressure on available grazing land and the appropriate equipment is available, it can make economic sense to reseed worn-out pastures after ploughing, as long as the land is potentially fertile. No permanent improvement would, however, result unless drainage, ditching, liming, fencing and hedging are first put in order. If there are large areas of pernicious perennial weeds in the old turf, they should first be destroyed with herbicides and 2–3 weeks should elapse before ploughing begins. For spring sowing, autumn ploughing of the old turf is desirable. Sometimes it may be necessary to use heavy discs or cultivators to break up the turf.

Table 10.17 Plants reported in the scientific literature to have caused poisoning in horses and ponies in the world (Hails & Crane 1982).

Species	Common name	Effects reported
Agrostemma githago	Corncockle	Hypersalivation, abnormal thirst, accelerated respiration
Allium validum	Wild onion	Haemolytic anaemia
Amsinckia intermedia	Tarweed	Walking disease, haemolytic anaemia, liver necrosis, possibly nitrate toxicity
Aristolochia clematitis	Birthwort	Tachycardia, weak pulse, decreased appetite, constipation, polyuria
Arum maculatum	Lords-and-ladies, Cuckoopint	Purgative, irritation of GI mucosa and kidneys, fatty liver
Astragalus spp.	Locoweed or milk vetch	Eye lesions, locoism, Se accumulation in several species
A. lentiginosus, A. mollissimus	Locoweed	Depression, anorexia, ataxia, hyperexcitability and espec. violent responses to stimuli; effects irreversible
Atalaya hemiglauca	Whitewood	Anorexia, dullness, irritability
Centaurea repens	Russian knapweed	Nigropallidal encephalomalacia, severe damage to nerve cells, muscular hypertonia
Centaurea solstitialis	Yellow star thistle	Nigropallidal encephalomalacia (chewing disease), hypertonia of muscles and muzzle
Cestrum diurnum		Calcinosis-hypercalcaemia (other *Cestrum* spp. cause severe gastritis and liver degeneration, but not reported in horse)

Table 10.17   *Continued*

Species	Common name	Effects reported
Coriandrum sativum	Coriander	(No signs given in Hails & Crane 1982)
Crotalaria spp.		Hepatotoxicity, jaundice, dyspnoea, weak pulse, collapse
C. crispata		Kimberley horse disease, anorexia, dullness, staggering gait
C. dissitiflora var. rugosa		(No signs given in Hails & Crane 1982)
C. retusa		Kimberley horse disease, liver and central nervous system lesions, anorexia, dullness, staggering gait
Cuscuta spp.	Dodder	Enteritis, anorexia, nervous symptoms
C. campestris	Peppery cuscus, dodder	Not described, but staggering, salivation,
C. breviflora,		increased pulse and respiratory rates reported in cattle
Echium lycopsis (or) *E. plantagineum*)	Purple viper's bugloss	Hepatotoxicity, blindness
Equisetum spp.	Horsetail, marestail	Thiamin deficiency – the horse is very
E. fluviatile	Water horsetail	susceptible
E. hyemale	Rough horsetail or Dutch rush	Akin to bracken poisoning owing to presence of thiaminase
E. palustre	Marsh horsetail	Loss of condition, some diarrhoea,
E. variegatum	Variegated horsetail	usually caused by eating hay containing horsetails – administer thiamin or dried yeast
Eupatorium spp. (incl. *E. rugosum*)	Incl. white snakeroot and hemp agrimony	Trembling, lethargy
Eupatorium adenophorum	Croftonweed	Pulmonary oedema, possible nitrate toxicity
Galeopsis spp.	Hempnettle	Pulmonary oedema, enteritis, anorexia
Glyceria maxima	Reed sweet-grass	Cyanide poisoning
Helichrysum cylindricum	Everlasting flower	Blindness, encephalopathy
Heliotropium europaeum	European heliotrope	Hepatotoxicity
Hypericum crispum		Photosensitivity
Indigofera dominii, I. enneaphylla	Birdsville indigo	Birdsville disease, drowsiness, immobility, discharges from eyes and nose, dragging of hind feet, liver lesions
Juglans nigra	Black walnut	Acute laminitis
Jussieia (Ludwigia) peruviana		
Lathyrus nissolia	Grass vetchling	Incoordination and collapse during exercise
Lupinus spp.	Lupin	(1) Chronic: lupinosis-congenital defects, liver damage (due to mould growing on lupins). (2) Acute: respiratory paralysis due to lupin alkaloids when large amounts of the plant eaten
Medicago sativa	Lucerne	Photosensitization, possible oestrogen toxicity; certain strains cause anaemia and hepatotoxicity
Morinda reticulata		Se toxicity
Nerium oleander	Oleander	Highly toxic – convulsions, diarrhoea, colic, petechial haemorrhages
Oxytropis campestris, O. sericea	Locoweed	Nervous signs, anorexia, ataxia; damage permanent

Table 10.17   *Continued*

Species	Common name	Effects reported
Papaver nudicaule	Iceland poppy	Ataxia, convulsions
Perilla frutescens		Lung disease
Persea americana	Avocado	(No signs given in Hails & Crane 1982)
Phaseolus vulgaris	Kidney, haricot or navy bean	Central nervous system lesions, GI disorder
Pimelea decora	Rice flower	Colic, diarrhoea, collapse, ulceration of mouth, tongue, oesophagus, gastritis (also St George disease in cattle)
Polygonum aviculare	Knotgrass, wireweed	Nitrite poisoning, GI irritation
Prunus laurocerasus	Cherry laurel	Cyanide poisoning (also caused by other *Prunus* spp. e.g. *P. serotina* – wild black cherry)
Pteridium aquilinum	Bracken	Thiamin deficiency
Quercus spp. (incl. *Q. rubra* var. *borealis*)	Oak	Acorn poisoning: hepatotoxicity; oak-leaf poisoning: dullness, diarrhoea, constipation, dark urine
Ricinus communis	Castor oil plant	Castor-bean poisoning: dullness, incoordination, sweating, tetanic spasms, watery diarrhoea
Robinia pseudo-acacia	False acacia	Acacia bark poisoning: colic, diarrhoea, irregular pulse, hyperexcitability, paralysis
Senecio spp.	Groundsel, ragwort	Hepatotoxicity
S. jacobaea	Common ragwort	Hepatotoxicity
Setaria sphacelata	Bristle grass	Oxalate poisoning, osteodystrophia fibrosa
Sinapis arvensis	Charlock	Cyanide poisoning, colic, gastroenteritis, diarrhoea
Solanum malacoxylon		Calcinosis – wasting, stiffness, hypercalcaemia, calcification of arteries
Sorghum almum	Sorghum	Cystitis, ataxia (growing grass)
Sorghum bicolor	Sudan grass	Cystitis, ataxia (growing grass)
Sorghum halepense	Johnson grass	Cyanide poisoning
Sorghum sudanense	Sudan grass	Possible lathyrism, equine cystitis and ataxia, ankylosis (growing grass)
Sphenosciadum capitellatum	Whitehead	(No signs given in Hails & Crane 1982)
Stipa viridula	Sleepy grass	Hypnosis
Swainsona spp.	Darling pea	Depression, emaciation, ataxia, blindness
Tanaecium exitiosum		Weakness, staggering, collapse, frequent micturition, inflammation of stomach and heart
Taxus baccata	Yew	Heart failure – trembling, dyspnoea, collapse
Taxus cuspidata	Japanese yew	Heart failure – trembling, dyspnoea, collapse
Trichodesma incanum		Severe liver damage in horses and other farm stock
Viscum album	Mistletoe	Incoordination, dilated pupils, salivation, hypersensitivity

Also, there is some justification for killing all the old turf by spraying with glyphosate, delaying ploughing for at least 2 weeks to ensure root kill. On very light soils it may be possible to plough during the late winter and sow in March or April in the northern hemisphere, but it is essential to achieve consolidation of the ground and a fine tilth after ploughing.

A pasture seed mixture should be sown as soon as a good seedbed can be effected, before the surface soil is subjected to periods of desiccation. In the UK, sowing should be accomplished by March or April, otherwise a delay until July or August may be inevitable, and undertaken then only if there are prospects of rain. It is also imperative that plant foods be provided; Table 10.10 gives suggested rates of application. In areas of low rainfall seeds should be sown 2.5 cm (1 in) deep; in areas of higher rainfall the sowing depth should be about 1.2 cm (0.5 in). Drilling deeper than 2.5 cm, which would be advisable in the height of summer, will result in seedlings of small seeds, such as poas and clovers, not reaching the surface; therefore for reasons of economy these should be omitted from the seed mixture. A fine firm tilth is then effected by harrowing and rolling in order to avoid a puffy, rapidly drying surface soil. Two seed mixtures are suggested in Table 10.18; these are composed of persistent strains that should not be given a cover crop and which suffer more than quicker growing strains from drought in the seedling stage if they have been sown in late spring in poorly formed seedbeds.

In formulating one's own seed mixtures the smaller the seed the lower the weight required per hectare. Thus, a lower weight of rough meadow grass, timothy and wild

Table 10.18 Suggested seed mixtures and minimum seeding rates (kg/ha) for permanent pastures when a good seedbed has been established (under adverse conditions the quantities may need to be increased up to double these amounts).

	kg/ha
Seed mixture 1	
Perennial ryegrass (Melle[1] or Contender)	10.00
Perennial ryegrass (Talbot, Parcour or Morenne)	3.50
Smooth meadow grass[2] (Arina, Dasas)	0.75
Creeping red fescue (Echo)	2.00
Rough meadow grass (VNS)	0.75
White clover (Blancho or NZ Huia)	0.75
Total	17.75
Seed mixture 2	
Perennial ryegrass (Trani or Springfield)	10.00
Perennial ryegrass (Melle)	5.00
Timothy (S48, S352, S4F)	2.00
White clover (Milkanova)	2.00
Total	19.00

[1] A mixture of strains similar to S23 and S24 could be used to give a range of heading dates.
[2] Kentucky bluegrass or smooth meadow-grass (*Poa pratensis*).

white clover would be required than of perennial ryegrass or meadow fescue and less is required of persistent leafy tillering strains of grasses for permanent pastures than would be used of aggressive strains in short-term leys. The better late-flowering strains of ryegrass and timothy should not be mixed with aggressive early flowering strains in the establishment of a permanent pasture. On light soils smooth meadow grass or Kentucky bluegrass (*Poa pratensis*) may be a helpful constituent of the mixture by virtue of its underground stems, which knit the surface soil together. On heavier soils this species might be replaced with tall fescue.

For some time a new sward will be more susceptible to destruction by treading than would an established turf and so ideally it should not be grazed in the first year, but should nevertheless be topped to prevent the grasses and any annual weeds from flowering. If the latter are present in excessive quantities, a selective weedkiller could be used, although this would check the establishment of clover and in any event it should not be used before the grasses have developed two or three leaves. Another option might be grazing by sheep late in the first year, but not when the surface is wet. A hay crop should not be taken from a newly established permanent pasture during the first 2 years at the very least.

Where ploughing and reseeding are impractical, or it is essential to establish a new pasture quickly, the author has realized considerable success in horse paddocks by spraying the old turf with glyphosate once or twice during a period of rapid growth and leaving it for at least 14 days. Following this, there are two alternative procedures for reseeding: (1) The surface is thoroughly disced (Plate 10.2), harrowed and rolled to effect a good tilth and seedbed before drilling the seed. This is delayed until the surface trash has decayed and if there are large amounts it can be burnt or raked off. The essential point is that seedlings should be surrounded by mineral soil in a firm seedbed. Burning will destroy seeds of exotic plants that may have remained dormant in permanent pastures for many years and that can yield a fascinating, but mildly bothersome, crop in the young sward. Moreover, excessive vegetative matter, which may contain some herbicide residues, can suppress the growth of the seedlings, but the cultivations allow soil microorganisms to contact and rapidly destroy residual herbicide. In areas of low rainfall the cultivations will dry out the surface soil during the summer, which also suppresses seed germination. Therefore (2), after the vegetation has died, surface moisture is retained and time is saved by direct drilling the seed immediately into the soil with a special drill that has heavy disc coulters for cutting through the dead turf. The same principles of top dressing and rolling at the appropriate times should be applied as for the more traditional husbandry described.

Another alternative is to graze a worn-out paddock intensively with steers, then disc and harrow heavily at a time when the surface soil is sufficiently moist to allow penetration of the implements. After a reasonable seedbed has been attained, sow, harrow, roll and fertilize as before. Heavy regrowth of the old turf can be controlled by regular topping until the new seedlings are well established. This procedure is the least expensive but unlikely to give complete satisfaction, especially where the previous sward was dense or its regrowth too rapid during the seedling stage.

Plate 10.2 Preparation of soil for resowing permanent pasture at Upend Stud, Newmarket, Suffolk. Discing began 8 days after spraying with glyphosate, a satisfactory seedbed was achieved 21 days after spraying and seed was drilled after 24 days, followed by rolling. The top and middle pictures show stages in seedbed preparation and the bottom picture shows pasture 1 year after sowing. Previously this paddock was rife with docks and thistles.

(a)

(b)

(c)

Plant disease control

Grass seedlings are sometimes killed soon after germination by soil-borne *Fusarium culmorum*. There is increasing reluctance to use pesticides that may be harmful to useful domestic insect and other animal species. Success has been achieved in the control of this fungus by using another fungus that is antagonistic to the *Fusarium*, but harmless to the grass seedlings. Indeed, a number of endophyte species (living entirely within the plant tissue and producing no spores) have been discovered that are synergic with grasses, that is, they even promote the health and growth of the grass species. Endophytes occur in both ryegrass and fescue pastures (see also 'Equine dysautonomia (grass sickness)', and 'Fescue toxicosis' caused by *Acremonium coenophialum*, this chapter). Endophytes are transmitted in infected seed and the resulting plants may be more resistant to a range of pests, grow more vigorously, producing a higher DM yield, and even may be more drought tolerant and winter hardy. Nevertheless, considerable care is now exercised by plant breeders in the use of these fungi as there is confirmed evidence that certain endophytes are a cause of disease in grazing stock. Ryegrass is frequently infected by *Acremonium lolii* which is likely to be the cause of sporadic disease (see 'ryegrass staggers', this chapter) in grazing animals, associated with unthriftiness and loss of coordination, observed in the UK, New Zealand and in several other countries, and *A. coenophialum* is the cause of fescue toxicity. Interestingly, it is thought that the tremorigenic toxin causing the animal disease is different from that afflicting insect pests, so useful progress should be possible in selecting endophyte strains for inoculating seeds that are benign to four-footed stock. Only time will tell.

TROPICAL GRASSLAND AND FORAGES

A characteristic of most tropical forages is a high yield of dry matter, which is relatively rich in fibre but impoverished of crude protein and P (Table 10.19) and from which the rewards of animal production are relatively meagre. There are, nevertheless, large differences between the products of wet and dry seasons. In one study (Kozak & Bickel 1981) the crude protein content of grasses decreased from a range of 6–10% in the rainy season to 4–5% in the dry season in Tanzania without any change in the crude fibre content, yet apparent digestibilities of the crude protein were 34–58% and 16–25%, respectively. Kozak and Bickel also noted that the digestibility of the dry matter of pasture forage decreased from 47–63% at heading to 30–53% at flowering. It has been asserted that the yield per hectare of digestible nutrients is at best only half what can be achieved from temperate grassland. The low digestibility of tropical grass is apparently an effect of a higher lignin content than that of temperate grass. Analyses reveal that as the environmental temperature rises there is a fall in the cellulose content of tropical grass but a proportionate rise in the hemicellulose and lignin contents. Malaysian experience has shown that horses fed as much cut tropical grass as they can eat, plus oats or other concentrates, lose weight. Their general health and their performance can be

Table 10.19 Dry matter composition of four tropical grasses fed to horses compared with a temperate (Newmarket, UK) grass sample (D. Frape unpublished observations).

	Young Napier grass	Napier hay	Mature Napier hay	Signal grass	Signal grass hay	Paragrass hay	Guatemala grass	Newmarket grass
Crude protein (%)	5.7–10	5–6	2.7–2.9	11–11.5	5–6	5.4–9.8	7.4–7.5	12.7
MAD fibre (%)	43–51	54–55	54–55	41–42	51–52	42–44	46–47	27.5
Ash (%)	3.2–4.5	3.6–3.8	2.1–2.3	7.3–7.8	3–3.2	3.4–8	6–7	9.95
Ca (%)	0.046–0.16	0.08–0.1	0.1–0.12	0.09–0.13	0.09–0.1	0.11–0.2	0.05–0.06	0.93
P (%)	0.06–0.16	0.11–0.13	0.1–0.12	0.17–0.18	0.14–0.15	0.15–0.23	0.17–0.175	0.31
Mg (%)	0.21–0.32	0.20–0.22	0.22–0.25	0.26–0.27	0.11–0.12	0.18–0.36	0.087–0.088	0.16
K (%)	0.9–1.3	0.8–0.9	0.38–0.4	2.8–3.3	0.95–1.05	0.66–2.6	2.1–2.2	1.9
Na (mg/kg)	170–690	200–300	170–190	510–650	170–180	1280–11300	150–170	1800
Zn (mg/kg)	21–38	34–36	20–22	27–29		35–43		38
Se (mg/kg)	0.055	—	—	—	0.03	0.03	0.04	0.1
F (mg/kg)	8.4	—	—	—	—	—	—	—

MAD, modified acid-detergent.

improved by limiting their access to grassland, or its products, and confining them for greater periods of time to the stable which affords shade, cool ventilation and a more satisfactory control of insect-borne disease. Furthermore, where the land is light, significant amounts of sand may be consumed by grazing stock and this may interfere with bowel function.

Plate 10.3a shows TB paddocks at 1350 m in Zimbabwe at the end of October, during the foaling season (surprisingly the hottest part of the summer, before the rains come). The grass is dry and bleached and the only green material appears to be *Indigofera* spp., Leguminosae, probably *I. setiflora* (Plate 10.3b). Species of this genus contain a hepatotoxic amino acid, indospicine, which competes with arginine, causing liver and kidney damage. Horses may be protected from Birdsville disease in Australia by supplementing their feed with arginine-rich substances, such as peanut meal, or gelatine. Signs of toxicity include loss of appetite, apathy, discharges from the eyes and nose, loss of flesh, laboured breathing and severe incoordination, with a dragging of the hindfeet. The horse may fall backwards during a canter. Many other potentially harmful plant species were identified in and around these TB paddocks (Table 10.20).

TB studs in Zimbabwe give mares 9 kg (barren mares) to 12 kg (lactating mares) relatively high protein concentrates daily, which incidentally will reduce the waste heat of fermentation to be disposed of which would be generated should otherwise large amounts of roughage be given. Another reason for the large concentrate allowances may be not only the low digestibility of the herbage, but also a recognition of adverse reactions if large amounts of herbage, containing hazardous plants, are consumed.

Oxalate poisoning

(See also 'Grass toxins', this chapter.) In the author's experience working horses introduced to the tropics and required to subsist on indigenous forages and cereal grains frequently exhibit decreased performance, lameness (changing from one leg to another and involving both creaking hip and shoulder joints), arching of the back, swelling of the facial bones and muscular wasting of croup and rump. Similar observations and the poisoning of grazing cattle have been recorded in various parts of South-East Asia, the Philippines, Brunei and North Australia (Seawright *et al.* 1970; Blaney *et al.* 1981a,b). The poor nutritional quality of tropical grass has frequently been overcome by the introduction of high-quality clovers, trefoils and *Medicago* spp. including lucerne, selecting those varieties low in tannins, as increased tannin concentrations are prevalent when legumes are grown on tropical soils deficient in P.

The signs described above are those of osteodystrophia fibrosa, big-head, found by the author (unpublished observations) and others (Blaney *et al.* 1981a) to be caused by large amounts of oxalate and small amounts of Ca and P in many tropical grasses. Most published reports have concerned species of the genus *Setaria* (Blaney *et al.* 1981b; Seawright *et al.* 1970), which may contain 30–70 g oxalate/kg DM. A

(a)

(b)

Plate 10.3   TB stud in Zimbabwe, 1350 m, during October. This is the foaling season and the hottest and driest month, when pastures can be of poor feeding value. (a) a general view of the paddock; (b) the only green herbage in the paddock, *Indigofera* spp., a legume said to contain substances toxic to ruminants.

Table 10.20   Hazardous plant species identified in and around paddocks in a subtropical TB stud.

Plant species	Comments
Eleusine indica, Panicum novemnerve	Thyrotropic cyanogenic glycosides
Setaria pumila, Portulaca spp., *Oxalis* spp.	Oxalates reducing bone calcification
Verbena bonariensis	Members of the Verbena family contain hepatotoxic icterogenin and other triterpenoids, e.g. rehmannic acid; identical to that causing *Lantana* poisoning: 'Beach disease' (see below).
Nicandra physalodes	A poisonous member of Solanaceae
Senecio venosus	Members of the genus *Senecio* contain hepatotoxic alkaloids
Sideranthus spp.	Se accumulators*
Amaranthus hybridus	Nitrates and the alkaloid lycorine, salivation, diarrhoea, paralysis
Richardia scabra	(False ipecacuanha) Very poisonous
Tagetes minuta	Very poisonous, roots poisonous to nematodes
Spermacoce senensis	Very poisonous
Lantana camara	Very poisonous, rehmannic acid a hepatotoxin, 'Beach disease': cholestasis, anorexia, haemorrhagic gastroenteritis, ataxia, photosensitization
Convolvulus sagittatus var. *aschevsonii*	Very poisonous, purgative, nervous signs, incoordination, tremors, pulmonary oedema, polyuria, blindness. Contains lysergic acid alkaloids
Cynodon dactylon	(Bermuda grass) Hepatotoxins causing secondary photosensitization
Pennisetum clandestinum	(Kikuyu grass) Salivation, thirst, distension and inflammation of GI tract, incoordination etc.
Polygala albida	Members of the genus *Polygala* cause tremors, gastroenteritis, incoordination, cerebral congestion
Leucas martinicensis	Tetracyclic polyol andromedotoxin causes hypotension, respiratory depression, attempts to vomit. The amino acid mimosine, a depilatory, causes haemorrhagic gastritis in horses
Indigofera spp.	Hepatotoxic amino acid, indospicine, causing liver and kidney damage: 'Birdsville disease'

*Over 2mg Se/kg was detected in representative pasture herbage samples.

simple method for the quantitative estimation of oxalate in tropical grasses has been proposed by Roughan and Slack (1973). Lesser but harmful quantitites of oxalates have been found in the widely used Napier (*Pennisetum purpureum*) and Signal (*Brachiaria* spp.) grasses and Paragrasses. For several reasons Napier is an inferior grass for horses; some chemical characteristics of each of these tropical grasses, for comparison with temperate grasses, are given in Table 10.19. The samples were shown to be deficient also in Se. Ponies given a diet containing 10g oxalic acid/kg plus 4.5g Ca/kg exhibit decreased urinary Mg and are in negative Ca balance through reduced Ca absorption (Swartzman *et al.* 1978). It has been concluded that problems in horses may arise where the dietary dry matter contains more than 5g total oxalates/kg with a Ca:oxalate ratio of less than 0.5 (Blaney *et al.* 1981a). Horses presenting typical signs, and which were in negative balances of both Ca and

P, supplemented once per week with limestone, rock phosphate or dicalcium phosphate and 50–60% molasses, consumed up to 200 g Ca and 50 g P per hour and changed to positive Ca and P balances (Gartner *et al.* 1981). Success has also been achieved by supplementing horses with 125 g limestone per day (providing 45 g Ca) until the initial abnormality has subsided and then reducing the amount to 100 g daily. If rock phosphate is introduced to the diet, the F intake should not be allowed to exceed 50 mg/kg of the total diet.

POISONOUS PLANTS

Poisonous plants are generally unpalatable. They may be consumed when there is feed scarcity, when the horse has not previously encountered them, or when present dried in hay. Poisonous weed seeds/moulds, as contaminants of grains, may include *Lolium temulentum*, *Ricinus communis* and *Claviceps purpurea*. Table 10.16 lists the plant species more likely to be a cause of horse illness in Europe.

Plant toxicosis caused by associated microorganisms in grasses

There are several microbial species growing in plants that are a cause of intoxication in horses. The subject of microbial toxins has received much attention in recent years, probably as a result of the development of laboratory methods in their analysis.

Fescue toxicosis

Perhaps one of the most pressing syndromes, at least in the USA, has been equine fescue toxicosis. Approximately 688 000 horses in the USA are kept on tall fescue (*Festuca arundinacea*) pastures, and for many years there have been reports of reproductive problems in mares grazing this species (Cross *et al.* 1995). Signs include increased gestation length, agalactia, foal and mare mortality, tough and thickened placentas, dystocia (abnormal labour at foaling), weak and dysmature foals, increased sweating during warm weather, reduced serum prolactin and progesterone and increased serum oestradiol-17β concentrations. Unlike the effects in many other species, horses consuming infected tall fescue do not exhibit increased body temperature. The abnormalities in gravid mares are caused by vasoconstrictive ergot peptide alkaloids (pyrrolizidine alkaloids have also been isolated) produced by endophytic fungi, principally *Acremonium coenophialum*, but also by *Balansia epichloe* and *B. henningsiana* identified in tall fescue (Table 10.21). These fungi infect other warm season weeds and grasses (including *Agrostis*, *Andropogon*, *Eragrostis*, *Paspalum*, *Sporobolu* and *Stipa*), in none of which are they pathological to the plant. Recent evidence showed that ergovaline, the primary ergopeptine isolated from *A. coenophialum*, at a dietary concentration of up to 308 μg/kg caused no adverse effect on nutritional, or reproductive, performance of mares.

Table 10.21 Ergot profiles in endophytic fungi isolated from grasses growing in central Georgia, USA (Bacon 1995).

Ergot alkaloids	*Balansia* spp.		*Acremonium coenophialum*
	B. epichloe	*B. henningsiana*	
Chanoclavine I	+	+	–
Isochanoclavine I	+	+	–
Agroclavine	+	–	+
Elymoclavine	+	–	+
Penniclavine	+	–	+
Ergoclavine	+	–	–
Ergonovine	+	–	±
Ergonovinine	+	–	–
6,7-Secoagroclavine	+	–	+
Dihydroelymoclavine	–	+	–
Festuclavine	–	–	+
Ergovaline	–	–	+
Ergovalinine			+

Endophyte-infected tall fescue hay is less digestible than endophyte-free hay, and young horses consuming infected pasture grow more slowly than do those on endophyte-free pasture. The alkaloids are apparently serving as D2 dopamine receptor agonists, explaining their prolactin-lowering effect. Cross *et al.* (1995) advise that domperidone, a dopamine receptor antagonist, is effective in preventing the signs of tall fescue toxicosis in horses without neuroleptic side effects. The minimum effective dose in gravid mares is 1.1 mg/kg BW daily given orally for 30 days before foaling, or 0.44 mg/kg BW given subcutaneously for 10 days before foaling.

Fescue toxicosis causes increased susceptibility to high environmental temperatures and light intolerance (Porter & Thompson 1992), and so recent hot summers in Europe may have increased the frequency there of this disease in horses, cattle and sheep. *Acremonium* grass endophytes are toxonomically aligned with the family Clavicipitaceae and they live their entire life within the aerial parts of their grass host, producing no spores.

Ryegrass staggers

Acremonium lolii is the endophyte of perennial ryegrass (*Lolium perenne*), which produces the indole-isoprenoid alkaloids paxilline and lolitrem (Porter 1995), causing ryegrass staggers, a neurological condition characterized by incoordination, staggering, head shaking and collapse in horses that have been disturbed. Lolitrem is in highest concentration in the leaf sheaths and lowest in the leaf blades. Thus, staggers is most frequently observed in closely grazed pastures and it has been reported in horses eating ryegrass straw. Reports from New Zealand and Oregon, USA are the most frequent. The ergopeptine alkaloids produced by this endophyte may be a cause of poor growth and poor reproduction on ryegrass pastures. The subject of the effect of natural toxins on reproduction in livestock has been reviewed by James *et al.* (1992) and by Cheeke (1995).

Pasture treatment against **Acremonium** *spp.*

At present there seems to be no effective prophylactic treatment for the pastures in which tall fescue or ryegrass is infected with *Acremonium* spp. The pasture infection improves the vigour and persistency of these grass species. In New Zealand prevention is afforded by providing protection to the grass from weevil damage and plant breeding is likely to be the route by which the problem is overcome. At present, by virtue of the increased vigour endowed, endophyte-infected cultivars of both ryegrass and tall fescue make up an increasing share of the total seed production.

Annual ryegrass toxicosis

Annual ryegrass toxicosis occurs particularly in South Africa and Australia and is caused by a group of highly toxic glycolipids called corynetoxins. The signs are neurological disturbances, presented as high stepping gait, ataxia and convulsions. It is a lethal condition causing damage to the cerebellum. The aetiology is complex. Seedlings may become infected by a soil nematode, *Anguina agrostis*, that leads to a gall in the flower, where the worm lays eggs. The nematode is nontoxic, but if it is, in turn, infected by a bacterium, *Clavibacter toxicus,* corynetoxins are produced and the galls are then toxic. There is some evidence that the toxin is produced only if the bacteria are themselves infected with a bacteriophage. Control of the toxicosis requires that nematode infection of the grass is prevented. Crop rotation, field burning, clipping immature seed heads, fallowing and avoidance of transfer of infected material to other fields are control methods.

Sleepygrass toxicosis

Sleepygrass (*Stipa robusta*) is a perennial bunch grass found on rangelands of the southwestern USA. Consumption of the grass causes a profound stuporous condition which may last for several days in horses. *Acremonium* endophytes, containing the ergot alkaloid lysergic acid amide, are probably the causative agents (see 'Fescue toxicosis', above).

Equine dysautonomia (grass sickness)

The clinical signs of equine dysautonomia (grass sickness) are muscular tremors, patchy sweating, difficulty in swallowing, salivation, dilation of the bowels and stomach and chronic loss of weight. It is more common in horses that have been on the premises for less than 2 months. The incidence may be reduced by stabling the horses for part of the time and giving them hay and supplementary feed.

It is, at the time of writing, a disease of unknown aetiology, but its incidence in the UK has been recognized and recorded since 1909. Horses at pasture in temperate zone countries are most susceptible. It occurs most frequently following drier spells of weather. It is neither contagious nor infectious. The lesions are prominent in the sympathetic ganglia. There seems to be resistance to the disease in some horses, but whether this involves immunity, genetic resistance or both is unclear. The characteristics of the disease point to a neuromycotoxicosis, and *Fusarium* spp. have been suspected. The species of this genus are stimulated to grow on grasses during dry weather and some produce cytotoxic trichothecenes. As *Fusarium* spp. occur on many pastures but grass sickness does not, it is likely that other specific environmental conditions are necessary for clinical expression of the disease. Objective advice on prevention will depend on a fuller understanding of the aetiology.

Chronic lupinosis

Chronic lupinosis is caused in sheep by the presence in lupin species of the fungus *Phomopsis leptostromiformis*, causing liver damage. The condition has not been reported for horses, in which acute lupin poisoning may occasionally occur (see 'Toxic legumes and other species', this chapter).

Plant toxins

In the plant world there is a vast array of substances that are toxic and hazardous to horses when consumed in disproportionate amounts. Many of these are known as alkaloids, which are organic, basic compounds, although the chemical diversity of all hazardous compounds is considerable, and their range of toxicity to the horse is wide. Numerous broad-leaved plant species on or around pastures are toxic to a greater or lesser degree as a consequence of their production of endogenous toxins. Access by horses to shrubs, trees and hedgerow plants is a typical cause, although minor intoxication can occur, for example, from widely abundant members of the Ranunculaceae. In the author's experience, heavy infestation of pastures by some members of this family causes buccal irritation in horses. Members of this family contain an irritant yellow volatile oil, protoanemonin, but in differing amounts. This substance can cause irritation to the mouth of grazing animals, but hay containing Ranunculaceae is safe, as far as *this* chemical is concerned, as the curing process causes precipitation of protoanemonin in a harmless form. More severe toxicosis is caused by contamination of upland hay by bracken which causes progressive ataxia. The whole plant is toxic and the principles include a cyanogenetic glycoside in relatively low concentrations, a thiaminase (causing the ataxia), an aplastic anaemia factor, a factor causing haematuria, and a carcinogen, although these last three may be identical. Water meadows are another source of hazardous plants and dry parched grassland, on which drought-resistant harmful shrubs survive, is yet another. In many cases conclusive diagnoses are unlikely.

Photosensitization

Photosensitization refers to the production of skin lesions caused by the interaction of sunlight with exogenous substances that are capable of activation by solar radiation to form free radicals. Primary photosensitization is caused by sunlight reacting directly with dietary substances in the skin after absorption. Examples include hypericin contained in St John's worts (*Hypericum* spp. especially *H. perforatum*), fagopyrin in buckwheat (*Fagopyrum esculentum*), and toxins in bog asphodel (*Narthecium ossifragum*), in species of *Vicia* and sometimes in lucerne (*Medicago sativa*), alsike (*Trifolium hybridum*) and red clovers. On other occasions therpeutic drugs may be incriminated. Lucerne photosensitization is caused by pheophorbide-α contained in the leaves of dried lucerne. The pheophorbide-α is formed by breakdown of chlorophyll under the influence of chlorophyllase, during processing. There is a higher activity of this enzyme in legume forages than in grass.

Many absorbed toxins are detoxified in the liver and excreted in bile. This process is obviously less efficient where there is a measure of liver dysfunction. Secondary photosensitization occurs when a damaged liver is unable to clear, for example, phylloerythrin, a photodynamic, chlorophyll metabolite, in the bile. Skin reactions are most likely to be observed following the consumption of large amounts of green forage by horses that have received some hepatotoxic agent. Numerous warm-season grass species cause secondary photosensitization, characterized by photophobia and severe dermatitis. These species include *Panicum* and *Brachiaria* which contain steroidal saponins, causing liver damage, possibly by interaction with mycotoxins.

Facial eczema in sheep is secondary photosensitization caused by the mycotoxin sporidesmin, contained in spores produced by the fungus *Pithomyces chartarum*. The toxin causes free radical liver damage. As Cu strongly catalyses the oxidation, protection is afforded by Zn supplementation that reduces Cu absorption and blood Cu level. Zn also binds with a sporidesmin metabolite, preventing its autoxidation. It is of concern in New Zealand, where sheep and cattle develop severe dermatitis of light-skinned areas of the body.

Toxic legumes and other species

Yew (*Taxus baccata*) is the most toxic plant (nonlegume) in the UK, and little more than 100 g of it will kill a horse by cardiac arrest. The second most poisonous is laburnum (*Laburnum*), a member of the legume family, which contains many plants known to cause damage to horses, among them broom (*Cytisus scoparius*) and lupins (*Lupinus*). The toxicity of lupins is principally confined to the seeds, and the various strains differ in their potency. Sweet lupins (*L. lutens*) have a low alkaloid content and are grown on poor land as a source of fodder. If horses eat them death is rare and is caused by respiratory paralysis, not by liver damage. An accumulative poisoning associated with progressive liver damage (chronic lupinosis) occurs in sheep and horses in Australia. Here the causative agent is a fungus growing on the lupins.

Sweet clover or melilot (*Melilotus*) contains coumarin which is broken down to dicoumarol in hay made under bad harvesting conditions, or during moulding. Dicoumarol prolongs blood clotting time. Both white and red clovers may contain toxic factors. Both species contain appreciable amounts of oestrogens which are also found (in much higher concentrations) in subterranean clover (*Trifolium subterraneum*) grown on light soils. Some reports indicate the presence of oestrogenic activity in moulds infecting the clover leaves. These hormone-like substances have been associated with infertility and increased teat length of sheep and it is an open question as to whether any comparable problem occurs in grazing mares. Two genera of vetches in the USA, milk vetches or locoweeds (*Astragalus*) and the related *Oxytropis* (see Table 10.17), are implicated in several abnormal conditions of horses. One caused by locoweeds is recognized as eliciting irreversible nervous signs. Although none of the poisonous members of these genera are found in the UK, certain pasture species of *Vicia* and *Lathyrus* found there are mildly toxic. In parts of the USA locoweeds begin their growth in late summer and remain green through the winter. They must be grazed for a period before poisoning is obvious. Some species are Se accumulators and may contain up to 300 mg/kg; they are poisonous for this reason.

Grass toxins

Toxins are rarely produced by grass tissues, although several species are known to do this.

Kikuyu grass poisoning
In addition to its soluble oxalates, Kikuyu grass (*Pennisetum clandestinum*) causes a poisoning of horses that is probably related to its saponin content, occurring during periods of rapid growth of the grass. Signs include false simulated drinking, anorexia, depression, pilo-erection, drooling, colic, grinding of teeth, cessation of intestinal movement and lack of faecal excretion.

Reed canary grass poisoning
Reed canary grass (*Phalaris arundinacea*) is a forage grown on wet, poorly drained soils. It contains at least eight different alkaloids which have been a cause of poisoning in farm livestock. Reed canary grass is relatively unpalatable and this may be one reason why there have been no reported cases of intoxication in horses.

Oxalate poisoning
Various tropical grass species, including buffel grass (*Cenchrus ciliaris*), pangola grass (*Digitaria decumbens*), setaria (*Setaria sphacelata*) and Kikuyu grass (*Pennisetum clandestinum*) contain soluble oxalates that react with Ca to form insoluble Ca oxalate, reducing Ca absorption. This causes mobilization of bone mineral and secondary hyperparathyroidism, or osteodystrophy fibrosa in horses. Cattle and sheep are less affected, but not unaffected, owing to degradation of oxalates in the rumen. Concentrations of 5 g or more soluble oxalates/kg DM in forage grasses induce the condition in horses. Oxalate content of these grasses is highest under conditions of rapid growth. Oxalate toxicity is also discussed under 'Tropical grassland and forages', this chapter.

Poisonous trees

Many tree species contain substances that are toxic to a greater or lesser degree (Table 10.17). The toxins may be present in the leaves, bark and/or fruits. A high proportion of horses bedded on black walnut (*Juglans nigra*) wood shavings develop laminitis. An aqueous extract of the heartwood given by stomach tube was shown to cause limb oedema, mild sedation and Obel grade 3 or 4 laminitis (see Chapter 11) within 12 h. The signs are inconsistent with those caused by carbohydrate overload. The toxin has not been identified.

Cyanogenic species

Hydrogen cyanide toxicity

Cyanide toxicosis is caused by the inhibition of cytochrome oxidase (EC 1.9.3.1), a terminal respiratory enzyme in all cells, depriving the cell of ATP. Thus, signs of acute intoxication include laboured breathing, excitement, gasping, staggering, convulsions, paralysis and death. In tropical and subtropical countries enzootic equine cystitis and ataxia occur in horses grazing fresh summer annual forages of the genus *Sorghum* (Johnson grass, Sudan grass, *S. sudanense* and common sorghum). However, acute toxicosis is more likely with the consumption of sorghum hay, or especially with ground and pelleted sorghum hay, owing to the rapid rate of intake and of cyanide release. Ensiling markedly reduces the risk of cyanide toxicosis. Sorghums contain a cyanogenic glycoside, dhurrin, from which free cyanide can be released by enzymatic action. The glycoside and the enzyme are contained in different plant cells, but damage to the plant from wilting, trampling, frost and drought result in the breakdown of cell walls, with the mixing of the juices and release of free cyanide. Arrow grass (*Triglochin*), wild black cherry (*Prunus serotina*), choke cherry (*P. virginiana*), pincherry (*P. pennsylvanica*) and flax (*Linum*) also contain cyanogenetic glycosides which are readily hydrolysed to hydrogen cyanide. They are most toxic during rapid growth immediately after a freeze. Heavy N fertilization, wilting, trampling and plant diseases may increase the hazard. Silage or haylage produced from the grasses is risky. Cystitis, or inflammation of the urinary tract, and incontinence are more common in mares than in stallions or geldings, but posterior ataxia is manifested in all horses, from which they seldom recover.

Thiocyanate

Cyanide is readily detoxified in animal tissues, during low rates of intake, by reacting with thiosulphate to form thiocyanate, the blood and urinary concentrations of which increase during chronic exposure:

$$S_2O_3^{2-} + CN^- \rightarrow SO_3^{2-} + SCN^-$$

However, the chronic production of thiocyanate (SCN⁻) can induce ataxia, degenerative lesions of the central nervous system, goitre and a deficiency of S, owing to urinary loss of S as thiocyanate.

Hepatotoxins

Malnutrition can increase the risks of many toxins, or one toxin may compound the effects of another. Although definite conclusions are usually unlikely there are several well-known interactions in the field. *Heliotropium europaeum* (Boraginaceæ) is a shrub that produces the pyrrolizidine alkaloid hepatotoxins heliotrine and lasiocarpine, causing lesions, associated with Cu accumulation and a subsequent haemolytic crisis of chronic Cu poisoning. *Heliotropium europaeum* would not normally be grazed if there was other more appetizing vegetation. A straight-forward case was presented once to the author in which severe liver damage and death occurred among sheep. These animals browsed in an environment where the pink flowers and green foliage of *H. europaeum* were about the only species not desiccated and these plants were set against the background of a blue streak of Cu salts in the bed of a dried-up wadi.

Seneciosis

In the UK and many temperate lands the most common source of pyrrolizidine alkaloids is ragwort (*Senecio jacobaea*), frequently consumed in hay. Seneciosis is a disease caused by the ingestion of certain plants of the genus *Senecio* (Compositae) which induce liver enlargement, degeneration, necrosis, cirrhosis and ascites. A constant feature has been occlusion of the centrilobular vein (veno-occlusive disease – VOD), briefly reviewed by Hill (1959). Although ragwort may be a hazardous member of this group, the inconspicuous groundsel (*S. vulgaris*) is a source, albeit a lesser one, of the toxins. There have been many reports of seneciosis from the USA, Europe and the other continents. The worldwide distribution of the disease is attested by the variety of names used to describe it: 'walking disease' of horses and cattle in Nebraska, 'walking about' disease of horses and cattle in Australia, 'Pictou' disease in Nova Scotia, 'Zdar' disease in former Czechoslovakia, 'Schweinberger' disease in Germany, 'Dunziekte' of horses and cattle in South Africa, and 'Winton' disease in New Zealand. It is also recognized in goats, chickens, quail, doves and pigs. *Crotalaria* (Leguminosæ) and *Heliotropium* have similar hepatotoxic properties so their effects are included under the umbrella term seneciosis.

Dry summers leave many paddocks and pastures in a poor state, with typical grass species dying back. On the other hand, deep rooted and drought-resistant weeds survive. In this weather ragwort and its seeds spread and an increase in the frequency of liver damage among horses and other herbivores can be expected to occur subsequently. The different resistances of grazing animals to this toxin reflect variation in the efficiency of its urinary elimination (Holton *et al.* 1983). Elevated plasma concentrations of GGT (EC 2.3.2.2) are a useful early indicator of the hepatic damage caused by ragwort toxins. Horses with ragwort-damaged livers should be given a well-balanced diet containing good-quality protein supplemented with B vitamins and trace elements.

The common comfrey (*Symphytum officinale*) contains at least nine potentially hepatotoxic pyrrolizidine alkaloids in its leaves and roots, which are less toxic than those in ragwort. Comfrey has been recommended for inclusion in horse feeds, but this recommendation cannot be supported, owing to the risk of liver damage.

Aflatoxicosis
Liver damage through aflatoxicosis, derived from *Aspergillus flavus* intoxication, to which the horse is very susceptible, is less likely now within the EU as a result of legislation. This toxin was first described in groundnuts, or peanuts (*Arachis hypogaea*), but has subsequently been detected at lower concentrations in other plant species, including some cereal grains.

HOMEOPATHY

The principle of homeopathy is to give a potentially toxic chemical, which in large doses causes the signs of a specific disease, but which in small doses is said to cure that disease. Unfortunately, much of the evidence supporting the claims is anecdotal and there is a need for these claims to be subjected to acceptable experimental methods of examination. The remedies used in homeopathy are extracted from materials in the animal and plant kingdoms and from natural minerals. These include:

- bryonia (wild hops)
- belladonna (deadly nightshade)
- silica (pure flint)
- sulphur
- sepia (Cuttlefish ink)
- apis (honey bee) (Evans 1995).

A potential development of some treatment procedures is hazardous, e.g. 'nosodes'. Several of the treatments involve use of irritant substances extracted from plants, e.g. pulsatilla from *Pulsatilla nigricans* (Ranunculaceae), arnica from flowers of *Arnica montana* and related Compositae, with well-known pharmacological properties. The general question is whether the very low doses typically used have any measurable effect, whereas high doses could be harmful.

STUDY QUESTIONS

(1) What would you propose should be done about worn-out, worm-infested pastures on (a) heavy clay and (b) light alkaline soils?
(2) Where (a) sheep or (b) cattle can be purchased, how would you propose a mixed pasture management system should be organized for a 50-ha stud with 30 mares and followers?
(3) What would be the sequence of decisions in planning to make haylage for a stud of 20 mares?

FURTHER READING

AFRC Institute for Grassland and Animal Production, Welsh Plant Breeding Station, Plas Gogerddan, Aberystwyth, Dyfed SY23 3EB, Wales. Various publications on grassland research.

Aiken, G.E., Potter, G.D., Conrad, B.E. and Evans, J.W. (1989) Growth performance of yearling horses grazing Bermuda grass pastures at different grazing pressures. *Journal of Animal Science*, **67**, 2692–7.

Andrews, A.H. and Humphreys, D.J. (1982) *Poisoning in Veterinary Practice*, 2nd edn. National Office of Animal Health, Enfield, Middlesex.

Anon (1986) Better use of nitrogen – the prospect for grassland. *National Agricultural Conference Proceedings*. Royal Agricultural Society of England and Agricultural Development and Advisory Service, National Agriculture Centre, Warwickshire.

Anon (1995a) *Compendium of Data Sheets for Veterinary Products, 1994–95*. National Office of Animal Health, Enfield, Middlesex.

Anon (1995b) *The UK Pesticide Guide* (ed. R. Whitehead). CAB International, Wallingford, and The British Crop Protection Council, Farnham.

Bacon, C.W. (1995) Toxic endophyte-infected tall fescue and range grasses: historic perspectives. *Journal of Animal Science*, **73**, 861–70.

Cheeke, P.R. (1995) Endogenous toxins and mycotoxins in forage grasses and their effects on livestock. *Journal of Animal Science*, **73**, 909–918.

Clarke, E.G.C. and Clarke, M.L. (1975) *Veterinary Toxicology*. Baillière Tindall, London.

Cross, D.L. Redmond, L.M. and Strickland, J.R. (1995) Equine fescue toxicosis: signs and solutions. *Journal of Animal Science*, **73**, 899–908.

Forbes, T.J., Dibb, C., Green, J.O., Hopkins, A. and Peel, S. (1980) *Factors Affecting Productivity of Permanent Grassland*. A National Farm Survey. Grassland Research Institute, Hurley, Maidenhead.

Frape, D.L. (1996) Sherlock Holmes and chemical poisons. *Equine Veterinary Journal*, **28**, 89–91.

Green, J.O. (1982) *A Sample Survey of Grassland in England and Wales 1970–1972*. Grassland Research Institute, Hurley, Maidenhead.

Hails, M.R. and Crane, T.D. (1983) *Plant Poisoning in Animals. A Bibliography From the World Literature, 1960–1979*. Commonwealth Agricultural Bureaux, Slough.

Hopkins, A. (1986) Botanical composition of permanent grassland in England and Wales in relation to soil, environment and management factors. *Grass and Forage Science*, **41**, 237–46.

Hopkins, A., Martyn, T.M. and Bowling, P.J. (1994) Companion species to improve seasonality of production and nutrient uptake in grass/clover swards. *Proceedings of the 15th General Meeting, European Grassland Federation* (ed. L. Mannetje), Wageningen, the Netherlands, pp. 73–6.

James, L.F., Panter, K.E., Nielsen, D.B. and Molyneux, R.J. (1992) The effect of natural toxins on reproduction in livestock. *Journal of Animal Science*, **70**, 1573–9.

Ministry of Agriculture, Fisheries and Food (1984) *Poisonous Plants in Britain and Their Effects on Animals and Man* (Reference Book 161). HMSO. London.

Murray, A. (1993) *The intake of a molassed mineral block by a group of horses at pasture*. MEqS thesis, Faculties of Agriculture and Veterinary Medicine, National University of Ireland, Dublin.

Porter, J.K. (1995) Analysis of endophyte toxins: fescue and other grasses toxic to livestock. *Journal of Animal Science*, **73**, 871–80.

Ricketts, S.W., Greet, T.R.C., Glyn, P.J., *et al.* (1984) Thirteen cases of botulism in horses fed big bale silage. *Equine Veterinary Journal*, **16**, 515–18.

Sheldrick, R.D., Lavender, R.H., Martyn, T.M. and Deschard, G. (1990) *Rates and frequencies of superphosphate fertiliser application for grass–clover swards*. Session 1: Poster 2. British Grassland Society, Research Meeting No 2, Scottish Agricultural College, Ayr.

Underwood, E.J. (1977) *Trace Elements in Human and Animal Nutrition*, 4th ed. Academic Press, New York.

Chapter 11
Pests and Ailments Related to Grazing Area, Diet and Housing

For a surfeited horse. Take a handful of pennyroyal, half a handful of hyssop, an handful of sage, an handful of elder leaves or buds, an handful of nettle tops, fix large sprigs of rue, and handful of celendine, cut small and boiled in three pints of stale beer, which must be boiled to a quart.

Sir Paulet St John 1780

ARTHROPOD PARASITES

Lice

There are two species of horse lice: *Haematopinus asini* is a blood sucker and *Damalinia equi* lives on skin scales. The females lay eggs on the hair and a greater problem of scratching or rubbing is frequently observed in the winter than in the summer so that many are lost when the winter coat is shed. Control is achieved by dipping, spraying or dusting with insecticides, but a second treatment should be undertaken to kill those that will hatch from eggs already laid.

Ticks

Grazing horses, particularly those sharing ground with wild grazing and browsing animals, are prone to infestation by ticks, which can transmit diseases. However, the cattle and sheep tick in the UK does not usually cause symptoms in horses. In the USA the soft tick (*Otobius megini*) lives deep in the ears, but the adults, which do not feed, live in cracks in stables, fences and under troughs where they also lay eggs. Larvae of hard ticks (*Dermacentor andersoni; Amblyomma americanum*) are found in various places on the horse, and the adults, after mating, fall to the ground where eggs are laid in secluded locations. The tropical horse tick (*Dermacentor nitens*), whose primary host is the horse, transmits equine piroplasmosis, a protozoan blood disease. Insecticide should be applied to all parts of the skin where the ticks may be attached, including the ears, and as the parasites spend long periods off the host, the grass and other areas around the stable should also be treated. With slight infestations, the ticks can be detached by application of chloroform to release their mouthparts.

Lyme disease was first described in 1977, following an outbreak in man of arthritis in Lyme, Connecticut. The disease is caused by *Borrelia burgdorferi* and antibodies to this spirochaete bacterium have been detected in the sera and synovial fluid of horses in the UK (there is, unfortunately, a lack of specificity owing to cross-reacting antibodies), the majority of which did not display clinical manifestations of Lyme disease. These signs include arthritis, myositis, weight loss, fever, laminitis and possibly meningoencephalitis. Diagnosis requires histological demonstration of silver-stained spirochaetes in skin biopsy specimens, or the difficult culture of the organism from blood or cerebrospinal fluid. The bacterium is transmitted by several species of the ixodid tick (*Ixodides ricinus and I. persulcatus* in Europe) in the northern hemisphere, which feed on many species of animal. The susceptibility of these animals to infection with *B. burgdorferi* is largely unknown, although they undoubtedly affect the prevalence of the infection. Acute infection responds to appropriate antibiotic treatment, but the chronic arthritis is often unresponsive.

Mites

Mites (*Psoroptes equi, P. cuniculi, Chorioptes bovis*) cause itch or scabs and may be controlled by dipping or spraying with insecticide. As a general rule, high-pressure sprays frighten horses so that low-pressure hand-pumped sprayers are preferable and the horse should be confined to a chute during treatment.

Biting midges

An intensely itching dermatitis called sweetitch, which occurs during the summer months and is quite common in Ireland, is probably caused by species of *Culicoides* (Baker & Quinn 1978), a blood-sucking midge, whose saliva induces an immediate-type hypersensitivity. Stock should not be grazed over wet areas where the midges are found and they should be stabled before dusk. Some control is achieved with antihistamines.

Flies

Several species of fly are more of a nuisance than a direct cause of trouble. The warble fly (*Hypoderma lineatum*) can cause some damage, mainly in young horses when the larva penetrates the skin of the legs and wanders under the skin to the back. When it is nearly ready to emerge it should be poulticed. The screw worm fly (*Callitroga hominivorax*) does not occur in Western Europe and has probably been eliminated from the USA. It causes wounds in the skin in which it lays eggs from which the larvae hatch. Direct treatment of the wounds with insecticide is appropriate. Several species of botfly (*Gastrophilus*) are widespread. The adult lays eggs on the breast and around the mouth and gums of the horse. The larvae are swallowed and attach by hooks to the stomach or small intestine, detaching when fully grown and pupating in the manure. Some control can be achieved by sponging areas of the skin on which eggs are attached to hair, using water at a minimum of 49°C (120°F), and where necessary ivermectin or other insecticidal anthelmintic can be given orally.

General stable hygiene is a major factor in the control of all flies, including the immediate removal of dung, contaminated feed and bedding. Routine grooming may assist not only by removing potential trouble but by ensuring that there is a regular scrutiny of the horse's coat.

Blister beetles

Blister beetles (*Epicauta* spp. and *Macrobasis* spp.) are lethal when ingested by livestock. Various species are distributed throughout Canada and the USA. They range from 0.8 to 2.7 cm in length and they may be black, black with grey hairs, black with red or yellow contrasting stripes, yellow with black stripes, metallic green or purple. They travel in swarms and feed on flowering plants, such as lucerne or clover. When lucerne hay is harvested, the insects can be crushed and incorporated in the bale. Upon ingestion cantharidin, an extremely stable toxin, is released for which there is no known remedy. It is claimed that 6 g of the beetle are sufficient to kill a horse. Cantharidin causes severe inflammation of the oesophagus, stomach and intestines and during urinary excretion causes severe irritation to the urinary tract. The horse develops colic and dies within 48 h.

Hay baled in July and August is more likely to be infested than that cut earlier. If the insects are detected, they may be present in considerable numbers as a consequence of their swarming nature. By knocking biscuits of hay before feeding at least one or two may fall out from infested material. The existence of the risk is, however, not a justification for the exclusion of lucerne hay from the diet.

WORM INFESTATIONS

In temperate countries helminth infection (helminthiasis) in horses is limited to GI nematodes, including lung worm, and to liver fluke (trematodes). In tropical countries, however, horses suffer spirurid and filarial infections as well. Foals may be heavily infected with migrating large strongyle nematode larvae with a prepatent period of 6–12 months and therefore injurious infections can be present for many months before eggs are detected. Adult horses may become severely parasitized by migrating larvae, even if wormed, when sharing pastures with horses that are not wormed.

The determination of the severity of worm infestation is no simple matter. Faecal egg counts simply reflect the presence of egg-laying worms. The only reliable means of establishing the degree of parasitism by GI nematodes is to analyse serum proteins. Alpha and beta globulins peak in concentration 6 months after infection and thereafter the latter of these proteins declines. There is a coincidental depression in serum albumin and eventually in haemoglobin. Worm egg counts are, however, of use in assessing anthelmintic efficacy in control schemes. Raised eosinophil counts reflect only migrating larvae so that these counts may not differ between treated and untreated animals. The counts tend to be highest in July and August in the northern hemisphere and so are not diagnostic. Both small and large infections with strongyles cause an elevation in the immunoglobulin IgG (T) concomitantly with depressed serum albumin before a patent infection occurs.

GI parasitic nematodes

The eggs of parasitic nematodes pass out in faeces from adults in the gut and hatch into the first and second larval stages while contained in the droppings on pasture, e.g. *Strongylus* spp. These stages are not infectious (some species undergo maturation through all three stages within the egg, e.g. ascarids (*Parascaris equorum*), and so are more resistant to deleterious environments, and may survive for years). Larvae of the third larval stage of *Strongylus* move out onto the surrounding blades of grass and are infectious. These developments require moisture. Once eaten, larvae of the third stage pass through two more stages before becoming adult, mating and beginning to lay eggs. The complete cycle takes about 8 weeks in the summer, the rate depending on the ambient temperature. As autumn approaches an increasing number of the parasites stop development and hibernate in the gut wall, to emerge again in the spring. The larvae of some nematode species migrate within the body, passing along blood vessels through the liver and lungs, causing damage. Larvae in the pasture can survive winter to infect horses the following spring. Most parasites are 'host specific', that is, parasites of sheep and cattle will not generally infect horses, but those of donkeys will infect horses.

Strongyloides westeri *(threadworm)*

Threadworms are very small worms that live in the small intestines of foals. Foals are infected shortly after birth, either by ingestion of colostrum and milk containing the larvae, or by the larvae penetrating the skin. Heavy infestations cause sufficient damage to the intestinal lining to precipitate diarrhoea, loss of appetite and dullness. Foals are susceptible up to 6 months of age, after which, usually, they will have developed immunity.

Parascaris equorum *(large roundworm)*

Large roundworms can reach a length of 50 cm and their lifecycle is 10–12 weeks. Eggs containing infective larvae are picked up from the pasture, or from contaminated stable bedding. The larvae migrate through the vascular system to the liver and lungs, before returning to the small intestine, where they develop into adults and lay eggs. Adult worms arrest growth and harm the appearance of foals. Heavy infestations can block the gut and migrating stages in the lungs cause 'summer colds' with fever, coughing and loss of appetite.

Cyathostome spp. *(small redworm)*

Small redworms are major parasites. They are very small and live in the large intestine. The life cycle is 5–18 weeks and infective larvae are eaten with the pasture during the summer. The larvae of these worms may be predominant in pastures in the UK during the summer. The seasonal emergence of huge numbers of larvae from the gut mucosa during the late winter and early spring can cause debilitating acute diarrhoea, weight loss, colic and even death. They may be the most common cause of diarrhoea among adult horses in the UK. Horses infected with larval cyathostomiasis, or *Strongylus* spp., frequently show elevated serum β-1-globulin concentrations. Inflammation caused by cyathostomes can also elevate α-2-globulin; and leakage from the gut causes depressed serum albumin levels (protein-losing enteropathy). These facts may be helpful in diagnosis of horses that have recently received anthelmintic treatment, and which may not be passing larvae. Most anthelmintic treatment will not affect larvae already encysted in the mucosa. A 5-day treatment with fenbendazole has shown efficacy. A good-quality dietary protein will assist compensation for blood albumin loss.

Strongylus spp. *(large redworm)*

Large redworms are reddish-brown worms, 2–5 cm long and have a life cycle of 6–11 months. The larvae have been detected in smaller numbers on pastures in the UK in recent years. Picked up from the pasture, the larvae of *Strongylus vulgaris* penetrate the gut wall and migrate to the main coeliac artery, where they are responsible for severe damage and blood clots. These thrombi can become dislodged and then block smaller branches of the artery. This interruption to blood supply typically causes colic.

Oxyuris equi *(pinworm)*

Pinworm females are up to 10 cm in length. The life cycle is 4–5 months. The adults migrate to, and lay eggs on, the skin surrounding the anus, causing irritation. Rubbing of the anal region causes the opening of wounds, the hair to be removed and the eggs to drop off to the stabling and pasture, from which they are picked up.

Dictyocaulus arnfieldi *(lungworm)*

The life cycle of lungworm is 2–4 months. Infective larvae are picked up from the pasture. The larvae are swallowed and migrate through the bloodstream to the lungs, where they develop into adults. Eggs are laid there, coughed up, swallowed and passed out in the faeces. Although donkeys frequently act as carriers, they and foals rarely show signs. The examination of faeces for larvae is useful in detecting carrier animals responsible for spreading the infection. Most horses possess some resistance and do not develop patent infections. However, where they do, the prepatent period before larvae may be found in the faeces is 3 months. The larvae may remain in a state of retarded development and hosts can elicit persistent coughing for periods exceeding 1 year in adults, during which time resort to veterinary lavage of the trachea for the detection of larvae in the washings is a rational move in seeking proof of the cause, even though the larvae are not readily demonstrated. Although lungworm evoke eosinophilia there is no detectable change in serum proteins. Management of lungworm infection requires location of the carrier which may be a donkey, or unhealthy mare, shedding faecal larvae without showing signs. This animal should then be removed and treated with effective anthelmintics. Fluke and lungworm are not as economically significant among horses in the UK as GI nematodes.

Trichostrongylus axei *(stomach hairworm)*

Stomach hairworms have a life cycle of 3 weeks, they live in the stomach causing damage and irritation and they are able also to infect cattle and sheep. Larvae of this parasite have been found to assume major importance on pastures in the UK from August until October.

Habronema muscae *(large-mouthed stomach worm)*

The adults of the large-mouthed stomach worm live in the stomach. Eggs are passed out in the dung, where they hatch and are picked up by fly maggots feeding in the dung. The larvae are carried in the mouthparts of the fly. As the fly feeds it passes the larvae to the feeding horse, which swallows the larvae. Larvae deposited on sores and wounds of the horse's skin do not complete their life cycle, but cause intense irritation and 'summer sores'.

Onchocerca *spp. (neck threadworm)*

Adults of the neck threadworm live in tendons and ligaments. The larvae (microfilariae) live under the skin and in eye tissue, and are taken up by feeding midges. Microfilariae in the eyes cause problems.

Other worm parasites

Gastrophilus *(bot fly)*

The bot fly lays eggs on the legs and face of the horse. These hatch and enter the mouth, where the larvae live in the tissues of the lining and the tongue for several weeks before entering the stomach. The larvae attach to the stomach wall, where they remain until the following spring. These larvae can cause ulceration and perforation of the stomach wall during this time. They then pass out in the faeces. The larvae mine underground and pupate. The adult flies emerge during the summer to lay eggs. The lifecycle is 1 year.

Anoplocephala perfoliata *(tapeworm)*

Horses with obstructions of the ileum and caecum are frequently found to harbour the tapeworm, and horses with concurrent infestation run an increased risk of ileo–caecal colic (Proudman & Edwards 1993). Tapeworm eggs are passed out, contained in proglottides (segments), in the dung. The eggs are consumed by free-living oribatid mites, which are, in turn, eaten with the grass by the horse. The adult worms attach to the wall of the intestines at the junction of the small and large gut.

Control of GI worm parasites

Control requires an effective worming programme and good pasture management. The essence of control is to reduce the number of infective larvae on pastures grazed by susceptible stock, particularly those under 3 years old. There is some evidence that a tolerance is developed to both strongyle and ascarid nematodes so that stock should not be kept entirely isolated from infective sources. Faecal egg counts reflect only the activity of adult worms in the intestines and may not give a good indication of the seriousness of a strongyle infection.

Horses should be treated orally with anthelmintics on arrival at a stable or as directed by the veterinary surgeon; however, an initial dose, much larger than normal, of an effective wormer may be prudent in cases of severe infection with strongyles as such doses of thiabendazole, or fenbendazole, can be larvicidal. Veterinary guidance is essential. Moderate but closely defined doses of oxfendazole or ivermectin (Dunsmore 1985) have been shown to possess efficacy against adult ascarids and small and large strongyles at all stages from egg to adults, including migrating larvae. Ivermectin also controls botfly lavae.

Young stock should always have access to the cleanest pasture until they have developed some tolerance to worms. Mares must therefore be properly treated so that they do not pass on any severe infection to their offspring; droppings can be removed expeditiously by the use of vacuum cleaner attachments to tractors. Stabled horses should also be treated routinely, particularly where they have been given access to pasture, even for short periods, in the summer. Table 11.1 gives a simple routine of treatment. The life histories of two species are indicated in Fig. 11.1.

Strongyles *(redworm)*

Strongyle eggs develop into infective larvae only in the period between March and October, especially in warm weather. Infective larvae can, however, survive the winter in the UK , but in the spring there is a rapid disappearance of these larvae from pasture with increasing ambient temperature. Overwintered larvae die out by June. Nevertheless, in the early grazing season this source augments that from eggs passed by other horses throughout early grazing. The high level of infectivity accumulating during the summer on pasture can be contained by regular anthelmintic

Table 11.1 Treatment programme for GI parasites in northern latitudes.

	Treatment	Purpose	Additional activity
January February March	C for 5 days B double dose	Encysted small redworm Tapeworm	Faecal examination of all stock
April	A, B or C every 6–10 weeks	Grazing season	
May June			Faecal examination of early foals
July August			Faecal examination of late foals
September October	B double dose A or C	Mid-October to end of December migrating large redworm	Faecal examination of all stock
November	C	Late October to end of December encysted small redworm	
December	A	Bot	

A, avermectins; B, pyrimidines; C, benzimidazoles.

dosing at 4–6-week intervals, which complements the management procedures given in Chapter 10. For strongyle control there is little point in dosing foals less than 2 months old as the prepatent period of small strongyles is 8–10 weeks and the developmental stages are not susceptible to most anthelmintics in the normal dose range. Badly infested pastures may need ploughing and reseeding, or at least should be rested till June by which time overwintered larvae will have largely gone. However, young stock should not be given access to them until July or August before which time grazing should be restricted to cattle or sheep.

Ascarids (large roundworms)

Ascarid infection is common in horses under 3 years old, by which time a considerable measure of resistance will have developed. Foals are especially susceptible and it is thought that nearly all become infected without necessarily developing signs, owing to anthelmintic control measures and increasing immunity. The migrating larvae damage successively the liver and lungs within 14 days of infection. Eggs occur in the faeces from 80 days of age. Eggs acquired by the young foal through coprophagia of the dam's droppings are normally immature and pass passively through the foal's intestines.

Clinical signs of severe infection include pyrexia, coughing, nasal discharge, nervousness, colic and unthriftiness. To preclude this, foals should be treated at 4-week intervals from 1 month of age for the control of the intestinal stages. Eggs can remain viable on pasture over winter and in suitable conditions some may

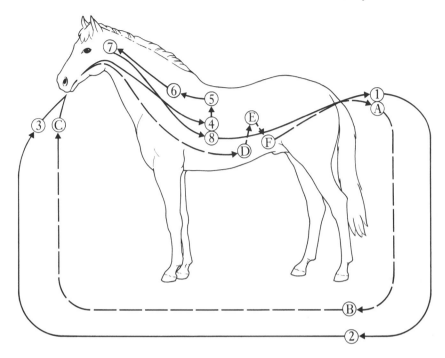

Ascarids

Parascaris equorum
(in young stock)

(1) Immature eggs appear in droppings from 12–15 weeks after infection
(2) Eggs may become infective in pasture or stable within 2 weeks, or remain dormant for up to several years
(3) Mature eggs ingested by susceptible foal or yearling
(4) Second stage larva hatches from egg and penetrates intestinal wall
(5) Larva reaches liver from 7 days after ingestion of infective egg
(6) Larva reaches lungs from 7–14 days after ingestion
(7) Larva coughed up and swallowed
(8) From fourth to approximately thirteenth week larva grows and matures in small intestine and begins shedding eggs 12–15 weeks after infection

Strongyles

Strongylus vulgaris
(in young and adult stock)

(A) Non-infective eggs shed on pasture
(B) Larva matures within egg case on the pasture during the grazing season
(C) Third-stage infective larva ingested
(D) Larva penetrates submucosa of intestine within a few days
(E) By 14 days after ingestion larva has reached the anterior mesenteric artery where it develops over a 4-month period.

Fig. 11.1 Life history of a roundworm (*Parascaris equorum*) and large redworm (*Strongylus vulgaris*).

persist in the environment for many years. The pasture management advised for strongyles is also applicable to the control of ascarids. For a more detailed discussion of parasitic worm control in the stud the reader is referred to Rossdale and Ricketts (1980).

Various wormers have a different spectrum of species against which they are effective, differences in activity against adults and larvae and differences in the number and frequency of dosing. Inadequate dosing can lead to the development of resistance to wormers, and small redworms are now widely resistant to the benzimidazoles. Three families of wormers are used:

(A)   Avermectins
(B)   Pyrimidines
(C)   Benzimidazoles.

There are several principles to worming:

- Know the weight of each horse so that correct dosages may be given.
- Treat any horse on arrival at a new establishment and keep it stabled separately for at least 2 days.
- Treat all horses grazing together at the same time and with the same product. If they are at pasture during the winter continue treatment during that period.
- Keep a diary of the date and product used for each horse.
- Do not overstock paddocks.
- Worm horses 48 h before moving them to clean pastures.
- Collect dung from the paddocks 2–3 times per week. This is probably *the most important procedure* in parasitic worm control. Moreover, strict adherence to this procedure can increase the grazing area by 50%, by eliminating the characteristic separation of horse pasture into roughs and lawns.
- To rest a pasture do not graze it with horses from autumn until the following mid-summer. Where possible use cattle or sheep to clear infected pastures of parasites.
- Rotate wormers used in the grazing season on an annual basis, not every time the horses are wormed. The rotation should be based on changing from *one to another* of the three chemical groups A, B and C listed above.

LIVER FLUKE

Reference to the infectivity of liver flukes (*Fasciola hepatica*) in horses was made in Chapter 10. Their presence may be detected by faecal egg counts and their influence by liver-function tests, indicating liver damage. Untreated cattle and sheep encourage their spread and snails are an essential intermediate host.

AILMENTS RELATED TO DIET

Microbial spoilage of feeds

(See also Chapter 5 for feed storage.) Bacteria grow in feeds where the moisture content is over about 16%. This may result from poor drying or from secondary water uptake in humid atmospheres and condensation on the surface of feeds. Cereal grains that have been ground or crushed, or feeds, such as bran, with a large surface area are more susceptible to bacterial and mould growth.

Endotoxaemia and laminitis

Evidence shows that endotoxaemia and laminitis are related in the horse. The dietary causes of acute laminitis and endotoxaemia are, to a considerable extent, limited to the consequences of an excessive consumption of readily fermentable carbohydrate by horses that have been inadequately adapted to the diet. However, equine laminitis is a local Shwartzman-type hypersensitivity reaction which may be provoked by several other agents that are neither antigenic nor dietary. These could include endotoxin.

Sensitization is a predetermining factor, so a history of laminitis, the endogenous release of corticosteroids in response to stress, previous exposure of tissues to endotoxin, followed by stress, are factors that increase the probability of laminitis occurring following grain overload or an overload of lush young grass. Platelet-activating factor (PAF) is a mediator of endotoxaemia, but whether PAF receptor antagonists play an important role in equine laminitis control is uncertain. Endotoxaemia and lactic acidosis are also implicated in obstructive bowel disease and equine colic precipitated by abrupt increases in the intake of starch and protein.

Endotoxaemia

The horse is particularly sensitive to endotoxins and endotoxic shock can be fatal. Endotoxins are lipopolysaccharides (LPS) which are a structural component of the outer bacterial cell wall of the gram-negative, nonsporing rods of the Enterobacteriaceae, including *Escherichia coli*, inhabiting the intestines. LPS are habitually present in the intestinal contents (as much as $80\,\mu g/ml$). In fact, repeated administration of sublethal doses of LPS results in attenuation of the host response. Moreover, both early- and late-phase endotoxin tolerance have been recognized (Allen *et al.* 1996). This tolerance not only provides protection in some individuals, but also may point to ways of providing prophylaxis. However, large doses of LPS are toxic.

During grain (soluble carbohydrate) overload there is a rapid increase in numbers of lactic acid-producing bacteria (species of anaerobic lactobacilli and streptococci) concomitant with a decline in intracaecal pH, which may fall from 7 to 4 within 12–24h. Starch-fermenting organisms grow much more rapidly than do those that ferment cellulose, and the starch fermenters proliferate at their expense. Organisms that use lactic acid as a source of energy are not present in sufficient numbers to cope with the surge and their numbers may decline as some are unable to withstand the very low pH values attained. The ciliate protozoa, which are much larger and slower growing than bacteria, normally engulf starch, fermenting it at a relatively slow rate. They then act as a starch reservoir, preventing an excessive rate of bacterial starch fermentation, but they are also killed by the acid environment and so no longer function as a buffer. As these normal homeostatic mechanisms are destroyed, acid production proceeds at an accelerating rate. The author's own evidence (Frape *et al.* 1982) showed that numbers of protozoa increased with increasing starch intake up to a threshold, beyond which there was a precipitous decline in numbers.

Following the death of many of the Enterobacteriaceae, large amounts of LPS are released into the gut lumen and the integrity of the colonic mucosa is frequently lost. This is likely to be caused by intestinal ischaemia, perhaps abetted by the lactic acidosis, or by previous parasitism. Concentrate overload has led to concentrations of endotoxin between 1 and 30 μg/ml in the small intestine, but in the large intestine the concentrations can be up to 160 μg/ml. Consequently there is a considerable transmural movement of LPS, leading to their presence in portal and systemic blood.

In health, endotoxin is cleared rapidly by the Kupffer's cells of the mononuclear phagocytic system within the liver and so plasma levels are normally <0.1 ng/l. The overwhelming of this system, leading to endotoxaemia (plasma levels of 2.5–82 ng/l with two peak concentrations 32 and 48 h after carbohydrate overload), is associated with early systemic vasoconstriction, tachycardia, arterial hypoxaemia, hyperpnea, respiratory alkalosis, pulmonary hypertension and fever, followed by an increase in vascular permeability, haemoconcentration, systemic hypotension, an alteration in mucous membrane colour, a prolongation of the capillary refill time, capillary thrombosis, thrombocytopenia and neutropenia with neutrophil sequestration in blood vessels. The decrease in capillary blood flow causes decreased blood perfusion of vital organs, but there is an increased perfusion of the GI tract. The decreased capillary flow and slow refilling of the capillary bed is associated with cold extremities. Incomplete perfusion of the lungs through capillary shunting causes incomplete oxygenation and lowered oxygen tension of the blood (hypoxaemia). This response, added to the restricted blood flow through other tissues and organs, including restricted extraction of lactate by the kidneys, aggravates the situation and contributes to anaerobic glycolysis and further production of lactic acid. A restricted hepatic blood flow reduces the removal and metabolism of lactic acid and diarrhoea causes dehydration, contributing to haemoconcentration.

Endotoxaemia is said to occur in at least 25% of horses with colic admitted to clinics (Moore 1991), and the effects of endotoxin on the colon are very different from their effects on the small intestine. The decreased perfusion of tissues, with a shift to anaerobic metabolism, causes blood lactic acid concentrations of >700 mg/l, which are associated with death in LPS shock. Experiments in which these toxins have been administered i.v., or intraperitoneally, at rates of 2–30 μg/kg BW, have led to an increase in inflammatory circulating eicosanoids (thromboxane, TXB_2, prostacyclin, 6-keto-prostaglandin $F_{1\alpha}$, and PGE_2) (Ward *et al.* 1987; King & Gerring 1991). The eicosanoids are derived from arachidonic acid, mobilized during damage by endotoxin to the endothelium of blood vessels (see Chapter 5, polyunsaturated fatty acid supplementation). Horses with intestinal ischaemia have a plasma endotoxin concentration in the range 30–100 ng/kg BW. Plasma endotoxin of 0.1 μg/kg BW (100 ng/kg) leads to an increase in body temperature of approximately 1°C. The synthesis and action of inflammatory mediators are central to the aetiology of endotoxaemia, and potent cyclo-oxygenase inhibitors (which inhibit conversion of arachidonic acid to eicosanoids) have been an effective therapy, if administered promptly. The feeding of fish oil (probably 500 g/day plus vitamin E) as a prophylactic may help.

Where grain overloading has occurred, and endotoxaemia is likely, early therapy is most rewarding. Treatment should also be directed towards preventing laminitis and may include:

- cyclo-oxygenase inhibitors;
- fluid replacement (this is important in treatment, including glucose, and correction of any bicarbonate deficit, after measuring acid–base status). Oral fluid therapy may be adequate in mild cases, but i.v. administration of physiological solutions will be necessary in horses severely affected;
- colloid solutions (increased vascular permeability and loss of blood colloid is a characteristic of endotoxaemia);
- evacuation of the starch overload from the GI tract. Mineral oil, administered by nasogastric tube, assists by decreasing bacterial fermentation, it may slow the absorption of endotoxin and it facilitates evacuation of colonic contents;
- virginiamycin treatment (see 'Lactic acid production and measurement', this chapter). This requires further investigation to exclude potential adverse outcomes.

Dietary prevention of grain overload
The underlying processes in grain overload indicate that feeding methods should be imposed to foster dietary health. By slowly increasing the concentrate ration, the bacteria that ferment lactic acid and the protozoa that engulf starch are encouraged to multiply (Table 11.2). These increased numbers of organisms act as a substantial buffer against a decline in pH of the large intestine. The concentrate portion of the ration should never be increased by more than 200 g/day for a 550-kg horse. This means that 40 days should elapse in raising the concentrate portion of the diet from 0 to 8 kg. Potter and colleagues (1992a) found that low intakes of almost any source of starch leads, primarily, to its digestion in the small intestine, but with larger meals starch spills over into the large intestine. In order to prevent digestive dysfunction resulting from starch overload to the small intestine, starch intake in horses, given two or three meals daily, should be limited to approximately 0.4% of bodyweight per feeding. Where the starch is relatively insoluble a lower percentage would apply. Processes that gelatinize the starch grain, such as micronization, enhance small intestinal digestion of that starch at moderate and high rates of intake and so those processes reduce the risk of overloading.

Table 11.2 Daily ration to be divided into at least three feeds for horses and ponies with laminitis.

	Mature weight (kg)	Concentrate	Concentrate per day (kg)	Grass hay per day (kg)
Pony	200		—	3.50
	300		—	4.25
Horse	500	Micronized barley	4.5[1]	*Ad libitum*
		Oatfeed or molassed chaff	1–2	
		Soya/micronized soya	0.5	
		Molasses (beet or cane)	0.4–0.5[2]	
		Limestone flour	0.05	
		Potassium chloride	0.03	
		Sodium bentonite[3]	0.05	
		Stabilized fish oil[3] + vitamin E	0.35	

[1] Micronized barley is digested precaecally to a greater extent than micronized oats or other cereals.
[2] If molassed chaff is given, provide only 0.4 kg molasses as well.
[3] These should be helpful in the prophylaxis of laminitis in horses and ponies prone to the disease. The use of fish oil should be considered experimental.

Lactic acid production and measurement

The shock of endotoxaemia results in increased anaerobic glycolysis in muscles. The product of this is L-lactic acid, associated with metabolic acidosis. Reliable measurements of this acid require careful problem-free sample collection and immediate mixing with cold perchloric acid or sodium fluoride. To measure blood D-lactic acid (see below) the samples should be treated with perchloric acid and analysed using D-lactate dehydrogenase. Alternatively, plasma anion gap [sodium – (chloride plus bicarbonate)] measurement is free from these problems and is a slightly better prognostic indicator than is lactic acid measurement. It is a good measure of the accumulation of acid anions (e.g. lactate, ketoacids, phosphate) generally (Gossett *et al.* 1987), and is suitable for routine application.

Lactic acid is, of course, normally produced in the muscles during anaerobic work, as described in Chapter 9. Why then is the lactic acid produced in the intestine potentially more lethal? First, anaerobic work can be sustained for only a few minutes, after which aerobic conditions lead to the complete metabolism of the lactic acid. In contrast, lactic acid fermentation may persist for 24–36 h. Second, at least ten species of *Lactobacillus* and *Streptococcus bovis* produce lactic acid as either a racemic mixture of the D(−) and L(+) forms, or in the D(−) form (a few produce the L(+) form), whereas that produced in the muscles is solely of the L(+) type. This latter type is dehydrogenated by lactic dehydrogenase, with the formation of pyruvic acid. However, lactic dehydrogenase in the muscle is unable to catalyse the dehydrogenation of the laevorotatory D(−)isomer, which therefore accumulates in tissues following absorption and so can exacerbate the effects of endotoxin, as its protracted existence leads to greater tissue damage. Horses need a fully functional liver to catabolize large amounts of lactic acid and those that have suffered liver damage through disease and infestation, and from bad feed management, are less able to cope with rapid increases in the dietary energy allowance. The lactic acid accumulating in the hindgut can also induce diarrhoea. A granular form of virginiamycin (4 or 8 g/kg feed, with 8 kg feed/day), which retains activity in the hindgut of horses, has been shown to suppress D(−) lactic acid production.

Clostridia

A comparable but less-frequent problem may also result from the excessive consumption of concentrates. It favours the rapid multiplication of clostridia and equine intestinal clostridiosis (enterotoxaemia) has been described, owing to the rapid growth of *Clostridium perfringens* type A or D in adults and *C. perfringens* type C in foals. *Clostridium perfringens* which secretes an enterotoxin is a normal inhabitant of the gut. In small amounts this enterotoxin is apparently harmless, but when there are large numbers of this species, excessive gas is produced and the toxin causes damage to the intestinal mucosa and precipitates diarrhoea. The toxin is neutralized by antibody, but immediate therapy involves the replacement of depleted tissue water and electrolytes and the relief of GI tympany (ballooning).

Laminitis

Laminitis may be caused by overwork, concussion of the feet, infections, abortion, high fever, drug-induced complications and the consumption of certain toxins, especially where liver function is abnormal. However, by far the most common cause (compounded by inadequate exercise) is the overconsumption of concentrates or of lush young grass by animals unaccustomed to them. After a grain overload, horses seem to be more likely to survive severe cardiovascular stress than are ponies. Ponies seem to have a greater risk of laminitis than do other equine animals. Geldings may be more prone than mares, according to a survey in England (J. Ridgeway personal communication). This survey recorded that 70% of the cases occurred among horses on pasture, particularly in the months of more rapid grass growth (April, May, June and October).

Garner *et al.* (1977) reported that following carbohydrate overload the greatest increase in blood lactate normally precipitates circulatory collapse and death with or without symptoms of laminitis. Lesser increases in blood lactate are more frequently associated with laminitis and the lowest increases with neither effect (circulatory collapse or laminitis). The development of laminitis is the most frequent outcome of lactic acidosis. In their study, 70% of the cases developed this condition, whereas only 15% suffered fatal cardiovascular collapse. Garner *et al.* (1977) concluded that the rate of increase in blood lactate, as determined by blood measurements at 8 and 16 h, gives a fair indication of whether the horse will die or will contract laminitis, and so it provides the basis for appropriate therapy. Other work has shown that where arterial plasma lactate exceeds 8 mmol/l, death is inevitable, whereas survival is probable with maximum values below 3 mmol/l (Coffman 1979c).

A survey of 108 laminitis cases by Slater *et al.* (1995) indicated that GI disease, occurring just prior to its onset, is a very frequent cause of acute laminitis. Of the 35 acute cases the predisposing health problems were: colic 23%, grain overload 23% and grass founder 8%. Generally, the principal associated causes are:

(1) fat ponies on lush pasture;
(2) carbohydrate overload;
(3) endotoxaemia;
(4) excessive weight bearing on a sound leg and postexhaustion myopathy;
(5) stress of exercise in overweight animals;
(6) excessive tube feeding of sick, aphagic horses with a high carbohydrate–protein diet.

Causes (1), (2) and (6) are very likely owing to the production of lactic acid of the D(–) or DL type in the large intestine from rapidly fermentable carbohydrate.

The first scientific investigation of laminitis was conducted by Obel in 1948, whose name remains linked to the scale describing the severity of lameness. Obel's four grading scales for laminitis are:

• Obel Grade I: feet lifted incessantly and alternately; no lameness at a walk, but at a trot horse moves with a short stilted gait.
• Obel Grade II: horse is willing to move, but gait is characteristic for laminitis. A forefoot may be lifted without difficulty.
• Obel Grade III: horse moves reluctantly and vigorously resists attempts to lift a forefoot.
• Obel Grade IV: horse moves only when forced.

Laminitis of Obel Grades I to IV occurs with an unerring frequency following endotoxaemia. Laminitis is the local manifestation of a serious metabolic disturbance. The front feet are frequently the most severely affected, when the animal adopts a straddled stance with the front feet pushed forwards. Often a strong digital pulse can be detected at the fetlock; sometimes petechial haemorrhages are observed in the buccal cavity and oedema seen along the underline of the belly. Abdominal pain (colic) accompanies a loss of intestinal borborygmi and ischaemia. Chronic inflammation of the hoof initiates a more rapid growth of the hoof wall than would otherwise occur so that the toe of the hoof extends and curls up, heavy rings developing on the wall in response to inflammation of the coronary band. Infections can focus on cracks developing between the hoof wall and the sole. Where the angle between the hoof wall and the surface of the third phalanx (pedal bone), measured on X-ray photographs, is excessive the prognosis is doubtful. The malformation can be partly overcome by rasping and corrective shoeing and any overweight condition should be rectified.

The primary lesion in laminitis is the opening of arteriovenous anastomoses, causing inadequate perfusion of the dermal laminae. The endotoxin-induced formation of microthrombi and venous thrombosis contribute to this failure in circulation. Increased blood flow to the foot, recognized by a bounding digital pulse, thus occurs with a shunting of blood away from the laminae. The ischaemia, and lack of nutrition, of the laminae eventually causes a degeneration of the bond between the hoof and the pedal bone (third phalanx), which sinks and rotates, owing to the horse's weight, initiating a chronic phase of the disease. In the acute phase it is therefore essential to correct the loss of blood to the laminae in order that chronic damage is arrested.

Treatment

In cases of grain overload, mineral oil administered by nasogastric tube may slow the absorption of endotoxin. Evidence of pedal bone (distal phalanx) displacement and rotation should be sought and the animal should be placed on sand, or mud, or have frog supports bandaged to the sole. Marked displacement is an indicator of a likely fatal outcome to laminitis. Shoeing, sole support and trimming were commonly part of the treatment protocol for chronic cases. The starch content of the diet should be reduced abruptly and replaced by good-quality hay to decrease hindgut fermentation rate. Total feed withdrawal should not occur, as this may cause hyperlipaemia, especially in ponies. Where the dietary Ca:P ratio is low, the addition of limestone to the diet may have both therapeutic and prophylactic effects.

Severe laminitis accompanying endotoxaemia and intracellular lactic acid accumulation is associated with loss of K from muscle cells into the plasma, decreasing the intra- to extracellular K concentration ratio. This causes membrane depolarization, but urinary K excretion may not necessarily increase (nevertheless, assessment of urinary K:creatinine clearance ratios may be useful). K depletion may lead to vasoconstriction of muscle capillaries, causing local ischaemia and hypoxia, anaerobic glycolysis and metabolic acidosis. This sequence may be a factor in equine exertional rhabdomyolysis syndrome (ERS), discussed in this chapter. The fact that both normal and abnormal plasma concentrations of K have been detected in ERS may simply reflect the difficulty experienced in assessing K status and its *in vivo* intra- to extracellular ratio. The ischaemic damage to muscle cell membranes results in excessive loss of cellular K. Thus, plasma K may be normal, raised or lowered. The maintenance of normal cellular K concentrations depends on the integrity of energy yielding systems. Intracellular K leaks from red blood cells into plasma in the absence of adequate glucose and analyses of whole blood should be conducted within 2 h of collection. In acute cases of K loss, i.v. dosing of limited amounts of K is appropriate, but must be carried out slowly while cardiac action is continuously monitored. Abnormal electrocardiographic changes can occur at plasma K concentrations of 6.2 mmol/l and severe cardiotoxic effects have been reported at 8–10.1 mmol/l. Molasses (cane or sugar beet) and potassium chloride are recommended to be given with the hay in cases of K depletion, as the vasodilatory effects of K could be helpful.

Fat ponies and horses with a history of laminitis should be given grass hay and should not be turned out onto lush pasture. Individuals presenting signs of laminitis should be removed from pasture, or concentrates, and the feed then restricted to coarse hay (Table 11.2). The hooves should be radiographed for evidence of pedal bone rotation and the individual placed on soft sand or wet soft earth, or frog supports may be applied. Severe restriction of the energy intake of ponies may precipitate hyperlipidaemia.

Nitric oxide

Fundamentally acute laminitis is a vascular disease associated with areas of ischaemia or haemostasis within the hoof. A key to this is a failure of the arginine-nitric oxide (NO) system. NO is produced by the action of NO synthase on its substrate, the amino acid L-arginine. NO relaxes vascular smooth muscle to cause vasodilatation. L-arginine administered i.v. at the rate of 0.42 g/kg BW, as a 10% saline solution, given, in turn, at the rate of 1 mg/kg BW/min, caused immediate reperfusion of laminal tissue in an acutely laminitic pony (Hinckley *et al.* 1996). The involvement of NO was further implied when glyceryl trinitrate paste applied topically to the pasterns of an acutely laminitic pony reduced the 'bounding pulse' in the treated limbs, reduced lameness and lowered systemic blood pressure. It is too early to suggest that a dietary supplement of L-arginine for animals at risk may be a preventive approach. L-arginine is likely to be a semidietary essential amino acid in the horse. This means that the horse may not be able to synthesize sufficient from other amino acids. The dietary requirement is probably in the region of 25 mg/kg BW daily, so that daily supplements of the order of 5–10 g for a 500-kg horse might be considered. This is only 3% of the dose given to a laminitic pony and so may have no preventive effect for this and other reasons.

Thyroxine (T₄)

Abnormal blood T_3 and T_4 concentrations have been observed in horses affected with laminitis. The values may be depressed during the 2 days prior to the onset of lameness and horses with chronic laminitis have elevated serum T_3 levels. These effects, and a reduced insulin sensitivity, are considered to be a consequence of laminitis and not a cause. In cases of equine goitre associated with depressed plasma T_4 levels the recommended approach is to replace existing feedstuffs with feeds of known quality and to treat affected animals with thyroxine. The reason for this is that any hypothyroidism may have resulted, among other causes, from excessive, or inadequate, dietary iodine. If either of these is a cause it should be established by dietary analysis before dietary changes are invoked.

Some other causes of lameness

Lameness during training is not clearly related to diet, although poor Ca status can increase the risk of stress damage to long bones. Hardness and other features of the surface on which the horse is trained seem to be critical. A good well-formed turf is protective, probably through a greater compliance compared with dirt. Moyer *et al.* (1991) reported that the incidence of dorsometacarpal disease (buck shins, sore shins, shin splints and stress fractures) was less in horses training on wood fibre than in those training on dirt. The wood fibre was a more compliant surface. There was a decrease of approximately 10% in stress and strain on the metacarpus (force per unit area) during fast work in horses on the wood fibre.

Osteochondrosis

Osteochondrosis (OC) is discussed in Chapter 8. The effect of OC in a joint on susceptibility to lameness depends greatly on the joint affected. Some reports on horses in training indicate that OC of the tarsocrural joint causes a degree of movement disturbance, whereas others have reported no relationship. OC of the stifle joint, on the other hand, is more frequently associated with clinical signs of lameness.

Colic (abdominal pain) and related disorders

Many colics involve the presence in the stomach, or intestines, of a thick sticky mass of fermenting feed or a compacted mass of roughage. Colic may wax and wane in concert with intestinal smooth muscle contractions and the pain is present in several abnormal conditions. As this implies no diagnosis it is apposite to discuss the various types and causes of colic and the management favouring a healthy prognosis. Probably all equine animals experience colic several times in their life, so that in various degrees of severity it is very common and in its most severe forms is associated with disorders which are the most common causes of death. Records show that 80% of cases recover spontaneously within 1–2 h, but in the remaining 20%, unless immediate action is taken, a disturbance that may initially be mild can become fatal. Colic usually accompanies a rise in blood lactate, and the severity and outcome are closely correlated with this increased value. Lactate concentrations in the peritoneal fluid are also typically higher than in the blood, except for cases of impaction.

Most colics are characterized by some of the following postures and reactions in various forms and intensities: tail twitching, pawing the ground and restlessness in which the horse may get up and down frequently, playing with its food and water, submersing the nostrils and blowing bubbles, and generally losing appetite, the head is frequently turned towards the flanks and, in the extreme, the horse rolls and thrashes about, risking further damage. However, one might enter the box to find the horse cast, with no intestinal sounds, no droppings, or a very few small ones, and a much distended abdomen. Frequent staling (urination) may be attempted in an endeavour to relieve pressure on the bladder. The rapidity of heart beat and respiration rate and the extent of sweating and fever will depend on the severity of the disorder. Normal heart rate lies generally between 38 and 40 beats/min but the rate may rise to 68–92/min in moderate colic and to over 100/min in severe pain. Similarly, respiration rate, normally 12–24/min, may exceed 72/min and the normal body temperature of $37.7 \pm 0.3°C$ ($100 \pm 0.6°F$) will be elevated. Other signs can include diarrhoea with undigested cereals in the faeces, a foul-smelling breath, ingesta in the nostrils, frequent stretching, and on some occasions skin changes in the form of a nettle rash. Capillary perfusion time is increased as measured by thumb pressure on the gum, after which the white patch regains its colour over a longer period than the normal 1–2 sec. Dehydration is also expressed as a delay in the return of the skin to its normal posture after being pinched.

Gastric ulcers

(See also gastric lesions in foals, Chapter 7.) More than half of the TBs in training may suffer from gastric ulceration. Lesions lie in the nonglandular pars proventricularis, particularly the region adjacent to the margo plicatus. Lesions were present in the gastric mucosa of nearly all TB horses in training examined by Murray *et al.* (1996) and those in the glandular mucosa were much less severe than those in the squamous mucosa, which became particularly severe as time in training and racing progressed. Signs generally include periprandial colic, bruxism, ructus and reflux. It is considered that gastric acid hypersecretion, gastric emptying disorders and disturbances in gastric mucosal blood flow are potentially involved in the initiation of gastric ulceration. Stress-associated catecholamine secretion may result in sufficiently frequent vasoconstriction, hypoxia and inanition of the mucosa to precipitate this lesion. After ulceration gastric acid prevents healing of the mucosa and therapy has successfully included H_2 receptor antagonists. Although there is some association of gastric ulceration with the consumption of large quantities of concentrate feeds, the absence of some environmental stresses among horses on pasture may contribute to a lower prevalence in that environment. The absence of frequent stress reactions, together with the lower dry matter and pH of gastric contents in horses receiving bulky feeds, could be important preventative factors. Gastric ulcers in human subjects have an association with the presence of *Helicobacter pylori*. The higher postprandial gastric pH in concentrate-fed horses may be more conducive to the survival of this, or a related, bacterium, although it tends to be protected from an acid environment by urease secretion. If survival is enhanced by a raised pH then the use of H_2 receptor antagonists could be counterproductive in the longer term.

Ill-health associated with physical and microbial quality of feeds

The physical and hygienic quality of all feeds used for horses is important in health maintenance. If the concentration of moulds, yeasts and bacteria in cereals is high then there is increased risk of digestive disturbances and of respiratory disease. A high feed yeast population may be associated with increased risk of gastric colic and tympany. A high feed level of lipopolysaccharides (Enterobacteriaceae) is associated with health disturbance, and *Salmonella* in silage can cause fatal colic (see Chapter 10 for botulism). Chopping hay and straw too short can increase the risk of colic, and excessive stickiness of concentrate feeds (e.g. high gluten content) seems to cause gastric colic. Moulds, including ergot, can cause severe respiratory disease (Meyer *et al.* 1986).

Oesophageal impaction (choke)

Greediness, poor dentition, inadequate water, foreign bodies and the consumption of coarse bedding may predispose the horse to oesophageal impaction, that is, a foreign object, or feed, lodging in the oesophagus. Normally the obstruction will clear after a while and it may be common in individuals possessing gullets with an abnormal structure. Long-standing impaction can result in chronic damage to the wall of the oesophagus and spasm, causing recurrence of impaction, dysphagia, coughing, and regurgitation of food from the nostrils and mouth. There may be enlargement of the cervical oesophagus. Food withdrawal and immediate veterinary action is vital for successful management. In simple cases sedation and nasal intubation with a small warm water lavage and external massage frequently works. Rehydration with isotonic fluids given i.v. and other veterinary treatment follow.

The obstruction is sometimes caused not by physical impaction of the oesophagus, but by an absence of adequate *free* saliva in the throat. Thus, the faster a horse eats, the more likely is choke to occur. The implication is that softer feed pellets and sugar beet pulp pellets are more likely to be causes owing to the absence of large amounts of free saliva, in comparison to the effects, in horses with good teeth, of a harder pellet that requires more chewing, or whole cereal grain that does not absorb much saliva.

It has been calculated that the intestinal tract of the average horse holds about 100 l fluid, which in water deprivation – as may be induced by choke – is drawn on to maintain homeostasis. If choke prevents feeding for more than 6–7 days, dehydration precipitates prerenal azotaemia (raised blood urea). Once the obstruction has been removed the horse should be watered and fed several times per day with small quantities of wetted nuts or other feed, and stones the size of tennis balls should be placed in the feed box to retard the rate of feed consumption. Some scarring of the oesophagus will have occurred, which may cause repeated trouble. These horses should be given soaked feeds.

Gastric impaction

There is little evidence concerning the causes of gastric impaction, although the ingestion of coarse roughage and inadequate water intake may contribute to it. A stomach tube is used to allow the expulsion of gases and to permit the administration of antifermentatives such as chloral hydrate or turpentine (an oil obtained from various species of *Pinus*) in raw linseed oil, although surgery may be needed. Liquid paraffin may be given by a nasogastric tube for impactions at rates of 2–6 l/500-kg horse once or twice per day for several days, and 0.5–1 l of raw linseed oil (acting as an emollient cathartic) may be used in obstinate cases together with warm salt water to stimulate thirst. For flatulent colic 30–60 ml turpentine may be added to the oil.

Intestinal impaction

The aetiology of ileal impactions is also unknown, although ascarid impactions in foals and tapeworm infestations are less common causes. Large amounts of coarse roughage feeds or excessive cereal intake may contribute to the risk, when ingesta may be present in the nostrils. Normal gut movement in cases of impactions of the large bowel is encouraged by cold-water enemas and massage of any impaction at the pelvic flexure via the rectum. The general use of tranquillizers and pain-killing drugs removes the need for continued forced exercise to prevent the animal from damaging itself and they reduce the risk of a simple impaction becoming a volvulus (twisting of the intestine on its mesenteric axis). Walking may only be necessary to distract the horse's attention where drugs are unavailable or ineffectual. Quick action in mild colic can forestall a more serious derangement precipitating endotoxin shock and death.

Spasmodic colic

In spasmodic colic there is an increase in bowel movement, which may be precipitated by a sudden change of feed, work and chilling. Spasms may last for a few minutes or up to half-an-hour, and may occur repeatedly over a period of hours, the signs being typically those already discussed under 'Colic', this chapter. Recovery occurs without treatment, but relief of pain and the use of spasmolytic drugs is helpful in amelioration. The colic is associated with increased parasympathetic tone and, from experience, flighty horses are subject to spasms of this kind with any abrupt change of feed which should, therefore, be made with extra caution.

Ileal impaction (ileus)

Ileal impactions are defined as intestinal obstruction, with the accumulation of fluid and gas, as a consequence of loss of smooth muscular action with its peristaltic movement of digesta. Solid food, or water, given orally in cases of small intestinal ileus will aggravate gastric distension and should be withheld. Therefore isotonic i.v. fluid treatment together with intestinal lubricants may be indicated. Surgical intervention is usually necessary where obstruction is complete.

Parasitic worms

Strongyle larvae cause damage to the lining of blood vessels, particularly that of the anterior mesentric artery and its branches, and this can lead to various degrees of occlusion and inhibition of blood flow (ischaemia). A thromboembolism can be a major contributory cause of the complete loss of blood flow to, and death of, a portion of the intestinal tract, leading to obstructive colic. Where the blockage is incomplete, recurrent colic will be experienced. In acute cases surgery is needed, but thorough worming at 30–60-day intervals will help to contain the situation and a defined dose of specific wormers (Table 11.1) under veterinary guidance will have some beneficial impact on the larval stages. (These stages cause a rise in intestinal alkaline phosphatase activity of peritoneal fluid; see Chapter 12). In young horses, impaction of the small intestine with ascarid worms can occur where management is bad, necessitating immediate anthelmintic treatment.

Large intestinal impaction

About 30% of all colics are in the form of intestinal impactions, and of these, most impactions occur in the large intestine. As briefly mentioned in the first chapter, these are located typically either at points where there is a change in the diameter of the colon, or at flexures where it turns acutely. More frequent sites are the pelvic flexure and where the right dorsal colon empties into the small colon, but occasionally the sternal and diaphragmatic flexures may be involved. Impaction may also occur at the ileo–caecal valve. The closer to the ileum the large intestinal blockage occurs, the more dangerous it is as it will severely restrict water resorption in the caecum and ventral colon, which can lead to dehydration and hypovolaemic shock. The horse suffering colonic impaction will frequently look at its flank, emit no intestinal sounds and void small mucus-covered droppings; palpation of the impaction is frequently possible. Inadequate water intake, excessive sweating and hard exercise together with excessive amounts of coarse roughage are also thought to contribute to colonic impactions. An association with positive *Salmonella* spp. faecal cultures has been observed, which may indicate a reaction to bowel inflammation. Impactions of the caecum seem to be related to several disease conditions including endotoxaemia. Impactions are fairly common in old horses with bad teeth restricted to poor-quality hay with little water after experiencing lush grass. Impaction in the small colon may also occur in foals between 2 and 6 months old when roughage feeding is initiated.

Sand colic

Intraluminal obstruction can also result from concretions of hair and of plant material, and combinations of these, and from the horse chewing objects in its environment. These concretions typically occur in the small colon and form as a mineral precipitate on the surface of the material during passage through the colon, over an extended period of time, before the obstruction occurs. The clinical signs include abdominal discomfort, distension and straining to defaecate. Treatment for small colon impactions includes the veterinary introduction of gentle warm water enemas with a lubricated nasogastric tube through the rectum.

Sand colic is probably the most common cause of colic in areas of very sandy soils. Some horses contract the bad habit of consuming large quantities of sand and soil. The problem therefore can frequently recur and is associated with periods of inappetence, diarrhoea, anxious pacing up and down with groans on lying down, pawing the ground or a crouched stance with a turned head. Sand may be present in the droppings. Treatment includes repeated large doses of liquid paraffin. Occasionally, however, enteroliths (large stones) are formed, apparently on a nidus of ammonium magnesium phosphate. This gradually enlarges and its removal requires surgery. Observation may indicate that the horse has a predilection for chewing some material in the surroundings. The solution could then lie in changing the environment and reducing boredom.

Foal colic

Foal colic is very common in the first 48h after birth and is caused by the meconium blocking the large intestine at various levels. Lubrication of the impacted mass with orally administered liquid paraffin (200ml) or glycerol, the use of enemata of soap and water, and the relief of discomfort are normally sufficient remedies. If no response is registered within a few days, volvulus or intussusception may be suspected. Abdominal pain can also occur at this age through rupturing of the bladder, which is effectively repaired by surgery, and in older foals discomfort may coincide with eruption of permanent teeth. Umbilical hernias can cause colic in young foals, but these usually correct themselves by 6–8 months of age. Inguinal hernias, especially in colts, also have similar effects.

Oral and dietary treatment

Unless one is very familiar with the sequence of events in a particular horse, veterinary help should be sought immediately signs of colic are initiated. Feed should be removed *until the cause has been determined,* but clean water should be provided. If the animal shows signs of injuring itself through violent actions the horse may be walked; otherwise leave it alone in a box free from projections or structures that might endanger it. In all colic cases the horse should be kept warm in cold weather and during recovery a warm bran mash is helpful. The general use of a mineral oil or a kaolin–pectin paste is safe and will accelerate the expulsion of the offending masses. General treatment also has the objective of preventing rupture of some part of the GI tract, or displacement of its parts by control of pain and tympany, by evacuation of the bowels, by arresting rapid bacterial fermentation and by re-establishment of normal peristalsis. An enema of 9–13 l (2–3 gallons) warm soapy water is frequently given in addition to the lubricant mineral oil for intestinal impactions. A mild soapy water enema is particularly helpful in cases of colonic impactions in foals. Where there is impaction of the large colon, or caecum, oral fluids, including solutions of electrolytes and glucose, may be allowed, but solid food should not be given until the impaction is passed.

From the dietary point of view therapy includes withholding normal feed, but allowing access to water if there is no nasogastric reflux. Rehydration of the horse both per os (in the absence of reflux) and i.v. is essential and laxatives are normally given by nasogastric tube. With severe intestinal impactions, a 500-kg horse should be given 6 l fluid every 2 h through an indwelling nasogastric tube. Mineral oil can be administered to facilitate passage after the impaction begins to resolve. Following resolution of the impaction, and in the absence of perforations, feeding should be resumed gradually to avoid an immediate recurrence. As soon as GI function resumes mashes of bran are preferable for the first 24–48 h. Following this, leafy hay may be introduced gradually, along with concentrate mashes, or short grazing spells on a halter should be offered. With prolonged dysfunction partial or total parenteral nutrition may be required.

Forage chopped very short is said to increase the risk of ileal obstruction. Grinding of forages reduces digestibility, as it tends to accelerate rate of passage through the GI tract. The main advantages of pelleting may be a reduction in dustiness, increased keeping quality of ground material and a reduction in bulk for storage.

Horses with sand impaction of the large colon respond to psyllium hydrophilic mucilloid, given at the rate of 400 g/500-kg BW daily for 3 weeks, divided among three daily feeds. Flavoured psylliums are available for reluctant patients. Over the period it should remove most of the sand. This treatment may require repeating every 4–12 months. Horses likely to eat sand should be kept on a thick sward and hay should not be offered on the ground, but from racks, in order to reduce the risk of recurrence.

Gas or flatulent colic and its treatment

Gas or flatulent colic may be secondary to an obstruction or an impaction and is extremely painful. Distension of the small bowel is rarely noticeable and if the abdomen is unusually large, tympany of the ventral colon may well be present. Sometimes on post-mortem examination several regions of the GI tract appear to be involved. An impaction and lack of movement of the intestines inhibit expulsion and minimize absorption of gas into the blood. The latter route of removal is more important than may be realized as about 150 l carbon dioxide and methane can be absorbed daily from the intestinal tract.

If gastric tympany occurs, the condition becomes evident within 4–6 h of eating. Intubation with a nasogastric tube is essential to relieve the pressure, which can otherwise cause rupture of the stomach. A warm 4% salt solution, which decreases the viscosity of the fermenting mass and encourages water consumption, is sometimes administered in small quantities at intervals by allowing it to drain into the stomach through the tube. The horse may have adopted a typical dog-sitting attitude, or it may stand without moving the feet, especially where a rupture or a serious intestinal infarct exists. After stomach rupture, in particular, ingesta are sometimes observed in the nostrils. With different intensities of gas colic the horse may feel cold, despite experiencing a fever, it may exhibit congested mucous membranes of the eyes and have a sour and vinegary breath. Powerful anodynes prevent violent rolling and self-inflicted trauma. Where there is risk of violent action the horse should, if possible, be kept on its feet, but quiet horses are best left alone.

Gas colic is frequently the sequel to a rich diet of cereals, or lush legumes, or even to the inadvertent consumption of a pile of grass cuttings. Quick action is essential, as again violent reactions on the part of the horse may lead to ruptures or to twisting of the gut, with a poor prognosis.

Colic associated with torsion, twists and rotations

Torsion (rotation on its own axis), volvulus (twisting on the mesenteric axis) and intussusceptions (infolding), usually of the terminal ileum into the caecum, all require immediate surgery with rather poor prospects. The membranes of the eyes and lips are typically dry and pain induces a rapid rise in pulse and respiratory rates so that the loss of carbon dioxide causes alkalosis despite a raised blood lactate.

Other colics

Abdominal pain is not the reserve of the intestinal tract and may result from bladder and kidney stones, urinary infections, pericardial effusions and sometimes liver disease.

Predisposing factors in colic

- 'Overheating' – sudden access to large quantities of cereals or stands of green clover and lush grass.
- Stress caused by changes in routine, changing stables, and mares and foals in new surroundings.
- Irregular work, horses standing idle on full feed or changes in the timing of feeds.
- Working a horse on a full stomach. Even during protracted slow work large feeds should not be given and bulk feeds should be excluded. Bulky feeds should be given in the evening, waiting until the digestive powers have been restored by rest.
- Work itself can precipitate colic, especially towards the end of an exhausting day – both feed and work should be regular. Extended work should be interrupted by short rests every 2–3 h when a few mouthfuls of concentrated feed and water are provided.
- The consumption of excessive amounts of cold water after severe and hot work before the horse has cooled down and/or providing heavy feed at this time.
- Insufficient good-quality roughage or mouldy corn and mouldy silage.
- Large quantities of cut green feed.
- Unfit horses changed abruptly to an increased work rate and a concentrate-rich diet.
- Failure to provide fresh clean water at all times.
- Lack of teeth care. 'Quidding' may be noticed in which small balls of partly chewed feed are dropped into the manger. This is usually associated with cheek teeth that require rasping.
- Greedy feeders that bolt feed, or greedy bullies in group-fed herds.
- Inadequate work control.

Prevention

- Each horse has its idiosyncrasies so that before a new horse is placed gradually onto a rich working diet its habits and particular requirements should be studied and understood.
- Hard-worked animals should receive a small feed of concentrates at frequent and regular intervals, and the time of meals should be constant, even at weekends.
- Increased demands for energy should be met by an increased feeding rate of concentrates of no more than 200 g/day in a 500-kg horse and a proportionately lesser rate of increase in smaller horses.
- A sensible and regular exercise programme should be instituted.
- Where a horse is changed from one stable to another, its old routine should go with it and if necessary be changed gradually.
- All animals should be checked last thing at night.
- Good-quality roughage free from contaminating weeds, and in the USA also free from blister beetles.
- Stores of horse feeds and dangerous chemicals should always be held in rooms, the doors of which cannot be opened by horses.
- No horse should be overworked and, following strenuous work, no substantial feed or water should be given until the animal is cool and rested and then should only be given in moderate amounts.
- Cribbers or wind suckers (swallowing of air into the stomach) (Plate 11.1) should be fitted with cribbing straps, or, if cribbing is known to precipitate colic in the individual, surgery may be necessary.
- Teeth should be inspected at regular intervals. These may require the filing of the upper and lower molars and premolars or removal of teeth with infected roots – decaying teeth may be inferred from excessive salivation.
- Individuals that bolt their concentrate ration should have it mixed with chaff or dry bran. The placement of stones the size of tennis balls in the manger may retard the rate of feed consumption.
- A proper worming programme is essential in all horses and a pasture rotation system should be instituted where horses have access to grass.

Overfeeding

In addition to the consequences discussed in this chapter, rank overfeeding can have a number of other deleterious effects in horses. Some of the more obvious are listed below:

- Obesity. This is said to reduce fertility and to present difficulties at foaling in mares, to affect the work rate of horses and to accelerate the onset of fatigue.
- Obese horses that suddenly experience food deprivation, as in a drought, may be subject to anorexia secondary to colic, which sometimes causes hyperlipidaemia (high concentration of blood lipids). Pony mares in late pregnancy or in peak

Plate 11.1 A 'wind-sucker' mare (cribber) – a vice in which the incisor teeth grip a solid object, the horse pulls down and swallows gulps of air. Sometimes a leather strap is fastened snugly around the neck just behind the jaw to deter the horse from the practice.

lactation and subject to pasture changes are prone to this problem. Treatment can be problematic and certainly requires veterinary advice. Anorexia may also be a sequel to acidosis.

- The overfeeding of young horses, (a) in particular of colts, (b) when the food is given in separate and discrete meals, and (c) where the ration is unbalanced in respect of its mineral content, may cause bone disorders. The principal one is epiphysitis – typically of the distal epiphyses of the radius, metacarpus, tibia and metatarsus – and recognized by bony enlargements and lipping of the physes.
- Contracted tendons, as previously discussed in Chapter 7, may be associated with overnutrition of energy-rich feeds in late pregnancy (producing too much milk), or in the foal and yearling.
- Enterotoxaemia in young horses fed in groups is occasionally precipitated in the largest and most aggressive individual. A flatulent colic occurs in which the intestines are loaded with rich feed and gas. Symptoms of dyspnea and subcutaneous oedema may be presented and the cause is apparently a rapid proliferation of the bacterium *Clostridium perfringens*.

Hyperlipaemia

Hyperlipaemia is a clinical condition of which its subclinical form is known as hyperlipidaemia. In the horse most of the circulating fat is in the form of very low density lipoprotein (VLDL, of hepatic origin), especially in the postabsorptive state of hyperlipaemia. The clinical disorder is characterized by depression, anorexia, elevated plasma TAG concentrations, lipid infiltration of the liver and hepatic failure. In large horses azotaemia also occurs. Various physiological stresses, fasting, obesity, pregnancy, lactation and depressed feed intake predispose horses to the disease. Ponies and the donkey are more susceptible to the condition, possibly owing to a greater likelihood of insulin resistance. Insulin is required for the activation of lipoprotein lipase, required for TAG clearance from the blood into adipocytes. Tube feeding with readily digestible carbohydrate in relatively small amounts, initially, can be instigated if the animal will not eat voluntarily. Insulin therapy (monitoring blood sugar) is frequently used to promote carbohydrate metabolism and fat clearance, and to reduce the activity of intracellular lipase which mobilizes stored fat to form VLDL (Table 11.3).

The disorder can lead to organ failure and mortality rates of 65–80%. The aetiology is apparently different from the condition in man, where the function of the enzyme lipoprotein lipase is impaired. In the horse, overproduction, rather than defective catabolism, of VLDL is the cause of hyperlipaemia. Treatment should therefore be directed towards use of lipid lowering agents that reduce VLDL synthesis. Hyperlipaemia in the pony is accompanied by an elevation in plasma FFAs, without ketosis, which may provide the stimulus for TAG synthesis and the secretion of the VLDL. Thus, there may be scope for the prevention and/or treatment of ponies by lowering this hepatic VLDL production with agents such as niacin, as this reduces adipose tissue lipolysis and FFA flux (Watson *et al.* 1992).

Botulism (forage poisoning in adult horses and shaker foal syndrome in foals)

(See silage/haylage, Chapter 10.) Characterized by dysphagia, weakness and progressive flaccid paralysis, botulism is caused by the exotoxin of *C. botulinum* which interferes with the release of acetylcholine at the neuromuscular junction. Wound botulism has rarely occurred in horses. In adult horses the disease is caused by the consumption of the toxin, but in young foals the toxin can be elaborated by organisms present in the GI tract.

Table 11.3 Glucose and insulin therapy for hyperlipaemia.

	Day 1	Day 2	Day 3	Day 4	Day 5
Insulin (iu/kg) intramuscularly twice daily	0.15	0.075	0.15	0.075	0.15
Glucose (g) per os twice daily	100		100		100
Glucose (g) per os once daily		100		100	
Heparin (iu/kg) intramuscularly twice daily[1]	40–150	40–150	40–150	40–150	40–150
Sodium bicarbonate and fluid[2] i.v.					

[1] Blood clotting ability should be ensured and maintained, especially if larger doses are used. Pharmaceutical doses of heparin release lipoprotein lipase into the blood.
[2] Blood acid–base status should be checked first for a base deficit. PCV and blood urea should be measured and elevated values rectified with Ringer's solution.

Treatment of infected foals and adults is difficult and requires intensive care with the use of a specific antitoxin. This treatment will not reverse the effects of toxin already bound at the presynaptic membrane and it should be introduced before recumbency occurs. Tube feeding will be necessary together with measures to prevent aspiration pneumonia, constipation, corneal ulcers and gastric ulceration. Mineral oil may be added to feed if constipation occurs. Foals are optimally protected by vaccinating the pregnant mare, with the annual booster given 1 month before foaling. This will protect the foal until 2–3 months of age. The foal should then receive the three-dose toxoid series starting at 2 months of age.

Idiopathic colitis (of unknown causation), malabsorption of nutrients and chronic diarrhoea

Severe colitis has been produced in ponies by oral treatment with the antibiotics clindamycin and lincomycin, followed by the intestinal contents of horses dying from naturally occurring idiopathic colitis (Prescott *et al.* 1988). Treatment of three further ponies with 25 mg lincomycin only/kg BW twice daily for 3–5 days caused death. A clostridium closely resembling *Clostridium cadaveris* was isolated from the colon of each of these ponies and from one of six horses dying from idiopathic colitis, but not from horses with nonfatal diarrhoea. These data may indicate the importance of maintaining a large mixed culture of symbiont bacteria in the large intestine by mixed feeding which prevents potential pathogens achieving rapid unrestricted growth.

Chronic diarrhoea is normally of large intestinal origin resulting from some upset to the normal balance of the intestinal flora. It can follow stress such as the prophylactic use of oxytetracycline or some other antibiotics. Where the diarrhoea is of small intestinal origin, it may be connected with a want of certain digestive enzymes detected by xylose and other tolerance tests (Chapter 12). It should also be remembered that the adult horse loses the ability to digest lactose when about 3 years old so that large intakes of milk sugar after this time may induce diarrhoea. Chronic diarrhoea can also be prompted by parasitism, mesenteric abscesses or some disorder of vital organs.

The loss of integrity of the gut mucosa associated with protein-losing gastroenteropathy commonly causes chronic diarrhoea. The loss will reduce the efficiency of net absorption of energy and protein sources as well as those of minerals, trace elements and vitamins. The diet should be of the highest quality and rich in these essential nutrients, including leafy hay, to compensate for a reduced efficiency of use. Water and electrolyte losses of young stock suffering from diarrhoea can lead to a rapid decline in vigour unless fluid loss is continually replenished. Salmonellosis, colitis and many other causes of diarrhoea are associated with rapid transit of digesta through the large intestine. This means that fibre digestion is impaired and that the efficiency of reabsorption of water and Na and K ions is depressed (see Chapter 1). Hyponatraemia, hypochloraemia and hypokalaemia may occur. Blood monitoring and a determination of blood acid–base balance are advised, so that the appropriate electrolyte drinks, and fresh water separately, can be given. In the absence of this and of inclement weather, it may be feasible to turn the horse out to sheltered pasture, as green herbage is a good source of electrolytes. Remember that bacterial diarrhoea may cause a contamination of the pasture for a period.

Acute diarrhoea

(See also Chapter 7.) Severe diarrhoea is frequently caused by salmonella infection precipitated by stresses of transport and particularly by strongyle worm infection. Antibiotic treatment to eliminate salmonella is of questionable value and in fact oxytetracycline may trigger the onset of the infection. Salmonellosis may occur in closely stocked groups after mild winters and is associated with heavy worm burdens. It may be transmitted to foals by adults that are asymptomatic carriers. Suspected cases should be isolated in a box with a very strict programme to ensure that contaminated faeces are not transmitted to other stock. The bacteria are not continuously excreted in the faeces, but careful veterinary examination may be necessary to detect any carriers that are shedding the organisms. Feed and rodents may also be suspect reservoirs of potential infection.

All cases of acute diarrhoea are associated with a critical loss of fluid, K, Na and Cl, and unless replacement therapy is quickly instituted the consequences are rapidly fatal in young stock. Appropriate fluids are listed in Table 9.3. Fluid losses of 20–50 l, Na deficits of 2000–6000 mmol, K deficits of 700–3000 mmol and bicarbonate deficits of 1000–2000 mmol may exist and should be made good in adults and young. Foals may experience absolute losses amounting to 15–20% of that of the adult. In acute diarrhoea where there is hypotonicity and dehydration, the use of hypertonic solutions yields an immediate response. Where the animal is dehydrated and hypertonic, then hypotonic solutions are given. The normal plasma values for Na, K, Cl and bicarbonate in horses are, respectively, 139, 3.6, 99 and 26 mmol/l.

Foal heat diarrhoea

Foal heat diarrhoea occurs typically at the time their dam's first postpartum oestrus is expected. The diarrhoea is normally self-limiting in 3 or 4 days, and is probably a secretory diarrhoea. Hypersecretion in the small intestinal mucosa may overwhelm an immature colon unable to compensate by increased fluid and electrolyte absorption. If a prolonged diarrhoea occurs, fluid and electrolyte losses should be replaced.

Dehydration and potassium status

Although the carcass of a 500-kg horse may contain by weight only 15% as much Na or K as of Ca, on average the 1100–1200 g K are subject to a much greater flux than is the Ca owing to its higher solubility in tissue fluids. The volume and water content of muscle cells and of red cells are modulated primarily by the control of their Na^+ and K^+ contents. The cell membrane is relatively impermeable to small cations, that is, they diffuse slowly, whereas small anions diffuse freely, but haemoglobin acting as a large anion remains as an intracellular entity. However, the equilibrium distribution of charged particles between red cells and plasma differs from that predicted by normal diffusion processes. The slow passive movements of Na^+ and K^+ are balanced by an active outward movement of Na^+ and an inward transport of K^+ in each cell, mediated by several hundred discrete pumps fuelled with ATP. If glycolytic mechanisms yielding ATP break down, or if the cell membrane is damaged such that diffusion leakage increases, then there is a decline in resting membrane potential and the pump mechanism is incapable of maintaining a physiological cellular environment. That an intact glycolytic pathway in red cells is necessary for the maintenance of a physiological cation distribution between cells and plasma has been amply confirmed in numerous experiments.

It is clear that the horse must maintain cellular K within strict concentration limits for normal health to prevail. The measurement of plasma or serum K^+ concentrations, as a guide to body K status, although frequently done, is misleading. Measurements of large numbers of horses have failed to detect any correlation between serum and cellular concentrations of K^+ (Frape 1984b; Muylle *et al.* 1984b) and serum contains on average only 3.7–4.3 mmol K^+/l, that is, 3.8–4.4% of the mean concentration in red cells. In fact, the extracellular fluid of the body in total contains only 1.3–1.4% of the total body K. In maximal anaerobic exercise, serum K^+ tends to increase, whereas it has a tendency to decrease in endurance work without comparable changes in the cells.

In severe diarrhoea, the bodily loss of K by a 500-kg horse may approach 4500 mmol (175 g), associated with a fall in red cell K^+ from 97.5 to 75 mmol/l, but without significant change in plasma K^+ (Muylle *et al.* 1984a). Protracted work in hot weather apparently leads in a horse of this size to losses of K^+ and Na^+ of 1500–1800 and 4000–5000 mmol, respectively. Thus, the measurement of red-cell K^+, the concentration of which appears to be well correlated with that of muscle-cell K^+ (Carlson 1983b), is recognized as a more reliable means of assessing K^+ status and possibly of understanding the underlying processes obtaining (existing) in setfast ('tying up') and azoturia.

A fall in red-cell K^+ concentration below 81 mmol/l is associated with weakness of skeletal and smooth muscles, tremors, and, in severe depletion, with recumbency, cyanosis and eventually respiratory and heart failure. The K^+ ions that are released from muscle cells during hard exercise act as potent arteriolar vasodilators and they stimulate cardiorespiratory reflex activity. Thus, there is a close correlation between the extracellular increase in K^+ and the increase in both muscle blood flow and oxygen consumption and therefore in performance. The blood flow is insufficient in K^+ depletion precipitating hypoxia, anaerobic glycolysis and metabolic acidosis. This turn of events is pathological, and the ensuing damage to muscle-cell membranes leads to further cellular K^+ loss, which cannot be restored by an Na^+/K^+ pump deficient in readily available energy. Investigations in Belgium (Muylle *et al.* 1984b) have revealed an anomalous situation in which about 10% of 436 horses examined possessed a normal red-cell K$^+$ concentration distributed independently of the remainder. Their mean red-blood cell K$^+$ concentration was 83.8 mmol/l, some 13.7 mmol less than the normal for the other 90% despite similarities in management and diet. Other studies by the same group (Muylle *et al.* 1983) indicated that in a smaller sample of 43 horses, 11 were in this low red-cell K$^+$ range, and of these 9 were performing unsatisfactorily on the racetrack and they expressed a more nervous temperament.

A rank dietary K deficiency is unlikely to be more than an occasional cause of K$^+$ depletion, although several investigations implicate dietary Mg deficiency as a cause. A 500-kg horse given a diet composed of a 50:50 mixture of grain and hay may absorb daily 90–100 g K$^+$, well above the maintenance level of requirement. Excess dietary K is rapidly excreted so that a generous dietary content is of no avail in the acute K$^+$ depletion of diarrhoea, or of abnormally high sweat loss, when in any event appetite is depressed. Dosing with an appropriate solution is the only reasonable approach in overcoming the worst of the deficit. K in excess is a moderately potent toxin to heart muscle so that only limited quantities may be given i.v. while cardiac action is monitored. I.v. infusion rates of 11.5–13.7 mmol KCl/min bring about plasma K$^+$ concentrations exceeding 8 mmol/l and consequential cardiac arrhythmias and abnormal electrocardiograms, through a transient but excessive alteration to the gradient of the transmembrane K$^+$ (Epstein 1984). Thus, the bulk of such dosage must be given orally or by nasogastric intubation. By this route Muylle and colleagues (1984a) administered a solution containing glucose (50 g/l), a commercial amino-acid mixture (0.05 l/l), KCl (5 mmol/l), CaCl$_2$ (3 mmol/l) made isotonic with NaCl, which was partially replaced by Na acetate according to the acid:base balance of the horse; the quantities given daily were proportional to the deficit calculated from red-cell K$^+$ values (see Chapter 12 for assessment of red-cell K).

Malabsorption of fat-soluble vitamins

Occasionally the efficiency of absorption of vitamins A, D, E and K is reduced and the most frequent cause may be an interruption to biliary flow by obstruction of the bile duct. The immediate effect is a failure in the blood-clotting mechanism, but this can be overcome by injections of vitamin K. Vitamin K administration is also successful in counteracting the bleeding syndrome of dicoumarol poisoning. Plant sources of this poison are referred to in Chapter 10.

Urticaria

(See also Chapters 5 and 12.) Feed proteins, or peptones, are absorbed from the gut in sufficient quantities to stimulate an immune response with circulating antibodies. However, such antibodies occur in the blood of horses without clinical signs. Feed allergies are expressed as respiratory and/or skin disorder.

Mycotoxins

Depending on the species of the incriminated mould, mycotoxicoses cause digestive disturbances, liver and kidney damage, nervous symptoms, infertility and abortions. The aetiology of several mycotoxins has been discussed in Chapters 5 and 10.

Grass sickness (equine dysautonomia)

Similar signs to choke may be present in grass sickness, but stomach distension is frequently involved, as failure of the normal function of the stomach and of the oesophagus results from neural impairment of the muscles controlling the contraction of these organs. Nasal return of ingesta indicates such oesophageal impairment. Neural toxins in herbage are apparently a cause (see Chapter 10 for current evidence).

Ammonia toxicity

(See also Chapter 2 for treatment.) Ammonia toxicity is much more likely to occur in the ruminant than in the horse, when the source of ammonia is urea taken orally. The reason for this is that the breakdown of urea to ammonia requires the intervention of urease, which is not found in the tissues of mammals, but is secreted by protein-degrading bacteria in the gut. In the horse any dietary urea would be absorbed into the blood before it was significantly degraded by these bacteria. With grossly excessive protein (or urea) intake, or when intestinal haemorrhage occurs, protein (or urea) may be bacterially deaminated to the extent that the portal system and the liver are overloaded, and ammonia spills over into the systemic circulation. Otherwise, the liver will 'mop up' ammonia in urea formation, or in transaminations. Where there is hepatic dysfunction, as in pyrrolizidine poisoning and portacaval shunting of blood, the ammonia will enter the systemic circulation. Its disposal through the kidneys may be arrested by renal pathology. The kidneys would otherwise tend to combat metabolic acidosis by secretion of H^+ ions and ammonia into the tubules, where they combine to form ammonium ions, with reabsorption of Na ions:

$$H_2CO_3 \rightarrow HCO_3^- + H^+ \rightarrow Na^+ + Cl^- + H^+ + NH_3 \rightarrow NH_4^+ + Cl^- \text{ excreted,}$$
$$\leftarrow Na^+ \text{ reabsorbed to neutralize } HCO_3^-$$

Ammonia toxicity has also been recognized in rare cases of unknown aetiology, where equine liver function is normal.

The signs of ammonia toxicity include head pressing, blindness, usually abdominal pain and varying degrees of maniacal behaviour, systemic ataxia and depression. Encephalopathy is the major cause of the behavioural disturbance, which can return to normal following early treatment. Appropriately collected blood samples reveal metabolic acidosis, low plasma bicarbonate (10–15 mmol/l) and hyperammonaemia (150–400 μmol/l). Where liver function is normal hyperglycaemia (15–24 mmol/l) and haemoconcentration have been observed; otherwise with liver dysfunction (detected by raised liver enzymes, bile acids and bilirubin and depressed blood albumin) hypoglycaemia is likely. Blood urea may, or may not, be elevated and blood albumin may be depressed when there has been protein-losing enteropathy, without liver dysfunction.

When the source of the ammonia is the intestines, bacterial action is partly arrested with nasogastric neomycin treatment, and polyionic fluids are given to counteract dehydration. Sodium bicarbonate should be reserved for the most severely acidotic cases and given by slow i.v. infusion, as rapid correction of acidosis may increase the intracellular movement of ammonia. K^+ treatment counteracts the toxicity of ammonia on neuronal cell membranes, and is given as i.v. potassium chloride (10 mEq/h) while monitoring heart rate.

Protein-losing gastroenteropathy

The condition of protein-losing gastroenteropathy is characterized by weight and muscle tissue loss, lethargy and diarrhoea, caused by leakage of plasma proteins into the lumen of the GI tract. There is a loss of integrity of the mucosa of the GI tract through gastric, or colonic, ulceration, GI parasitism and enteritis caused by various bacterial infections. It is crucial that the cause of the protein-losing condition is determined and that the appropriate veterinary therapy is instituted. Of the plasma proteins, virtually all the albumin and fibrinogen and 60–80% of the globulins are synthesized in the liver (the remaining γ-globulins are mainly formed in the plasma cells of lymph tissue). Globulin synthesis is faster than that of the albumin, for which the half-life is longer, i.e. halflife ($t_{1/2}$) 19–21 days. Thus, in chronic cases, especially with hepatic malfunction, the albumin:globulin ratio and the colloid osmotic pressure of plasma both decline, owing to a failure of liver protein synthesis to keep pace with the loss (normal values are given in Table 11.4).

The rate of plasma protein synthesis can be very high, but it depends, apparently, on the level of amino acids, and critically of dietary essential amino acids, in the blood (however, normal postabsorptive plasma free amino acid levels, even with adequate intakes, are extremely variable: lysine is about 15–100, threonine 100–250 and methionine 40–60 µmol/l). The ratio of tissue proteins to plasma proteins re-

Table 11.4 Mean plasma protein concentrations (g/l).

	Birth	3 Weeks	Yearlings	2 Years	3+ Years	Mares at stud
TB						
Albumin	25	25	27–28	32	34	27
Total globulins	20	22	31–35	25–28	25–35	31
Non-TB						
Albumin			29	27	28	34–37
Total globulins			38–44	44	47–49	37–40

mains relatively constant at about 33:1. Amino acids from muscle tissues and from the diet are used by the liver in albumin, etc. synthesis. Thus, losses are buffered, but the net loss will depend to a large extent on the provision of good-quality dietary protein. Diets containing 140g protein derived from soya and legume leaf are recommended.

Protein-losing nephropathy

In severe kidney disease there can be a large loss of plasma protein in the urine. Dietary protein should be of high quality. Until the kidney disease has been re-solved by veterinary treatment there has to be a quantitative balance between the protein intake necessary to stem the renal loss, and the amount that would tax the kidneys in urea disposal, causing azotaemia. Dietary protein with an ideal balance of amino acids will minimize urea production.

Urinary calculi (uroliths)

The chemical composition of urinary calculi varies with diet. A high content of oxalates in pasture (and possibly silica) is a major contributor, and a rise in urinary pH induces precipitation of Ca phosphates, carbonates and oxalates and occasionally of Ca sulphates, particularly where water intake is restricted. When high urinary pH is an immediate cause, the dissolution of calculi is facilitated by decreasing the pH with acid salts such as ammonium chloride (NH_4Cl), or with ascorbic acid. Daily doses of NH_4Cl may be between 45 and 100g. A dose of 0.33g/kg BW has been shown to lower urinary pH from 8 to 6.2. A dose of 500g ascorbic acid on each of 2 days, or 1kg once, both by gastric intubation, effectively lowered the pH to 4.7.

Intestinal stones

(See also 'Sand colic and abnormal appetite', this chapter.) Excessive dietary Mg and P may encourage stone formation in the gut and feeding 200–300g NaCl daily assists in preventing the precipitation of Mg and PO_4 around the nidus of a calculus by stimulating fluid intake.

LIVER DISEASE

(See also seneciosis, pyrrolizidine alkaloid toxicity, Chapter 10.) The liver is central to the intermediary metabolism of both nutrients and nonnutrients, and it is the main line of defence in the detoxification of ingested substances. During browsing and grazing the horse ingests many chemicals requiring detoxification, some of which may damage the liver, leading to clinical liver disease. A frequent example is hepatic cirrhosis, resulting from the ingestion of pyrrolizidine alkaloids (seneciosis) present in *Senecio jacobaea* (see Chapter 10). The risk of damage is increased by several antecedent causes of injury. These include:

- hepatic necrosis caused by *Strongylus equinus* and *S. edentatus* larvae and mouldy hay;
- aflatoxicosis and hepatic lipidosis (fatty change of liver, hyperlipaemia and amyloidosis);
- hepatic metastases from carcinoma of kidney, stomach and pancreas;
- GI disease (obstructive conditions, strongylosis, malabsorption and grass sickness; West 1996) infections; and
- biliary lithiasis.

Liver disease (Table 11.5) causes an inability to metabolize and excrete photodynamic substances in the bile. Reactions to ultraviolet light then occur in unpigmented skin and skin not covered by hair, e.g. the muzzle, eyelids and ears, especially in horses grazing green pasture (also see lucerne, chlorophyll, Chapter 5). Liver disease occurs typically in pregnant mares of 5–14 years of age and Welsh and Shetland ponies are particularly prone. Hepatic lipidosis is a feature of pituitary neoplasia with hirsuitism, laminitis and high plasma cortisol.

Table 11.5 Clinical signs and abnormalities of liver disease (West 1996). (Note: characteristics within rows, i.e. signs and clinical abnormalities, are not necessarily related.)

Signs	Clinical abnormalities
Weight loss	Increased serum glutamate dehydrogenase (EC 1.4.1.3)
Anorexia	Increased protein and neutrophil counts in abdominal paracentesis samples
Dullness and depression	Intravascular haemolysis and haemoglobinuria
Jaundice	Raised haematocrit, owing to dehydration
Tachycardia	Leucocytosis, owing to neutrophilia
Intermittent pyrexia	Elevated serum bile acids[1]
Abdominal pain	High serum GGT (EC 2.3.2.2) in biliary tract damage[2]
Ventral oedema	Elevated plasma ammonia[2]
Muscle fasciculations	Increased prothrombin time[3]
Diarrhoea or constipation	Low plasma fibrinogen and platelet count
Dysphagia	Frequently normal values of:
Photosensitization	plasma urea[4], but may be low, or high in renal failure
Encephalopathy	plasma glucose, but severe hypoglycaemia terminally
Mucosal petechiation and nasal bleeding	hyperbilirubinaemia[5]

[1] Serum bile acids are sensitive measures of early *S. jacobaea* toxicity and with glutamate dehydrogenase, GGT and liver biopsy are useful in diagnosing different types of liver disease.
[2] May be raised in both intra- and extrahepatic cholestasis.
[3] Delayed coagulation is a definite contraindication for liver biopsy, as haemorrhage into the abdominal cavity is likely to result.
[4] Normal ammonia:urea ratio (μmol:mmol) is 3:1. A ratio of 40:1 carries a poor prognosis.
[5] Frequently with acute hepatic failure. High concentrations are a terminal change. The conjugated form rarely exceeds 25% as the equine kidney is only permeable to conjugated pigment. A conjugated fraction of greater than 35% is normally terminal, whereas high total bilirubin, when most is unconjugated, may follow a few days of starvation.

Table 11.6   Composition of wet mash (g/kg diet), 4–6 times daily, for acute liver disease.

Crushed barley	326
Crushed oats	400
Beet pulp	150
Beet molasses	120
Lysine-HCl	2
Vitamins* plus trace elements	0.5
Limestone	1.5
Grass hay	1 kg maximum

*Should include choline chloride, 1 g daily.

Table 11.7   Amino acid supplement (g/500 kg BW daily) for anorectic horses.

L-leucine	6
L-isoleucine	4
L-valine	5

The liver is able to regenerate and the objective in cases of acute liver failure is to provide supportive care until its function is restored. Depending on the cause and development, SDH and AST liver enzymes are at increased concentration in the blood. If biliary obstruction occurs (chronic cases) GGT and ALP are increased and conjugated bilirubin exceeds 25% of the total. In acute cases oral, or i.v., glucose administration is frequently helpful and imbalances in electrolyte, acid–base and hydration status should be gradually rectified. In cases of hypoglycaemia 10% glucose i.v. should be administered to establish normoglycaemia and then 5% glucose, at the rate of 2 ml/kg/h, given for the initial 24 h. Blood glucose should be monitored and the postabsorptive value maintained approximately between 4.5 and 6 mmol/l plasma. Hyperlipaemia may be moderated by oral dosing with methyl donors such as choline chloride, or betaine and a general B-vitamin supplement. If signs of hepatoencephalopathy occur a reduction in the absorption of toxic metabolites from the gut is aided by giving mineral oil (contraindicated in horses with GI reflux) carefully by nasogastric tube, as a laxative, at the rate of 2–4 l/500 kg BW once daily. Horses should be kept out of intense sunlight, to reduce the risk of photosensitization. High protein diets, legume hays and haylage, or silage, should be avoided. A mash, mixed with water, of the composition shown in Table 11.6 and allowed to stand for 30 min before feeding is recommended. Anorectic horses can be force-fed a gruel by nasogastric tube.

Thiamin and folic acid may be given parenterally once per week. There may be a case for raised, or pharmaceutical, doses of niacin to reduce adipose tissue lipolysis and FFA flux (see 'Hyperlipaemia', this chapter). Some benefit may be obtained by including synthetic sources of the branched chain amino acids leucine, isoleucine and valine in the gruel (Table 11.7).

CHRONIC WEIGHT LOSS

Apart from the effects of an inadequate diet and bullying, other causes of weight loss are many, but GI diseases are among the most common. At feeding time the activity of the horse should be closely observed. If the reaction is atypical the mouth should be examined and the teeth checked. If the buccal cavity is normal the problem may still lie in other regions of the GI tract. Increased energy expenditure can result from chronic infections, neoplasia, or chronic obstructive pulmonary disease (COPD) (see this chapter). Therapy for most causes includes correction of fluid deficits, hypoglycaemia and acid–base imbalances and the provision of a high-quality diet given in small amounts at frequent intervals, supplemented with micronutrients. One of the following causes might be suspected in cases of chronic weight loss (protein-losing syndromes are discussed above in 'Protein-losing gastroenteropathy' and 'Protein-losing nephropathy'):

- jaw and dental abnormalities, including sharp points on the upper and lower molars and premolars, or abscesses below the teeth;
- roundworm infestation damaging the gut, which has interrupted the uptake of nutrients, and larval stages injuring mesenteric blood vessels that supply the intestines;
- diarrhoea;
- tuberculosis which very occasionally involves the digestive system;
- liver disease or the presence of some chronic septic focus in the body;
- windsucking;
- shy horses, low in the social order and subjected to group feeding;
- heart abnormalities and anaemia;
- arthritis and other causes of chronic low-grade pain;
- cancer, especially in older horses. Grey horses are prone to internal melanomas.

THE MATURE SICK OR GERIATRIC HORSE

The following points are a guide to the nutritional management of the adult sick horse.

- Provide fresh clean pasture which is sheltered and within observation distance.
- Where concentrate feeds form a major portion of the diet remember that it is necessary primarily to meet the energy requirement. To this end those feeds offered should be edible without difficulty. Grains should be crushed and made palatable with a little molasses. The horse is likely to be fussy and so a range of feeds should be offered to find those most acceptable.
- Include readily consumed bulky treats, such as apples and sliced mould-free carrots.
- Vegetable oil is a useful addition to the feed as it will not cause colic.
- Give B vitamins, including vitamin B_{12} by injection.

- If digestive function is compromised, partial parenteral nutrition (PN) is useful to reduce the load on the GI tract. Total PN may be adopted for a few days. This should be introduced, and subsequently removed, gradually over several days. Where maintenance of fluid status is necessary acid–base status should be assessed and Ringer's, or lactated Ringer's, solution should be given i.v. through a separate catheter.

Many of these points may apply to the old horse. Teeth are likely to be in poorer condition than in younger animals and therefore small quantities of high-quality leafy forage will be required supplementing larger quantities of digestible concentrates. These can usefully include micronized cereals. Ralston and Breuer (1996) reported that a commercial feed containing 85 g protein, 2.7 g Ca and 2.2 g P/kg was inadequate when it represented from 20 to 100% of the total feed, with the remainder being timothy and lucerne hay. A compound feed containing 140 g protein, 6 g Ca and 4–6 g P, given in similar proportions with the hay and providing similar amounts of DE, led to greater bodyweight gains, better condition, higher plasma total protein and P concentrations, and higher haematocrit and blood haemoglobin values. Renal function was not compromised by the larger protein intake, as assessed by plasma creatinine concentration. Where older horses are known to have poor renal function the dietary protein quality, i.e. amino acid balance, should be ideal and the dietary protein allowance should prevent any excessive rise in plasma urea level.

MUSCULAR AILMENTS

There are several metabolic abnormalities connected with muscular ailments, for which many of the signs are shared and attached to which the nomenclature is frequently confused. Previous terms include 'tying-up', myositis, setfast and azoturia.

Exercise-associated myopathy

There is a range of degrees of exercise-associated myopathy, from severe rhabdomyolysis with azoturia, to milder 'tying-up', although the terminology is variable. The onset of clinical signs of muscle disease usually occurs within 5–20 min of commencing exercise which may be either mild or strenuous. Signs after racing are also seen in racehorses. Muscle damage is accompanied by an elevation in serum CK and AST.

Postexhaustion syndrome

Muscle problems occur some time after severe, prolonged exercise which causes exhaustion. Signs may be presented 2–4 days after completion of the exercise. Damage may involve both skeletal and cardiac myopathy, hepatic lipidosis, renal damage, laminitis and GI ulceration. The onset of muscular malfunction may occur within a few initial strides, when the horse falters and stops, generally refusing to move. The muscles are not palpably abnormal, but are stiff and sore. The horse may adopt a cramped stance and if willing to move it does so reluctantly and slowly. Lesser attacks may occur following cooling off. Muscular stiffness, which passes within 1–3 days, is associated with the muscular accumulation of lactate. The activity of serum CK returns to normal after 4–6 days, but that of AAT may take 4–5 weeks to do so. In addition to a depletion of muscle glycogen (principally in the fast twitch fibres), there is also a depletion of ATP and CP. However, muscle glucose as well as lactate concentration is raised, suggesting local hypoxia (insufficient oxygen reaching the muscle cells from the blood). A leakage of the muscle enzymes AAT, CK and LDH is the cause of their raised blood concentration (note: determination of iso-enzyme concentrations may help locate the origin, although, for example, LDH_5 is present in both liver and locomotor muscles).

The horse should be kept warm; effective treatment normally includes the i.v. administration of calcium gluconate, magnesium and phosphate ions and vitamin D. Prevention requires training to increase condition, more frequent rest periods, and the administration of electrolyte solutions during and after physical activity. There is, moreover, some evidence that 'tying-up' is associated with elevated phosphate clearance and that feeding calcium carbonate is helpful. The condition is also occasionally noticed in fillies coming into season (oestrogens increase the activity of 1-hydroxycholecalciferase, which further implicates vitamin D).

After hard work it is an advantage to trot or canter slowly as this will stimulate the transport of lactic acid from the muscles to the liver in healthy horses so that muscle and blood pH return to normal more rapidly. Moreover, this light exercise stimulates the flow of oxygen to the muscles, accelerating the conversion of lactic acid back to glycogen. Accordingly the static changes are avoided.

Equine exertional rhabdomyolysis syndrome

The signs of exertional rhabdomyolysis syndrome (ERS) include apparent anxiety, profuse sweating, elevated pulse and respiration rates, reluctance to move and gait abnormalities, with affected muscle groups often being hard and painful upon deep palpation. There is a severe disintegration of skeletal muscle, associated with the fall in muscle pH which appears to be a cause of coagulation of muscle protein and liberation of myoglobin, which escapes in the urine, colouring it dark red–brown. This in turn can lead to nephrosis and uraemia. However, simple diagnostic methods used to distinguish myoglobin from any urinary haemoglobin contain inaccuracies. The muscle enzyme CK attains a peak concentration in the blood 6–12h after a single severe episode of muscle damage and returns to normal frequently within a week, if no further damage occurs. AST reaches a peak about 24h after the episode, but normal values are not regained for 2–4 weeks. Very high plasma concentrations of CK and raised AST are also observed following strenuous exercise, although they may not represent muscle damage and normal values may be achieved after 24h.

The aetiology of ERS is still a mystery. It may indicate poor management of training and it commonly occurs within the first hour of exercise. The ailment usually does not follow a rest period of only 1 day, or rest as long as 14 days, but commonly occurs following 2 days' rest on full rations. Some forms of the disease that succeed extended exercise may be related to a depletion of glycogen stores and an increased shift to mitochondrial β-oxidation. An interruption in regular exercise, with the maintenance of an immoderate dietary starch intake, seem to be frequently associated factors. Horses may be in a state of dehydration and metabolic alkalosis, or acidosis, and therefore solutions of sodium bicarbonate should not be given unless the acid–base and electrolyte status (see 'Laminitis', this chapter, for K status) has been established. In the absence of this knowledge, isotonic neutral solutions, such as Ringer's, should be given i.v. Diuresis should be induced (and anyway would do no harm) to diminish the risk of nephrotoxic effects of myoglobinuria. It is vital that the horse is not allowed to move at all as recovery requires that complete rest is immediately instituted. It may remain on its feet, or become completely recumbent, and severe pain and distress are often accompanied by repeated attempts to rise. The horse should be removed to its stable as quickly as possible in a low-loading trailer, where every effort should be made to keep it standing by slinging, or other means, as this may prevent the development of uraemia. Not only racehorses but also draught horses on heavy cereal rations and those kept on lush pastures during the week and ridden only at weekends may be susceptible.

The affliction does not seem to be initiated by lactic acidosis, but there does seem to be a loss in the ability of the muscle to maintain appropriate concentrations of free Ca, and intracellular concentrations of this increase, associated with damage to mitochondria and other organelles. One study (Valberg *et al.* 1993) showed that horses suffering from recurrent ERS had a higher ratio of type IIA to type IIB muscle fibres, but this may have reflected better training which caused a lower accumulation of lactate and ammonia during near-maximal exercise (also see 'Ammonia toxicity', this chapter). Blood cortisol and glucose concentrations tend to be higher in ERS horses, although this may be an effect of the stress, rather than a cause of the ailment.

If exercise and movement have been stopped and treatment instituted immediately the horse may recover in 2–4 days. Treatment has included narcotic drugs and corticosteroids administered i.v. to control the swelling and to stimulate energy metabolism, and i.v. or oral administration of electrolytes in large quantities to maintain a high rate of urine flow at an alkaline pH, preventing myoglobin precipitation in the renal tubules. Although acidosis is a normal feature of azoturia this should not be assumed without analysis of plasma acid–base and electrolyte status. Some horses have been found to be alkalotic, when, of course, treatment with sodium bicarbonate would be harmful. Intramuscular injections of 0.5 g thiamin repeated daily seem to be warranted, and the inclusion of pantothenic acid and riboflavin, also involved in oxidative energy metabolism, may be justified. If the painful swelling of the muscles is not reduced, pressure on the sciatic nerve can induce secondary degeneration of other muscles. The maintenance of proper kidney function is vital if health is to be restored.

The risk of ERS is reduced if horses are always warmed-up slowly, their concentrate allowance is halved at weekends, or at other times when they are not worked, and their work is reinstated gradually with a gradual increase in the consumption of starchy and high-protein feeds. The addition of dimethylglycine to the feed, as a prophylactic measure, has been recommended, but without any convincing supportive published evidence. The thyroid hormones, T_4 and T_3, are intimately involved in resting energy metabolism and it has been suggested that transient hypothyroidism may occur in ERS, so that nutritional supplementation with thyroxine may assist recovery.

Recurrent exertional rhabdomyolysis

Horses in training with recurrent exertional rhabdomyolysis (RER) have moderately elevated AST and/or CK concentrations in blood samples taken at rest, even if they have been free of clinical signs for weeks. There appear to be subclinical episodes of rhabdomyolysis after exercise. Muscle fibre necrosis and associated increases in plasma AST, CK and myoglobin occur with exercise more frequently than can be detected clinically (Valberg *et al.* 1993).

Infectious myopathies

Myopathy may result from bacterial, viral or parasitic infection.

Hyperkalaemic periodic paralysis

Recurrent muscle cramping, fasciculations and weakness are associated with hyperkalaemia which is a disease for which horses have a genetic predisposition. The condition is confirmed by oral potassium challenge. There is debate as to whether this is associated with a breakdown in normal postexercise potassium uptake into skeletal muscle under β_2-adrenoreceptor stimulation by epinephrine.

Nutritional myopathy

Horses with nutritional myopathy present with depression, weakness, dysphagia and dropping of the head and neck (also see Chapters 3 (selenium) and 4 (vitamin E) and equine degenerative myeloencephalopathy). Muscle enzymes are elevated in serum and there is severe degenerative myopathy with hyalinization and fragmentation of muscle cells. The condition occurs typically in neonatal foals, but also in adult horses, and is associated with deficiencies of selenium and/or vitamin E.

Stress tetany (hypocalcaemic tetany)

Extended exertion, particularly in hot weather, leads to dehydration, loss of electrolytes and to energy depletion. The signs of fatigue presented may reflect a combination of these losses even though the diet is quite satisfactory. Ca losses in the sweat can amount to 350–500 mg Ca/h, and continued hyperventilation is associated with alkalosis (discussed in Chapter 9). An acute life threat in horses suffering from stress tetany may be posed by hypocalcaemia in which plasma levels can fall to 1.5 mmol/l, whereupon muscular twitching and cramps are manifested. The fall is largely the result of alkalaemia. Occasionally hypomagnesaemia may be present. As normal muscle function, including that of cardiac muscle, requires the concentration of Ca in the blood to remain within strict limits, the most immediate need is for careful i.v. administration of a Ca gluconate solution (Table 9.3) while heart function is monitored, as an excessive rate of administration can be fatal. Where the clinical signs of hypocalcaemia are clear, a greater risk is entailed in awaiting confirmation by laboratory determination so that immediate veterinary treatment of this kind is indicated.

A fall in plasma Ca and of Mg also occasionally occurs in lactating mares, although clinical signs are quite rare (see Chapter 7). Where tetany does occur this is usually precipitated by additional stresses of exertion, transport, weather or disease. Ca gluconate treatment is instituted. In the dairy cow hypocalcaemia causes paresis rather than tetany.

Electrolyte losses in extended exercise

Several days' rest are required for the regeneration of muscle cell glycogen reserves after extended work. The exhaustion of this and loss of cellular potassium in sweat contribute to a sense of fatigue and, as suggested in Chapter 9, recovery during the next few days may be accelerated by providing potassium chloride as well as common salt in the feed (10 g NaCl plus 5 g KCl/kg total feed is recommended). Isotonic dehydration depresses thirst so that rehydration is also stimulated by the provision of these electrolytes. Salt licks are supplied in most horse boxes, but many horses will not consume sufficient in this form and the licks normally contain sodium chloride only. Thus, a powdered feed supplement available from feed merchants containing both salts yields a more satisfactory outcome. In fact, during summer weather generally many horses in training for flat races do not consume enough salt from licks and therefore present a higher PCV and plasma viscosity than should be the case. By encouraging water consumption, electrolytes can have the effect of improving performance. Similarly, during long endurance events those horses that do not drink fatigue more easily and are less likely to finish. Thus, by satisfying a need for electrolytes at rest points, a thirst response is induced, water consumption is increased and dehydration is deferred.

Losses of the electrolytes calcium, chloride and potassium throughout long rides may cause 'thumps' or synchronous diaphragmatic flutter during or after the exercise. Losses of these electrolytes and 'thumps' may also be brought on by severe diarrhoea. A decrease in the plasma concentration of calcium, chloride and potassium is thought to change nerve irritability, initiating a contraction of the diaphragm muscles in unison with that of the heart beat. This is seen as sudden movements of the horse's flanks. Treatment consists of replacing lost electrolytes, which are always beneficial and never detrimental as long as water is available. The quantities to be administered are given in Chapter 9. The exhausted endurance horse is usually alkalotic, contributing to the hypocalcaemia, so that Ringer's solution, which is slightly acid, is preferred for immediate use. Sodium bicarbonate should not be used in these circumstances unless metabolic acidosis has been demonstrated.

LAMENESS

In one survey of 314 TBs, 53% suffered lameness at some time and in 20% of the cases lameness prevented subsequent racing (Jeffcott *et al.* 1982b). Undoubtedly, the condition represents a considerable embarrassment to the industry and is a problem in horses and ponies of all types. Lameness can be defined as a disturbance of gait, which reduces the weight on the affected limb. Although there are a multitude of causes, one study carried out by the author (unpublished observations) on horses in the Far East revealed that faulty mineral nutrition, as estimated by phosphate clearance, all too frequently was associated with vertebral fractures, and probably with fractures of other kinds. Unsoundness of joints may result from sprains, strains and jarring forces causing inflammation, which may also result from osteoarthritis. Although this and many bone disorders, for example, splints, ringbone, osselets, bone spavin, curb, capped hocks and thoroughpin, are unlikely to have a major dietary involvement in their causes, the severity of response may well have dietary implications (see Chapter 5, polyunsaturated fatty acids). Abnormalities in mineral, trace-element and vitamin nutrition are frequently associated with various types of lameness, but little research work (apart from investigations of OCD, see Chapter 8) has been undertaken to achieve any objective assessment of the scale of that involvement. Other reasons for lameness include bruised feet, bowed tendons, navicular disease and spinal lesions, which may in part implicate poor hoof care; inspection of hooves daily frequently avoids long-term problems.

Physitis has already been discussed in Chapter 7 and some reference was made to contracted tendons (Plate 7.1). The latter often occurs in foals that are doing well with mares having ample milk on good grass. The speed of onset in a foal is surprising in that within 24–48 h the heel will rise and a slight concavity develops on the front wall of the hoof. Wear at the heel is decreased and increased tension occurs in the extensor tendons. These do not contract, but apparently fail to develop at a rate commensurate with bone growth, and the fetlock joint also tends to enlarge. It is essential to spot the aberration in the early stages so that it can be counteracted by weekly rasping of the heels, exercise, the removal of concentrates and reduction in milk intake. Ultimately, surgical remedy may be the only means of rectifying severe angular deformities of leg joints. Exercise is important in the prevention of leg-growth abnormalities, which may be more prevalent where mares and foals spend long periods in their boxes during adverse weather without any cut in feed intake. It may be preferable to allow mares and foals to remain out all the time in the summer regardless of weather, so long as some form of shelter is available.

Navicular disease

Lameness and damage to the navicular bone of the hoof are possibly caused by thrombosis of the navicular arteries, abnormal stresses on the bone or by degenerative changes. Treatment includes corrective trimming of the hoof and fitting a wide webbed egg-bar shoe. Drug treatment with isoxsuprine hydrochloride is a common practice as it avoids the risk of haemorrhage encountered with warfarin. Warfarin (dicoumarol) may be helpful, although its use is declining. This drug interrupts blood coagulation by extending prothrombin time. The dose has to be carefully titrated to extend clotting time by 2–4 s from the equine standard of 14 s. An excessive dose will lead to bleeding. If colic inadvertently occurs during treatment, an extended prothrombin time resulting from depressed liver function necessitates a removal of warfarin and the institution of vitamin-K treatment.

It is essential that horses treated with warfarin are fed consistently, particularly in respect of the amount of green feed, which is rich in vitamin K, and as the level of work affects clotting time any change in activity must be imposed gradually. If practical, regular work should be instituted and blood samples taken at intervals of at least one per month during rest periods, but immediately after exercise. Sampling should also occur 5–7 days after any change in work, or feed, routine. Veterinary treatment with warfarin must always err on the side of caution as excess can be fatal. The drug seems to act by reducing the viscosity and increasing the flow of blood, which improves the nutrition of the navicular bone; this is probably more relevant than the prevention of thrombus formation.

HOUSING

Any detailed consideration of housing is beyond the scope of this book, but the environment provided by housing has a profound impact on the wellbeing of horses. The aspects that impact on feed are discussed here. A series of studies by Houpt and colleagues (Houpt & Houpt 1988) at Cornell University, Ithaca, New York has revealed environmental factors important to the wellbeing of horses. For example, mares in visual contact with other mares are less active and spend more time eating than do those without visual contact. Mares were also found to prefer an artificially lit environment to a dark one. Environmental temperature studies with *ad libitum*-fed weanling horses (Cymbaluk *et al.* 1989a) showed that feed intake increased by 0.2% for each 1°C decrease in barn temperature below 0°C.

Ventilation

For reasons of savings in finance and labour there is a trend for housing racehorses in American-style barns (Townson 1992). Less attention to the detail of ventilation is required in the construction of individual horse loose boxes. On the other hand, the communal air and the aerial dust and ammonia levels that can occur at feeding and bedding-down times in barns dictate strict rules that should be followed for ventilation rates, as this is the principal route for evacuating the small particles that are the agents of respiratory distress.

In recommending ventilation rates, allowance has to be made for the fact that air volume per horse in barns tends to be more than double that in loose boxes (mean values 98 cf. 43 m^3/horse; Townson 1992). Natural systems of air flow (Fig. 11.2) should have a controllable inlet area of up to 0.3 m^2 per horse (Sainsbury 1981). If building design impedes natural air flow, fan assistance should be provided at the base of the outlet chimney. This will function as an aid to natural ventilation and is a more desirable solution than the installation of a pressurized system for which the costs are greater and the numbers of suspended air particles increased by greater air turbulence. There should, for similar reasons, be no recirculation of air. Abrupt changes in air flow as a result of 'on–off' regulators are to be avoided. Bottom-hinged air inlets are recommended, which deflect cold air up in cold weather and may be fully open in hot for cross-draughts. Their height above the floor should be such as not to interfere with the stock and they must, of course, be of safe construction. Additional inlets will probably be necessary for very hot weather. Where horses are grouped in barns and covered yards a copious air flow is essential at all times. This will be facilitated by a wide open ridge, 0.3–0.6 m wide with a covering flap. Space boards at the top half of the wall are successful (150-mm wide boarding with 25-mm gaps; Fig. 11.2), or on exposed sites, narrow sliding boards in which one set slides over the other.

The air space in barns is shared, whereas that in loose boxes is either shared or isolated. Although it should be isolated, the tendency is for sharing to be more frequent in modern loose boxes. Probably the most successful housing in the UK consists of various forms of monopitch lean-to buildings (Figs 11.3 and 11.4), which contain single open-fronted boxes facing the warmest wind and the greatest sunlight. These buildings have a low back and high front with hopper-flap air inlets such that the air flows from back to front. An extension to the roof at the front gives cover and cross-partitions act as load bearers. These divisions should be solid as the partly open ones allow crossflow of air, the incidence of wall kicking increases and where the division contains iron bars, even at quite high levels, shod horses can become ensnared.

Dust loads vary considerably, not only among stables, but also within a stable at various times in the routines of daily husbandry. The horse should be out of the environment when conditions are worst and ventilation should be encouraged at that time for the safety of the grooms. Dust can contain pathogens, allergens, irritants and nonpathogenic nuisance microbes. Townson (1992) found that 65% of racehorses in Ireland were bedded on straw, and all were given hay as forage. When this is provided in large round bales it adds dust to the barn atmosphere if shaken out in the feed passage before feeding. Where alternatives to hay are not available the hay should be of very top quality, or soaked prior to distribution. Clarke (1987) showed that the number of respirable particles released from moulded ryegrass hay was reduced from 45 000 to 1650/mg fresh material after 5 min of soaking and to 525/

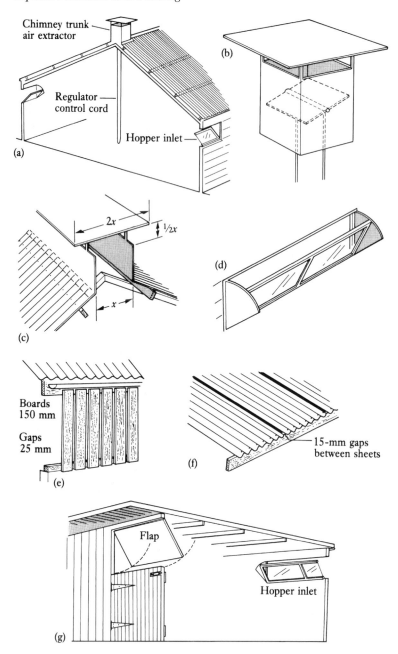

Fig. 11.2 Arrangements for the natural ventilation of stables. (a) Ventilation suitable for all stables using extractor chimney trunks and hopper inlets for fresh air. (b) Detail of the extractor chimney trunk, which may have a regulator or electric fan placed in the base. (c) Simple open ridge suitable for extraction ventilation of covered yards; x is normally about 300 mm in yards up to a width of 13 m and 600 mm in yards over 13 m and up to 25 m wide. (d) Hopper window suitable as a fresh-air inlet. (e) Spaced boards, giving draught-free ventilation: normally 25-mm gaps between 150-mm wide boarding. (f) 'Breathing roof' – corrugated roof sheets fixed with 15-mm gaps between for extractor ventilation. (g) Mono-pitch house showing hopper inlet at back and ventilating flap at front. Note overhang on roof to protect horses from rain, sun and wind (Sainsbury 1981).

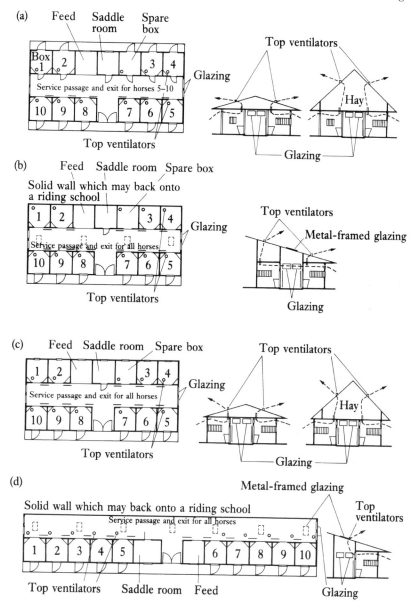

Fig. 11.3 Four layouts of ten horse boxes, which combine the advantages of indoor and outdoor boxes without draughts. (a) All the boxes open to the outside: boxes 1–4 have large exterior service doors; boxes 5–10 have interior sliding doors and shutters opening to the outside. (b) Stable with a solid back wall and interior service; boxes 5–10 have shutters opening to the outside. (c) Stable with a side exposed to bad weather and with interior service; boxes 5–10 have shutters opening to the outside. (d) Stable with a solid back wall and with interior service; all boxes have shutters opening to the outside (Ministère de l'Agriculture 1980).

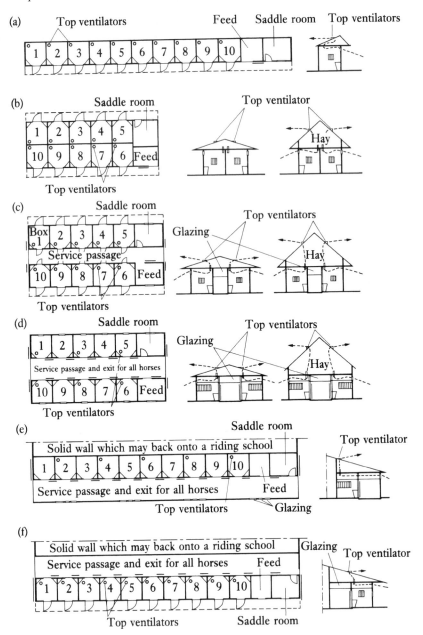

Fig. 11.4 Simple layouts for ten horse boxes. (a) A row of outside boxes with external services. (b) Back-to-back boxes with external services. (c) Boxes with external horse access and internal service; solid doors and walls in the passage keep draughts to the minimum. (d) Boxes with internal (central) servicing without access for the horses' heads to the outside (a less satisfactory arrangement). (e) Boxes with internal (lateral) servicing, without access for the horses' heads to the outside (a less satisfactory arrangement). (f) Boxes with internal (lateral) servicing and with access for the horses' heads to the outside; solid doors and walls onto the passage keep draughts to the minimum (Ministère de l'Agriculture 1980).

mg after 24 h of soaking (cf. 44/mg from haylage). Wood shavings should be used as bedding in barns, as they were found to release the lowest amount of dust of the sources compared (Table 11.8).

Too frequently an adequate ventilation rate depends on whether the top half of loose box doors is open and whether barn doors are open. The problem tends to be greater with barns as the top half of loose box doors is normally left ajar, but should have the facility for not being fully closed. The need for good ventilation should be gauged against the evidence that the thermoneutral zone for a mature and fed quarter horse is −10 to +10°C. Loss of body heat through concrete floors by conduction, when a horse is lying down, is greatly reduced by bedding, which also reduces floor draughts. Generally, ventilation rates should be between 4 and 8 air changes/h. These rates should keep the number of aerial particles of >0.5 µm diameter to less than 33 particles/cm³ (the threshold limiting value, TLV). However, this conclusion depends on the release rate of dust (Table 11.9). Hence, the importance of avoiding the creation of dust is self-evident.

Recommended dimensions for natural ventilation in boxes and barns are given in Table 11.10. Natural ventilation depends on both the area of inlets and outlets and the temperature difference between inside and outside, i.e. the temperature gradient. Insulation of a stable not only increases that gradient, and therefore increases the natural ventilation rate (stack effect), for a given area of inlets and outlets, but

Table 11.8 Number of dust particles/cm³ stable air in well-ventilated boxes (Webster *et al.* 1987) and mean weight of dust/m³ air in 1166 stables of 26 Irish racing establishments (Townson 1992).

	Number of airborne particles (diameter >0.5 µm/cm³ air)*		Weight of dust (mg/m³ air)
	Quiet	Bedding down	
Wood shavings	8.8	30.5	0.19
Paper	10.0	40.0	0.37
Straw	11.7	75.9	0.57

*Particles of up to 5 µm diameter may enter the alveoli of the lungs. These can include antigenic *Aspergillus fumigatus* and *Micropolyspora faeni*.

Table 11.9 Ventilation rates required to achieve a TLV of 33 particles/cm³ air at three rates of release of dust into the stable air (Townson 1992).

Dust release rates (particles/cm³/h)	Air changes required/h
60	3
300	10
600	22

Table 11.10 Requirements for natural ventilation of a typical box and barn (Sainsbury 1981; Webster *et al.* 1987; Townson 1992).

Dimensions per horse	Box	Barn
Volume (m³)	50	85
Surface area of building* (m²)	41	43
Ventilation inlet area (m²)	0.34	0.46
Ventilation outlet area (m²)	0.17	0.23
Height from inlets to outlets (m)	1	2
Ventilation rate (m³/s)	0.055	0.094
Air movement (m/s)	0.15–0.5	0.15–0.5
Ventilation heat loss (W/°C)	67	114
Ambient temperature (°C)	0–30	0–30
Relative humidity (%)	30–70	30–70

*Assuming 10 horses in a row of boxes or single barn. W, water vapour (kg).

also decreases the fluctuations in temperature of the building. Moreover, roof insulation can decrease the solar heating effect nearly ten-fold through a single skin roof.

Ailments related to housing

The importance of hygiene where foals are concerned cannot be overemphasised. This includes the removal of pests. Salmonellosis has been discussed already in 'Acute diarrhoea', this chapter. Severe liver damage and a lethal infection of young foals (from a few days to 6 weeks old) is caused by Tyzzer's bacillus (*Clostridium piliformis*). This bacterium is carried in the GI tract of many species of rodents and lagomorphs (rabbits and hares) and infection may occur by consumption of their carcasses, faeces or litter. The endospores can survive for very long periods in rodent litter and are sensitive to heating to 80°C for 15 min, or exposure to 0.015% sodium hypochlorite, to 1% iodophol and to 5% phenol.

Chronic obstructive pulmonary disease and other respiratory diseases

Viruses, bacteria (*Streptococcus equi*) and lungworm cause coughing, but the most frequent equine respiratory disease in the UK is chronic obstructive pulmonary disease (COPD). In a population of 300 adult horses referred for pulmonary examination in the north of the UK, Dixon *et al.* (1995) found that of 270 with pulmonary disease, 16.7% of the cases presented with infectious or postinfectious pulmonary disease, 2.6% with *Streptococcus zooepidemicus* pulmonary infection, 2.6% with lungworm infection, 5.9% with exercise-induced pulmonary haemorrhage, 3.3% with chronic idiopathic hypoxaemia, 14.1% with miscellaneous identified, or undifferentiated, pulmonary disorders and 54.8% with COPD. COPD is a hypersensitivity, inflammatory, reactive response of the lower respiratory tract, that is, the small airways (with fibrosis of the alveoli), to dust and mould (fungus and thermophilic actinomycete, especially two moulds, *Micropolypora faeni* and *Aspergillus fumigatus*) that causes respiratory hypersensitivity. Some horses are sensitive to other organisms and even to ryegrass pollen in the atmosphere of the stable, leading to a range of clinical reactions from mild exercise intolerance to severe dyspnoea, even at rest. Breathing is characterized by a 'double lift', seen in the abdomen on expiration, when the normal abdominal contraction is followed by a second lift as the horse endeavours to expel more air. This recurrent airway obstruction is known as 'broken wind' or 'heaves' in which parts of the lung tissue lose their elasticity. Individuals between 6 and 10 years old are most frequently affected.

Evidence indicates that generally more precipitins to mould antigens occur in the sera of horses housed in barns than in those in boxes. However, this does not necessarily relate to the risk to, or severity of, COPD. Consequently, some researchers hold the view that COPD horses may demonstrate precipitins to these fungi as a consequence of impairment to pulmonary function. The confusion may partly arise from the evidence that hypersensitivity has a rather strong genetic basis. Sensitization and the presence of serum-precipitating antibodies to mould antigens do not necessarily lead to clinical disease in a particular horse, but when clinical signs of dyspnoea occur the IgA:albumin ratio of bronchoalveolar lavage (BAL) fluid increases and pulmonary neutrophilia of this fluid is frequently found. Relief is achieved by veterinary intervention with a cyclo-oxygenase blockade. Whether dietary fats rich in n-3 fatty acids would reduce the risk of inflammatory responses (see Chapter 5) has not to the author's knowledge been examined. Susceptible horses should be removed from an environment of high allergen loads, otherwise the forced abdominal breathing, caused by bronchospasm, excessive mucus secretion and inflammation of the airway mucosa is likely to occur during periods of raised atmospheric concentrations of allergens.

The environmental causes are therefore indoor accommodation with inadequate ventilation and poor hay and bedding. There is probably too great a concern that horses will catch a chill at night unless doors and windows are closed. Horses and ponies prefer boxes and stalls that are bedded, especially from the viewpoint of lying down. There is no general preference for straw, or wood shavings, but from the point of view of avoidance of COPD the latter is to be preferred and good ventilation is the most important environmental characteristic.

COPD typically presents with a chronic cough, nasal discharge and expiratory dyspnoea. The widespread endoscopic examination of airways, with cytological examination of respiratory secretions and use of other diagnostic tests has led to the conclusion that COPD is an important cause of poor performance in the absence of any other overt clinical signs. In addition, there has been the recognition, with increasing incidence, of summer pasture-associated obstructive pulmonary disease. Many practitioners in the UK are of the opinion that the prevalence of COPD is actually increasing, although cases of lower severity may now be identified. Earlier cases were characterized by elevated maximal intrapleural pressure changes (max dPpl), whereas Dixon *et al.* (1995) found elevated max dPpl values in fewer than half of their 300 referred cases. There are many similarities of COPD to asthma in man, which has shown an increase in prevalence over the last two decades. It has been proposed that air pollution with nitric oxide and related hydrocarbon combustion products predisposes patients to respiratory disease. The mechanism probably involves the induction of airway inflammation and increased epithelial permeability, with reduced clearance of allergens and other inhaled particles that may trigger the development of COPD.

COPD occurs in individual horses and is not infectious, but there are other respiratory diseases that spread from horse to horse and cause impairment and damage to pulmonary function. The principal transmissible diseases are caused by bacteria, parasites and viruses. Viral infections seem to be of increasing incidence among horses and ponies and include equine influenza, equine herpes viruses I and II, rhinoviruses I and II and adenovirus, particularly in foals. It is thought that the widespread occurrence of viral infections may be associated with the trend to have larger numbers of horses in close proximity in totally enclosed buildings, greater national and international traffic in horses and, it must be said, an increasing awareness and understanding of viruses. Lung damage caused by these infections may leave horses more subject to the allergic responses of COPD; certainly one aggravates the other and both are influenced by building design and atmospheric pollution.

It is concluded that horse boxes should be not only well ventilated, but also sited distant from road traffic. Where haylage or damped hay is impractical to use, some benefit may be obtained from the use of mould spore extracted hay, or from the installation of ventilated hay racks, into which the air is drawn and extracted from the building by a fan. The system can be used as the ventilation system for the box, which should be bedded with wood shavings or paper. Prophylactic use of sodium cromoglycate has been found effective in sensitive individuals. However, prevention is better and less expensive than a partial cure.

All hay contains moulds, but some is visibly mouldy or musty and therefore presents a greater hazard than hard, stemmy, clean and shiny hay. Thus, there is reasoned justification for the use of such safe, though nutritionally poor, hay for horses, the lack of nutrients and energy being made up with a high-quality compounded nut or other concentrate. Where horses are affected they should be turned out to grass. If this is impracticable, hay should be soaked rather than merely dampened prior to feeding, or it should be replaced by silage, haylage or high-fibre compounded nuts. Straw bedding should be replaced by shredded paper or by softwood shavings and the ventilation of the box should be improved. Symptomatic horses may become asymptomatic in 4–24 days when horses are bedded on shredded paper and routinely fed a complete cubed diet. The pulmonary function values of asymptomatic horses may then not differ significantly from those of normal horses. The ingestion of mould spores rather than their inhalation, as would occur with the feeding of soaked hay or nuts, is *not* a cause of the problem as it depends on a direct reaction between the inhaled particle and the lung alveoli. Meal feeding should also be avoided as some evidence indicates that allergy may be caused by oat dust and some other sources of feed dust.

Crib-biting

Crib-biting is probably mediated by neurotransmitters and where there is a high frequency of the vice among horses it may indicate stress caused by bad management. A similar vice apparently does not occur among feral horses. Ways to palliate it include increased handling and exercise, a change of environment and feeding strategies.

STUDY QUESTIONS

(1) How would you set about minimizing the risks of GI disturbances in a stable of working horses?
(2) Where hyperlipaemia has been diagnosed in several pony mares, what is the appropriate course of action?
(3) What are the causes of wasting disease in an elderly horse and what management should be instituted?

FURTHER READING

Allen, G.K., Campbell-Beggs, C., Robinson, J.A., Johnson, P.J. and Green, E.M. (1996) Induction of early-phase endotoxin tolerance in horses. *Equine Veterinary Journal*, **28**, 269–74.

Bogan, J.A., Lees, P. and Yoxall, A.T. (1983) (eds) *Pharmacological Basis of Large Animal Medicine*. Blackwell Science, Oxford.

Clarke, A.F. (1987) A review of environmental and host factors in relation to equine respiratory disease. *Equine Veterinary Journal*, **19**, 435–41.

Dixon, P.M., Railton, D.I. and McGorum, B.C. (1995) Equine pulmonary disease: a case control study of 300 referred cases, Parts 1–4. *Equine Veterinary Journal*, **27**, 416–39.

Gossett, K.A., Cleghorn, B., Martin, G.S. and Church, G.E. (1987) Correlation between anion gap, blood L-lactate concentration and survival in horses. *Equine Veterinary Journal*, **19**, 29–30.

Hinckley, K.A., Fearn, S., Howard, B.R. and Henderson, I.W. (1996) Nitric oxide donors as treatment for grass induced acute laminitis in ponies. *Equine Veterinary Journal*, **28**, 17–28.

King, J.N. and Gerring, E.L. (1991) The action of low dose endotoxin on equine bowel motility. *Equine Veterinary Journal*, **23**, 11–17.

Ministère de l'Agriculture (1980) *Aménagement et Équipement des Centres Équestres. Section Technique des Équipements Hippiques*. Fiche Nos CE. E. 4, 5, 13 and 14. Service des Haras et de l'Equitation, Institut de Cheval, Le Lion-d'Angèrs.

Moore, J.N. (1991) Rethinking endotoxaemia in 1991. *Equine Veterinary Journal*, **23**, 3–4.

Moyer, W., Spencer, P.A. and Kallish, M. (1991) Relative incidence of dorsal metacarpal disease in young Thoroughbred racehorses training on two different surfaces. *Equine Veterinary Journal*, **23**, 166–8.

Prescott, J.F., Staempfli, H.R., Barker, I.K., Bettoni, R. and Delaney, K. (1988) A method for reproducing fatal idiopathic colitis (colitis X) in ponies and isolation of a clostridium as a possible agent. *Equine Veterinary Journal*, **20**, 417–20.

Proudman, C.J. and Edwards, G.B. (1993) Are tapeworms associated with equine colic? A case control study. *Equine Veterinary Journal*, **25**, 224–6.

Slater, M.R., Hood, D.M. and Carter, G.K. (1995) Descriptive epidemiological study of equine laminitis. *Equine Veterinary Journal*, **27**, 364–7.

Townson, J. (1992) *A survey and assessment of racehorse stables in Ireland.* MEqS thesis, Faculties of Agriculture and Veterinary Medicine, National University of Ireland, Dublin.

Valberg, S., Jönsson, L., Lindholm, A. and Holmgren, N. (1993) Muscle histopathology and plasma aspartate amino transferase, creatine kinase and myoglobin changes with exercise in horses with recurrent exertional rhabdomyolysis. *Equine Veterinary Journal*, **25**, 11–16.

Ward, D.S., Fessler, J.F., Bottoms, G.D. and Turek, J. (1987) Equine endotoxaemia: cardiovascular, eicosanoid, hematologic, blood chemical, and plasma enzyme alterations. *American Journal of Veterinary Research*, **48**, 1150–56.

Webster, A.J., Clarke, M.T.M. and Wathes, C.M. (1987) Effects of stable design, ventilation and management on the concentration of respirable dust. *Equine Veterinary Journal*, **19**, 448–53.

West, H.J. (1996) Clinical and pathological studies in horses with hepatic disease. *Equine Veterinary Journal*, **28**, 146–56.

Chapter 12
Laboratory Methods for Assessing Nutritional Status and Some Dietary Options

If the urine of a horse be somewhat high coloured, bright and cleare like lamber, and not like amber, or like a cup of strong March beere, then it sheweth the horse hath inflammation in his blood.... Now for the smell of his dung, you must understand, that the more provender you give, the greater will be the smell, and the lesse provender the lesse the smell.

G. Markham 1636

Many methods of nutritional assessment are only appropriate for large stables where normal variation in values can be overcome by determinations carried out in a number of horses fed in a similar manner. Nonetheless, whether one is dealing with individuals or groups, it is essential to establish normal values for the individuals or groups concerned. The normal value of a particular parameter in an individual may differ from the breed mean, as age, sex, time of year, system of management, stage of training, diet and breed all influence the norm. Thus, by establishing some routine practice it is possible to assess quickly whether a particular value has shifted from its norm, which enables one to uncover a disturbance in its early stages. Furthermore, observation and a range of tests are undoubtedly necessary to truly understand a situation.

Metabolic profiles have been adopted as measures of the nutritional and physiological health of dairy herds and some progress has been made towards similar techniques in stables. In assessing nutritional status, it is essential to have a full knowledge of the ingredient and chemical composition of all the feeds and the actual weights of each type fed over an extended period. Records and the retention of feed samples assist subsequent solution of problems. This procedure will facilitate an objective diet evaluation, which, with the monitoring of management and disease, should indicate the most appropriate laboratory measurements to be undertaken. Direct and appropriate information should then come to hand and unnecessary expenditure would be avoided.

The adequate nutrition of the horse implies the normal nutrition of the tissues and cells of the body, but nutrient-related abnormality of these units does not of necessity imply incorrect feeding. A knowledge of the physiological and health status of the horse must be accommodated in any assessment. An improvement in health may not be achieved by nutritional adjustment, or it may require enteral, or parenteral, nutrition to supplement or supplant normal diet for a limited period. A purpose of this chapter is to aid decisions in this direction.

427

METABOLIC TESTS

Many metabolic tests are now available for gauging nutritional status and some of the more commonly used measures are discussed in this chapter. It should, however, be emphasized that single measurements, let alone single determinations, are practically valueless. A salient reason for this is that few, if any, of these methods are specific in determining the status of a particular nutrient.

Variability of measured values

Repeat samples obtained from an individual horse normally yield variable values. This variability is caused by the inherent variability in any analytical method, in variable sample handling and variation within the horse while there may be no apparent variation in its health. Different analytical procedures for the same determinant yield different means, so it is important to quote the method employed with any set of values. Normal ranges for a particular analyte are quoted to accommodate this variability. Nonetheless, the important criterion is not the accepted normal range but the change in value from one day to another for an individual, as it could point to a trend in health. In order to detect such trends it is critical to control measurements so that real trends can be distinguished from cyclic variation.

Circadian variation in physiological characteristics has been well established as a phenomenon in man and other animals. Some control of this can be achieved by, for example, sampling at the same time in the daily routine, preferably at a time when the rate of change is least and when environmental and metabolic factors of no concern have least influence. Meal-fed horses, like humans, show a fairly marked 24h circadian rhythm in the plasma concentration of PCV, plasma total protein, urea, glucose, insulin, neutral fat, NEFAs, cholesterol, Ca, P and several, frequently measured, enzymes (Greppi *et al.* 1996). For consistency of data, fixed daily times reduce variability in repeat samples for these characteristics, assuming the daily husbandry does not change.

Enzymes

Nomenclature

Over the years the name of many enzymes has changed, sometimes more than once. So that it is known which enzyme is referred to in any particular statement, the international Enzyme Commission has classified each identified and specific enzyme with its own code number, known, not surprisingly, as the EC number (International Union of Biochemistry and Molecular Biology on the Nomenclature and Classification of Enzymes 1992). There are about 3500 enzymes on the list, which is still increasing in length. The EC number is a four-part system, the first number of which defines the class, the second the subclass of that class, the third the sub-subclass and the fourth the number it is given in the list of that sub-subclass, e.g. the oxidoreductase alanine dehydrogenase is defined as EC 1.4.1.1 and alanine amino-transferase as EC 2.6.1.2, as indicated in Table 12.1.

Specific tissue isoenzymes

The activity of many specific tissue enzymes is frequently measured as an aid to diagnosis. The plasma enzyme concentration may rise during leakage from the cells in which the enzyme is active, either as a result of maximum activity of those cells, e.g. muscle cells during galloping, or as a consequence of damage to the tissue, e.g. toxin damage of liver parenchymal cells. The location of the relevant tissue/organ is helped by measuring the iso-enzyme concentration relative to a total raised activity. For example, there are at least 5 iso-forms of lactic dehydrogenase (EC 1.1.1.27). Isozymes are physically distinct, e.g. immunologically so, but they all catalyse the same reaction.

Diet and enzyme activity

Tissue enzymes are proteins that function only when present as the holoenzyme, which consists of the protein (apoenzyme that on its own has no catalytic activity) plus a heat-stable, dialyzable nonprotein coenzyme. This dialyzable fraction contains nutrients in the form of certain water-soluble vitamers or vitamins A and K. Frequently also a metallic cofactor, e.g. zinc, is needed. A dietary deficiency of one of these nutrients therefore depresses the activity of the specific enzyme(s) in which the nutrient functions.

The degree of saturation of several enzymes (the fraction present in the holoenzyme form), measured *in vitro* as the enzyme activity coefficient, can indicate the nutritional status in respect of the vitamin component of the cofactor (Table 12.2). It should be noted that the addition of the cofactor vitamer to the *in vitro* system will increase the activity regardless of the nutritional status, as a portion of the enzyme is normally present in the unsaturated apoenzyme form. Therefore the extent of the response is critical in assessing whether a dietary deficiency exists.

Table 12.1 Classification of enzymes.

Group	Name
1	Oxidoreductases
2	Transferases
3	Hydrolases
4	Lyases
5	Isomerases
6	Ligases

Table 12.2   Cofactor vitamers of several tissue enzymes, with their EC numbers, of particular significance in equine nutrition and those* for which enzyme saturation measurements have been used to assess
adequacy.

Enzyme	Cofactor vitamer	Nutrient in cofactor
Erythrocyte glutathione reductase, 1.6.4.2*	Flavin adenine dinucleotide	Riboflavin
Erythrocyte transketolase, 2.2.1.1*	Thiamin pyrophosphate	Thiamin
Pyruvate dehydrogenase, 1.2.4.1*	Cocarboxylase (TPP)	Thiamin
Fatty acid synthase[1]	Acetyl-CoA	Pantothenic acid
Glucose-6-phosphate dehydrogenase, 1.1.1.49	NADP(H)	Niacin
β-Hydroxyacyl-CoA dehydrogenase, 1.1.1.35	NADP(H)	Niacin
Lactic dehydrogenase, 1.1.1.28	NADH and FAD	Niacin and riboflavin
Pyruvate carboxylase, 6.4.1.1	Biotin	Biotin
Erythrocyte alanine aminotransferase, 2.6.1.2*	Pyridoxal phosphate	Pyridoxine
Erythrocyte aspartate aminotransferase, 2.6.1.1*	Pyridoxal phosphate	Pyridoxine
Methionine synthetase, 4.2.99.10	5-Methyltetrahydrofolic acid	Folic acid
Methylmalonyl-CoA mutase, 5.4.99.2	Adenosylcobalamin	Vitamin B_{12}
α-Carboxylase, 4.1.1.1	K hydroquinone	Vitamin K
Succinic dehydrogenase, 1.3.99.1	CoQ_{10}	Ubiquinone[2]

[1] Multi-enzyme complex which differs among species.
[2] Not generally considered to be a vitamer because the molecular species shares an antioxidant role with
other nutrients.
Also see vitamin status in 'Fat- and water-soluble vitamins', this chapter.

Serum enzymes and liver function

Clinical signs of hepatic insufficiency include loss of appetite, icterus, depression,
weight loss, lethargy, yawning and sometimes wandering. Haemorrhages in oral
mucous membranes have been reported and photosensitization may be observed,
especially in horses with unpigmented skin and hair who are exposed to bright light.
Photodynamic substances in the feed are incompletely metabolized by the compromised liver and cause hypersensitivity reactions in the skin.

Measurement of the serum activities of several hepatic enzymes is normally
included in reaching a conclusion concerning the function and dysfunction of hepatic tissue. Increased serum activities of hepatic ALP, AST [formerly serum
glutamic—oxaloacetic amino-transferase (SGOT)] and SDH (EC 1.1.1.14) may
result from reversible changes in hepatocellular membranes, structural injury of
hepatic tissue, caused by ischaemic necrosis or cholestasis, or from microsomal
enzyme induction. Determinations of GGT (EC 2.3.2.2), serum bilirubin, serum
proteins, the clearance of bromsulphthalein (BSP), excreted in the bile, and the
histological examination of liver biopsy samples will also be appropriate in suspected cases of hepatobiliary dysfunction. However, hepatobiliary disease is rare in
horses, unless the common bile duct is obstructed.

Liver function is also assessed by plasma coagulation tests, including prothrombin
time and partial thromboplastin time (platelet function tests, for which several

instruments are available, should be evaluated in cases of bleeding problems). Increased liver enzyme activity in serum can arise from decreased hepatic perfusion, causing hypoxia. Cardiac failure, endotoxaemia, septicaemia, hypothyroidism and hyperthermia may contribute to hypoxia of hepatic cells, but generally serum enzyme values do not rise more than 2–3 times the normal value, as a result of these extrahepatic causes.

Aspartate aminotransferase (AST, AAT, EC 2.6.1.1)

AST is a cytoplasmic and mitochondrial enzyme present in several tissues – liver and skeletal and cardiac muscles. The activity of this enzyme in serum/plasma is elevated most rapidly following acute hepatic necrosis and it may reach values 10–40 times normal, attaining a peak in 12–24 h and declining over a 2-week period. The enzyme may increase after larvicidal treatment with thiabendazole and in the hyperlipaemic syndrome of ponies and adult horses, but in chronic hepatic fibrosis the serum activity may be normal. The author (unpublished observations) has found values chronically 2–10 times normal in Miniature Shetland mares with bile duct injury, but without hyperlipidaemia.

Alkaline phosphatase (ALP, EC 3.1.3.1)

(See also 'Bone metabolism', this chapter.) ALP is a membrane-bound enzyme synthesized in many tissues, notably active bone tissue, liver, kidney and intestinal mucosa (highest specific activity in the duodenum), although the intestinal source does not contribute significantly to total serum ALP in the horse. During ischaemic colic of the bowel, intestinal ALP is released into body fluids. A measure of this aids both diagnosis and a decision over immediate action in acute cases. Intestinal ALP is not bound by L-phenylalanine, whereas other sources are, thus providing a means of differentiation and detection of ischaemic colic when this source is shown to be elevated in peritoneal fluid (Davies *et al.* 1984). Serum values are elevated during bone growth as a consequence of osteoblastic activity, and in horses with renal hyperparathyroidism or with NSHP in Ca or even in P deficiency. When hepatic injury accounts for the increase, it is frequently caused by cholestasis or primary bile duct injury and the serum level often remains elevated with severe chronic disease.

Sorbitol dehydrogenase (SDH, EC 1.1.1.14)

SDH is concentrated in hepatocytes and so is used to demonstrate hepatic necrosis, but the serum activity declines in 2–4 h following hepatic necrosis, and it has relatively low stability in stored samples. Nevertheless, elevated values indicate acute hepatic necrosis, and along with a raised AST they point to an hepatic origin of the disease.

Gamma-glutamyltransferase (GGT, EC 2.3.2.2)

GGT is membrane-bound in hepatic cells, but it also occurs in kidney and pancreatic tissue. However, elevated serum concentrations are accounted for almost entirely by release from bile duct cells and serum increases result from intra- or extrahepatic cholestasis. Cholelithiasis is such a cause. The source of urinary GGT is probably the brush border of the proximal tubular renal epithelium. High urinary concentrations may indicate proximal tubular dysfunction. Rudolph and Corvalan (1992) found urinary GGT concentrations of 47.6 ± 27.3 iu/l in horses positive for proteinuria and 27.2 ± 17.2 iu/l in those negative (note: some proteinuria is a physiological consequence of a high protein diet).

Serum enzymes and exercise

Aminotransferase and creatine kinase (CK, EC 2.7.3.2)
AST, ALT (EC 2.6.1.2) and CK can be detected in the serum and tissues of the horse, although different isoenzymes of AST and ALT are found in the mitochondria and in the soluble portion of the cytoplasm of cells. In the absence of severe tissue injury the cytoplasmic forms predominate in serum. These two enzymes require pyridoxyl 5'-phosphate as a cofactor for activity.

There is a large individual variation in the normal serum activity of these enzymes and individuals with apparently high circulating levels may achieve a good running performance. Nevertheless, where the intention is to determine whether the concentration within the peripheral circulation of, for example, a transferase is within physiological limits, and not elevated into pathological concentrations, a wide margin of uncertainty must be allowed, unless there is good previous data available on a particular horse. It is also necessary to ensure that circulating apoenzyme is included in the total. For this, exogenous pyridoxal phosphate should be included in the assay medium (Rej *et al.* 1990). Serum concentrations increase slightly postexercise and are also elevated in disorders of muscle soreness, azoturia or 'tying-up' myositis. Equine rhabdomyolysis is a muscle disorder resulting in elevated serum levels of muscle enzymes, stiffness, loss of performance and/or lameness, with a slightly higher frequency in 2- and 3-year-old fillies than in other horses. Serum concentrations of muscle enzymes 24h following exercise may more easily allow differentiation between muscle damage and the postexercise physiological elevation. The 24-h upper limits of normality may be taken as 100 and 300iu/l, respectively for CK and AST. CK is relatively muscle-specific and has a $t_{1/2}$ of about 2h, whereas AST has a $t_{1/2}$ of 7–8 days.

Plasma proteins

Albumin is the major plasma protein (Table 12.3) synthesized in the liver and it determines plasma colloidal osmotic pressure. Low serum values may indicate that over half the liver has been nonfunctional for several weeks, as albumin has a long half-life in plasma. Causation includes trauma, malnutrition, parasitic-worm infestation, protein-losing enteropathy and renal dysfunction, with elevated urinary levels, or poor hepatic circulation through hypotension, hypovolaemia or chronic inflammatory processes. Insufficient dietary protein is more likely to be the cause of

Table 12.3 Blood serum albumin concentrations in TBs determined by the sulphate/sulphite method of Reinhold (1953).

	Number sampled	Range (g/l)	Mean (g/l)
Foal	140	12–45	25.8
Yearling	70	12–42	28.5
Two-year-old	120	9.5–42	26.8
Three-year-old	90	13.5–49	30.3
Four-year-old	52	18–47	29.8
Mare	32	18.5–52	29.6

depressed blood haemoglobin than is a relatively scarce quantity of dietary iron, where normal dietary ranges are concerned.

Plasma minerals

The plasma concentration of minerals is used, with considerable reservations, to assess the status of nutrients. These include Mg, K (but see p. 401), P, Zn, Se, Cu, and T_3 and T_4, reflecting I status. Cr is a component of the glucose tolerance factor and there is speculative evidence that depressed plasma Cr, or a measure of its renal excretion rate, may relate to glucose tolerance and insulin sensitivity. Serum ferritin provides a good index of hepatic and splenic Fe and can be used to evaluate Fe storage in horses. Whereas a deficiency of Se is characterized by a depression in serum concentration of the element and a depressed activity of Se-containing serum GSH-Px (EC 1.11.1.9), the situation with Zn is unclear. In growing children, dietary deprivation of Zn causes a depression in the serum concentration of the element only where there is sufficient dietary protein to promote a normal growth rate. In other species, at least, the activity of serum ALP (EC 3.1.3.1) is sensitive to Zn status, as Zn is a cofactor. The relationship in the horse has not been studied in detail, and the activity per se is not specific for Zn status (see page 431). Blood-plasma concentrations of inorganic phosphate reflect the dietary intake of P and in the deficient state there is an *increase* in the actvity of plasma ALP. Plasma measurements of inorganic phosphate and Ca are necessary, together with the urinary measurement of P, in the clearance test for Ca adequacy to be discussed in 'Urinary fractional electrolyte excretion (FE) test (creatinine clearance)', this chapter.

Cellular content of minerals

When there is, for example, a dietary protein deficiency some success has been achieved in detecting Zn inadequacy by measuring the more labile Zn content of leucocytes (Table 12.4). The reason for this is that Zn is a cofactor of cellular superoxide dismutase (SOD) (EC 1.15.1.1). There are two known forms of this enzyme: one contains Cu–Zn and is found in the cytoplasm of most cells and the other contains Mn and is present in the mitochondrial compartment of cells. The

Table 12.4   Range in leucocyte Cu and Zn contents in seven Shetland ponies receiving poor hay only, before and after supplementation with Cu and Zn for 50 days (author's unpublished observations, by method of Williams *et al.* 1995).

	Cu (μg/10^{-9} cells)	Zn (μg/10^{-9} cells)
Before supplementation	0.11–0.18	2.57–6.25
After supplementation	0.40–2.86	2.57–10.57

measurement of the leucocyte activity of SOD (EC 1.15.1.1), or the cellular content of Zn and Mn which tends to reflect the activity of the enzyme, is a measure of the adequacy of these elements. For similar reasons the measurement of leucocyte or platelet Cu, required as a cofactor in cytochrome c oxidase (CCO, EC 1.9.3.1), or the cellular activity of CCO, is informative as a measure of Cu status. Plasma Cu is largely present in caeruloplasmin (EC 1.16.3.1), an acute phase protein, the concentration of which relates to the inflammatory process, so plasma Cu concentration is unrelated to status.

Fat- and water-soluble vitamins

Among the fat-soluble vitamins the measurement of plasma α-tocopherol in relation to plasma lipid is a measure of storage status, and elevated breath alkanes may indicate deficiency. A deficiency of vitamin E and/or Se can cause muscle and liver damage that nonspecifically leads to an increase in the activity of certain plasma enzymes, particularly AST and CK, owing to leakage from the tissue cells concerned. A more specific measure of adequacy is, however, the determination of red-cell fragility in the presence of dialuric acid or hydrogen peroxide. (See Chapter 4 for a fuller discussion.)

The homeostatic control of many blood components and the interaction of many nutrients imply that variations in blood concentrations require careful interpretation. Plasma retinol (vitamin A) concentration is only marginally informative. However, the concentration of vitamin A in the liver varies more closely with the dietary intake of vitamin A. Liver tissue is not easily accessed and an alternative sensitive measure of vitamin A adequacy is the RDR, determined in the horse's jugular blood and discussed in Chapter 4. Vitamin D deficiency occurs only in horses confined indoors for long periods, e.g. pit-ponies, or those in the higher northern latitudes in winter. Status may be assessed by plasma concentration of 25-(OH)-D$_3$, possibly in conjunction with serum ALP (EC 3.1.3.1) activity. The determination of blood prothrombin time as a measure of vitamin K status is discussed in Chapter 11. Elevated times would generally indicate antibiotic misuse, causing depressed intestinal synthesis of the vitamin.

Among the B vitamins, quick and routine methods are now available for assessing whole blood vitamin B$_{12}$ (cyanocobalamin), or erythrocyte B$_{12}$, and serum or leucocyte folic acid, which clearly reflect diet in the experience of the author. In the human subject the most sensitive indicators used in detecting, and separating, B_{12} and folate deficiencies are the measurements of methylmalonate and homocysteine concentrations in urine and blood. Elevation of both these metabolites indicates a B_{12} deficiency, whereas an elevation in homocysteine only, indicates a folate deficiency. Among other B vitamins, methods have been successfully developed for assessing thiamin, riboflavin and pyridoxine adequacy through measuring the activity coefficient of the appropriate enzymes discussed in 'Diet and enzyme activity', this chapter. The activity per se (not the coefficient) of a transaminase in red blood cells of sow pigs has been proposed as a good indicator of pyridoxine status. The *serum* activity of AST (AAT, EC 2.6.1.1), routinely measured in horses, is greatly influenced by the extent of leakage from liver and muscle cells and therefore the activity in horse red cells may reflect pyridoxine status. Biotin, a cofactor in carboxylases, is assessed in humans from whole blood or urinary biotin concentrations.

Tests for liver and kidney disease

Blood urea

Urea is derived from ammonia and although the liver has a large over-capacity for its synthesis, blood urea values can be half normal with chronic severe liver disease. Blood values will also be greatly elevated when daily protein intakes exceed NRC requirements by a large margin, a not infrequent occurrence in TBs during training.

Blood ammonia

Blood ammonia may derive from the action of bacterial urease on urea in the gut, although much of this ammonia will be reutilized in bacterial protein synthesis. Ammonia also results from the deamination of amino acids. Ammonia is converted to urea in the liver, but raised plasma levels have been noted in hepatic encephalopathy. Normal values range from 80 to 160 µg/l. However, as blood samples must be placed on ice (not dry ice) immediately following collection and the plasma separated within 30 min, this measurement is problematic from a practical viewpoint.

Hepatic encephalopathy

Hepatic encephalopathy is a neurological dysfunction resulting from acute, or chronic, liver disease, characterized by depression, frequent yawning, muscle twitching, poor coordination, head pressing and loss of strength and posture. It is associated with high concentrations of blood ammonia, low blood glucose and increased levels of blood VFA.

Bile acids

About 75% of bile acids in the horse are accounted for by chenodeoxycholic acid. Approximately 85% of these bile acids are conjugated to taurine and the remainder to glycine. The plasma content of total bile acids determined by HPLC rises slightly 2–6 h after feeding, primarily owing to an increase in glycocholic acid production. Liver failure, obstruction of bile flow, or vascular shunting causes a rise in plasma bile acids, as a proportion of those absorbed from the gut is not removed by the liver for resecretion. Normal serum bile acid concentrations can be up to approximately 12 µmol/l, whereas in horses with clinical signs of liver disease the serum concentration may be more than double this level.

Bilirubin

Bilirubin is synthesized from the degradation of the haemoproteins of red blood cells in the reticuloendothelial cells of the spleen and liver. Bilirubin conjugated with glucuronic acid is excreted in the bile, except where hepatobiliary obstruction occurs and jaundice develops. Icterus is apparent when total plasma bilirubin concentration exceeds 34 µmol/l (20 mg/l). ALP and GGT are also usually elevated. Fasting hyperbilirubinaemia can occur in the horse, possibly caused by a reduction in hepatic blood flow.

Reticuloendothelial system function

The cells of the reticuloendothelial system (RES) have several functions. They act as a sieve protecting the systemic blood circulation from some products carried to the liver by the portal system. These products include bacterial antigens of gut origin and both enterotoxins and endotoxins resulting from grain overload. It is thought that where there is a systemic acid:base imbalance and liver perfusion may be compromised, the sieve is then partly bypassed and clinical endotoxaemia may result.

Crystalline urine deposits

Urolithiasis is uncommon in the horse despite the supersaturation of equine urine with calcium carbonate. The most common site of calculus formation is the bladder and cystic and urethral calculi have been reported, whereas renal and ureteral calculi are rare. Calcium carbonate is the predominant mineral of equine urinary calculi, although oxalate and phosphate types exist. Calculi in the horse may result from the mineralization of a nidus, possibly provided by a prior disease, such as pyelonephritis, renal papillary necrosis or tubular necrosis. In other species, urolithiasis typically develops with diets with low Ca:P ratios and excessively high, or low, Mg contents. Low Ca:P ratios activate calcitonin secretion, bone mobilization and soft tissue calcification, leading to both nephrocalcinosis and urolithiasis. The situation in the horse differs in that the urine is normally supersaturated with Ca salts and persistent stone formers may lack some inhibitor of crystallization.

Urinary fractional electrolyte excretion test (creatinine clearance)

Electrolyte status is a function of intestinal absorption, renal tubule reabsorption, tissue deposition and mobilization, sweat loss and renal excretion. Serum or plasma concentrations of electrolytes may not be used to detect electrolyte imbalance, owing to efficient homeostatic mechanisms which maintain relatively normal blood concentrations despite extreme body depletion. Homeostasis is primarily mediated by the kidney, so that the amount of an electrolyte excreted daily varies with the body's status. However, the daily urine volume varies considerably between individuals so the concentration of an electrolyte in the urine is an unreliable guide to nutritional status. The collection of urinary losses over a period of several days would indicate the quantities surplus to requirement, but such extended urinary collection is impractical. Urine concentration of an electrolyte is therefore related to that of a control substance. This substance should (1) have an excretion rate similar to the glomerular filtration rate, and (2) not be secreted or reabsorbed by the renal tubules. Creatinine, the excretion product of creatine metabolism, fulfils these requirements reasonably well and the fractional electrolyte excretion (FE) is measured as the creatinine clearance ratio (renal creatinine averages 1.15 ± 0.41 mg/kg BW/h; Meyer 1990). Urine sample creatinine concentrations of <9000 mmol/l may indicate contamination of the sample or excessive salt consumption, causing polydipsea/polyuria, so the FE values obtained should not be accepted, although low values in horses of less than 18 months old may indicate a physiological abnormality; P.A. Harris 1996 personal communication).

A urine sample and a serum sample are required. The urine sampling should be achieved without resort to the use of diuretics because they affect Na and, to a lesser extent, K and chloride losses. The clearance of an electrolyte equals the concentration in the urine times the urine volume divided by the concentration in the serum. The clearance ratio is the clearance of the electrolyte divided by the clearance of creatinine. In this ratio, the urine volume cancels out and so need not be measured. The equation for which the determinants are required is given below:

$$\text{Per cent electrolyte clearance}(\% \text{ creatinine}) = \frac{[E]_u \times [Creat]_s}{[E]_s \times [Creat]_u} \times 100$$

where $[E]_u$ is the concentration of electrolyte in urine, $[E]_s$ is the concentration of electrolyte in serum, $[Creat]_s$ is the concentration of creatinine in serum and $[Creat]_u$ is the concentration of creatinine in urine.

High-concentrate rations tend to give a raised phosphate clearance and a depressed K clearance, the reverse being the case for high-roughage rations. These effects are quite normal. K clearance, together with measurements of blood and urinary pH, is useful in evaluating the type of acidosis and in assessing K depletion in exhausted horses, as the urinary excretion of K and H ions tends to display a reciprocal relationship. K clearance is depressed in chronic laminitis. Note that even slight blood haemolysis may increase plasma K and therefore yield falsely low FE values. A raised Na clearance may indicate excessive Na intake in the form of common salt, Addison's disease, dehydration or tubular malfunction.

Owing to the circadian rhythm in excretion of electrolytes, referred to in 'Variability of measured values', this chapter, the FE ratio should ideally be measured at least three times during a 24-h period. The serum and urine samples do not have to be taken simultaneously, but the samples should be taken pre-meal and pre-exercise. Decreased urinary Mg and Ca concentrations occur shortly before and after feeding with maximum values 4–8h after feeding. The serum sample should be sterile and that of urine as close to sterility as is possible. The urine sampling may benefit from some advice and it should comply with the following restrictions (P.A. Harris, Animal Health Trust, Newmarket, personal communication; Meyer 1990):

(1) Samples should be freely voided using a collection harness before exercise (entire product of micturition). Walking the horse slowly 'first thing' in the morning while its stable is being cleaned and rebedded may promote urination on return. If catheterization is necessary, fillies and mares should first be bled and then given a short brisk trot before sampling. Catheterization will decrease the risk of contamination, and so be helpful. Plasma K concentrations also decrease during eating and rise to a peak several hours postingestion.

(2) Meyer (1990) found that the mean urinary concentration of creatinine tended to be lower during the first 2h after eating. Thus, for the assessment of dietary adequacy by measurement of the ratio, the measurements should be made before feeding or at the following times after a meal:

    3h   for Na
    6–10h   for Cl
    3–8h   for K
    4–8h   for Ca

(3) There is little difference between the early, middle or late portions of a voided urine stream in the FE of Na, K, PO_4 and Cl, but for Ca collect the entire voided sample. Owing to the supersaturation of urinary Ca, reliable urine sampling for Ca clearance may not be achieved satisfactorily. The amount of Ca precipitate voided at any one moment is unpredictable. However, where a low urinary Ca value is supported by a low Mg value there is strong evidence of a dietary Ca deficiency and possibly a deficiency of Mg. The reason for this is that Mg is more soluble so that urinary values are more reliable and Mg responds in a parallel fashion to that of Ca. The FE test may not be used to monitor low phosphate intakes.

(4) If a delay between collection and analysis is unavoidable there is little change in the Na, K or Cl concentration in plasma, or urine, with storage at 18, 4 or −20°C; however, considerable changes in Ca, PO_4 and creatinine concentrations occur, especially at 18°C. Hygiene is important. Samples should be transported in capped sterile containers and stored at 3–4°C for short periods, or at −20°C for longer periods. Ca, PO_4 and creatinine should be analysed as soon as possible, i.e. within 4 days.

Table 12.5 Interquartile (25–75%) normal reference ranges for FE % (or creatinine clearance, % of electrolyte) of adult horses at rest and in work from spot samples (P. A. Harris personal communication)* and a mean published range.

Electrolyte	Grain based diet* FE %	Balanced compound and hay diet* FE %	Published FE %
PO_4	0.0–0.5	0.0–0.2	0.04–1.19
Na	0.02–1	0.04–0.52	0.01–1
K	15–65	35–80	15–75
Cl	0.04–1.6	0.7–2.1	0.04–1.65
Ca		>7 (8–24)	
Mg		>15	

(5) For urinary Ca determinations mix well and use a flame atomic absorption spectrophotometer, as colorimetric methods used for serum or plasma samples are unsuitable.

(6) Samples collected during, or shortly after, an equine rhabdomyolysis episode do not reflect electrolyte status, owing to circulatory disturbances and to raised plasma myoglobin concentration which may affect renal function. Samples with a pH of 6 or below or that are positive for glucose, blood or myoglobin/haemoglobin are unsuitable. A low urinary pH is accompanied by an increase in Ca and Mg excretion, giving falsely high Ca and Mg FE values. Blood and urinary pH may be decreased on a high starch diet, yielding raised urinary Ca and P clearances, unrelated to changes in DCAB (see Chapter 9).

Effect of diet on FE

Diets with a Ca:P ratio of 1:1, or lower, produce an elevated FE PO_4, whereas inadequate Ca and low Na intakes lead to low FE Ca and low FE Na. Normal clearance ratio values are given in Table 12.5. Frequently TBs in hard training in the UK appear to be receiving inadequate Ca and Na, reflected in low FE values for Na and high values for PO_4. However, it is unclear whether there is any relationship between these inadequacies and the frequency of muscle stiffness, as the appropriate definitive experiments do not seem to have been conducted.

DIETS FOR LIVER DISEASE

Dietary causes of liver disease include aflatoxin, from mould-infected grains and protein concentrates (especially groundnuts), and pyrrolizidine alkaloid poisoning from *Senecio* and *Crotalaria* species. Supportive therapy has the objective of allowing time for regeneration of hepatocytes. This may initially involve i.v. glucose administration, followed by enteral feeding of glucose. For horses suffering severe hepatic dysfunction dietary management should ensure the following:

- The ration should be divided into at least three daily meals.
- The ration should contain the highest quality protein in adequate, but not excessive amounts.
- An amino acid supplement of the glucogenic branched chain amino acids isoleucine and valine (1 g/kg diet of each) may help.
- Soluble fibre sources, such as citrus pectin and beet pulp, are useful, together with wheat bran and other insoluble fibre sources, and a moderate level of several cooked starch sources.
- Fat supplements should not be added to the diet (although there may be a case for fats containing n-3 fatty acids, e.g. fish oils).
- Vitamin E (1500 iu/day) and a water-soluble B vitamin supplement, including 1000 mg choline/kg diet, are advisable.
- A supplement of 0.5 kg DL-methionine/t, has been recommended. There is, however, some evidence that excessive dietary methionine is converted by gut bacteria to mercaptans. Following absorption a diseased liver is unable to adequately clear these bacterial metabolites and acting with ammonia they may cause signs of encephalopathy. Unfortunately mercaptans are also derived from cystine, so, if the evidence is reliable, the basal diet should be relatively low in total sulphur-amino acids, i.e. not more than 3.5 g/kg diet. If methionine is to be added it should ideally be the L-α-isomer (preferable to DL-), and included at a concentration of 0.25 kg/t. (The author has never observed encephalopathy to result from methionine supplements in horses with compromised livers and so the risk may be slight.)

DIETS FOR KIDNEY DYSFUNCTION

Renal failure may lead to metabolic acidosis; therefore the diet should have as an objective the prevention of a low urinary pH. Dietary protein should be as described for liver disease (see 'Diets for liver disease' above). Dietary P should not be excessive and the Ca:P ratio should be 2:1. Sodium chloride supplements should not be given in excessive amounts and diuretics should be avoided. Grasses and other raw materials known to contain oxalates should be excluded and a pyridoxine supplement might be given. Fresh water should be available at all times. Blood urea and ammonia should be monitored in both kidney and liver dysfunction.

BONE METABOLISM

Skeletal ailments are not infrequent in young growing stock and chronic nonspecific lameness in adults is sometimes an expression of NSHP, by far the most common cause of which is faulty Ca and P nutrition and an improper balance between these two minerals. Thus, a simple means of evaluating Ca adequacy is required. Ca concentrations in serum vary to a small degree in relation to Ca intake, and where they are measured in a number of horses, significant differences can be detected between deficient and normal groups (Fig. 3.2). However, the method is insufficiently sensitive to be of practical use. Greater sensitivity is achieved with creatinine clearance ratios. The horse seems to regulate serum Ca more by renal excretion than by controlling intestinal absorption. Intact proximal tubules are required for P reabsorption and Ca excretion and in renal failure serum Ca is increased and P is depressed. The fully functioning kidney, on the other hand, excretes excess Ca in the urine and therefore an inadequate intake might be revealed by a reduced urinary excretion, if it were not for the fact that Ca sediments in horse urine, owing to its alkaline nature. Repeatable estimates of the Ca content of urine cannot therefore be readily attained.

Phosphate clearance may fall to zero where the diet is marginally adequate in P and the clearance will be increased when the intake is greatly in excess of need, or when the horse is suffering from NSHP. An inadequate dietary intake of Ca stimulates the secretion of parathormone, which increases the reabsorption of Ca by the renal tubules, mobilizes bone Ca, and increases the loss of phosphate through the tubules by decreasing tubular reabsorption of phosphate. The net effect of this is to stabilize the ionized concentration of serum Ca and to depress serum phosphate. Concentrations of Ca in serum tend to be below average, but still within the normal range. An increased phosphate clearance with a normal or slightly depressed serum Ca, is therefore frequently indicative of insufficient dietary Ca when other causes have been eliminated from consideration. The method, nevertheless, has its limits.

Measurement of changes in the rate of bone metabolism would be helpful in assessing bone modelling and remodelling, during training, and in the detection of cases of OCD, or early stages of other bone disorders and during bone fracture repair. Bone ALP is an enzyme located on the cell surface of osteoblasts and it plays an important role in bone formation. It accounts for approximately 60% of circulating ALP activity in horses of less than 1 year of age, whereas in those over 5 years it accounts for about 20%. Separation of the bone isoform is therefore necessary in the detection of changes in bone metabolism.

Price *et al.* (1995) reported procedures for the measurement of bone ALP in serum by precipitating it with wheat germ lectin (note: liver ALP binds with wheat germ in man, but not in the horse, and two-thirds of caecum ALP binds in the horse, but does not appear in the blood). This binding depends on the number of carbohydrate units in the side chain of the enzyme and although there are two or three isotypes within each tissue origin the procedure still segregates tissue sources. Using radioimmunoassay Price *et al.* (1995) also measured serum carboxyterminal propeptide of *type I* collagen (PICP) and serum pyridinoline cross-linked telopeptide domains of *type I* collagen (ICTP). ICTP is liberated into serum during the collagen degradation of bone resorption in man and it is a precursor of pyridinoline cross-links in urine. These determinants are all associated with bone modelling and are quantitatively and inversely correlated with age, at least up to 5 years in the horse (Table 12.6). Deviations from these ranges probably indicate bone metabolism, related to bone disease or repair, and with the parallel

Table 12.6 Reference ranges in serum from healthy horses for PICP, ICTP, bone ALP and total ALP (after Price *et al.* 1995).

Age group	PICP (µg/l)	ICTP (µg/l)	Bone ALP (u/l)	Total ALP (u/l)
<1 year	1216–2666	14–27	134–288	223–498
1–2 years	550–1472	8–23	33–125	134–238
3–4 years	248–925	6–15	25–70	101–203
5–20 years	136–394	0–9	13–47	91–352

determination of other parameters, particularly Ca and P clearance, the data should be of considerable diagnostic value. Type I collagen also occurs in the skin, tendons and ligaments, but the proportion in the serum pool, derived from these sources, seems to be small, and moreover inflammation of tendons in the horse is frequently related to bone resorption.

Osteocalcin has been used in measurement of bone metabolism, but the molecule is extremely labile and assay values in different laboratories vary considerably, so data derived from its measurement may be unreliable.

OTHER TESTS

Hair analysis

The chemical analysis of hair samples has been suggested as a means of measuring intakes of proteins, minerals, trace elements and toxic heavy metals. The author has found increased amounts of lead in hair in lead toxicosis. However, the mineral composition is influenced by hair colour (Hintz 1980b) and the intake of both minerals and trace elements may be more reliably quantified by other means. The bulb diameter of hair among horses under range conditions has been used successfully in the assessment of dietary protein adequacy (Godbee *et al.* 1979).

Sugar tolerance (as distinct from the glucose tolerance test, GTT)

Glucose and xylose absorption tests have been used to measure the gross function of the small intestine. The recommended procedure is to give 0.5 g D(+)xylose/kg BW as a 10% solution by stomach tube. By measuring the peak plasma-xylose concentration after 90 min it is possible to discriminate between normal and abnormal absorption. A normal control animal should also be measured under the same conditions.

Feed allergen tests

Feed allergen tests have been used successfully by the author in horses for determining whether feed entities are causal in certain oedematous skin and respiratory irregularities. The horse seems to be prone to such reactions, but cross reactions between feeds must be expected. Blood and skin tests are appropriate in the detection of dietary and mould allergens.

Haematology

The measurement of numbers of red and white blood cells has been used in folic acid, vitamin B_{12} and protein assessments. For consistency, samples should be drawn immediately after strenuous exercise; lower erythrocyte parameters should be expected for ponies and coldblooded horses. Normal values are also influenced by sex, age, stable and season, and, among racehorses, by training level. More sensitive methods of assessing vitamin B_{12} and folic acid status are suggested in 'Fat- and water-soluble vitamins', this chapter.

Potassium status

(See also Chapter 11.) The measurement of $[K^+]$ in red blood cells should be undertaken within 2h of removal of the blood from the horse. The metabolic energy required for the active transport of K^+ is not generated in blood samples stored in the refrigerator and consequently K^+ leaks by diffusion from the cells and Na^+

Table 12.7 A sequence of procedures for investigating possible causes of signs indicating possible nutritional anomalies in horses, recorded descriptively and by photograph, with times and dates.

Sequence of initial actions	Records to be kept	Laboratory investigation sequence
(1) Examination of horses, social interaction, collection of tissue samples, immediate therapy	Abnormal signs and age distribution	Routine haematology and biochemistry, preserve samples for later examination
(2) Examine 'co-grazing' ruminants	Associated signs	If necessary routine blood values
(3) Examine stable/paddocks	Water, weeds, pollutants, bedding, poisons, surface materials, dust	Identification of plants, pollutants and poisonous substances
(4) Sample all feeds, including pasture	Physical examination, ingredient make-up, calculate nutrient levels-intakes	Feed analysis for heavy metals, drugs, trace elements, fats, natural toxins
(5) Collect previous feed samples and records	Calculate presumed nutrient intakes	Calculate chronic intakes of nutrients and toxicants
(6)		Specific nonroutine determinations on initial samples
(7) Test presumptive corrective treatments in sequence or in selected sub-groups of horses	Clinical results	Determine any change in fresh sample values, previously abnormal

penetrates their membranes. This movement can be reversed by incubating the blood with glucose, or in fact by reinjecting the cells into the circulation.

PROCEDURES FOR DETERMINING CAUSES OF SUSPECTED NUTRITIONAL AND DIETARY PROBLEMS

Procedures for determining causes of suspected nutritional and dietary problems must vary according to the amount and reliability of the information at hand and whether there is a justification for determining causes rather than simply replacing an inadequate feed system by a proven system. A proposed procedure is given in Table 12.7.

STUDY QUESTIONS

(1) Construct a 'decision tree' and procedure you would adopt for determining the causes of a diet-induced nutritional problem.
(2) How would you propose a horse with poor liver function should be fed and managed?

FURTHER READING

Greppi, G.F., Casini, L., Gatta, D., Orlandi, M. and Pasquini, M. (1996) Daily fluctuations of haematology and blood biochemistry in horses fed varying levels of protein. *Equine Veterinary Journal*, **28**, 350–53.
Price, J.S., Jackson, B., Eastell, R., *et al.* (1995) Age related changes in biochemical markers of bone metabolism in horses. *Equine Veterinary Journal*, **27**, 201–207.

Appendix A
Example Calculation of Dietary Composition Required for a 400-kg Mare in the Fourth Month of Lactation

(1)   To calculate the proportions of hay and concentrate in the total diet, divide the required daily energy intake (MJ/day) by the desirable daily total feed intake (88% DM) to give the average energy density of the total diet. Then form a simple equation containing the energy densities of the roughage and of the cereal available (88% DM).

Required daily energy	84.5 MJ DE/day
Desirable total feed intake	9.0 kg
Oats	12.1 MJ DE/kg
Hay	7.3 MJ DE/kg

Let x be the proportion of cereal in the diet and $1 - x$ be the proportion of hay.

$$12.1x + 7.3(1-x) = 84.5/9.0$$
$$12.1x + 7.3 - 7.3x = 9.39 \text{ MJ DE/kg}$$
$$4.8x = (9.39 - 7.3) \text{ MJ DE/kg}$$
$$x = 2.09/4.8 = 0.435 \text{ or } 43.5\% \text{ oats } (435\,\text{g/kg})$$

If 43.5% of the diet is oats then 56.5% is hay.

(2)   Now calculate the protein, Ca and P contents of this simple mix from the information given in Appendix C.

Oats (12.1 MJ DE) contains (per kg):

    95 g   crude protein
    0.8 g Ca
    3.3 g P

Hay (7.3 MJ DE) contains (per kg):

    55 g   crude protein
    3.5 g Ca
    1.7 g P

Initial dietary composition (g)

	Crude protein	Ca	P
Contributed by oats:			
$\dfrac{43.5}{100}$ × each oat value above	41.3	0.35	1.44
Contributed by hay:			
$\dfrac{56.5}{100}$ × each hay value above	31.1	1.98	0.96
Total initial composition (g/kg)	72.4	2.33	2.40
Requirement (g/kg)	120.0	5.00	3.00
Deficit of minerals (g/kg)	—	2.67	0.60

Now calculate how much soyabean meal is required to make good the protein deficit (other protein concentrates could be used in amounts that are inversely proportional to the lysine contents of the protein source under consideration and that of soya). About 2–3 g more protein than is required should be allowed for because minerals will also be added to the diet. Soyabean meal from Appendix C contains 440 g crude protein/kg. As before:

$$72.4x + 440(1 - x) = 120.0 + 3 \text{g crude protein/kg}$$
$$72.4x + 440 - 440x = 123 \text{g/kg}$$
$$-367.6x = -317 \text{g/kg}$$

Change signs on both sides. $x = 0.862$ or 86.2% oats plus hay, and therefore (100 – 86.2) soya forms the remainder, i.e. 13.8% soyabean meal.

The soya contains more energy than oats and hay, but this will be approximately compensated for by the complete absence of energy in the mineral and vitamin supplement.

(3) Now 0.6 g P/kg is still required, being the deficit shown above. Dicalcium phosphate contains 188 g P/1000 g (from Appendix C). Therefore an addition of:

$$\frac{1000}{188} \times 0.6 \text{g} = 3.19 \text{g dicalcium phosphate (CaHPO}_4\text{)/kg}$$
total feed will provide the necessary P.

Dicalcium phosphate from Appendix C also contains 237 g Ca/1000 g. Therefore 3.19 g provides:

$$\frac{237}{1000} \times 3.19 \text{g} = 0.76 \text{g Ca}$$

The original deficit was 2.67 g Ca and it is now:

$$2.67 - 0.76\,\text{g} = 1.91\,\text{g Ca}$$

This can be made up with limestone flour ($CaCO_3$) containing 360 g Ca/1000 g. Therefore:

$$\frac{1000}{360} \times 1.91 = 5.31\,\text{g } CaCO_3/\text{kg total diet}$$

(4)   The diet should also contain 5 g common salt (NaCl)/kg and a proprietary vitamin/trace-element mixture suitable for horses.

(5)   The complete diet is now as follows:

	g/kg	Per cent
Oats	283	28.35 (43.5 − 13.8 − 0.32 − 0.53 − 0.5)
Soyabean meal	138	13.80
Dicalcium phosphate	3.2	0.32
Calcium carbonate	5.3	0.53
Salt	5.0	0.50
Vitamins/trace elements	+*	+
Hay	565	56.50
Total		100

* A supplement may be provided as described in Chapters 6 and 8.

The composition of the concentrate portion of the ration is as follows:

		Per cent
Oats	$\frac{28.3}{43.5} \times 100$	65.1
Soyabean meal	$\frac{13.8}{43.5} \times 100$	31.7
Dicalcium phosphate	$\frac{0.32}{43.5} \times 100$	0.74
Calcium carbonate	$\frac{0.53}{43.5} \times 100$	1.22
Salt	$\frac{0.5}{43.5} \times 100$	1.15
Vitamins/trace elements		0.1
Total		100

(6)   The total daily ration is to be 9 kg and of this the above concentrate would form 43.5/100 × 9 = 3.9 kg daily. The remaining 5.1 kg hay could be given in excess

as some will be lost and horses would naturally consume their ration of concentrate before filling up on hay.

The concentrate ration should be divided into a minimum of two feeds per day and introduced gradually in increasing amounts until the full ration is provided. Small adjustments of the quantity can be made to allow for differences in condition between individuals. Where growing horses are being fed, particularly of faster growing breeds, the overriding concern must be the condition of the legs and if there is any tendency towards contracted flexor tendons, epiphysitis or crooked legs the concentrate allowance should be reduced until the condition subsides (see Ch. 8 for details).

Appendix B
Common Dietary Errors in Studs and Racing Stables

Range of dietary composition for foals and yearlings in ten studs where home mixes have been prepared and when restricted access to pasture of moderate quality is provided (% of total diet, air-dry basis; dashes imply none used).

	Dietary range		Typical poor-quality diets at specific studs			
	Weaned foals	Yearlings	Foals (1)	Foals (2)	Yearlings (1)	Yearlings (2)
Oats	37–70	11–58	48	—	57	9.5
Boiled barley	0–1.6	0–26	1.6	—	—	20
Flaked maize	0	0–6	—	—	—	2.5
Bran (wheat)	0–12	5–28	5	—	28	16
Coarse mix (sweetfeed) of low quality	0–43	0–43	—	42	—	—
Barley chaff	0–7	0–7	—	7	—	—
Cubes of moderate quality	0–12	0–60	12	—	—	23
Soyabean meal	0–4	0	—	—	—	—
Linseed (boiled)	0–7	0–9	1.6	—	5	7
Palm kernel cake	0	0–3	—	—	—	—
Milk pellets	0–13	0	—	—	—	—
Locust beans	0	0–3	—	—	—	—
Egg	0	0–+	—	—	—	—
Molasses	0	0–2	—	—	—	—
Carrots	0	0–3	—	—	—	—
Grass hay	6–18	14–29	16	18	10	11

	Dietary range		Typical poor-quality diets at specific studs			
	Weaned foals	Yearlings	Foals (1)	Foals (2)	Yearlings (1)	Yearlings (2)
Lucerne hay	0–9.6	0	—	9.5	—	—
Limestone flour	0–2.2	0–2.2	—	2.2	—	—
Dicalcium phosphate	0–0.5	0	—	—	—	—
Vitamins, trace elements and minerals	+*	0–+	—	—	—	—
Pasture	12–36	14–40	16	22	—	11

* Some sort of supplement was used in each of the ten stables.

Some chemical characteristics

Crude protein (%)	10–16	10–17	11.1	10.0	11.8	11.7
Total lysine (%)	0.4–0.6	0.4–0.7	0.48	0.40	0.45	0.45
Ca (%)	0.32–0.9	0.22–1.2	0.33	0.65	0.23	0.44
P (%)	0.2–0.46	0.2–0.65	0.42	0.19	0.65	0.57

General comments

- Widespread use of poor-quality hay for young stock.
- Failure to rectify this by adjustment of concentrate composition.
- Insufficient checks on rates of growth.
- Failure to compensate for inadequacies of pastures.

Common errors

- Very variable protein and lysine intakes exacerbated by variable pasture quality and availability.
- Suboptimum protein: energy ratios for weaned foals.
- Extreme variation in Ca intake.
- Suboptimum Ca:P ratios.
- Excessive intakes of vitamins A and D, but possible inadequacies of several other vitamins.
- Lack of control of growth curve leading to poor conformation, epiphysitis and abnormal alignment of legs.
- Use of poorly formulated micronutrient supplements.
- Use of more than one micronutrient supplement lacking complementary effects.
- Pasture trace-element and other deficiencies, which contribute to metabolic and conformational problems.

Faults in diets (1) and (2) specified above

Foals:

(1)   Insufficient protein, lysine and Ca.
      Excessive use of vitamins A and D_3.
      Deficiency of selenium and marginal zinc status.
(2)   Insufficient protein and lysine.
      Excessively wide Ca:P ratio.
      Selenium depletion, zinc inadequacy and suspected induced manganese deficiency.
      Poor conformation.

Yearlings:

(1)   Marginal protein and lysine inadequacy.
      Rank Ca deficiency with adverse Ca:P ratio.
      Epiphysitis evident.
      Several vitamin inadequacies and suspect trace-element status.
(2)   Lack of objectivity in ration formulation and unnecessary complexity.
      Adverse Ca:P ratio.

Range of dietary composition for TBs in eight racing stables where home mixes are prepared (% of total diet, air-dry basis; dashes imply none used, or analytical value unavailable).

	Dietary range	Typical poor-quality diets at specific stables	
		(1)	(2)
Oats	32–59	49	51
Bran (wheat)	0–16	—	1.1
Coarse mix (sweetfeed)	0–20	—	—
Chaff	0–3.7	—	—
Cubes	0–24	—	2.0
Soyabean meal	0–2.2	0.8	0.7
Linseed (boiled)	0–1.5	0.8	1.5
Molassed peat	0–0.5	—	0.4
Carrots	0–2	—	—
Grass pellets	0–6	—	—
Molasses	0–1.5	—	1.4
Grass hay	28–53	49	40.6
Limestone flour	0–1.1	—	1.1
Salt (sodium chloride)	0–0.1	—	0.07
Corn oil	0–0.57	0.11	—
Vitamins, trace elements and minerals	0–4	0.2	0.1
Some chemical characteristics			
Crude protein (%)	7.2–11.5	7.3	7.4

	Dietary range	Typical poor-quality diets at specific stables	
		(1)	(2)
Total lysine (%)	0.35–0.5	0.36	0.40
Ca (%)	0.15–1.38	0.15	0.68
P (%)	0.24–0.43	0.25	0.28
K (%)	—	1.5	—
Na (%)	—	0.16	—
Mg (%)	—	0.18	—
Zn (mg/kg)	—	—	24
Mn (mg/kg)	—	—	46

General comments

- Oats and poor-quality grass hay frequently constitute over 90% of the diet and their composition is variable and unknown.
- Several noncomplementary micronutrient supplements are frequently used in the same diet.
- Frequently insufficient common salt is consumed in hot weather.
- Frequency of feeding is often insufficient.
- Rate of ration change and of energy intake frequently inappropriate.
- Notion that a rest on Sundays with changes in management and in feeding benefits the horse is misguided, in contrast to human benefits.

Common errors

- Variable protein and lysine intakes.
- Suboptimum protein:energy ratios.
- Extreme variability in Ca intake.
- Suboptimum Ca:P ratios.
- Excesses of vitamins A and D_3.
- Inadequate allowances of folic acid and possibly of other water-soluble vitamins and of salt.
- Incorrect quantities of vitamins, trace elements and minerals provided by most supplements.

Faults in diets (1) and (2) specified above

(1) Lameness, metabolic upsets (e.g. azoturia, setfast).
    Dietary faults include excesses of vitamins A and D_3 and excess iodine.
    Insufficient Ca and adverse Ca:P ratio.
    Insufficient sodium and folic acid, marginal protein and probably too wide an energy:protein ratio.
(2) Abnormal blood characteristics.
    Marginally low protein intake.
    Insufficient allowances of zinc, folic acid and salt.

Appendix C
Chemical Composition of Feedstuffs Used for Horses

Values for feedstuffs (a) assume 880 g DM/kg (dashes imply no value available); values for forages (b) are typical rather than average values. Equations in (c), (d) and (e) can be used to estimate UFC values from the chemical composition of feeds.

(a) Feedstuffs

	Crude protein (g/kg)	MADC (g/kg DM)	MADC (88% DM)	Oil (g/kg)	Crude fibre (g/kg)	MAD fibre (g/kg)	Ash (g/kg)	Ca (g/kg)	P (g/kg)	K (g/kg)
Oats	96	103	91	45	100	170	40	0.7	3.0	5.0
Barley	95	92	81	18	50	70	25	0.6	3.3	5.0
Wheat	100	98	86	15	22	40	19	0.4	3.2	4.2
Maize	85	79	70	38	25	30	15	0.2	3.0	3.1
Sorghum (white)	10.6			25	27	60	18	0.3	2.7	4.1
Rice (rough)	73			17	90	—	52	0.4	2.6	—
Millet	88			13	277	—	81	—	—	—
White fishmeal	660			80	0	0	215	57.0	34.0	8.3
Dried skimmed milk (spray)	340			6	0	0	80	10.5	9.8	16.0
Linseeds	219			316	76	135	45	2.4	5.2	9.4
Exp. linseed meal	320	323	284	60	100	170	60	3.0	7.3	11.0
Extr. soyabean meal	440	437	385	10	62	100	60	2.4	6.3	23.5
Extr. sunflower seedmeal	280–450	344	303	18	230	—	77	2.9	8.0	14.0
Exp. cottonseed cake	410			37	140	220	65	2.0	10.5	15.0
Groundnut meal extr. 48–50% CP	470	—	—	13	140	220	50	2.0	7.1	12.0
Maize gluten feed	210	191	168	25	70–80	—	70	2.5	7.5	9.7
Maize gluten feed (40%)	410	427	376	25	50	—	40	2.0	5.1	1.0
Field beans	255	272	239	90	74	114	29	0.8	4.8	11.0
Peas	229	223	196	50	57	82	27	0.7	4.0	11.0
High-protein grassmeal	160			32	220	360	70	6.0	2.3	21.0
Lucerne meal	170	101–122	89–107	30	250	400	100	15.0	2.0	22.0
Brewer's yeast	450			10	30	—	65	4.4	13.3	18.0
Rapeseed meal extr.	350			24	130	—	70	6.5	11.0	—
Wheatfeed	155	148	130	35	85	100	50	1.0	10.0	12.0
Wheat bran	155	130	114	30	110	120	70	1.0	12.1	14.0
Extr. rice bran	135			15	120	190	125	1.1	19.0	19.0

Lysine (g/kg)	DE (MJ/kg)	UFC (g/kg DM)	UFC (88% DM)	α-Tocopherol (mg/kg)	Free folic acid (mg/kg)	Available biotin (μg/kg)	Thiamin (mg/kg)	Riboflavin (mg/kg)	Pantothenic acid (mg/kg)
3.2	10.9–12.1	0.99	0.87	9	0.12	50	17.0	1.7	12.0
3.1	12.8	1.16	1.02	7	0.11	12	5.0	1.8	14.0
2.8	14.1	1.26	1.11	9	0.12	4	4.0	1.1	11.0
2.6	14.2	1.33	1.17	9	0.06	65	2.0	1.5	6.0
2.4	13.0	—	—	7	0.13	—	3.0	1.1	12.0
2.5	11.1	—	—	—	0.20	15	2.5	0.9	—
—	—	—	—	—	—	—	—	—	—
48.0	14.1	—	—	13	0.22	100	5.0	5.0	8.8
29.0	15.1	—	—	10	0.60	330	4.0	20.0	30.0
7.7	18.5	—	—	2	—	—	7.0	2.5	—
11.3	13.9	0.92	0.81	1	—	—	8.0	3.0	12.0
26.0	13.3	1.06	0.99	2	0.57	280	6.0	3.3	14.0
13.0	9.5	0.79	0.70	10	—	415	34.0	3.0	—
14.0	12.8	—	—	5	0.22	250	7.0	4.0	9.5
15.0	12.5	1.01	0.89	2	0.5	300	8.0	3.0	40.0
5.8	12.8	0.96	0.85	7	0.20	85	2.0	2.2	14.0
11.2	13.1	1.14	1	7	0.2	90	0.4	1.8	12.0
17.0	13.1	1.04	0.92	—	—	—	—	—	—
15.8	14.1	1.15	1.01	8	—	—	—	1.4	—
8.0	9.6	—	—	25	1.80	300	3.0	15.0	25.0
8.2	9.0	0.57–0.68	0.5–0.6	30	3.00	400	3.7	14.0	25.0
29.3	12.2			2	2.31	300	100.0	39.0	105.0
20.0	11.5			—	—	575	1.0	3.6	9.0
6.1	11.0	1.09	0.96	23	0.67	15	13.0	4.5	16.0
6.0	10.8	0.86	0.75	21	0.30	15	7.3	5.4	26.0
5.3	10.8	—	—	—	—	20	23.0	2.6	23.0

	Crude protein (g/kg)	MADC (g/kg DM)	MADC (88% DM)	Oil (g/kg)	Crude fibre (g/kg)	MAD fibre (g/kg)	Ash (g/kg)	Ca (g/kg)	P (g/kg)	K (g/kg)
Oatfeed	50			22	250	400	70	0.8	1.2	7.0
Malt sprouts (culms)	250	257	226	20	140	—	65	2.0	7.0	—
Dried brewer's grains	180–250	251	221	62	140–170	180	38	2.6	5.1	1.0
Beet pulp	100	45	40	10	174	340	50	7.0	0.8	2.0
Molassed beet pulp	90–120			1–6	130	—	60–80	6.0	0.7	16.0
Grass hay	45–80	40–60	35–53	25	330	380	50	2.9	1.7	17.0
Clover/grass hay	60–100	50–70	44–62	27	330	380	65	4.0	1.7	19.0
Maize silage in cob	—	30	26	—	—	—	—	—	—	—
Spring cereal straw, barley	30	0	0	19	410	590	70	2.0	0.4	19.0
Vegetable oil	0	—	—	100	0	0	0	0	0	0
Molasses (cane)	30	34	30	0	0	0	85	7.2	1.0	27.0
Molasses (sugar beet)	—	83	73	—	—	—	—	—	—	—
Nutritionally improved straw	45	—	—	12	340	490	—	4.0	1.0	18.0
Limestone flour	0	0	0	0	0	0	990	365	4	0
Dicalcium phosphate	0	0	0	0	0	0	100	238	187	0
Steamed bone flour	0	—	—	0	0	0	980	323	133	0
Pasture:										
1. First growth		103	91							
2. Second growth		130 to 92	114 to 81							
Pure clover		112	99							
Pure grass		98	86							
3. Blooming										
Pure clover		96	84							
Pure grass		40	35							
4. Winter after close grazing until July, and free growth from July to December										

* Contains 30 g Na/Kg.
Exp., expeller; extr., extracted; MAD, modified acid detergent.

Lysine (g/kg)	DE (MJ/kg)	UFC		α-Toco-pherol (mg/kg)	Free folic acid (mg/kg)	Available biotin (μg/kg)	Thiamin (mg/kg)	Riboflavin (mg/kg)	Pantothenic acid (mg/kg)
		(g/kg DM)	(88% DM)						
1.9	7.7			—	—	—	—	—	—
12.0	10.0	0.92	0.81	—	—	—	—	—	—
8.3	10.0	0.79	0.7	10	—	—	0.6	1.0	10.0
5.6	11.0	0.86	0.75	—	—	—	0.4	0.6	1.5
2.6	12.0	—	—	1	—	—	0.4	0.4	2
3.0	7.0	0.41–0.55	0.36–0.48	7	—	—	1.5	10	—
5.2	7.5	0.51–0.57	0.45–0.50	8	—	—	2.0	15	—
—	—	0.85	0.75	—	—	—	—	—	—
0	6	0.36	0.32	—	—	—	—	—	—
0	35	—	—	—	0	0	0	0	0
—	11.4	1.07	0.94	0	0.08	100	0.8	1.0	35
—	—	1.06	0.93	0	—	—	—	—	—
—	—	0.46	0.40	0	—	—	—	—	—
0	0	0	0	0	0	0	0	0	0
0	0	0	0	0	0	0	0	0	0
0	0	—	—	0	0	0	0	0	0
		0.88	0.77						
		0.80 to	0.70 to						
		0.78	0.69						
		0.80	0.70						
		0.88	0.77						
		0.69	0.61						
		0.67	0.59						

(b) Forages

	Crude protein (g/kg)	MADC (g/kg DM)	MADC (88% DM)	Oil (g/kg)	Crude fibre (g/kg)	MAD fibre (g/kg)	Ash (g/kg)	Ca (g/kg)	P (g/kg)	K (g/kg)	DE (MJ/kg)	UFC (g/kg DM)	UFC (88% DM)	pH	NH$_3$N as % of total N
Lucerne hay mid-bloom	150	84	74	17	270	350	70	11.4	1.9	16	7.6	0.5–0.63	0.44–0.55	—	—
Pasture:															
(1) First growth	167			38	176	194	97	5.3	3.1	26	9.3			—	—
(2) Second growth	194			35	195	215	90	5.3	1.9	21	9.6			—	—
Pure clover	220			31	176	229	97	15.8	1.9	18	9.0			—	—
Pure grass	176			54	150	211	85	5.7	1.8	21	10.0			—	—
(3) Blooming															
Pure clover	150			26	211	308	105	14.1	2.0	17	8.8			—	—
Pure grass	79			13	264	290	92	3.1	1.8	15	8.4			—	—
(4) Winter after close grazing until July, and free growth July–December	136			26	194	—	70	—	—	—	—			—	—
Clamp silage	108	40–60	35–53	28	299	334	62	5.3	2.6	15	9.2	0.6–0.68	0.53–0.6	4.2	12.4
Big-bale silage	100	36–55	32–48	35	232	273	62	—	—	—	—	0.65–0.72	0.57–0.63	5.1	8.9

(c) Some factors necessary for the estimation of km, used in the calculation of UFC values and the estimation of UFC values from chemical composition of feeds. Gross energy (GE, kJ/g) and km % of nutrients

	GE	km
glucose (GL)	15.65	85
acetate (C_2)	14.60	
propionate (C_3)	20.76	
butyrate (C_4)	24.94	
amino acid (AA)	23.44	70*
long chain fatty acids (LCFA)	39.76	80

* amino acid ME

So

$$km = 0.85E_{GL} + 0.80E_{LCFA} + 0.70E_{AA} + (0.63 \text{ to } 0.68) \, E_{VFA} - 0.14 \, (76.4 - ED),$$

where E is the % of absorbed energy from glucose or lactate (GL) etc., ED is energy digestibility (%) of the feed and the last term of the equation corresponds to the cost of eating, a term not included for concentrate feeds. The percentage of absorbed energy from a typical horse diet of roughage and concentrate is represented by 9–41% GL, 45–82% VFA, 7–17% AA, and 2–6% LCFA (Vermorel *et al.* 1997).
(d) Variations in molar proportions, energy content (E, kJ/g VFA mixture) and efficiency (km) of E utilization of VFA mixtures absorbed from the colon of the horse receiving diets of three crude fibre concentrations (Vermorel and Martin-Rosset 1997).

Crude fibre % DM:	15	20	30
C_2 %*	65	68	73
C_3 %	21	19	16
C_4 %	14	13	11
E, kJ/g	17.91	17.70	17.24
km %	66.6	65.8	64.5

* The molar proportion of acetate (C_2) in the VFA mix increases with increasing dietary crude fibre (CF), C_2 % = 0.54 CF (% DM) + 57. The energy expended in mastication also increases with increasing dietary crude fibre, $\Delta km = -0.20$ CF% + 2.50, so the above km values decrease further from 15% to 30% CF. Both the heat of fermentation and the energy wasted in metabolism are greater with C_2 metabolism than with the other VFA. Thus increasing fibre is associated with increased waste heat production and poor meadow hay has a km % of only 61, whereas that of maize is 80.

(e) Prediction of UFC value per kg dry matter (DM) of forages and concentrates from their cytoplasmic carbohydrate (CC), crude protein (CP), crude fibre (CF) and digestible organic matter (DOM), kg/kg DM, or digestible energy (DE, MJ/kg DM) content (Vermorel and Martin-Rosset 1997).
(i)   Forages

$$UFC = -0.124 + 0.254CC + 1.330DOM, \text{ RSD } 0.012, R^2 \text{ } 0.988.$$
$$UFC = -0.056 + 0.562CC + 0.0619DE, \text{ RSD } 0.007, R^2 \text{ } 0.996.$$

(ii)   Concentrates

$$UFC = -0.134 + 0.274CF - 0.362CP + 0.316CC + 0.0755DE, \text{ RSD } 0.017, R^2 \text{ } 0.995.$$

Note: See Martin-Rosset (1996c) for CC values, ie water-soluble carbohydrates, of feeds and reference to a comprehensive data source.
(iii)   Compound feeds

$$UFCo = 1.333 - 1.684ADFo - 0.096CPo, \text{ RSD } 0.060, R^2 \text{ } 0.958.$$

Note: Martin-Rosset *et al.* (1996c) define UFCo as a UFC per kg organic matter to allow for the high mineral content of some compound feeds. UFCo values of compound feeds must be increased by 0.02 UFCo for each 1% ether extract above 3.5% of feed.
(iv)   Organic matter digestibility of forages (OMD %)
Three alternative methods are proposed for the determination of OMD % of horses, of which the second and third are the preferred methods.

(1)   OMD % = $67.78 + 0.07088CP - 0.000045NDF^2 - 0.12180ADL$, RSD 2.5, R^2 0.878, Martin-Rosset *et al.* (1996a); where NDF is neutral detergent fibre, g/kg DM and ADL is acid detergent lignin, g/kg DM.

(2)   OMD % in horses predicted by near infrared spectrophotometry (NIRS); RSD 1.80, R^2 0.93, Andrieu *et al.* (1996).

(3)   OMD % = $-29.38 + di + 2.3032CDMD - 0.01384CDMD^2$, RSD 1.90, R^2 0.927, Martin-Rosset (1996b); where di = +4.12 for green forages, di = 0 for grass hays and di = −2.61 for legume hays; CDMD is the pepsin cellulase degradability (%) of DM.

Appendix D
Estimates of Base Excess of a Diet and of Blood Plasma

ESTIMATE OF BE OF A DIET FROM ITS FIXED ION CONTENT

(See Chapter 9 for further details.)

$$(\text{Cations} - \text{anions})_{\text{absorbed}} - (\text{cations} - \text{anions})_{\text{excreted in urine}} - H^+ \text{ endogenous}$$
$$= BE \tag{1}$$

Account here is taken only of fixed ions absorbed from the diet and Eq. (2) shows the principal ones involved. (Note: fixed ions are those that cannot be degraded by metabolism.)

$$(\text{Cations} - \text{anions})_{\text{absorbed}} = \text{mEq}(0.95 \text{ Na} + 0.95 \text{ K} + 0.5 * \text{Ca} + 0.5 \text{ Mg})$$
$$- \text{mEq}(0.95 \text{ Cl} + 0.95 \text{ S} + 0.5 * \text{P}) \tag{2}$$

* Approximate values which will be inversely related to dietary concentration.

In order to avoid a degree of arbitrariness a simplified balance of ions has been proposed and these are shown in Eq. (3) in relation to their optimum range in the diet of a horse in light work.

$$(\text{Na} + \text{K} - \text{Cl})_{\text{absorbed}} = 200\text{--}300 \text{ mEq/kg diet} \tag{3}$$

(Note: $Na^+ + K^+ = 95\%$ of all cations in extracellular fluid and $Cl^- + HCO_3^- = 85\%$ of all anions.)

ESTIMATE OF BE OF BLOOD PLASMA FROM ITS BICARBONATE CONCENTRATION

Plasma bicarbonate (HCO_3^-) (mEq/l) at pH 7.4:

$$\cong [HCO_3^-]_{\text{measured}} - 10(7.4 - \text{pH measured})$$

BE of plasma at pH 7.4:

$$\cong [HCO_3^-] \text{ at pH } 7.4 - 24$$

(Normal bicarbonate of venous blood at pH 7.4 = 24 mEq/l.)

If the venous plasma of a horse was found to have a pH of 7 and $[HCO_3^-]$ was 30 mEq/l then $[HCO_3^-]$ at pH 7.4 would

$$= 30 - 10(7.4 - 7)$$
$$= 26 \text{ mEq/l}$$

Therefore

$$BE = 26 - 24$$
$$\cong 2 \text{ mEq/l}$$

Note: other organic acid anions could also be included in a BE calculation.

Glossary

aboral	Away from or remote to the mouth.
acidaemia	An increased hydrogen ion concentration (acidity) and lowered blood bicarbonate, or decreased pH (q.v.) of the blood (see p. 292).
acidosis	See **acidaemia**.
ACTH	Adrenocorticotrophic hormone (corticotropin) is secreted by the anterior pituitary gland, so controlling corticol secretion by the adrenal cortex. The release of ACTH is in turn controlled by corticotropin releasing hormone (CRH) secreted in the hypothalamus.
acute	Applied to a metabolic upset, or a disease, which progresses rapidly to a climax followed by death or rapid recovery. Contrasts with a chronic (q.v.) condition or disease.
acute phase proteins	Specific serum proteins which increase rapidly in concentration (up to 100-fold) following infection, e.g. C-reactive protein, and which increase during inflammatory response, e.g. caeruloplasmin (ferroxidase), which also transports Cu (contains about 3% of the body's Cu).
adipose tissue	The cells of this tissue readily store fat which is drawn on as a source of energy, especially when the blood levels of glucose and VFAs are low.
ad libitum **feeding**	A system in which feed supply is unrestricted at all times except during exercise. However, usually applies only to growing horses.
adrenal glands	A pair of ductless glands, one situated near each kidney, and consisting of an internal medulla, which secretes the hormones epinephrine and norepinephrine (adrenaline and noradrenaline), and an external cortex, which secretes corticosterone, cortisol (glucocorticoids) and aldosterone.
aerobic	In aerobic respiration energy-yielding nutrients are broken

down with the consumption of dissolved oxygen that has reached the tissue cells from the lungs. In anaerobic respiration the breakdown of these nutrients is incomplete, it yields less energy and occurs in the absence of oxygen.

afferent
Afferent nerve fibres conduct impulses centripetally, e.g. from sense organs to the central nervous system.

agalactia
Failure of the secretion of milk.

agglutination
The clumping together of particulate antigen (foreign substance) in the presence of homologous antibody (defence substance).

air dry
Under UK regulations, feed that has been allowed to dry without heating in the air contains 100–140 g moisture/kg.

aldosterone
The main mineralocorticoid hormone secreted by the adrenal cortex, promoting the reabsorption of Na and therefore water by the kidney tubules.

alfalfa
See **lucerne**.

alimentary canal/tract
See **gastrointestinal tract**. In addition, it includes the buccal cavity and oesophagus.

alkalaemia
A decreased hydrogen ion concentration, or elevated pH (q.v.), of the blood, irrespective of changes in blood bicarbonate. Normal arterial blood pH is 7.5 (see p. 288).

allergy
A condition of exaggerated susceptibility, or sensitivity, to a specific substance, usually but not necessarily containing a specific protein. Exposure, especially to large amounts of the allergen, through inhalation, ingestion or injection, or even by skin contact, may cause respiratory difficulties, sneezing, a skin rash or diarrhoea of increasing severity through repeated contact.

ALT (also **GPT**)
Alanine amino-transferase (EC 2.6.1.2), formally called glutamic pyruvic-transaminase. The activity of this enzyme in the blood plasma shows a similar reaction to that of AST (q.v.), particularly in respect of exercise and muscle damage.

amino acids
These N-containing compounds are the building-blocks of proteins. There are some 25 different kinds in protein, 10 of which are known as dietary indispensable (essential) nutrients, the most critical of which is lysine.

α-amylase
An important digestive enzyme in the digestion of starch and other polysaccharides containing three or more α-1,4-linked D-glucose units. It hydrolyses the α-1,4-glucan links.

anabolism
The process of synthesis of complex organic molecules in the body from simpler precursors (cf. catabolism, q.v.).

anaemia
A condition in which there is a reduced number of red cells and/or a reduced haemoglobin content of the blood. The volume of packed red cells is reduced when the equilibrium

between blood loss, through bleeding or destruction, and blood production is disturbed.

analgesic   A pain-relieving substance.

angiogenesis   The development of vessels; in the text it refers to blood vessels.

anodyne   A drug used for relieving pain.

anorexia   Lack, or loss, of appetite for feed.

anthelmintic   A substance used to destroy parasitic worms.

antibiotic   A chemical substance produced by and obtained from living cells, especially of lower plants such as moulds, yeasts or bacteria, that is antagonistic to, or destroys, some other form of life. It may be so used to destroy infectious organisms.

antibody   A specific substance (immunoglobulin) found in the blood, or in certain secretions, in response to the antigenic stimulus of bacteria, viruses, worm parasites and certain other foreign substances. An antibody has a specific amino acid sequence and can combine specifically with the inducing foreign entity (antigen), helping to inactivate it.

antigen   Any substance that is capable under appropriate conditions of inducing the formation of antibodies and of reacting specifically with those antibodies.

antihistamine A drug that counteracts the effects of histamine or certain other amines that cause inflammation.

antitoxin   A substance found in blood serum or other body fluid that is antagonistic to a particular toxin (q.v.). For therapeutic, or protective, use it may be injected into horses to neutralize the toxin of a particular disease, but as it does not stimulate the horse to produce its own antitoxin, its (passive) effects may last for only a few weeks.

arrhythmia   Any variation from the normal rhythm of the heart beat.

artery   A vessel containing smooth muscle through which blood passes away from the heart to the various parts of the body.

ascarids   Roundworms. A group of large parasitic intestinal parasites in the phylum Nematoda. They are 15–20 cm long, white and infest primarily young horses, as those over 3–4 years of age have usually developed considerable immunity (for life history see Chapter 11). Large numbers of adult worms in the intestines can cause impactions, intestinal perforations and colic.

ascites   Effusion and accumulation of serous fluid in the abdominal cavity.

ash   The ash content of feed is determined by ignition of a known weight of it at 500°C until the carbon has been removed. The residue represents approximately the inorganic constituents of the feed – principally Ca, K, Mg, Na, S, P and Cl. Some feeds, particularly those contaminated with soil, may contain a significant amount of silica.

AST (also **AAT** and **GOT**)
Aspartate amino-transferase (EC 2.6.1.1), formally called glutamic oxaloacetic-transaminase. This enzyme is released into the blood following damage to liver or muscle cells so that the blood level rises sharply. Plasma activity of the enzyme usually increases after hard exercise. The normal maximum blood plasma level in adult horses is 250 iu/l.

ataxia
Failure of muscular coordination, or irregularity of muscular action, resulting in a staggering gait. It may result particularly from exhaustion or from pathological change in the nerves.

ATP
Adenosine triphosphate mediates the transfer of energy, from the breakdown of glucose and fatty acids (exergonic reactions), for the synthetic processes of growth, milk secretion, etc. (endergonic reactions) and for muscular action. ATP is split by the enzyme ATPase, with the liberation of inorganic phosphate.

autogenous
Self-generated. It refers particularly in the text to antibodies produced by the dam to blood proteins of the fetus that happen to be circulating in her blood. If colostrum is then taken by the foal within the first 12 h, there will be an antibody reaction with the foal's blood proteins.

autonomic nervous system
This is the self-controlling part of the nervous system – it is not subject to direct influence by the conscious brain. The sympathetic system (thoracolumbar outflow) and the parasympathetic (craniosacral flow) have largely antagonistic actions. The combined system is of importance in regulating the activities of many of the glands, the smooth musculature of the GI tract and elsewhere, and the heart and blood vessels.

availability of minerals
The proportion of a nutrient mineral supplied in the feed which, at a stated level of inclusion and level of feeding, can be absorbed and utilized by the horse to meet its net requirement. It is not necessarily synonymous with true digestibility.

azotaemia
An excess of urea and other nitrogenous compounds in the blood.

azoturia
An excess of nitrogenous compounds in the urine. Considered synonymous with exertion myopathy. It frequently occurs within a short interval of beginning exercise after a rest of 2 or 3 days. A reluctance to move and muscular spasms are witnessed and excessive lactic acid accumulates in the

muscles. The muscles of the hindquarters become tense and there is a tendency to knuckle-over at the fetlocks. At some stage the horse will pass quantities of urine from light-Burgundy-wine to dark-coffee colour. Sometimes the urine is retained and requires relief with a catheter.

β-carotene   See **carotene**.

big-head   A condition seen in horses and ponies given a ration based on cereals, bran and poor hay without adequate mineral supplementation. The bones of the upper jaw and face are particularly enlarged owing to a replacement of their normal mineralized structure by fibrous connective tissue.

bile duct   The duct carrying bile, synthesized by the liver, to the duodenum, where the bile facilitates emulsification of fats and contributes to the production of an alkaline reaction of the intestinal contents.

biological values (BV)   The amount of N retained by an animal per unit of N absorbed from the GI tract and from a given feed protein.

biuret   A simple organic compound containing three N atoms, which has been put forward as a source of dietary NPN. The structurally related compound urea contains only two N atoms per molecule.

bleeders   Broken blood vessels. This usually refers to the loss of blood through the nose after hard exercise. Blood is lost from small broken blood vessels in the lungs, or nasal passages, of about half the population of healthy TBs during hard exercise and the condition requires no treatment unless large blood vessels are involved.

blood counts   These normally include the numbers of red blood cells (erythrocytes) per unit volume, the packed cell volume (haematocrit), or proportion of cells by volume in the total blood measured after the blood has been centrifuged, and the haemoglobin content of the total blood. Three other characteristics of the red cells (McV, mean corpuscular volume; McH, mean corpuscular haemoglobin and McHc, mean corpuscular haemoglobin concentration) are calculated from these basic data. The numbers of white blood cells (leucocytes) per unit volume and the differential count (proportion of each type of white cell) may also be measured.

borborygmus   Rumbling noise caused by the propulsion of gas through the intestines.

botfly   The larvae, or maggots, of *Gastrophilus* spp. cause chronic gastritis and loss of condition in grazing animals. *Gastrophilus intestinalis* lays eggs of pin-head size on the horse's hair in the summer. When the horse licks itself, the eggs are taken into the mouth and hatch there or in the stomach. On rare occasions they cause perforation of the stomach and death (see Chapter 11).

botulism
A rapidly fatal motor paralysis caused by the ingestion of the toxin of *Clostridium botulinum*, a spore-forming anaerobic bacterium, which proliferates in decomposing animal tissue and sometimes in plant material. The toxin (a di-chain protein, MW 140 000) is the most neurotoxic substance known. In the text (Chapter 10) botulism is referred to in cases in which horses have consumed silage (q.v.) that has been subject to an abnormal fermentation. The toxin seems to inhibit irreversibly the release of acetylcholine from peripheral nerves, and so impedes neuromuscular transmission. A flaccid descending paralysis develops.

bradycardia
Slow heart beat.

broken-wind, heaves
These are outdated expressions applied to longstanding respiratory diseases in which a double expiratory effort is a feature. The causes include bacteria, viruses and, rarely, lung tumours. A common cause is an allergic reaction, and that related to the inhalation of spores and particles of the moulds *Micropolyspora faeni* and *Aspergillus fumigatus* is discussed in Chapter 11.

bruxism
Rhythmic or spasmodic grinding of teeth.

buccal cavity
The cavity of the mouth between the cheeks bounded at one end by the lips and at the other by the pharynx.

buffers
Substances that in aqueous solution increase the amount of acid or alkali that may be added without changing the pH, or degree of acidity or alkalinity. In general, a buffer is made up of two components: (1) a weak acid, e.g. H_2CO_3, and (2) its corresponding base, HCO_3^-. Arterial blood is well buffered around a pH of 7.5.

caecum
A great cul-de-sac intercalated between the small intestine and the colon. In the adult horse it is about 1.25 m long, with a capacity of 25–30 l (see Chapter 1).

calcitonin
A hormone synthesized by the parafollicular cells of the thyroid gland (q.v.). It is secreted when the serum concentration of Ca ions rises, promoting the deposition of Ca in bones, and so counteracting the action of parathyroid hormone (q.v.).

calculi
Urinary calculi consist of accumulations of mineral substances in the urinary tract. They form in the bladder, less frequently in the urethra and rarely in the kidneys. They are commonly rough and yellow–brown and are composed of

calcium carbonate. These are seen more commonly in horses on high-roughage diets or on pasture. The less common phosphate calculi are smooth and white and occur with high-cereal rations. A low intake of water may predispose the horse to calculi and the signs include difficulty in urination and incontinence.

calorie A unit of energy, being the amount of heat required to raise 1 g water 1°C. 1000 calories = 1 kilocalorie (kcal). The joule (J) has now been adopted as the unit of energy in nutrition; 4.184 kJ = 1 kcal.

carbohydrates Compounds composed of carbon, hydrogen and oxygen and including the sugars, starches and other storage carbohydrates and the structural (fibre) carbohydrates – cellulose and hemicelluloses; also pectins, gums and mucilages. Lignin is included in the structural fibre but is not strictly a carbohydrate.

cardiac output The volume of blood expelled by each ventricle per minute. The stroke volume is the volume of blood discharged by each ventricle at each beat (i.e. the cardiac output divided by the heart rate). The cardiac output of the left ventricle after birth is 2–8% higher than that of the right, so the value quoted represents the mean of these two.

carotene Green plants contain a number of yellow carotenoid pigments, the most important of which are α, β and γ-carotenes and hydroxy-β-carotene. The most potent of these, β-carotene, is converted to vitamin A by the intestinal wall.

carpus The 'knee' of the forelimb of the horse between the radius (q.v.) above and metacarpus (q.v.) below.

catabolism The breaking down of tissue nutrients and components to less complex molecules (cf. anabolism. q.v.).

catecholamines A group of similar compounds having sympathomimetic action. They include epinephrine, norepinephrine and dopamine. The latter two serve as neurotransmitters and the former two are hypertensive, stimulating smooth muscles.

cathartic A purgative, or medicine, that quickens the evacuation of the bowels (intestines).

cation The positively charged elements of electrolytes, which include all metals and hydrogen ions.

cellulolysis The breaking down, or digestion, of cellulose. Horses and other mammals depend on certain bacteria in their intestinal tracts to carry this out as they do not secrete enzymes capable of it.

cellulose A structural carbohydrate (fibre) of plant cells.

choke Obstruction to the passage of food through the pharynx and oesophagus, either partial or complete. This is frequently caused by a mass of dry impacted feed.

cholestasis Suppression of the flow of bile. This may be unusually caused by gallstones and is then termed cholelithiasis.

chondrocyte A mature cartilage cell embedded within the cartilage matrix.

chronic Long-continued, the opposite of acute (q.v.).

CK (or **CPK**) Creatine kinase (EC 2.7.3.2). This enzyme is measured as an indicator of muscle damage. The normal maximum blood-plasma level in adult horses is 105 iu/l.

coenzyme A nonprotein organic compound, which may be a vitamer, the presence of which is required by an enzyme for the catalysis of the particular reaction to occur.

coldblooded horse In Europe two types of horse evolved – the light short-legged Celtic pony and the Great Horse of the Middle Ages (the large powerful but slow heavy horse). Present-day breeds whose major blood lines derive from either of these types are termed 'coldblooded' (cf. hotblooded horse, q.v.).

colic Abdominal pain. Where this originates from part of the GI tract it is termed 'true colic', but where it derives from one of the other vital organs or muscles it is defined as 'false colic'.

colon Made up of the great colon, which originates from the caecocolic orifice and terminates where it joins the small colon, and the latter, which continues to the rectum. In the adult horse the great colon is 3–3.7 m long with an average diameter of 20–25 cm, whereas the small colon has a diameter of 7.5–10 cm and a length of about 3.5 m.

colostrum Secreted by the mammary gland of the mare shortly before foaling (parturition) and for about the first 24 h after the birth of the foal. It is rich in gamma globulins (q.v.), which comprise the antibodies that the foal absorbs undigested into its blood during approximately the first 12–18 h of life, providing it with a measure of protection from disease.

compounded feeds Balanced mixtures of ground or otherwise processed feedstuffs, to which appropriate supplements of vitamins, minerals and trace elements have been added.

concentrates The portion of the horse's ration, or a feedstuff that is rich in starch, protein or both, and that contains less than 15–17% crude fibre (q.v.).

contagious Transmissible from one horse to another.

contracted tendons Hyperflexion or flexural deformity of limbs. The condition in foals ranges from uprightness of fore- or hindlegs to knuckling over at the fetlock and/or inability to extend knee joints. Not uncommon in TB foals. Correction of slight abnormalities may include the fitting of boots and in more severe cases severance of fibres of the flexor muscles and tendons or desmotomy of the superior check ligament.

convulsions A violent involuntary (uncontrollable) contraction, or series of contractions, of the voluntary (skeletal) muscles.

coprophagy The eating by an animal of its faeces. Within 3 weeks of birth, foals will eat their dam's faeces and thereby acquire the species of bacteria and protozoa necessary for the rapid development of a normal microbial population in their GI tract so that invasion by harmful microorganisms is partially inhibited.

cornea The transparent structure that forms the anterior part of the eyeball.

coronary band, **coronary matrix** Runs around the horse's foot just below the hair line and forms part of the sensitive structures from which the wall grows. A permanent defect in the hoof wall usually follows injury to the coronary band.

corpus luteum 'Yellow body.' A yellow glandular mass in the ovary, formed by an ovarian follicle that has matured and discharged its ovum. It contains carotenoids and secretes progesterone.

cortical bone Compact bone forming the cylinder of the shaft, or diaphysis of long bones, and resulting mainly from periosteal ossification.

corticosteroids Comprise the natural glucocorticoids, cortisone and hydrocortisone hormones secreted by the adrenal cortex (q.v.) and synthetic equivalents, e.g. prednisone, prednisolone and fluoroprednisolone, used in the treatment of inflammation, shock, stress and, in other animals, ketosis.

CP Creatine phosphate (phosphocreatine). The fixation of energy in the form of ATP is a transitory phenomenon, and any energy produced in excess of immediate requirements is stored more permanently in compounds such as muscle phosphocreatine. As ATP becomes depleted, more is generated from phosphocreatine by a reverse reaction.

creatinine The normal excretory breakdown product of muscle creatine found in horses' urine. As the quantity produced daily is relatively constant for a particular horse, being proportional to muscle mass, its concentration in urine is used in the assessment of the level of other substances excreted in urine (see creatinine clearance tests, Chapter 12). Blood levels of creatinine rise dramatically following renal failure.

creatine kinase See **CK**.

creep feed A feed, normally dry pellets, offered to nursing foals behind a barrier, which does not allow access to the mare but permits entry to the foal.

cribbing
(aerophagia)
An outdated expression for 'wind-sucking', which is a vice of domesticated horses and ponies. This consists of the habitual swallowing of air while the animal bites or pulls down with its upper incisor teeth on some solid object such as a fencing rail or gate. The neck is slightly arched and gulps of air are swallowed into the stomach with emission of a grunt. The term 'wind-sucking' is also unfortunately used to describe mares that aspirate air, and frequently faecal material, into the vagina. This is rectified by Caslick's operation.

crimping
A term used for the pressing of cereal grains between corrugated rollers to rupture the kernels and to increase digestibility slightly.

croup
That part of the horse's hindquarters lying immediately behind the loins. The 'point of the croup' is its highest part and corresponds to the internal angles of the ilia.

crude fibre
The feed residue identified after subjecting the residual feed from ether extraction to successive treatments with boiling acid and alkali of defined concentrations. The crude fibre contains cellulose, hemicellulose and lignin, but it is not an accurate measure because it underestimates the structural components of vegetative matter (see **NFE**).

curb
A swelling about 100 cm below the point of the hock, owing to undue strain causing a sprain of the calcaneocuboid ligament, or of the superficial flexor tendon.

cyclic AMP
An intracellular hormonal mediator formed from ATP under the influence of the stimulating hormone.

DE
Digestible energy. The gross energy (or heat of combustion) of a feed minus the gross energy of the corresponding faeces, expressed as MJ or kJ/kg total feed. Synonymous with apparent digestible energy.

deamination
When amino acids are present in excess of needs for tissue protein synthesis, or when the horse is forced to catabolise tissue to maintain essential functions, amino acids may be degraded to provide energy. This occurs mainly in the liver and to some extent in the kidneys. The first step in the oxidative degradation of amino acids is the removal of the amino group, a process called deamination. This group is then either transferred to a keto acid to produce another amino acid or is incorporated into urea.

decarboxylation
In the text, this term is principally restricted to reactions of amino acids. Bacteria – for example, in the intestinal tract and in silage during the early stages of fermentation of the

clamp – elaborate enzymes (called carboxylases), which act upon amino acids to yield amines and carbon dioxide. This implies a loss of dietary protein value and the amines, including histamines and tryptamine, may have toxic effects following absorption.

dehydrated
Feed from which most of the moisture has been removed. This extends shelf life and greatly retards the rate of, or inhibits, mould spoilage.

dialysis
The process of separating crystalloids from colloids in solution by the difference in their rates of diffusion through a semipermeable membrane; colloids pass very slowly or not at all.

diaphragm
A thin muscular partition or membrane that separates the thorax (chest cavity) from the abdomen. During its contraction air is drawn into the lungs and when it relaxes air is expelled.

diaphysis
The shank of a long bone between the ends, or epiphyses, which are usually wider and articular.

dicoumarol
An anticoagulant with similar properties to warfarin except that its action has a slower onset, a longer duration and a less predictable response. When ribbed melilot or yellow sweet clover (*Melilotus officinalis*) or white melilot or white sweet clover (*M. alba*) plants are damaged by weather, badly harvested, or when as hay they become mouldy, coumarin contained in the sweet clover is broken down to dicoumarol. Both yellow and white sweet clover are found in pastures in the UK, and white sweet clover is grown as a forage crop in North America and the former USSR.

digestible energy
See **DE**.

dispensable amino acids
Amino acids that are synthesized in the tissues of horses, and/or are made available from synthesis by gut microorganisms, in amounts sufficient to meet tissue requirements of horses without a dietary source.

distal
Remote, farthest from the centre, or origin, as opposed to proximal (q.v.).

diuresis
Increased secretion of urine. A diuretic drug induces diuresis.

dopamine
A neurotransmitter and an intermediate in the synthesis of norepinephrine.

duodenum
The first (proximal) part of the small intestine and connected to the stomach. In the adult horse it is 1 m long with a diameter of 5–10 cm.

dyschondroplasia
Disordered, or abnormal, cartilage formation.

dysphagia
Difficult in swallowing.

dysplasia	Abnormality in the development of cells.
dyspnea	Difficult or laboured breathing.
dystrophy	Faulty nutrition. Dystrophy of muscles causes their atrophy and degeneration.
electrolytes	Substances that in water solution break up into particles carrying electrical charges. The principal electrolytes from a nutritional point of view are Na^+, K^+, Cl^-, HCO_3^-, Ca^{2+}, Mg^{2+} and HPO_4^{2+}.
emphysema	A swelling, or inflation, of the chest which is caused principally by the presence of air in the intra-alveolar tissue of the lungs following rupture of the alveoli.
endochondral ossification	Bone formation within cartilage, cf. periosteal ossification.
endocrine (hormone) system	Glands dispersed in various parts of the body, the function of which is to liberate into the blood, lymph or neuro-secretory channels specific substances that influence metabolism in other organs and tissues.
endogenous	Arising from within the horse, i.e. excluding the lumen of the GI tract.
endotoxaemia	Presence in the blood of endotoxins, which are nonprotein, lipopolysaccharide fragments of the cell wall of gram-negative bacteria. Per milligram they are much less toxic than are exotoxins, but appear to play a crucial role in certain diet-related disorders of bowel origin, including forms of colic (q.v.) and founder (q.v.).
enema (pl. enemata)	A fluid for injection into the rectum or small colon for cathartic or diagnostic purposes.
enterotoxaemia	Presence in the blood of toxins, produced and secreted in the intestines by certain bacteria; e.g. those of *Clostridium perfringens*, which are referred to in the text (Ch. 11). Enterotoxin produced in large quantities by this organism is also specific for cells of the intestinal mucosa and causes enteritis. This anaerobic organism is found in soil and the consumption of relatively small numbers of the spores in herbage is usually without remarkable effect. (See also **botulism.**)
epinephrine	See **catecholamines.**
epiphysis	A head of a long bone joined to the shaft, or diaphysis, during growth by a cartilaginous growth plate – a metaphysis.
epiphysitis	Inflammation of an epiphysis or of the cartilage that separates it from the shaft during growth.
epistaxis	A nose bleed. Blood may be present in the nostrils through being coughed up from broken blood vessels in the lung.

Small losses can be a normal phenomenon after a race, but persistent losses may occur in chronic bronchitis.

epithelial cells   All the body surfaces, including the external surface of the skin, the internal surfaces of the digestive, respiratory and genito-urinary tracts, the inner coats of vessels and ducts of all secreting and excreting glands are covered by one or more layers of cells, called epithelium or epithelial cells.

ergot   A fungus that infects and finally replaces the seed of a cereal or grass, especially the sclerotium of *Claviceps purpurea* (ergot of rye). This ergot is small, hard, black and resembles mouse droppings. Ergot contracts the arteriolar and other unstriped muscle fibres. Its toxins are used to arrest haemorrhage after parturition and internal injury. The persistent consumption of ergot of rye sufficiently decreases blood flow to cause gangrene (typically of the ears, tail and legs). Several ergot toxins cause abortions. After eating large amounts of ergotized hay, horses become dull and listless, a cold sweat breaks out on the neck and flanks, the breathing is slow and deep, the body temperature is subnormal and the pulse weak. Death occurs during a deep coma within the first 24 h. Lesser amounts over a longer period may cause diarrhoea, colic, trembling and loss of condition.

erythrocytes   See **red blood cells**.

essential amino acids   See **indispensable amino acids**.

ether extract   Chemical substances that are soluble in, and extracted by, ether. Whether diethyl ether, or 40:60 petroleum ether, is used should always be specified.

exertion myopathy   Azoturia (q.v.) myohaemoglobinuria. An acute condition in which affected muscles, especially of the hindquarters, become hard to the touch, and the hindlimbs rapidly become stiff and weak or staggering. There is a tendency to knuckle-over at the fetlocks. The risk is greatest in horses that have been in continuous work, abruptly rested for a few days on full feed and then returned to work.

expansion, extrusion   This relies on the cooking effects of super-heated steam injected into a slurry compressed against a die face by a revolving worm and the subsequent rapid fall in pressure during extrusion. Material is subjected to a temperature of around 120°C for about a minute.

extracellular space   The fluid space, or volume, within the body that is external to the cells.

fasciculation   Local contraction, or bundling, of muscles.

fatty acids   Composed of a hydrocarbon chain of one to more than twenty units attached to a carboxyl group. In the formation of storage fats, fatty acids are neutralized by the trihydric alcohol glycerol. Both neutral fat and fatty acids circulate in the blood (see also **free fatty acids** and **volatile fatty acids**).

feed That which is given to the animal to consume, whether a single feedingstuff or a mixture, but excluding water. Not synonymous with ration (q.v.).

fermentation Decomposition of organic substances by microorganisms. In the horse's GI tract this refers especially to bacteria, yeasts and ciliate protozoa; the first are the most significant.

fertilizer Inorganic fertilizers are plant nutrients prepared synthetically or mined as minerals. Organic fertilizers are sources of the same nutrients of animal and vegetable origins bound in organic (q.v.) form.

fetlock The horse's 'ankle' joint in fore- and hindlimbs between the metacarpus or metatarsus (cannon bones) and the first phalanx (long pastern (q.v.) bone).

FFA See **free fatty acids**.

fibre See **crude fibre**.

fibrin The insoluble protein formed from fibrinogen by the proteolytic action of thrombin during normal clotting of blood.

filled legs Oedema or puffiness of the legs. Abnormally large amounts of fluid (exudate) in the intercellular tissue spaces beneath the skin. The most common cause in healthy horses is a period of inactivity in a box following a season of hard exercise, particularly where corn or other concentrated feeds are being given. However, it is not protein poisoning. Oedema can also arise from diseases of the heart, liver or kidneys, or from longstanding malnutrition involving diets impoverished in protein.

flexor tendons Muscles are attached to the long bones of the legs by extensor tendons, which extend or straighten the leg at that point when the muscle contracts, and by flexor tendons, which flex joints by contraction of the appropriate muscles.

founder, laminitis A painful disease of the feet in which there is apparently a transitory inflammation followed by congestion of the laminate of the hooves. It is most frequent in ponies but can be induced readily in both horses and ponies by a sudden increase in the starch or protein content of the diet. Sometimes all four feet are affected, sometimes only the forefeet, and occasionally only the hindfeet, or a single foot. Affected feet feel hot and the body temperature may rise. The stance is unnatural, affected forefeet are thrust forwards and the horse is reluctant to move.

free fatty acids (FFA)
During work storage fats are mobilized when lipase enzymes catalyse the production of fatty acids, splitting them from glycerol and leading to a rise in the blood plasma concentration of both components.

frog
Horny central part of the lower surface of the foot, subject to thrush, an infection by bacteria and fungi occurring in wet unhygienic stables when there is poor routine foot care.

fundus gland region of stomach
A large region of the stomach containing glands of two types of cells. The region lies distal to the oesophageal and cardiac regions and proximal to the pyloric gland region.

fungal units
Fungi contaminating feed produce a mycelial mat of fine threads and fruiting bodies from which spores are shed. When badly affected feed is disturbed, the mycelial threads may break up into small particles and new growth can be initiated by each of these particles or by germination of spores. The fungal unit is any particle from which new growth can be started.

furlong
Forty poles, one-eighth of a mile. Originally the length of the furrow in the common field. The side of a square of 10 statute acres. Equals the Roman stadium, one-eighth of a Roman mile.

β-galactosidase
Neutral or brush-border lactase. This enzyme is present in the intestinal juice of normal young horses. It is necessary for the cleavage of milk sugar (lactose) to glucose and galactose, which can then be absorbed into the blood.

gamma globulin
A protein fraction of the plasma globulin, which has a slow moving electrophoretic mobility. Most antibodies are gamma globulins.

gastrointestinal tract
Stomach and intestines. Sometimes used synonymously with alimentary tract, which is the entire tube extending from the lips to the anus.

Gastrophilus
See **botfly**.

GGT gamma-glutamyltransferase
EC 2.3.2.2 is released into the blood following liver damage. The blood activity of this enzyme is elevated in liver cirrhosis, pancreatitis and renal disease. The normal range of plasma activity is 0–41 iu/l and seneciosis may cause a rise up to 80–280 iu/ol.

glucagon
Polypeptide hormone secreted by the alpha cells of the islets of Langerhans in response to hypoglycaemia. It stimulates glycogenolysis in the liver and opposes the action of insulin (q.v.).

glucogenic
Giving rise to, or producing, glucose. According to a

long-used classification, amino acids are ketogenic if (like leucine) they are converted to acetyl-CoA and, when fed to a starved animal, produce ketones in the blood. Glucogenic amino acids such as valine, when fed to a starved animal, promote the synthesis of glucose and glycogen.

gluconeogenesis   The formation within the body of glucose from amino acids by various routes that include pyruvic acid and lactic acid, but not via acetyl-CoA.

glutathione peroxidase (GSH-P$_x$)   The activity of this enzyme (EC 1.11.1.9) in horse erythrocytes is used as an indication of the horse's selenium status. The activity is determined as μmol NADPH oxidized in 1 min by 1 ml erythrocytes. In TBs the evidence (Blackmore *et al.* 1982) indicates that the activity should be approximately 25–35 units/ml red cells.

glycolysis   The major pathway whereby glucose is metabolized to give energy is a two-stage process; the first stage, called glycolysis, can occur anaerobically and yields pyruvate.

glycosuria   The presence of abnormal amounts of glucose in the urine. It arises from failure of renal tubular reabsorption or from an abnormality of hormone status as in diabetes mellitus.

goitrogenic   A term applied to substances in certain feeds, derived, for example, from members of the plant genus *Brassica* (family Cruciferae) which, if consumed persistently in large quantities, cause goitre (an enlargement of the thyroid gland, q.v.). A deficiency of iodine in the diet will cause a similar condition, especially in young horses.

gossypol   A toxin found with two other toxic pigments in the pigment glands of cotton seeds. It is a polyphenolic binaphthalene derivative. It is inactivated by heating, but in this procedure it combines with lysine, reducing the protein value. Selective breeding has produced glandless seed, not widely available.

GOT   See **AST**.

GPT   See **ALT**.

grass sickness   A disease of horses, ponies and donkeys that seems to be noncontagious. It occurs mainly in the summer months among grazing animals on certain pastures in Europe in a peracute form in which death occurs within 8–16h of initiation after some periods of great violence. In a subacute form the horse is dull, listless and salivates. The bowel becomes impacted and distended. Food material may appear in the nostrils. There seems to be some loss of motor control of the GI tract. Muscles of the back are hard, the horse is 'tucked-up' and twitching over the shoulders occurs. Recovery is uncommon (See Chapter 10).

growthplate	See **metaphyseal plate**.
haemagglutinins	Now called lectins, contained in *Phaseolus* spp., including *P. vulgaris*. Severe GI disorders can be caused by these substances when given in high dietary concentrations. They are proteins of which concanavalin A from jack beans is used medically as a mitogenic agent and to preferentially agglutinate cancer cells.
haematocrit	Packed cell volume (PCV). The proportion of blood by volume made up of cells, especially red cells, and expressed as a percentage. It is determined by centrifugation of blood samples containing an anticoagulant.
haematopoiesis	The formation and development of blood cells.
haemoglobin	The oxygen-carrying red pigment of the red cells (erythrocytes) of blood. A conjugated protein consisting of the protein globin combined with an iron-containing prosthetic group (heme).
haemolysis	The rupture of red blood cells.
haemolytic icterus of foals	Caused by the absorption of colostral antibodies that destroy the foal's red blood cells (see Ch. 7).
haustral	Referring to the haustra, or sacculations, of the colon.
haylage	Originally registered trade name for a silage containing a high proportion (35–50%) of dry matter made from wilted forage, precision-chopped to approx. 12–25 mm nominal length and ensiled in a Harvestore tower silo. However, the name has acquired a more liberal definition to include unchopped vacuum-packed material.
heating feed	A concentrated feed, which is readily digestible and fermentable and which leads to a rapid rise in blood metabolites, waste heat production and probably to some increase in metabolic rate.
heaves	See **broken-wind**.
hepatic encephalopathy	A degenerative disease of the brain resulting secondarily from advanced disease of the liver, or where a portacaval shunt occurs.
hindgut	The large intestine consisting of the caecum and colon.
hives	Large numbers of small bumps, or raised areas, about 0.5 cm diameter under the skin, which eventually form scabs. They appear suddenly and are caused by an allergic response to specific components of feed, to drugs or to insect bites. The cause (antigen, q.v.) can usually be determined by allergy tests, and recovery follows removal of the source from the diet or general environment. The condition sometimes arises when 'rich' feeds are suddenly introduced to the diet.

hock	The tarsal joint between the tibia (q.v.) and metatarsus (q.v.) (cannon bone) of the hindlimb.
homeostasis	Stability of the normal body states. Refers frequently to the constancy of pH and chemical composition of extracellular fluids.
hormone	A discrete chemical substance secreted into the body fluids by an endocrine gland and which influences the action of a tissue, or organ, other than that which produced it.
hotblooded horse	Hot and warmblooded horses (cf. coldblooded horses, q.v.) are those derived to a significant extent from breeds originating in Mediterranean countries, which came to be called Arabian, Barb and Turk. Modern breeds of this type include TB, Arabian, standardbred, American saddle horse, Morgan, quarter horse and Tennessee Walker. The principal North European warmblooded breeds are the Hanovarian, Trakehner, German Holstein, Dutch warmblooded and Oldenburg breeds.
hyper-, hypo-	Prefixes: hyper signifying above normal or excessive and hypo meaning below.
hyperaemia	Engorgement with blood.
hypercalcuria	Abnormally large amounts of Ca in the urine; hypocalcuria is abnormally small quantities of Ca.
hypercapnia	Excess carbon dioxide in the body fluids.
hyperglycaemia	An elevated blood glucose concentration.
hyper- and hypokalaemia	Abnormal level of plasma potassium. Hypokalaemia may be caused by a combination of inadequate dietary K with excessive losses from the body. The causes of both hypo- and hyperkalaemia can be metabolic derangements, when there are abnormal shifts of K^+ between intracellular and extracellular space, as in acidosis.
hyperlipidaemia	A general term for elevation of any, or all, of the lipids in the plasma, including hyperlipoproteinaemia and hypercholesterolaemia, whereas hyperlipaemia refers specifically to an elevation of the triacylglycerols, previously known as triglycerides.
hyperparathyroidism	Abnormally increased activity of the parathyroids (q.v.), causing loss of Ca from the bones.
hyperplasia	The abnormal multiplication, or increase, in the number of normal cells in normal arrangement in a tissue.
hyperplasmia	An excess in the proportion of plasma to cells in the blood.
hyperpnea	Abnormal increase in the depth and rate of the respiratory movements.

hypersensitivity	An exaggerated reaction to a foreign agent. These immune responses are classified as immediate or delayed, or as Types I–IV.
hypertension	Usually refers to high arterial blood pressure, which may be confined to a specific circulation such as pulmonary or renal.
hyperthermia	Abnormally elevated body temperature.
hypertonic	A body fluid with a concentration, or osmotic pressure, above normal (more than isotonic).
hypertrophy	The enlargement or overgrowth of an organ or tissue by the increase in size of its individual cells.
hypocalcaemia	A reduction of blood Ca below the normal range of concentrations.
hypocupraemia	A subnormal concentration of blood Cu.
hypoglycaemia	Concentration of blood glucose below the normal limit for the breed.
hypoglycaemic shock, insulin shock	Occurs when blood glucose falls below the normal range, causing nervousness, trembling and sweating.
hypomagnesaemia	A reduction of blood Mg below the normal range of concentrations.
hyponatraemia	Abnormally low plasma concentration of Na.
hypotension	Abnormally low blood pressure.
hypovolaemia	Abnormally decreased volume of circulating blood plasma.
hypoxia	A reduction below physiological limits of the oxygen supply to tissues despite adequate perfusion by blood.
icterus	Jaundice. A yellowish discoloration of the visible mucous membranes – eyes, mouth, nostrils and genital organs. It can also be detected in blood plasma of stabled animals receiving a diet low in pigments and can be caused by certain infections resulting in destruction of the red blood cells and release of the haem pigment (haemolytic jaundice) or by liver damage.
ileum	The distal portion of the small intestine extending from the jejunum to the caecum.
immunity	Resistance to infection by an organism, or to the action of certain poisons. Immunity can be inherited, acquired naturally or acquired artificially.
immunoglobulin	Specific proteins, found in blood, colostrum (q.v.) and in most secretions, produced by plasma cells in response to stimulation by specific antigens, which in turn are inactivated. The antigens may be carried by, or released from, bacteria, viruses or even certain parasites.
imprinting	An inborn tendency of a neonatal foal to attach itself to a set group of objects, or a single object, such as its mother.

indispensable (essential) amino acids	Those amino acids that are not synthesized in the tissues of the horse, or otherwise made available from, for example, synthesis by gut microorganisms, in amounts sufficient to meet the requirements of tissues and which must therefore be present in the feed.
infarct	An area of tissue necrosis (q.v.) due to local anaemia resulting from obstruction of the blood circulation. In the text this refers to the effects of migrating strongyle (q.v.) larvae on the mesenteric blood vessels, which causes death of a segment of the intestines.
inorganic	Not a precise term but can be taken to refer to the ash content of the body remaining after it has been incinerated, which removes H, C and N as oxides. The minerals (q.v.), in an oxidized form, remain, together with a very small proportion of carbonates.
insulin	A protein hormone synthesized by the islet cells of Langerhans in the pancreas and secreted into the blood where it regulates glucose metabolism. It is deficient in diabetes mellitus.
international unit (iu)	As applied to vitamins, is the internationally agreed unit of potency for a particular vitamin that may have several molecular forms. The unit is now being displaced in favour of gravimetric measurements of each molecular form or vitamer.
intracellular space	The fluid volume within the cells of the body as distinct from the extracellular fluid space. Movement of ions, water, glucose, etc. from one to the other is under metabolic control.
intussusception	Prolapse of one part of the intestine into the lumen of an immediately adjoining part.
ischaemia	A deficiency in the blood supply to a tissue, organ or part of the body.
isoantibody	An antibody (q.v.) generated in the body in reaction to an isoantigen (q.v.). An example is one found in the dam's blood and colostrum in response to a fetal protein that has entered her bloodstream.
isoantigen	An antigen in the body that will induce the production of an antibody against itself.
isoenzymes	Isoenzymes, or isozymes, are physically distinct, e.g. immunologically so, but they all catalyse the same reaction.
isotonic solutions	Those solutions that have the same concentration or, more specifically, the same osmotic pressure.
iu	See **international unit**.
jaundice	See **icterus**.
joint-ill	A disease in which organisms enter the body by way of the

	unclosed navel causing abscesses to form at the umbilicus and in some of the joints.
joule	SI (Système Internationale d'Unités) unit of energy; 4.184 J = 1 calorie. The joule is defined as the energy expended when 1 kg is moved 1 m by a force of 1 newton (1 N m), 1 J = 1 kg m^2 S^{-2}, or 1 watt second (1 Ws). The kilojoule (kJ) = 10^3 joules and the megajoule (MJ) = 10^6 joules.
keratinization	To become horny. Keratin, the principal protein of epidermis, hair and the hoof, is very insoluble and contains a relatively large amount of sulphur. Increased keratinization of epithelia can occur under physiological conditions and under pathological conditions, e.g. vitamin A deficiency.
knuckling-over	Usually refers to flexing of the fetlock (q.v.) joint owing to contraction of muscles and tendons or ligaments behind the cannon.
Kupffer's cells	Phagocytic cells lining the walls of the sinusoids of the liver and which form a part of the reticuloendothelial system.
labile	Chemically unstable.
lactase	An enzyme that hydrolyses the milk sugar lactose to form glucose and galactose. The enzyme accomplishing this in the intestine of the horse is β-galactosidase.
laminitis	See **founder**.
latent period	The period or state of seeming inactivity between the time of stimulation and the start of the response, e.g. the interval between the injection or absorption of antigen and the first appearance of antibody.
lathyrism	A condition characterized by sudden and transient paralysis of the larynx, with near suffocation of the horse, caused by β-aminopropionitrile in *Lathyrus sativus* (Indian pea) and other *Lathyrus* spp., including *L. odoratus* (sweet pea).
LDH	Lactic dehydrogenase (EC 1.1.1.27). A tissue enzyme that has several different forms or isoenzymes. It catalyses the transfer of H$^+$ with the formation of pyruvic acid. The blood activity of LDH is elevated during and following strenuous exercise and following tissue damage.
legumes	Plants of the family Leguminosae or Fabaceae, which include useful forage species (e.g. the clovers and lucerne, q.v.) and seed species (e.g. soyabeans, peas and field beans)
leukocytosis	A transient increase in the number of white cells in the blood.
level of feeding	Weight of complete dry diet eaten daily, not confined to energy. Strictly it should be given as a proportion of metabolic body size (BW$^{0.75}$), or more crudely per 100 kg bodyweight.

ley
Grass and clover seeds sown as part of a cropping rotation, subsequent to which it is usually ploughed-up after 1–6 years. The distinction from permanent pasture is not absolute.

ligament
A tough fibrous band supporting viscera, or which binds bones together. In different situations ligaments are cord-like or in flat bands, or, in forming the joint capsule, they are in sheets, preventing dislocation.

light horse
A loose term for small and large lady's hacks, usually of mixed breeding, which may include TB and English Arabian. An alternative definition includes all riding horses of the present day in this classification, which therefore excludes heavy draught horses and ponies.

lipases
A class of enzymes, members of which are found in the digestive secretions and in body tissues. Individual lipases have as major functions the hydrolysis of fats to yield fatty acids, monoglycerides, glycerol and cholesterol, and the hydrolysis of phospholipids.

lipids
A group of substances found in plant and animal tissues, insoluble in water but soluble in common organic solvents, including petroleum, benzene, ether and chloroform. The crude fat of feed is the material extracted using light petroleum.

lipolysis
The decomposition of fat into glycerol and fatty acids and, in the case of, for example, the phosphoglyceride lecithin, into phosphoric acid and choline as well.

lucerne (alfalfa)
A legume, *Medicago sativa*. A perennial forage crop with a strong tap root, which grows well on light alkaline soils in warm climates. Weeds must be kept at bay during its establishment.

lymphatics
A system of vessels that drains lymph from various body tissues and conveys it to the bloodstream and that conveys neutral fats from the small intestine after they have been absorbed.

lysine
An indispensable basic amino acid, the concentration, or frequency, of which in most vegetable proteins limits their biological value. L-Lysine hydrochloride is sometimes used as a feed additive to make good any deficit in the dietary protein.

MAD fibre
Modified acid detergent fibre. A fibre fraction of feed determined by the MAD fibre procedure isolates principally the lignocellulose complex. This complex is the fraction of plant material that has most influence on energy digestibility of feed. In comparison, during the chemical procedure for

crude fibre (q.v.) determination a considerable amount of the lignin may become soluble and hence lost from the residue, so leading to an overestimation of the digestible fraction of feed.

maintenance At the maintenance level of feeding, the requirements of the horse for nutrients for the continuity of vital processes within the body, including the replacement of obligatory losses in faeces and urine and from the skin, are just met so that there is no net gain or loss of nutrients and other tissue substances by the animal.

maize (corn) *Zea mays*. A member of the grass family, the Gramineae, the seeds, both cooked and uncooked, of which constitute an excellent high-energy cereal grain for horses. The above-ground vegetative parts, at the milky grain stage, can be made into a good silage for horses.

mandible Lower jawbone. In the adult horse this bone has sockets for three incisors, one canine, three premolars and three molars on each side in the male (in the female the canines are usually absent, or rudimentary). Its grinding movements are controlled by powerful muscles.

ME Metabolizable energy. The digestible energy (DE) of a unit weight of feed less the heats of combustion of the corresponding urine and gaseous products of digestion.

meconium A dark-brown, viscid, semi-fluid or hard material which accumulates in the intestines of the foal prior to birth. It should be discharged soon after birth. The colostrum (q.v.) has a natural purgative action on it.

mesentery Membranous peritoneal fold attaching the intestines to the dorsal wall of the abdomen.

metabolism A term embracing the chemical processes of anabolism (q.v.) and catabolism (q.v.) in the body.

metacarpus The part of the forelimb lying between the carpus (wrist) and the digit (cfr. metatarsus, q.v.). There are three metacarpal bones: the central, or third, metacarpal ('cannon') is the largest; the other two are rudimentary metacarpals (splint bones).

metalloproteinase Enzyme proteins binding a metal, including caeruloplasmin, cytochrome oxidase, lysyl oxidase and the superoxide dismutases.

metaphyseal plate The region of linear growth of the long bones of growing horses lying between the epiphysis, or head, and the diaphysis, or shank.

metaphysitis Inflammation of the metaphysis, seen as swelling or bumps adjacent to the joints of leg bones.

metatarsus   The part of the hindlimb lying between the tarsus, or hock, and the digit. It is similar in layout to the metacarpus (q.v.), with a central large metatarsal bone (cannon) and rudimentary metatarsals (splint bones).

methaemaglobin   A modified form of haemoglobin found in the blood in which the iron has been converted from the ferrous to the ferric state when it can no longer combine with, and transport, oxygen. The lesions can occur following the administration of large doses of certain drugs or after the consumption of nitrites in feed or water.

micronization   The cooking of cereal and legume seeds under ceramic burners that emit infrared irradiation in the 2-6-μm waveband, resulting in a rapid internal heating of the seed and a rise in water vapour pressure, during which the starch grains swell, fracture and gelatinize.

minerals   The essential elements in the diet other than carbon (C), hydrogen (H) and nitrogen (N). They include the macrominerals calcium (Ca), phosphorus (P), magnesium (Mg), potassium (K), sodium (Na), chlorine (Cl) and sulphur (S), and the microminerals iron (Fe), zinc (Zn), manganese (Mn), copper (Cu), cobalt (Co), iodine (I), selenium (Se) and fluorine (F). The microminerals are also known as trace elements. Some other elements, such as chromium (Cr), nickel (Ni) and molybdenum (Mo), are required in very small amounts. The mineral elements are required for incorporation into compounds different in the main from those in which they may appear in the diet whereas C, H and N are required as constituents of preformed organic nutrients. A crude approximation to the mineral content of the diet is obtained from the measurement of the ash (q.v.) content of the feed.

mitochondria   Minute bodies occurring in the cytoplasm of cells (except bacteria and blue–green algae, or cyanobacteria). They exert important regulatory functions both on catabolic and biosynthetic sequences. They are the seat of the citric acid cycle, the β-oxidation pathway and of oxidative phosphorylation.

MJ   1000 kJ or 10^6 joules.

mycotoxins   Substances produced under specific conditions by certain fungi or moulds. Their chemical form and their effects on animals are wide ranging. These include hormone-like effects, disruption of intestinal, renal and hepatic function, tumour induction and antibacterial effects. Most antibiotics are mycotoxins. The best-known include aflatoxin produced by *Aspergillus flavus*, T-2 toxin produced by *Fusarium* and *Myrothecium* spp., ergotoxin produced by *Claviceps purpurea*, zearalenone (F-2) produced by *Fusarium* spp., dicoumarin produced on melilots or sweet clover by *Aspergillus* and *Penicillium* spp., and vomitoxin produced by *Fusarium* spp. Several mycotoxins are harmful to the horse in normal husbandry conditions.

myoglobin A small oxygen-carrying protein of muscles containing the pigment haem with an atom of iron at its centre. Severe muscle damage leads to its appearance in the urine in azoturia (q.v.). Being a much smaller molecule than haemoglobin, it passes the glomerular filter much more readily, causing a dark-brown stain to the urine. Precipitation of myoglobin in the renal tubules, as with haemoglobin, may contribute to terminal uraemia (q.v.).

myopathy Disease of a muscle.

N-balance The net gain of nitrogen (N) by the animal. N in feed – (N in
(N-retention) faeces + N in urine) – loss of ammonia expired in air per unit time.

NE Net energy, the energy value of animal product formed, or of body substance saved at or below maintenance, per unit weight of feed consumed. NE = metabolizable energy – heat increment.

necrosis Death of a cell, or of a group of cells, which is in contact with living tissue.

NEFA Nonesterified, or free, fatty acids that are produced from the hydrolysis of triacylglycerols, or triglycerides.

nematode The Nematoda are a class of tapered cylindrical helminths, the roundworms, of the phylum Aschelminthes. Not all are parasitic.

nephritis Inflammation of the kidney; a focal or diffuse proliferative, or destructive, process which may involve the glomerulus, tubule or interstitial renal tissue (cf. nephrosis, q.v.).

nephrosis Any disease of the kidney, especially one characterized by degeneration of the renal tubules (cf. nephritis, q.v.).

nephrotoxic Toxic or destructive to kidney cells.

neurotransmitter A chemical, e.g. norepinephrine, acetylcholine, dopamine, etc., released from the axon terminal of a presynaptic neuron on excitation, that travels across the synaptic cleft to excite or inhibit the target cell.

NFE Nitrogen-free extractives. Measured in grams per kilogram of feed, this is numerically evaluated as 1000 – (moisture + ash + crude protein + ether extract + crude fibre). NFE includes some of the feed cellulose, hemicellulose, lignin, sugars, fructans, starch, pectins, organic acids, resins, tannins, pigments and water-soluble vitamins if each of these components was present in significant amounts in the original feed.

nonprotein nitrogen (NPN) The determination of the crude protein content of a feed assumes that the protein content is generally 6.25 times the determined N content. However, there are many compounds in feed, including nucleic acids, creatine and others, that have no protein value, but which are included in the calculation. Also, young herbage is rich in amino acids and nitrates that are components of NPN, although amino acids have an equivalent protein value.

nuts, cubes Compounded horse feed mixtures that have been compressed into solid cylinders 3–10 mm in diameter and two to three times as long by forcing the mixture through the holes of a metal die. The mix may or may not have been previously steamed in a kettle.

obligatory loss Usually refers to the minimal inevitable loss of a nutrient, by excretion from the body, at low dietary intakes of that nutrient.

oedema The presence of abnormally large amounts of fluid in the intercellular tissue spaces of the body. Applied in the text especially to an accumulation in the subcutaneous tissue.

oesophagus The gullet, a muscular membranous tube or canal extending from the pharynx to the stomach.

oestradiol The most potent naturally occurring ovarian and placental oestrogen which prepares the uterus for the implantation of the fertilized ovum and which induces and maintains the female secondary sex characteristics.

oestrous cycle A cycle in the mare typically lasts 21 days in the breeding season in which there is a pattern of physiological and behavioural events under hormonal control. The cycle, which forms the basis of sexual activity, has two components: oestrus (heat) in which the mare is receptive to the stallion and the egg is shed, and dioestrus, a period of sexual quiescence.

oncotic pressure As used in the text this refers to oncotic pressure of plasma, which is the osmotic pressure owing to the colloids present, principally albumin, that counterbalance the capillary blood pressure.

open knees A dished concave appearance to the front of the 'knee' or carpus joint caused by epiphysitis (q.v.) immediately above the knee.

orad Toward the mouth.

oral	Of the mouth.
organic	Complex molecules containing at least C and H and synthesized by living tissue.
osmolality	A solution that has 1 osmole solute dissolved in 1 kg water has an osmolality of 1 osmole/kg (1/1000 osmole dissolved/kg water has an osmolality of 1 mOsmole/kg; cf. osmolarity, the more practical measure, which is the osmolar concentration expressed as osmoles per litre solution rather than per kilogram water. For dilute solutions, as in the body, the quantitative difference is less than 1%.
osmolarity	The total number of dissolved particles, or osmoles, in water solution. Fluids with an osmolarity greater than that of body fluids are hypertonic and those for which it is lower are hypotonic. Osmolarity depends on molar concentration and not upon equivalents per litre (mEq/l). For instance, if Mg^{2+} concentration is 30 mEq/l, this is 15 mmol/l, or 15 mOsmol/l. 100 ml 0.6 M-$NaHCO_3$ provides 60 mmol Na^+ and 60 mmol HCO^-_3, or 120 mOsmole, on the assumption that the salt is completely dissociated. The osmolarity of body fluids is about 285 mOsmole/l.
osmotic pressure	The pressure that can be exerted when water moves from one solution to another of higher concentration through a semipermeable membrane, such as that of red cells. Solutions that have the same osmotic pressure, i.e. are isotonic, with the osmotic pressure of red cells are used for injections; otherwise with the use of water the cell membrane would become distended and burst (haemolysis). The energy of high energy compounds is consumed when any nutrient is required to move against an osmotic pressure gradient.
osteoarthritis	Chronic multiple degenerative joint disease characterized by degeneration of the articular cartilage, hypertrophy of the bone at the margins and changes in the synovial membrane.
osteoarthrosis	Chronic arthritis of a noninflammatory character.
osteochondritis	Inflammation of both bone and cartilage.
osteoid	The organic matrix of young bone.
osteomalacia	Softening of bones in adults from a deficiency of vitamin D or minerals. There is an increased amount of uncalcified bony matrix (osteoid).
oxalate	An organic acid anion that combines with Ca and some other positively charged dietary minerals to form a very insoluble precipitate that inhibits digestion and absorption (see tropical grassland, Ch. 10). Oxalates circulating in the

	blood also react with ionized blood Ca and form a crystalline deposition in the kidneys.
Oxyuris equi	Pinworm. A nematode intestinal worm, which is not a serious hazard to horses, but it causes intense irritation of the anal region and this encourages tail-rubbing and biting. Piperazine compounds are effective.
pancreas	A gland in the abdominal cavity that secretes a juice containing digestive enzymes into the duodenum (see Ch. 1) and secretes the hormones insulin (q.v.) and glucagon (q.v.) into the bloodstream.
Parascaris equorum	A nematode intestinal worm that commonly infects foals, yearlings and horses under 3 years old. It causes intestinal problems, colic, coughing and nasal discharge (see Ch. 11).
parathormone	Parathyroid hormone synthesized by the parathyroid gland.
parathyroid	An endocrine gland located in the upper neck adjacent to the thyroid gland. When serum concentration of calcium ions falls, parathyroid hormone is secreted into the blood. It induces mobilization of Ca in the bones, increases Ca absorption in the intestines, reabsorption of Ca by the renal tubules and the urinary excretion of P. Its effects are counteracted by that of thyrocalcitonin. See also **secondary nutritional hyperparathyroidism**.
paresis (general)	A condition short of complete paralysis in which certain muscles are relaxed and weak. If it is more generalized the animal cannot support itself or stumbles. Sometimes there is slight paralysis – that is, there has been some injury to or effect on certain motor nerves and an inability to make purposeful movements.
paresis (parturient)	This clinical sign, associated with hypocalcaemia (q.v.), is *not* characteristic of mares. Lactation tetany (q.v.) was commonly observed in draught horse mares. Hypocalcaemia is a consistent characteristic, although hypomagnesaemia (q.v.) has been associated with recent transport. Mares grazing lush pasture and with a heavy milk flow are particularly prone to tetany. For other causes see Chapters 3, 9, 10 and 11.
parturition	The act, or process, of foaling.
pastern	The region of the leg between the fetlock and the hoof in both fore- and hindlimbs, formed by the long and the short phalanges which create the pastern joint. The third phalanx is in the hoof.
peristalsis	The rhythmic involuntary muscular contractions of the alimentary canal by which it mixes and propels ingesta towards the rectum.

petechial haemorrhage	Small blood spot caused by the effusion of blood from a capillary. Frequently numerous spots are present in tissue reacting to toxins, such as endotoxin.
pH	The symbol used in expressing the H ion concentration of water solutions. It signifies the negative logarithm of the H ion concentration in gram-molecules (moles) per litre (the logarithm of the reciprocal of the H ion concentration). pH 7 is neutral. Progressively above this value alkalinity increases and below it acidity increases.
photosensitization	The development of abnormally heightened reactivity of the skin to sunlight. In the text, reference is made to a skin reaction of horses following the consumption of certain plants. Reports exist of domestic animals reacting to St John's worts (*Hypericum* spp.) or bog asphodel (*Narthecium ossifragum*). White, grey or piebald horses or those with liver damage are most susceptible.
phytase	Phytase EC 3.1.3.8 is produced by *Aspergillus niger* (CBS 114.94) and is listed as a permitted enzyme that hydrolyses phytates. Phytase is produced by many microbial cells, but, as far as it is known, does not occur in any of the digestive secretions of the horse.
phytates	Much of the phosphorus present in cereal grains and other seeds occurs as phytates, which are Ca, Mg, and other salts of phytic acid, a phosphoric acid derivative, composed of a six-carbon ring structure with a phosphate group bonded indigestibly to each carbon atom. The horse can utilize the phosphate phosphorus only to the extent of about 30%, partly through the activity of bacterial phytase present in the intestines, but probably also as a result of some phytase activity which has been recognized in the wall of the intestines of several species. Phytates reduce the availability of dietary zinc. Wheat bran contains 1% of P present as phytin.
piloerection	Erection of the hair.
pinworm	See *Oxyuris equi*.
pituitary gland (hypophysis)	A gland lying at the base of the brain and connected with the hypothalamus by the pituitary stalk. Physiologically it consists of two parts, the anterior and posterior pituitary. Six hormones are secreted by the anterior portion: growth hormone, adrenocorticotropin, thyroid stimulating hormone, prolactin, follicle-stimulating hormone and luteinizing hormone. The posterior portion secretes antidiuretic hormone (vasopressin) and oxytocin.
pK	The symbol used in expressing the dissociation constant of weak acids (or bases) in the form of a negative legarithm.

The larger the value the weaker, or less dissociated, is the acid. When equal concentrations of the salt of an acid and the acid are mixed, the pK = pH (q.v.) and the buffering capacity of the mixture is maximal. Thus, for the primary ionization of carbonic acid to bicarbonate, as in blood, the pK = 6.36, and at pH 6.36 half the molecules of carbonic acid are dissociated forming bicarbonate. At the normal pH of venous blood (7.4) the mixture is an even more effective buffer to acid produced during anaerobic muscular activity with the formation of undissociated carbonic acid.

placenta   An organ that develops within the uterus in early pregnancy and that establishes communication between the dam and the developing fetus. It is composed of a maternal portion and a fetal portion attached to the fetus by the umbilical cord. Following parturition, it is passed as the afterbirth.

progesterone   A hormone liberated by the corpus luteum, adrenal cortex and placenta. It prepares the uterus for the reception, development and maintenance of the fertilized ovum.

prolactin   A hormone secreted by the anterior pituitary gland that stimulates and sustains lactation.

prostaglandins   Grouped in six main series of cyclic compounds derived from unsaturated fatty acids such as arachidonic acid – itself a derivative of the dietary indispensable (essential) linoleic acid – and from fatty acids with one less and one more double bond. Prostaglandins were first recognized in seminal fluid and the prostate gland. They show a variety of biological actions that influence smooth muscle contraction (as in contraction of uterine muscle and in blood-pressure control) and they are mediators in the regulation of the dilatation and permeability of arterioles, capillaries and venules in the inflammatory response. They are involved in immune mechanisms and are used for oestrus synchronization and for abortions in cases of twin fetuses. Prostaglandin $F_{2\alpha}$ ($PGF_{2\alpha}$) which terminates the life of the corpus luteum has been the one most commonly used.

protease   An enzyme that digests proteins by hydrolytically splitting off amino acids. Several kinds are secreted into the alimentary canal (see Ch. 1).

protein   True proteins are chains in which the links are amino acids. All amino acids possess at least one N-containing amino-group. The crude protein content of the diet is defined as the N-content $\times$ 6.25 as it is assumed that protein contains 16% N. However, this product includes dietary nucleic acids, nitrogenous glycosides, amines, nitrates, etc. and so overestimates the true protein content.

proteolysis
The enzymatic digestion of protein, which, if carried out by the horse's own secretions, yields proteoses, peptones and amino acids, but if carried out by intestinal bacteria it includes deamination with a loss of protein value.

prothrombin time
The synthesis of prothrombin occurs in the liver and requires vitamin K. It is essential for the clotting of blood, and any defect in prothrombin formation, or in the activity of other substances involved in clotting, extends the interval between the initiation of the process and the formation of fibrin from fibrinogen. Fibrin spontaneously coagulates. Dicoumarol (q.v.), which arises from the activity of a mould on coumarin in spoiled sweet clover or melilots, interferes with the metabolism of vitamin K causing an extension of prothrombin time and consequential extensive haemorrhaging (see Ch. 11).

proximal
Nearest to the centre or origin, and opposed to distal (q.v.). The duodenum is proximal to the jejunum and the 'knee' is proximal to the fetlock.

purgative
Cathartic, a medicine that stimulates peristaltic action and evacuation of the intestines.

pylorus
The distal or duodenal aperture of the stomach. It is controlled by a sphincter muscle and through it stomach contents enter the small intestine.

pyrexia
Abnormal elevation of body temperature.

quarter horse
(American quarter
horse)
A breed devised mainly from dams of Spanish origin, for long bred by American Indians, and from Galloway sires introduced by early settlers of North America.

quidding
The expulsion of partially chewed feed from the mouth. The habit may arise from injuries to the tongue, or cheek, resulting from molar teeth which are too sharp, irregular in height or in alignment, or even from permanent teeth pushing the temporaries out from the gums. Causes also include infections of the mouth or teeth and paralysis of the throat and consequent inability to swallow.

radius
One of the two long bones of the 'fore-arm', between the point of the 'elbow' and the 'knee'. The other long bone is the ulna.

ration
The amount of daily feed rather than its composition, but it should include all the constituents of the diet apart from water.

rectum
The distal portion of the large intestine extending from the small colon to the anus and holding the faeces.

red blood cells (erythrocytes)
The most numerous cells in the blood ($6.8–12.9 \times 10^{12}/l$), there being only $5.4–14.3 \times 10^9/l$ white blood cells (leucocytes). The cellular portion of blood makes up 32–53% of the total, the remainder being the plasma. About 35% of each red cell is the protein haemoglobin, which transports oxygen from the lungs to the various body tissues.

renal clearance
The ratio of the concentration of a substance in the urine to that in the blood times the rate of urine formation. As the latter is normally unknown, the creatinine clearance ratio is measured (see Ch. 12). To calculate the glomerular filtration rate, when the rate of urine formation is known, creatinine (q.v.) or insulin may be used; these substances are readily filtered by the glomerulus but not secreted or absorbed by the renal tubules.

renal tubules
These run from the glomerulus through the cortex of the kidney, as convoluted tubules, then through the medulla as collecting tubules, and they open into the pelvis of the kidney at the apices of the renal pyramids. Their main function is the reabsorption of water and various solutes – glucose, chloride, Ca, P, etc. – required by the animal.

renin
A proteolytic enzyme synthesized by the juxtaglomerular cells of the kidney that plays a role in blood pressure control by catalysing the conversion of angiotensinogen to angiotensin when renal arterial pressure falls.

reproductive cycle
The time from the conception of one foal to the conception of the next.

requirement (nutrient)
The requirement for any given nutrient is the amount of that nutrient that must be supplied in the diet to meet the net requirement of a normal healthy animal given a completely adequate diet in an environment compatible with good health. The net requirement is the quantity of that nutrient that should be absorbed to meet the needs of maintenance (q.v.), including the replacement of obligatory losses, and of any work, growth, production or reproduction taking place.

rhinopneumonitis (equine)
A mild viral disease of the upper respiratory tract of horses, which also commonly causes abortion.

rickets
Defective calcification of the epiphyseal cartilage of growing horses owing to inadequate dietary vitamin D, Ca and P or an incorrect proportion of Ca to P in the diet.

ringbones
Any bony exostosis affecting the interphalangeal joints of the horse's foot, or any bony enlargement in the same region.

Ringer's solution
An isotonic solution, devised by Sydney Ringer, containing sodium chloride, potassium chloride and calcium chloride.

However, it is but little more physiological than physiological saline as its chloride concentration is even higher and therefore it can cause metabolic acidosis of the same magnitude (see Ch. 9).

roughage

There are several types of roughage, which fall broadly into the following categories: (1) long and dry, e.g. hay and straw; (2) ground and pelleted hay, straw and oatfeed; (3) ensiled long grass and comparable succulent forages; and (4) chopped succulent ensiled material. Within each category only feeds analysed to contain more than 20% crude fibre on an air-dry basis should be included. Roughages tend to reduce the intake of dry matter in horses fed *ad libitum* and they decrease net energy intake in these animals in comparison with those also receiving concentrates (q.v.). The fibre is useful in maintaining the microbial populations of the large intestine in a steady state.

ructus

Eructation or belching of gas from the GI tract.

ruminant

Herbivorous species that possess an enlarged forestomach (rumen) and that chew the cud by regurgitation of ingesta from the forestomach to the mouth.

sclerosis

An induration or hardening, especially resulting from persistent inflammation.

scours

Diarrhoea.

secondary nutritional hyperparathyroidism

The increased secretion of parathyroid hormone (q.v.) as a compensatory mechanism directed against a disturbance in mineral homeostasis induced by nutritional imbalances. A loss of Ca from the bones is induced, resulting in a condition marked by pain, spontaneous fractures, muscular weakness and osteofibrosis (see Chs. 3 and 11).

septicaemia

A serious condition in which bacteria circulate in the bloodstream and become widely distributed throughout practically every organ. The horse becomes distressed in severe septicaemia, respiration rate and heart action are accelerated and body temperature is elevated.

serum (blood)

The clear liquid that separates from the clot and the corpuscles in the clotting of blood.

set-fast

See **tying-up**.

shank (of long bone)

See **diaphysis**.

silage (ensilage)

Succulent feed preserved either by adding acid or allowing natural fermentation to occur under anaerobic conditions in compacted material. The pH achieved is approximately 4–4.2. Too low a pH limits intake and too high a pH provokes protein breakdown. Materials ensiled include fresh grass, forage crops, the above-ground growth of young cereal crops and a variety of byproducts from beet pulp to fish waste. The dry matter content is 30–50%.

spavin
Bone spavin is a disease of the hock, or tarsus, in which changes occur in the small bones on the inner aspect of the joint resulting in the deposition of new bone. Bog spavin is a puffy swelling of the same joint.

spleen
A glandlike but ductless organ in the anterior part of the abdominal cavity on the left side. Its functions are at least threefold. First, it disintegrates red cells, setting free the haemoglobin, which the liver converts to bilirubin, conserving the iron. Second, it acts as a storehouse of red cells, which it releases into the blood during times of higher oxygen demand. Third, evidence (mainly from other species) indicates a role for the spleen in immunological responses. The spleen is an antibody-forming tissue and macrophages constitute a predominant cell form in it.

splints
Bony enlargements that occur on the cannon bones or in connection with the small metacarpals or metatarsals (splint bones) as the result of localized inflammation of the bone or periosteum (periostitis or osteitis).

stifle
The joint corresponding to the human knee at the top of the hindlimb.

stocking up
Swelling, or oedema (q.v.), of the legs, owing to the accumulation of fluid beneath the skin frequently caused by a period of inactivity on rich feed immediately after an extended period of activity. Exercise, purging and a reduction in energy intake normally bring rapid relief. The condition is also occasionally seen in horses with damaged livers, especially where the diet is of low quality and deficient in protein.

strangles
An acute contagious fever of horses, donkeys and mules caused by the bacterium *Streptococcus equi*, characterized by catarrhal inflammation of the mucous membranes of the nasal passages and pharynx, and frequently accompanied by abscess formation in the submaxillary or pharyngeal lymphatic glands noticeable under the jaw.

stroke volume
See **cardicac output**.

strongyles
A group of strongyloid nematodes or roundworms widely distributed in the intestinal contents of mammals. In horses, strongyles are commonly called redworm (see Ch. 11).

sucrase (invertase)
A digestive enzyme secreted by the small intestine, which hydrolyses sucrose (cane and beet sugar), forming glucose and fructose, both of which are readily absorbed.

sweetfeed	An American term for a concentrate mix containing molasses.
synchronous diaphragmatic flutter	Contraction of the diaphragm in synchrony with the heart beat. It is observed in fatigued horses following severe exercise in hot weather when excessive sweating may precipitate a large decrease in the plasma concentrations of ionized Ca, Cl and K.
tachycardia	Excessive rapidity of heart action and pulse.
tapeworm	A parasitic intestinal cestode composed of numerous flattened segments and attached to the gut wall by a head. Not infrequently occurring in horses, but generally causes no major problem. Occasionally it may block the ileo–caecal sphincter and cause severe colic. This can be accomplished by *Anoplocephala perfoliata*, the only species to have been observed infecting horses in the UK.
tarsus	Bones of the hock. There are usually six bones of the hock, which join the tibia above to the two metatarsal bones below in the hindleg.
tetany	A condition in which there are localized spasmodic contractions, or twitching, of muscles (see Ch. 11 for stress tetany).
Thoroughbred (TB)	A hotblooded (q.v.) breed of about 1.625 m (16 hands), which originated in the UK and which has been used to improve many other breeds.
thoroughpin	Distension of the sheath of the deep flexor tendon where it passes over the arch of the tarsus (hock) resulting from a sprain, but it is not usually serious.
thrombophlebitis	Inflammation of a vein associated with thrombus formation.
thrush	A degenerative condition of the horn in the central cleft of the frog (q.v.) of the horse's foot caused by a bacterial infection and often resulting from the horse standing in dirty wet boxes with insufficient clean dry bedding.
thumps	See **synchronous diaphragmatic flutter**.
thyroid gland	Situated in the neck in connection with the upper extremity of the trachea. The gland secretes two hormones: an iodine containing thyroxine and calcitonin (thyrocalcitonin) (see Ch. 3).
tibia	The major long bone between the stifle (q.v.) and the hock (q.v.) of the hindlimb.
tidal volume	The volume of air inspired or expired with each normal breath.
tillering	The process of forming side shoots from the base of the stem of graminaceous plants (cereals and grasses). Grazing, or cutting, in the vegetative phase of growth encourages this process, so thickening the 'bottom' or base of young pasture

swards and increasing their suitability for grazing and exercising horses.

α-tocopherol The principal tocopherol with vitamin E potency (see Ch. 4).

torsion As described in the text, the twisting, or rotation, of part of the intestine, causing an obstruction.

toxin Any poisonous substance of microbial, vegetable or animal origin.

trace elements See **minerals**.

tricarboxylic acid (TCA) cycle The series of metabolic reactions by which acetylcoenzyme A (acetyl CoA) is oxidized to carbon dioxide and water. The energy released is stored as ATP (q.v.) and the process occurs only in the mitochondria of cells (see Ch. 9). Acetyl-CoA is generated by the catabolism (q.v.) of fatty acids, carbohydrates and amino acids.

trypsin This enzyme acts on peptide linkages that involve the carboxyl groups of lysine and arginine and ruptures protein chains at these points. It is one of the proteolytic enzymes secreted by the pancreas, but in the inactive form of trypsinogen. This is activated by the enzyme enterokinase liberated from the duodenal mucosa.

turgor The normal consistency of soft tissue, as opposed to the flaccid condition of dehydration.

tying-up (set-fast) A condition in racehorses in which stiffness, blowing and sweating occur after a period of hard extended exercise, said to be caused by a depletion of muscle glycogen (see Chs. 9 and 11).

tympany Distension of the stomach, or intestines, by gas, usually caused by rapid microbial fermentation of ingesta, leading to a drum-like condition of the abdomen and colic.

ulna A bone behind the radius in the foreleg that forms the point of the elbow. The shaft of the ulna is vestigial and the tapered end is fused to the radius.

umbilical cord The nourishment of the fetus mainly passes to it through the cord from the placenta. After birth, the cord should not be interfered with as early severance deprives the newborn foal of 1000–1500 ml placental fetal blood, whereas under 'normal' conditions the amount concerned is under 200 ml.

uraemia The presence of urinary constituents in the blood and the toxic condition produced thereby. Normal blood urea level is greatly exceeded, indicating a failure of normal renal function.

urea The chief nitrogenous wasteproduct synthesized in the liver, discharged from the body in the urine and also secreted into the small intestine. It is highly soluble in water and the amount produced by the healthy horse receiving regular meals is proportional to the crude protein content of the diet.

urticaria   See **hives**.

uterus   A hollow muscular organ lying in the abdominal cavity below the rectum. In the mare it has a large body and small horns. It carries the fetal foal and nourishes it during pregnancy through the placenta attached to its wall.

vagus (pneumogastric) nerve   A major parasympathetic nerve (tenth cranial) of the autonomic nervous system (q.v.) possessing both efferent and afferent fibres distributed to the larynx, lungs, heart, oesophagus, stomach, liver, intestines and in fact most of the abdominal viscera. It therefore plays a considerable part in digestion and physical exercise.

vein   Thinner walled vessels than arteries (q.v.), in which deoxygenated blood under lower pressure is carried back to the heart. Veins possess a system of valves that control the direction of blood flow. Muscular contraction and relaxation of the limbs therefore provides the main force by which blood is lifted back up the legs against the force of gravity.

vertebrae   A chain of bones running from the base of the skull to the tip of the tail and which carries in the spinal canal the spinal cord – the posterior part of the central nervous system.

VFA   See **volatile fatty acids**.

villus (pl. villi)   Small vascular processes covering the mucous epithelium of the small intestine. They greatly enlarge the surface area for the absorption of nutrients into branches of the portal blood system and into the lymphatic system.

viscera   In this text the term refers to the abdominal viscera – the organs of the abdominal cavity.

vitamer   Any of a number of compounds that possess a given vitamin activity.

vitamins   A group of unrelated organic substances which occur in many foods in small amounts and which are necessary for normal metabolism of the body. They have been arbitrarily divided into a group of four major fat-soluble vitamins and at least ten water-soluble ones (see Ch. 4).

volatile fatty acids   Short-chain steam-volatile acids, principally acetic, propionic, butyric and smaller quantities of higher acids, which are the microbial wasteproducts of fermentation of dietary polysaccharides and protein within the alimentary canal. They are absorbed into the bloodstream and constitute a major energy source to the horse.

volvulus   Intestinal obstruction due to a knotting and twisting of the intestines.

warmblooded	See **hotblooded**.
wasting disease	A state of chronic emaciation.
windsucking (crib-biting)	In the English literature these terms refer to the act of swallowing gulps of air into the stomach. They are habitual vices. The crib-biter effects the action by grasping the edge of the manger, fence, etc. with the incisor teeth and by the coincidence of the raising of the floor of the mouth, opening the soft palate and a swallowing action, a gulp of air passes into the stomach (Plate 11.1). The wind-sucker achieves the same end without a resting place for the teeth. Young idle animals are said to acquire the habit from individuals in their company with a confirmed habit, which can initiate repeated bouts of mild colic. Excessive wear of the incisor teeth may compromise the individual's grazing powers. Sometimes a cribbing strap is fitted snugly around the upper neck of persistent offenders. The American literature restricts the term 'windsucker' to mares that aspirate air and faecal material into the vagina. This is corrected by Caslick's operation. The term 'cribbing' is then reserved for the aspiration of air into the stomach.
withers	The ridge on the back of the horse over the dorsal processes of the thoracic vertebrae and the shoulder blades and directly in front of the saddle.
wobbler	The name given to a horse showing a slight swaying action of the hindquarters, or stumbling, occurring mainly between 1 and 3 years of age. The signs can become progressively worse over 6–9 months when the horse is unable to trot without rolling from side to side and falling. The condition apparently results from damage to the spinal cord in the neck through injury and/or nutritionally induced abnormalities of the vertebrae caused by imbalances or inadequacies of Ca and P.
wood chewing	A habit developed by many horses probably as a result of boredom. The horse normally does not swallow the wood and the habit is unlikely to have any dietary implications.
zone of thermal neutrality	The range of environmental temperatures over which heat production by the animal is minimized. Below this range the animal must increase heat production by shivering and other means in order to maintain a normal body temperature. Above the range normal cooling mechanisms prove inadequate, body temperature rises and with it metabolic rate.

References and Bibliography

Abrams, J.T. (1979) The effect of dietary vitamin A supplements on the clinical condition and track performance of racehorses. *Bibl. Nutr. Dieta*, **27**, 113–20.

Adam, K.M.G. (1951) The quantity and distribution of the ciliate protozoa in the large intestine of the horse. *Parasitology*, **41**, 301–311.

Agricultural and Food Research Council (AFRC) Institute for Grassland and Animal Production, Welsh Plant Breeding Station, Plas Gogerddan, Aberystwyth. Various publications on grassland research.

Agricultural Research Council (1980) *The Nutrient Requirements of Ruminant Livestock*, p. 293. Commonwealth Agricultural Bureaux, London.

Agricultural Research Council (1981) *The Nutrient Requirements of Pigs*, p. 146. Commonwealth Agricultural Bureaux, London.

Aherne, F.X. and Kennelly, J.J. (1983) Oilseed meals for livestock feeding. In: *Recent Advances in Animal Nutrition 1982* (ed. W. Haresign), pp. 39–89. Butterworths, London.

Ahlswede, L. and Konermann, H. (1980) Experiences with oral and parenteral administration of β-carotene in the horse. *Praktische Tierärztliche*, **68**, 56–60.

Aiken, G.E., Potter, G.D., Conrad, B.E. and Evans, J.W. (1989) Growth performance of yearling horses grazing Bermuda-grass pastures at different grazing pressures. *Journal of Animal Science*, **67**, 2692–7.

Aitken, M.M., Anderson, M.G., Mackenzie, G. and Sanford, J. (1974) Correlations between physiological and biochemical parameters used to assess fitness in the horse. *Journal of the South African Veterinary Association*, **45**, 361–70.

Aldred, T., Fontenot, J.P. and Webb, K.E., Jr (1978) Availability of phosphorus from three sources for ponies. *Virginia Polytechnic Institute State University Research Division Report*, **174**, 152–7.

Alexander, F. (1963) Digestion in the horse. In: *Progress in Nutrition and Allied Sciences* (ed. D.P. Cuthbertson), pp. 259–68. Oliver & Boyd, Edinburgh.

Alexander, F. (1972) Symposium (1). Certain aspects of the physiology and pharmacology of the horse's digestive tract. *Equine Veterinary Journal*, **4**, 166–9.

Alexander, F. (1978) The effect of some anti-diarrhoeal drugs on intestinal transit and faecal excretion of water and electrolytes in the horse. *Equine Veterinary Journal*, **10**, 229–34.

Alexander, F. and Benzie, D. (1951) A radiological study of the digestive tract of the foal. *Quarterly Journal of Experimental Physiology*, **36**, 213–17.

Alexander, F. and Hickson, J.C.D. (1969) The salivary and pancreatic secretions of the horse. In: *Physiology of Digestion and Metabolism in the Ruminant. Proceedings of the 3rd International Symposium*, August 1969, Cambridge (ed. A.T. Phillipson), pp. 375–89. Oriel Press, Stocksfield.

Alexander, F., Macpherson, J.D. and Oxford, A.E. (1952) Fermentative activities of some members of the normal coccal flora of the horse's large intestine. *Journal of Comparative Pathology*, **62**, 252–9.

Alexander, J.C. (1977) Biological effects due to changes in fats during heating. *Journal of the American Oil Chemists' Society*, **55**, 711–17.

Allen, G.K., Campbell-Beggs, C., Robinson, J.A., Johnson, P.J. and Green, E.M. (1996) Induction of early-phase endotoxin tolerance in horses. *Equine Veterinary Journal*, **28**, 269–74.

Aluja, A.S., de Gross, D.R., McCosker, P.J. and Svendsen, J. (1968) Effect of altitude on horses. *Veterinary Record*, **82**, 368–72.

Anderson, C.E., Potter, G.D., Kreider, J.L. and Courtney, C.C. (1981) Digestible energy requirements for exercising horses. *Journal of Animal Science*, **53** (Suppl. 1), 42, abstract 101.

Anderson, C.E., Potter, G.D., Kreider, J.L. and Courtney, C.C. (1983) Digestible energy requirements for exercising horses. *Journal of Animal Science*, **56**, 91–5.

Anderson, M.G. (1975a) The effect of exercise on blood metabolite levels in the horse. *Equine Veterinary Journal*, **7**, 27–33.

Anderson, M.G. (1975b) The influence of exercise on serum enzyme levels in the horse. *Equine Veterinary Journal*, **7**, 160–65.

Anderson, M.G. (1976) The effect of exercise on the lactic dehydrogenase and creatine kinase isoenzyme composition of horse serum. *Research in Veterinary Science*, **20**, 191–6.

Anderson, P.H., Patterson, D.S.P. and Berrett, S. (1978) Selenium deficiency. *Veterinary Record*, **103**, 145–6.

Anderson, R.A., Bryden, N.A., Polansky, M.M. and Patterson, K.Y. (1983) Strenuous exercise: effects on selected clinical values and chromium, copper and zinc concentrations in urine and serum of male runners. *Federation Proceedings*, **42**, 804, abstract 2998.

Andrews, A.H. and Humphreys, D.J. (1982) *Poisoning in Veterinary Practice*, 2nd edn., pp. 1–114. National Office of Animal Health, Enfield.

Andrieu, J., Jestin, M. and Martin-Rosset, W. (1996) Prediction of organic matter digestibility (OMD) of forages in horses by near infrared spectrophotometry (NIRS). *47th European Association of Animal Production Meeting*, Lillehammer, Norway, 25–29th August. Horse Commission Session-H4: Nutrition, pp. 1–6.

Angsubhakorn, S., Poomvises, P., Romruen, K. and Newberne, P.M. (1981) Aflatoxicosis in horses. *Journal–American Veterinary Medical Association*, **178**, 274–8.

Anon (1976a) Drugs and doping in horses. *Veterinary Record*, **98**, 453.

Anon (1976b) Salmonellosis in horses. *Veterinary Record*, **99**, 19–20.

Anon (1984) Tables des apports alimentaires recommandés pour le cheval. In: *Le Cheval. Reproduction, Sélection, Alimentation, Exploitation* (eds R. Jarridge and W. Martin-Rossett), pp. 645–89. INRA, Paris.

Anon (1986) Better use of nitrogen – the prospect for grassland. *National Agricultural Conference Proceedings*. Royal Agricultural Society of England and Agricultural Development and Advisory Service, National Agriculture Centre, Warwickshire.

Anon (1995a) *Compendium of Data Sheets for Veterinary Products, 1994–95*, pp. 1–755. National Office of Animal Health, Enfield.

Anon (1995b) *The UK Pesticide Guide* (ed. R. Whitehead), pp. 1–589. CAB International, Wallingford and the British Crop Protection Council, Farnham.

Anon (1995c) The Feeding Stuffs Regulations 1995, pp. 1–100. *Statutory Instruments, 1995, No. 1412, Agriculture*. The Stationery Office, London.

Answer, M.S., Chapman, T.E. and Gronwall, R. (1976) Glucose utilization and recycling in ponies. *American Journal of Physiology*, **230**, 138–42.

Answer, M.S., Gronwall, R., Chapman, T.E. and Klentz, R.D. (1975) Glucose utilization and contribution to milk components in lactating ponies. *Journal of Animal Science*, **41**, 568–71.

Archer, M. (1973) Variations in potash levels in pastures grazed by horses: a preliminary communication. *Equine Veterinary Journal*, **5**, 45–6.

Archer, M. (1978a) Studies on producing and maintaining balanced pastures for studs. *Equine Veterinary Journal*, **10**, 54–9.

Archer, M. (1978b) Further studies on palatability of grasses to horses. *Journal of the British Grassland Society*, **33**, 239–43.

Argenzio, R.A. (1975) Functions of the equine large intestine and their interrelationship in disease. *Cornell Veterinarian*, **65**, 303–29.

Argenzio, R.A. and Hintz, H.F. (1970) Glucose tolerance and effect of volatile fatty acid on plasma glucose concentration in ponies. *Journal of Animal Science*, **30**, 514–18.

Argenzio, R.A. and Hintz, H.F. (1972) Effect of diet on glucose entry and oxidation rates in ponies. *Journal of Nutrition*, **102**, 879–92.

Argenzio, R.A., Lowe, J.E., Hintz, H.F. and Schryver, H.F. (1974) Calcium and phosphorus homeostasis in horses. *Journal of Nutrition*, **104**, 18–27.

Argenzio, R.A., Southworth, M., Lowe, J.E. and Stevens, C.E. (1977) Interrelationship of $NaHCO_3$ and volatile fatty acid transport by equine large intestine. *American Journal of Physiology*, **233**, E469–78.

Argiroudis, S.A., Kent, J.E. and Blackmore, D.J. (1982) Observations on the isoenzymes of creatine kinase in equine serum and tissues. *Equine Veterinary Journal*, **14**, 317–21.

Art, T. and Lekeux, P. (1993) Training-induced modifications in cardiorespiratory and ventilatory measurements in Thoroughbred horses. *Equine Veterinary Journal*, **25**, 532–6.

Asmundsson, T., Gunnarsson, E. and Johannesson, T. (1983) 'Haysickness' in Icelandic horses: precipitin tests and other studies. *Equine Veterinary Journal*, **15**, 229–32.

Austic, R.E. (1980) Acid–base interrelationships in nutrition. *Proceedings of the Cornell Nutrition Conference for Feed Manufacturers*, 12–17. Cornell University, Ithaca, New York.

Baalsrud, K.J. and Øvernes, G. (1986) Influence of vitamin E and selenium supplement on antibody production in horses. *Equine Veterinary Journal*, **18**, 472–4.

Bachman, S.E., Galyean, M.L., Smith, G.S., Hallford, D.M. and Graham, J.D. (1992) Early aspects of locoweed toxicosis and evaluation of a mineral supplement or clinoptilolite as dietary treatments. *Journal of Animal Science*, **70**, 3125–32.

Bacon, C.W. (1995) Toxic endophyte-infected tall fescue and range grasses: historic perspectives. *Journal of Animal Science*, **73**, 861–70.

Baker, H.J. and Lindsey, J.R. (1968) Equine goiter due to excess dietary iodide. *Journal of American Veterinary Medical Association*, **153**, 1618–30.

Baker, J.P. (1971) Horse nutritive requirements. *Feed Management*, September, 10–15.

Baker, J.P., Lieb, S., Crawford, B.H., Jr and Potter, G.D. (1972) Utilization of energy sources by the equine. *Proceedings of the 27th Distillers Feed Conference*, **27**, 28–33. Distillers Co.

Baker, J.P. and Quinn, P.J. (1978) A report on clinical aspects and histopathology of sweet itch. *Equine Veterinary Journal*, **10**, 243–8.

Baker, J.P., Sutton, H.H., Crawford, B.H., Jr and Lieb, S. (1969) Multiple fistulation of the equine large intestine. *Journal of Animal Science*, **29**, 916–20.

Baker, J.R. (1970) Salmonellosis in the horse. *British Veterinary Journal*, **126**, 100–105.

Baker, J.R. and Leyland, A. (1973) Diarrhoea in the horse associated with stress and tetracycline therapy. *Veterinary Record*, **93**, 583–4.

Baker, J.R., Wyn-Jones, G. and Eley, J.L. (1983) Case of equine goitre. *Veterinary Record*, **112**, 407–408.

Baker, L.A., Topliff, D.R., Freeman, D.W., Telter, R.G. and Breazile, J.W. (1992) Effect of dietary cation–anion balance on acid–base status in horses. *Journal of Equine Veterinary Science*, **12**, 160–3.

Balls, D. (1976) Notes on equine toxicology. *Veterinary Practice*, **8**, 5–6.

Banach, M.A. and Evans, J.W. (1981a) The effects of energy intake during gestation and lactation on reproductive performance in mares. *Proceedings of the Western Section of the American Society of Animal Science*, Vancouver, British Columbia, 23–25 June 1981, **32**, 264–7.

Banach, M.A. and Evans, J.W. (1981b) The effects of energy intake during gestation and lactation on reproductive performance in mares. *Journal of Animal Science*, **53** (Suppl. 1), 500, abstract 94.

Baranova, D. (1977) Vitamins in the feeding of weaned foals. *Konevodstvoi Konnyi Sport*, No. 10, 29–30.

Barclay, M.N.I. and MacPherson, A. (1992) Selenium content of wheat for bread making in Scotland and the relationship between glutathione peroxidase (EC 1.11.1.9) levels in whole blood and bread consumption. *British Journal of Nutrition*, **68**, 261–70.

Barratt, M.E.J., Strachan, P.J. and Porter, P. (1979) Immunologically mediated nutritional disturbances associated with soya-protein antigens. *Proceedings of the Nutrition Society*, **38**, 143–50.

Bartel, D.L., Schryver, H.F., Lowe, J.E. and Parker, R.A. (1978) Locomotion in the horse: a procedure for computing the internal forces in the digit. *American Journal of Veterinary Research*, **39**, 1721–7.

Barth, K.M., Williams, J.W. and Brown, D.G. (1977) Digestible energy requirements of working and non-working ponies. *Journal of Animal Science*, **44**, 585–9.

Bartley, E.E., Avery, T.B., Nagaraja, T.G., *et al.* (1981) Ammonia toxicity in cattle v. ammonia concentration of lymph and portal, carotid and jugular blood after the ingestion of urea. *Journal of Animal Science*, **53**, 494–8.

Basler, S.E. and Holtan, D.W. (1981) Factors affecting blood selenium levels in Oregon horses and association of blood selenium level with disease incidence. *Proceedings of the Western Section of the American Society of Animal Science*, **32**, 399–400.

Batt, R.M. (1991) Oral sugar tests for diagnosis of small intestinal disease. *Equine Veterinary Journal*, **23**, 325–6.

Battle, G.H., Jackson, S.G. and Baker, J.P. (1988) Acceptability and digestibility of preservative-treated hay by horses. *Nutrition Reports International*, **37**, 83–9.

Baucus, K.L., Ralston, S.L., Rich, G. and Squires, E.L. (1987) The effect of dietary copper and zinc supplementation on composition of mares' milk. *Proceedings of the 10th Equine Nutrition and Physiology Society*, Colorado State University, Fort Collins, 11–13 June 1987, pp. 179–80.

Bauer, J.E. (1983) Plasma lipids and lipoproteins of fasted ponies. *American Journal of Veterinary Research*, **44**, 379–84.

Belko, A.Z., Barbieri, T.F. and Wong, E.C. (1986) Effect of energy and protein intake and exercise intensity on the thermic effect of food. *American Journal of Clinical Nutrition*, **43**, 863–9.

Belko, A.Z. and Roe, D.A. (1983) Exercise effects on riboflavin status. *Federation Proceedings*, **42**, 804, abstract 2995.

Benamou, A.E. and Harris, R.C. (1993) Effect of carnitine supplement to the dam on plasma carnitine concentration in the sucking foal. *Equine Veterinary Journal*, **25**, 49–52.

Bendroth, M. (1981) A survey of reasons for some trotters being non-starters as 2-, 3-, and 4-year olds. *Proceedings of the 32nd Annual Meeting of the European Association of Animal Production*, Zagreb, Yugoslavia, 31 August–3 September, IIA–1.

Bentley, O.E., Burns, S.J., McDonald, D.R., *et al.* (1978) Safety evaluation of pyrantel pamoate administered with trichlorfon as a broad-spectrum anthelmintic in horses. *Veterinary Medicine and Small Animal Clinician*, **73**, 70–73.

Bergsten, G., Holmbäck, R. and Lindberg, P. (1970) Blood selenium in naturally fed horses and the effect of selenium administration. *Acta Veterinary Scandinavica*, **11**, 571–6.

Berliner, V.R. (1942) Seasonal influences on the reproductive performance of mares and jennets in Mississippi. *Journal of Animal Science*, 63–4.

Beuchat, L.R. (1978) Microbial alternations of grains, legumes and oil seeds. *Food Technology*, May, 193–6.

Bird, A.R., Chandler, K.D. and Bell, A.W. (1981) Effects of exercise and plane of nutrition on nutrient utilization by the hind limb of the sheep. *Australian Journal of Biological Sciences*, **34**, 541–50.

Blackmore, D.J. and Brobst, D. (1981) *Biochemical Values in Equine Medicine*. Animal Health Trust, Newmarket, Suffolk.

Blackmore, D.J., Campbell, C., Dant, C., Holden, J.E. and Kent, J.E. (1982) Selenium status of thoroughbreds in the United Kingdom. *Equine Veterinary Journal*, **14**, 139–43.

Blackmore, D.J. and Elton, D. (1975) Enzyme activity in the serum of Thoroughbred horses in the United Kingdom. *Equine Veterinary Journal*, **7**, 34–9.

Blackmore, D.J., Henley, M.I. and Mapp, B.J. (1983) Colorimetric measurement of albumin in horse sera. *Equine Veterinary Journal*, **15**, 373–4.

Blackmore, D.J., Willett, K. and Agness, D. (1979) Selenium and gamma-glutamyl transferase activity in the serum of thoroughbreds. *Research in Veterinary Science*, **26**, 76–80.

Blaney, B.J., Gartner, R.J.W. and McKenzie, R.A. (1981a) The effect of oxalate in some tropical grasses on the availability to horses of calcium, phosphorus and magnesium. *Journal of Agricultural Science, Cambridge*, **97**, 507–514.

Blaney, B.J., Gartner, R.J.W. and McKenzie, R.A. (1981b) The inability of horses to absorb calcium from calcium oxalate. *Journal of Agricultural Science, Cambridge*, **97**, 639–41.

Blaxter, K.L. (1962) *The Energy Metabolism of Ruminants*. Hutchinson, London.

Bochröder, B., Schubert, R. and Bödeker, D. (1994) Studies on the transport *in vitro* of lysine, histidine, arginine and ammonia across the mucosa of the equine colon. *Equine Veterinary Journal*, **26**, 131–3.

Boening, K.J. and Leendertse, I.P. (1993) Review of 115 cases of colic in the pregnant mare. *Equine Veterinary Journal*, **25**, 518–21.

Bogan, J.A., Lees, P. and Yoxall, A.T. (eds) (1983) *Pharmacological Basis of Large Animal Medicine*. Blackwell Science, Oxford.

Bolton, J.R., Merritt, A.M., Cimprich, R.E., Ramberg, C.F. and Streett, W. (1976) Normal and abnormal xylose absorption in the horse. *Cornell Veterinarian*, **66**, 183–90.

Bonhomme-Florentin, A. (1988) Degradation of hemicellulose and pectin by horse caecum contents. *British Journal of Nutrition*, **60**, 185–92.

Boren, S.R., Topliff, D.R., Freeman, D.W., Bahr, R.J., Wagner, D.G. and Maxwell, C.V. (1987) Growth of weanling Quarter horses fed varying energy and protein levels. *Proceedings of the 10th Equine Nutrition and Physiology Society*, Colorado State University, Fort Collins, 11–13 June 1987, pp. 43–8.

Bouwman, H. (1978) Digestibility trials with extruded feeds and rolled oats in ponies. *Landbouwkundig Tijdschrift*, **90**, 2–6.

Bouwman, H. and van der Schee, W. (1978) Composition and production of milk from Dutch warmblooded saddle horse mares. *Zeitschrift für Tierphysiologie Tierernährung und Futtermittelkunde*, **40**, 39–53.

Bowland, J.P. and Newell, J.A. (1974) Fatty acid composition of shoulder fat and perinephric fat from pasture-fed horses. *Canadian Journal of Animal Science*, **54**, 373–6.

Bowman, V.A., Meacham, T.N., Dana, G.R. and Fontenot, J.P. (1978) Pelleted complete rations containing different roughage bases for horses. *Va Polytech Inst State University Research Div Reports*, **174**, 179–82.

Bracher, V., von Fellenberg, R., Winder, C.N., Gruenig, G., Hermann, M. and Kraehenmann, A. (1991) An investigation of the incidence of chronic obstuctive pulmonary disease (COPD) in random populations of Swiss horses. *Equine Veterinary Journal*, **23**, 136–41.

Brady, P.S., Ku, P.K. and Ullrey, D.E. (1978) Lack of effect of selenium supplementation on the response of the equine erythrocyte glutathione system and plasma enzymes to exercise. *Journal of Animal Science*, **47**, 492–6.

Brady, P.S., Shelle, J.E. and Ullrey, D.E. (1977) Rapid changes in equine erythrocyte glutathione reductase with exercise. *American Journal of Veterinary Research*, **38**, 1045–7.

Braverman, Y. (1988) Preferred landing sites of *Culicoides* species (Diptera: Ceratopogonidae) on a horse in Israel and its relevance to summer seasonal recurrent dermatitis (sweet itch). *Equine Veterinary Journal*, **20**, 426–9.

Breuer, L.H. (1970) Horse nutrition and feeding. *Feedstuffs*, **42**, 44–5.

Breukink, H.J. (1974) Oral mono- and disaccharide tolerance tests in ponies. *American Journal of Veterinary Research*, **35**, 1523–7.

Bridges, C.H. and Harris, E.D. (1988) Experimentally induced cartilaginous fractures (osteochondritis dissecans) in foals fed low-copper diets. *Journal of the American Veterinary Medical Association*, **193**, 215–21.

British Equine Veterinary Association (c. 1978) *Veterinary Guidelines on Equine Endurance Competitions*, working party report. BEVA, London.

British Horse Society (n.d.) *Grassland Management for Horse and Pony Owners*. BHS, Kenilworth, Warwickshire.

Brobst, D.F. and Bayly, W.M. (1982) Response of horses to a water deprivation test. *Journal of Equine Veterinary Science*, **2**, 51–6.

Brook, D. and Schmidt, G.R. (1979) Pre-renal azotaemia in a pony with an oesophageal obstruction. *Equine Veterinary Journal*, **11**, 53–5.

Brophy, P.O. (1981) Assessment of the immunological status of the newborn foal. *Proceedings of the 32nd Annual Meeting of the European Association of Animal Production*, Zagreb, Yugoslavia, 31 August–3 September, IIIa–5.

Brown, R.F., Houpt, K. and Schryver, H.F. (1976) Stimulation of food intake in horses by diazepam and promazine. *Pharmacology Biochemistry and Behavior*, **5**, 495–7.

Brown, R.H. (1978) Horses can digest high fat levels, Georgians told. *Feedstuffs*, **50**(10), 15.

Browning, G.F., Chalmers, R.M., Snodgrass, D.R., *et al.* (1991) The prevalence of enteric pathogens in diarrhoeic Thoroughbred foals in Britain and Ireland. *Equine Veterinary Journal*, **23**, 405–409.

Brownlow, M.A. and Hutchins, D.R. (1982) The concept of osmolality; its use in the evaluation of 'dehydration' in the horse. *Equine Veterinary Journal*, **14**, 106–110.

Buffa, E.A., Van Den Berg, S.S., Verstraete, F.J.M. and Swart, N.G.N. (1992) Effect of dietary biotin supplement on equine hoof horn growth rate and hardness. *Equine Veterinary Journal*, **24**, 472–4.

Bunch, T.D., Panter, K.E. and James, L.F. (1992) Ultrasound studies of the effects of certain poisonous plants on uterine function and fetal development in livestock. *Journal of Animal Science*, **70**, 1639–43.

Buntain, B.J. and Coffman, J.R. (1981) Polyuria polydypsia in a horse induced by psychogenic salt consumption. *Equine Veterinary Journal*, **13**, 266–8.

Burke, D.J. and Albert, W.W. (1978) Methods for measuring physical condition and energy expenditure in horses. *Journal of Animal Science*, **46**, 1666–72.

Burridge, J.C., Reith, J.W.S. and Berrow, M.L. (1983) Soil factors and treatments affecting trace elements in crops and herbage. In: *Trace Elements in Animal Production and Veterinary Practice* (eds N.F. Suttle, R.G. Gunn, W.M. Allen, K.A. Linklater and G. Weiner), Occasional Publication No. 7, pp. 77–85. British Society of Animal Production, Edinburgh.

Burrows, G.E. (1981) Endotoxaemia in the horse. *Equine Veterinary Journal*, **13**, 89–94.

Burton, J.H., Pollack, G. and de la Roche, T. (1987) Palatability and digestibility studies with high moisture forage. *Proceedings of the 10th Equine Nutrition and Physiology Society*, Colorado State University, Fort Collins, 11–13 June 1987, pp. 599–602.

Butler, K.D., Jr and Hintz, H.F. (1977) Effect of level of feed intake and gelatin supplementation on growth and quality of hoofs of ponies. *Journal of Animal Science*, **44**, 257–63.

Butler, P. and Blackmore, D.J. (1982) Retinol values in the plasma of stabled thoroughbred horses in training. *Veterinary Record*, **111**, 37–8.

Butler, P. and Blackmore, D.J. (1983) Vitamin E values in the plasma of stabled thoroughbred horses in training. *Veterinary Record*, **112**, 60.

Cabrera, L., Julliand, V., Faurie, F. and Tisserand, J.L. (1992) Influence of feeding roughage and concentrate (soybean meal) simultaneously or consecutively on levels of plasma free amino acids and plasma urea in the equine. *First Europäische Konferenz über die Ernährung des Pferdes,* Institut für Tierernährung, Tierärzliche Hochschule, Hannover, 3–4 September 1992, pp. 144–9.

Callear, J.F.F. and Neave, R.M.S. (1971) The clinical use of the anthelmintic membendazole. *British Veterinary Journal*, **127**, xli–xliii.

Cameron, I.R. and Hall, R.J.C. (1975) The effect of dietary K^+ depletion and subsequent repletion on intracellular K^+ concentration and pH of cardiac and skeletal muscle in rabbits. *Journal of Physiology*, **251**, 70–71P.

Campbell, J.R. (1977) Bone growth in foals and epiphyseal compression. *Equine Veterinary Journal*, **9**, 116–21.

Campbell, J.R. and Lee, R. (1981) Radiological estimation of differential growth rates of the long bones of foals. *Equine Veterinary Journal*, **13**, 247–50.

Candau, M. and Bueno, L. (1977) Motricité caecale et transit chez le poney: influence de l'état de réplétion du caecum et des fermentations microbiennes. *Annals Biological Animal Biochemistry Biophysic*, **17**, 503–508.

Cape, L. and Hintz, H.F. (1982) Influence of month, color, age, corticosteroids, and dietary molybdenum on mineral concentration of equine hair. *American Journal of Veterinary Research*, **43**, 1132–6.

Caple, I.W., Edwards, S.J.A., Forsyth, W.M., Whiteley, P., Selth, R.H. and Fulton, L.J. (1978) Blood glutathione peroxidase activity in horses in relation to muscular dystrophy and selenium nutrition. *Australian Veterinary Journal*, **54**, 57–60.

Cardinet, G.H., Fowler, M.E. and Tyler, W.S. (1963) Heart rates, respiratory rates for evaluating performance in horses during endurance trial ride competition. *Journal–American Veterinary Medical Association*, **143**, 1303–309.

Cardinet, G.H., Littrell, J.F. and Freedland, R.A. (1967) Comparative investigations of serum creatine phosphokinase and glutamic-oxaloacetic transaminase activities in equine paralytic myoglobinuria. *Research in Veterinary Science*, **8**, 219–26.

Carlin, J.I., Harris, R.C., Cederblad, G., Constantin-Teodosiu, D., Snow, D.H. and Hultman, E. (1990) Association between muscle acetyl-CoA and acetylcarnitine levels in the exercising horse. *Journal of Applied Physiology*, **69**, 42–5.

Carlson, G.P. (1975) Hematological alterations in endurance-trained horses. *Proceedings of the 1st International Symposium on Equine Hematology*, Michigan State University, 28–30 May 1975, pp. 444–9. American Association of Equine Practitioners, Golden, Colorado.

Carlson, G.P. (1983a) Response to saline solution of normally fed horses and horses dehydrated by fasting. *American Journal of Veterinary Research*, **44**, 964–8.

Carlson, G.P. (1983b) Thermoregulation and fluid balance in the exercising horse. In: *Equine Exercise Physiology, Proceedings of the 1st International Conference*, Oxford 1982 (eds D.H. Snow, S.G.B. Persson and R.J. Rose), pp. 291–309. Granta Editions. Cambridge.

Carlson, G.P. and Mansmann, R.A. (1974) Serum electrolyte and plasma protein alterations in horses used in endurance rides. *Journal–American Veterinary Medical Association*, **165**, 262–4.

Carlson, G.P. and Ocen, P.O. (1979) Composition of equine sweat following exercise in high environmental temperatures and in response to intravenous epinephrine administration. *Journal of Equine Medicine and Surgery*, **3**, 27–31.

Carlson, G.P., Ocen, P.O. and Harrold, D. (1976) Clinicopathologic alterations in normal and exhausted endurance horses. *Theriogenology*, **6**, 93–104.

Carlson, L.A., Fröberg, S. and Persson, S. (1965) Concentration and turnover of the free fatty acids of plasma and concentration of blood glucose during exercise in horses. *Acta Physiologica Scandinavica*, **63**, 434–41.

Carrick, J.B., Morris, D.D. and Moore, J.N. (1993) Administration of a receptor antagonist for platelet-activating factor during equine endotoxaemia. *Equine Veterinary Journal*, **25**, 152–7.

Carroll, C.L., Hazard, G., Coloe, P.J. and Hooper, P.T. (1987) Laminitis and possible enterotoxaemia associated with carbohydrate overload in mares. *Equine Veterinary Journal*, **19**, 344–6.

Carroll, C.L. and Huntington, P.J. (1988) Body condition scoring and weight estimation of horses. *Equine Veterinary Journal*, **20**, 41–5.

Carroll, F.D., Goss, H. and Howell, C.E. (1949) The synthesis of B vitamins in the horse. *Journal of Animal Science*, **8**, 290–96.

Carson, K. and Wood-Guch, D.G.M. (1983) Behaviour of Thoroughbred foals during nursing. *Equine Veterinary Journal*, **15**, 257–62.

Chachula, J. and Chachulowa, J. (1969) The use of the concentrates in feeding arden and fjord stallions. *Rocz. Nauk rol.*, **91B–4**, 635–56.

Chachula, J. and Chachulowa, J. (1970) Further investigations on the use of mixed feeds in feeding of various groups of breeding horses. *Rocz. Nauk Rol.*, **92B–3**, 351–75.

Chachula, J. and Chrzanowski, S. (1972) Investigations upon several factors affecting the results of breeding of the fjording horse at the Nowielice state stud. *Instytut Genetyki i Hodowli Zwierzat Biuletyn*, **26**, 71–85.

Chapman, D.I. Haywood, P.E. and Lloyd, P. (1981) Occurrence of glycosuria in horses after strenuous exercise. *Equine Veterinary Journal*, **13**, 259–60.

Cheeke, P.R. (1995) Endogenous toxins and mycotoxins in forage grasses and their effects on livestock. *Journal of Animal Science*, **73**, 909–918.

Chong, Y.C., Duffus, W.P.H., Field, H.J., *et al.* (1991) The raising of equine colostrum-deprived foals; maintenance and assessment of specific pathogen (EHV-1/4) free status. *Equine Veterinary Journal*, **23**, 111–15.

Chrichlow, E.C., Yoshida, K. and Wallace, K. (1980) Dust levels in a riding stable. *Equine Veterinary Journal*, **12**, 185–8.

Clark, H., Alcock, M.B., Harvey, A. (1987) Tissue turnover and animal production on perennial ryegrass swards continuously stocked by sheep to maintain two sward surface heights. In: *Efficient Sheep Production* (ed. G.E. Pollott), pp. 149–52. British Grassland Society, Occasional Symposium, No. 21.

Clark, I. (1969) Metabolic interrelations of calcium, magnesium and phosphorus. *American Journal of Physiology*, **217**, 871–8.

Clarke, A.F. (1987) A review of environmental and host factors in relation to equine respiratory disease. *Equine Veterinary Journal*, **19**, 435–41.

Clarke, A.F. (1993) Stable dust – threshold limiting values, exposures variables and host risk factors. *Equine Veterinary Journal*, **25**, 172–4.

Clarke, A.F. and Madelin, T.M. (1987) The relationship of air hygiene in stables to lower airway disease and pharyngeal lymphoid hyperplasia in two groups of Thoroughbred horses. *Equine Veterinary Journal*, **19**, 524–30.

Clarke, E.G.C. and Clarke, M.L. (1975) *Veterinary Toxicology*, pp. 1–438. Baillière Tindall, London.

Clater, F. (1786) *Every Man His Own Farrier, or, the Whole Art of Farriery Laid Open*. J. Tomlinson for Baldwin and Bladon, Newark.

Clay, C.M., Squires, E.L., Amann, R.P. and Nett, T.M. (1988) Influences of season and artificial photoperiod on stallions: luteinizing hormone, follicle-stimulating hormone and testosterone. *Journal of Animal Science*, **66**, 1246–55.

Codazza, D., Maffeo, G. and Redaelli, G. (1974) Serum enzyme changes and haematochemical levels in Thoroughbreds after transport and exercise. *Journal of the South African Veterinary Association*, **45**, 331–4.

Coenen, M. (1992a) Chloridkonzentrationen und Mengen im Verdauungskanal des Pferdes. *First Europäische Konferenz über die Ernährung des Pferdes*, Institut für Tierernähung, Tierärztliche Hochschule, Hannover, 3–4 September 1992, pp. 73–6.

Coenen, M. (1992b) Observations in the occurrence of gastric ulcers in horses. *First Europäische Konferenz über die Ernährung des Pferdes*, Institut für Tierernähung, Tierärztliche Hochschule, Hannover, 3–4 September 1992, pp. 188–91.

Coffman, J. (1979a) Blood glucose 1 – factors affecting blood levels and test results. *Veterinary Medicine and Small Animal Clinician*, **74**, 719–23.

Coffman, J. (1979b) Blood glucose 2 – clinical application of blood glucose determination per se. *Veterinary Medicine and Small Animal Clinician*, **74**, 855–8.

Coffman, J. (1979c) Plasma lactate determinations. *Veterinary Medicine and Small Animal Clinician*, **74**, 997–1002.

Coffman, J. (1979d) The plasma proteins. *Veterinary Medicine and Small Animal Clinician*, **74**, 1168–70.

Coffman, J. (1980a) Calcium and phosphorus physiology and pathophysiology. *Veterinary Medicine and Small Animal Clinician*, **75**, 93–6.

Coffman, J. (1980b) Adrenocortical pathophysiology and consideration of sodium, potassium and chloride. *Veterinary Medicine and Small Animal Clinician*, **75**, 271–5.

Coffman, J. (1980c) Acid: base balance. *Veterinary Medicine and Small Animal Clinician*, **75**, 489–98.

Coffman, J. (1980d) Percent creatinine clearance ratios. *Veterinary Medicine and Small Animal Clinician*, **75**, 671–6.

Coffman, J. (1980e) Urology – 2. Testing for renal disease. *Veterinary Medicine and Small Animal Clinician*, **75**, 1039–44.

Coffman, J. (1980f) Hemostasis and bleeding disorders. *Veterinary Medicine and Small Animal Clinician*, **75**, 1157–64.

Coffman, J. (1980g) A data base for abdominal pain – 1. *Veterinary Medicine and Small Animal Clinician*, **75**, 1583–8.

Coffman, J.R., Hammond, L.S., Garner, H.E., Thawley, D.G. and Selby, L.A. (1980) Haematology as an aid to prognosis of chronic laminitis. *Equine Veterinary Journal*, **12**, 30–31.

Coger, L.S., Hintz, H.F., Schryver, H.F. and Lowe, J.E. (1987) The effect of high zinc intake on copper metabolism and bone development in growing horses. *Proceedings of the 10th Equine Nutrition and Physiology Society*, Colorado State University, Fort Collins, 11–13 June 1987, pp. 173–7.

Colles, C.M. (1979) A preliminary report on the use of warfarin in the treatment of navicular disease. *Equine Veterinary Journal*, **11**, 187–90.

Comben, N., Clark, R.J. and Sutherland, D.J.B. (1983) Improving the integrity of hoof horn in equines by high-level dietary supplementation with biotin. *Proceedings of the Annual Congress of the British Equine Veterinary Association*, University of York, York, 5 September 1983, pp. 1–17.

Comben, N., Clark, R.J. and Sutherland, D.J.B. (1984) Clinical observations on the response of equine hoof defects to dietary supplementation with biotin. *Veterinary Record*, **115**, 642–5.

Comerford, P.M., Edwards, R.L., Hudson, L.W. and Wardlaw, F.B. (1979) Supplemental lysine and methionine for the equine. *Technical Bulletin of the South Carolina Agricultural Experimental Station*, No. 1073.

Comline, R.S., Hall, L.W., Hickson, J.C.D., Murillo, A. and Walker, R.G. (1969) Pancreatic secretion in the horse. *Proceedings of the Physiology Society*, **204**, 10–11P.

Comline, R.S., Hickson, J.C.D. and Message, M.A. (1963) Nervous tissue in the pancreas of different species. *Proceedings of the Physiology Society*, **170**, 47–8P.

Cook, W.R. (1973) Diarrhoea in the horse associated with stress and tetracycline therapy. *Veterinary Record*, **93**, 15–17.

Cooper, J.P., Green, J.D. and Haggar, R. (1981) *The Management of Horse Paddocks. A Booklet of Instructions*. Horserace Betting Levy Board, London.

Cooper, S.R., Kline, K.H., Foreman, J.H., *et al.* (1995) Effects of dietary cation–anion balance on blood pH, acid–base parameters, serum and urine mineral levels, and parathyroid hormone (PTH) in weanling horses. *Journal of Equine Veterinary Science*, **15**, 417–20.

Corbally, A.F. (1995) *The contribution of the sport horse industry to the Irish economy*. MEqS thesis, Faculty of Agriculture, National University of Ireland, Dublin.

Corke, M.J. (1986) Diabetes mellitus: the tip of the iceberg. *Equine Veterinary Journal*, **18**, 87–8.

Corn, C.D., Potter, G.D. and Odom, T.W. (1993) Blood buffering in sedentary miniature horses after administration of sodium bicarbonate in single doses of varying amounts. *Proceedings of the 13th Equine Nutrition and Physiology Society*, University of Florida, Gainesville, 21–23 January 1993, No. 504, pp. 108–112.

Cornell, C.N., Garner, G.B., Yates, S.G. and Bell, S. (1982) Comparative fescue foot potential of fescue varieties. *Journal of Animal Science*, **55**, 180–84.

Cornwell, R.L. and Jones, R.M. (1968) Critical tests in the horse with the anthelmintic pyrantel tartrate. *Veterinary Record*, **82**, 483–4.

Cross, D.L., Redmond, L.M. and Strickland, J.R. (1995) Equine fescue toxicosis: signs and solutions. *Journal of Animal Science*, **73**, 899–908.

Crowell-Davis, S.L., Houpt, K.A. and Carnevale, J. (1985) Feeding and drinking behavior of mares and foals with free access to pasture and water. *Journal of Animal Science*, **60**, 883–9.

Cuddeford, D. (1994) Artificially dehydrated lucerne for horses. *Veterinary Record*, **135**, 426–9.

Cuddeford, D., Pearson, R.A., Archibald, R.F. and Muirhead, R.H. (1995) Digestibility and gastrointestinal transit time of diets containing different proportions of alfalfa and oat straw given to Thoroughbreds, Shetland ponies, Highland ponies and donkeys. *Animal Science*, **61**, 407–417.

Cuddeford, D., Woodhead, A. and Muirhead, R. (1992) A comparison between the nutritive value of short-cutting cycle, high temperature-dried alfalfa and timothy hay for horses. *Equine Veterinary Journal*, **24**, 84–9.

Cunha, T.J. (1969) Horse feeding and nutrition. *Feedstuffs*, **41**(28), 19–24.

Cunha, T.J. (1971) The mineral needs of the horse. *Feedstuffs*, **43**(46), 34–8.

Cushnahan, A. and Gordon, F.J. (1995) The effects of grass preservation on intake, apparent digestibility and rumen degradation characteristics. *Animal Science*, **60**, 429–38.

Cushnahan, A. and Mayne, C.S. (1995) Effects of ensilage of grass on performance and nutrient utilization by dairy cattle. 1. Food intake and milk production. *Animal Science*, **60**, 337–45.

Cushnahan, A., Mayne, C.S. and Unsworth, E.F. (1995) Effects of ensilage of grass on performance and nutrient utilization by dairy cattle. 2. Nutrient metabolism and rumen fermentation. *Animal Science*, **60**, 347–59.

Custalow, S.E., Ferrante, P.L., Taylor, L.E., *et al.* (1993) Lactate and glucose responses to exercise in the horse are affected by training and dietary fat. *Proceedings of the 13th Equine Nutrition and Physiology Society*, University of Florida, Gainesville, 21–23 January 1993, No. 504, pp. 179–84.

Cutmore, C.M.M., Snow, D.H. and Newsholme, E.A. (1985) Activities of key enzymes of aerobic and anaerobic metabolism in middle gluteal muscle from trained and untrained horses. *Equine Veterinary Journal*, **17**, 354–6.

Cutmore, C.M.M., Snow, D.H. and Newsholme, E.A. (1986) Effects of training on enzyme activities involved in purine nucleotide metabolism in Thoroughbred horses. *Equine Veterinary Journal*, **18**, 72–3.

Cygax, A. and Gerber, H. (1973) Normal values of and the effect of age on haematocrit, total bilirubin, calcium, inorganic phosphates and alkaline phosphatase in the serum of horses. *Schweizer Archiv für Tierheilkunde*, **115**, 321–31.

Cymbaluk, N.F. and Christison, G.I. (1989) Effects of dietary energy and phosphorus content on blood chemistry and development of growing horses. *Journal of Animal Science*, **67**, 951–8.

Cymbaluk, N.F., Christison, G.I. and Leach, D.H. (1989a) Energy uptake and utilization by limit- and *ad libitum*-fed growing horses. *Journal of Animal Science*, **67**, 403–413.

Cymbaluk, N.F., Christison, G.I. and Leach, D.H. (1989b) Nutrient utilization by limit- and *ad libitum*-fed growing horses. *Journal of Animal Science*, **67**, 414–25.

Cymbaluk, N.F., Christison, G.I. and Leach, D.H. (1990) Longitudinal growth analysis of horses following limited and *ad libitum* feeding. *Equine Veterinary Journal*, **22**, 198–204.

Cymbaluk, N.F., Fretz, P.B. and Loew, F.M. (1978) Amprolium-induced thiamine deficiency in horses: clinical features. *American Journal of Veterinary Research*, **39**, 255–61.

Cymbaluk, N.F., Schryver, H.F. and Hintz, H.F. (1981a) Copper metabolism and requirement in mature ponies. *Journal of Nutrition*, **111**, 87–95.

Cymbaluk, N.F., Schryver, H.F., Hintz, H.F., Smith, D.F. and Lowe, J,E. (1981b) Influence of dietary molybdenum on copper metabolism in ponies. *Journal of Nutrition*, **111**, 96–106.

Cysewski, S.J., Pier, A.C., Baetz, A.L. and Cheville, N.F. (1982) Experimental equine aflatoxicosis. *Toxicology and Applied Pharmacology*, **65**, 354–65.

Darlington, F.G. and Chassels, J.B. (1960) The final inclusive report on a 5-year-study on the effect of administering alpha-tocopherol to Thoroughbreds. *The Summary*, **12**, 52.

Datt, S.C. and Usenik, E.A. (1975) Intestinal obstruction in the horse. Physical signs and blood chemistry. *Cornell Veterinarian*, **65**, 152–72.

Davie, A.J., Evans, D.L., Hodgson, D.R. and Rose, R.J. (1994) The effects of an oral polymer on muscle glycogen resynthesis in standardbred horses. *American Institute of Nutrition, Journal of Nutrition*, **124**, 27405–415.

Davies, A. and Jones, D.R. (1988) Changes in the grass/clover balance in continuously grazed swards in relation to sward production components. *European Grassland Federation Proceedings*, 12th Meeting, Dublin, pp. 297–301.

Davies, J.V., Gerring, E.L., Goodburn, R. and Manderville, P. (1984) Experimental ischaemia of the ileum and concentrations of the intestinal isoenzyme of alkaline phosphatase in plasma and peritoneal fluid. *Equine Veterinary Journal*, **16**, 215–17.

Davies, M.E. (1968) Role of colon liquor in the cultivation of cellullolytic bacteria from the large intestine of the horse. *Journal of Applied Bacteriology*, **31**, 286–89.

510   *References and Bibliography*

Davies, M.E. (1971) The production of vitamin B_{12} in the horse. *British Veterinary Journal,* **127**, 34–6.

Davies, W. (1952) *The Grass Crop; Its Development, Use and Maintenance.* E. and F.N. Spon, London.

Dawson, W.M., Phillips, R.W. and Speelman, S.R. (1945) Growth of horses under Western range conditions. *Journal of Animal Science,* **4**, 47–51.

Deegen, E., Ohnesorge, B., Dieckmann, M. and Stadler, P. (1992a) Ulcerative gastritis in horses. *First Europäische Konferenz über die Ernährung des Pferdes,* Institute für Tierernahrüng, Tierärzliche Hochschule, Hannover, 3–4 September 1992, pp. 183–7.

Deegen, E., Ohnesorge, B. and Harps, O. (1992b) Therapy in typhlocolitis. *First Europäische Konferenz über die Ernährung des Pferdes,* Institut für Tierernahrüng, Tierärztliche Hochschule, Hannover, 3–4 September 1992, pp. 207–208.

De Gray, T. (1639) *The Compleat Horseman and Expert Farrier. In Two Bookes.* Thomas Harper, London.

Demarquilly, C. (1970) Feeding value of green forages as influenced by nitrogen fertilization. *Annals of Zootechnology,* **19**, 423–37.

Denman, A.M. (1979) Nature and diagnosis of food allergy. *Proceedings of the Nutrition Society,* **38**, 391–402.

DePew, C.L., Thompson, D.L., Jr, Fernandez, J.M., Southern, L.L., Sticker, L.S. and Ward, T.L. (1994a) Plasma concentrations of prolactin, glucose, insulin, urea nitrogen, and total amino acids in stallions after ingestion of feed or gastric administration of feed components. *Journal of Animal Science,* **72**, 2345–53.

DePew, C.L., Thompson, D.L., Jr, Fernandez, J.M., Sticker, L.S. and Burleigh, D.W. (1994b) Changes in concentrations of hormones, metabolites, and amino acids in plasma of adult horses relative to overnight feed deprivation followed by a pellet-hay meal fed at noon. *Journal of Animal Science,* **72**, 1530–39.

Derksen, F.J. (1993) Chronic obstructive pulmonary disease (heaves) as an inflammatory condition. *Equine Veterinary Journal,* **25**, 257–8.

Diekman, M.A. and Green, M.L. (1992) Mycotoxins and reproduction in domestic livestock. *Journal of Animal Science,* **70**, 1615–27.

Divers, T.J., Mohammed, H.O., Cummings, J.F., *et al.* (1994) Equine motor neuron disease: findings in 28 horses and proposal of a pathophysiological mechanism for the disease. *Equine Veterinary Journal,* **26**, 409–415.

Dixon, P.M. and Brown, R. (1977) Effects of storage on the methaemoglobin content of equine blood. *Research in Veterinary Science,* **23**, 241–3.

Dixon, P.M., McPherson, E.A. and Muir, A. (1977) Familial methaemoglobinaemia and haemolytic anaemia in the horse associated with decreased erythrocytic glutathione reductase and glutathione. *Equine Veterinary Journal,* **9**, 198–201.

Dixon, P.M., Railton, D.I. and McGorum, B.C. (1995) Equine pulmonary disease: a case control study of 300 referred cases. Parts 1–4. *Equine Veterinary Journal,* **27**, 416–39.

Doige, C.E. and McLaughlin, B.G. (1981) Hyperplastic goitre in newborn foals in Western Canada. *Canadian Veterinary Journal,* **22**, 42–5.

Dollahite, J.W., Younger, R.L., Crookshank, H.R., Jones, L.P. and Peterson, H.D. (1978) Chronic lead poisoning in horses. *American Journal of Veterinary Research,* **39**, 961–4.

Domingue, B.M.F., Wilson, P.R., Dellow, D.W. and Barry, T.N. (1992) Effects of subcutaneous melatonin implants during long daylength on voluntary feed intake, rumen capacity and heart rate of red deer (*Cervus elaphus*) fed on forage diet. *British Journal of Nutrition,* **68**, 77–88.

Donoghue, S., Kronfield, D.S., Berkowitz, S.J. and Copp, R.L. (1981) Vitamin A nutrition of the equine: growth serum biochemistry and hematology. *Journal of Nutrition,* **111**, 365–74.

Doreau, M. (1978) Comportement alimentaire du cheval à l'écurie. *Annals of Zootechnology,* **27**, 291–302.

Doreau, M., Martin-Rosset, W. and Boulot, S. (1988) Energy requirements and the feeding of mares during lactation: a review. *Livestock Production Science,* **20**, 53–68.

Doreau, M., Moretti, C. and Martin-Rosset, W. (1990) Effect of quality of hay given to mares around foaling on their voluntary intake and foal growth. *Annals of Zootechnology,* **39**, 125–31.

Doreau, M., Boulot, S. and Martin-Rosset, W. (1991) Effect of parity and physiological state on intake, milk production and blood parameters in lactating mares differing in body size. *Animal Production,* **53**, 111–18.

Doreau, M., Boulot, S., Bauchart, D., Barlet, J.-P. and Martin-Rosset, W. (1992) Voluntary intake, milk production and plasma metabolites in nursing mares fed two different diets. *Journal of Nutrition*, **122**, 992–9.

Dorn, C.R., Garner, H.E., Coffman, J.R., Hahn, A.W. and Tritschler, L.G. (1975) Castration and other factors affecting the risk of equine laminitis. *Cornell Veterinarian*, **65**, 57–64.

Doxey, D.L., Gilmour, J.S. and Milne, E.M. (1991a) The relationship between meteorological features and equine grass sickness (dysautonomia). *Equine Veterinary Journal*, **23**, 370–73.

Doxey, D.L., Milne, E.M., Gilmour, J.S. and Pogson, D.M. (1991b) Clinical and biochemical features of grass sickness (equine dysautonomia). *Equine Veterinary Journal*, **23**, 360–64.

Drew, B., Barber, W.P. and Williams, D.G. (1975) The effect of excess dietary iodine on pregnant mares and foals. *Veterinary Record*, **97**, 93–5.

Driscoll, J., Hintz, H.F. and Schryver, H.F. (1978) Goiter in foals caused by excessive iodine. *Journal of American Veterinary Medical Association*, **173**, 858–9.

Drolet, R., Laverty, S., Braselton, W.E. and Lord, N. (1996) Zinc phosphide poisoning in a horse. *Equine Veterinary Journal*, **28**, 161–2.

Drudge, J.H. and Lyons, E.T. (1972) Critical tests of a resin-pellet formulation of dichlorvos against internal parasites of the horse. *American Journal of Veterinary Research*, **33**, 1365–75.

Drummond, R.O. (1981) Biology and control of insect pests of horses. *Pony of the Americas*, **26**, August, 30–32.

DuBose, L.E. (1987) The effect of urea and lysine in alfalfa and coastal Bermuda grass hay rations on the growth of young equines. *Proceedings of the 10th Equine Nutrition and Physiology Society*, Colorado State University, Fort Collins, 11–13 June 1987, p. 605.

Duncan, J.L., McBeath, D.G., Best, J.M.J. and Preston, N.K. (1977) The efficacy of fenbendazole in the control of immature strongyle infections in ponies. *Equine Veterinary Journal*, **9**, 146–9.

Duncan, J.L., McBeath, D.G. and Preston N.K. (1980) Studies on the efficacy of fenbendazole used in a divided dosage regime against strongyle infections in ponies. *Equine Veterinary Journal*, **12**, 78–80.

Duncan, J.L. and Reid, J.F.S. (1978) An evaluation of the efficacy of oxfendazole against the common nematode parasites of the horse. *Veterinary Record*, **103**, 332–4.

Dunnett, M. and Harris, R.C. (1992) Determination of carnosine and other biogenic imidazoles in equine plasma by isocratic reversed-phase ion-pair high-performance liquid chromatography. *Journal of Chromatography*, **579**, 45–53.

Dunsmore, J.D. (1985) Integrated control of *Strongylus vulgaris* infection in horses using ivermectin. *Equine Veterinary Journal*, **17**, 191–5.

Duren, S.E., Manohar, M., Sikkes, B., Jackson, S. and Baker, J. (1992) Influence of feeding and exercise on the distribution of intestinal and muscle blood flow in ponies. *First Europäische Konferenz über die Ernährung des Pferdes*, Institut für Tierernahrung, Tierärztliche Hochschule, Hannover, 3–4 September 1992, pp. 24–28.

Dybdal, N.O., Gribble, D., Madigan, J.E. and Stabenfeldt, G.H. (1980) Alterations in plasma corticosteroids, insulin and selected metabolites in horses used in endurance rides. *Equine Veterinary Journal*, **12**, 137–40.

Dyce, K.M., Hartman, W. and Aalfs, R.H.G. (1976) A cinefluoroscopic study of the caecal base of the horse. *Research in Veterinary Science*, **20**, 40–6.

Eckersall, P.D., Aitchison, T. and Colquhoun, K.M. (1985) Equine whole saliva: variability of some major constituents. *Equine Veterinary Journal*. **17**, 391–3.

Edens, L.M., Robertson, J.L. and Feldman, B.F. (1993) Cholestatic hepatopathy, thrombocytopenia and lymphopenia associated with iron toxicity in a Thoroughbred gelding. *Equine Veterinary Journal*, **25**, 81–4.

Edwards, G.B. and Proudman, C.J. (1994) An analysis of 75 cases of intestinal obstruction caused by pedunculated lipomas. *Equine Veterinary Journal*, **26**, 18–21.

Egan, D.A. and Murrin, M.P. (1973) Copper concentration and distribution in the livers of equine fetuses, neonates and foals. *Research in Veterinary Science*, **15**, 147–8.

El Shorafa, W.M. (1978) Effect of vitamin D and sunlight on growth and bone development of young ponies. *Dissertation Abstracts International B*, **30**, 1556–7, No. 7817436.

El Shorafa, W.M., Feaster, J.P., Ott, E.A. and Asquith, R.L. (1979) Effect of vitamin D and sunlight on growth and bone development of young ponies. *Journal of Animal Science*, **48**, 882–6.

Ellis, R.N.W. and Lawrence, T.L.J. (1978a) Energy under-nutrition in the weanling filly foal. I. Effects on subsequent live-weight gains and onset of oestrus. *British Veterinary Journal*, **134**, 205–11.

Ellis, R.N.W. and Lawrence, T.L.J. (1978b) Energy under-nutrition in the weanling filly foal. II. Effects on body conformation and epiphyseal plate closure in the fore-limb. *British Veterinary Journal*, **134**, 322–32.

Ellis, R.N.W. and Lawrence, T.L.J. (1978c) Energy under-nutrition in the weanling filly foal. III. Effects on heart rate and subsequent voluntary food intake. *British Veterinary Journal*, **134**, 333–41.

Ellis, R.N.W. and Lawrence, T.L.J (1979) Energy and protein under-nutrition in the weanling filly foal. *British Veterinary Journal*, **135**, 331–7.

Ellis, R.N.W. and Lawrence, T.L.J. (1980) The energy and protein requirements of the light horse. *British Veterinary Journal*, **136**, 116–21.

Elsden, S.R., Hitchcock, M.W.S., Marshall, R.A. and Phillipson, A.T. (1946) Volatile acid in the digesta of ruminants and other animals. *Journal of Experimental Biology*, **22**, 191–202.

Epstein, V. (1984) Relationship between potassium administration, hyperkalaemia and the electrocardiogram: an experimental study. *Equine Veterinary Journal*, **16**, 453–6.

Erickson, H.H., Erickson, B.K., Landgren, G.L., Hopper, M.K., Butler, H.C. and Gillespie, J.R. (1987) Physiological characteristics of a champion endurance horse. *Proceedings of the 10th Equine Nutrition and Physiology Society*, pp. 493–8.

Essén-Gustavsson, B. and Lindholm, A. (1985) Muscle fibre characteristics of active and inactive standardbred horses. *Equine Veterinary Journal*, **17**, 434–8.

Essén-Gustavsson, B., Karlström, K. and Lindholm, A. (1984) Fibre types, enzyme activities and substrate utilisation in skeletal muscles of horses competing in endurance rides. *Equine Veterinary Journal*, **16**, 197–202.

Essén-Gustavsson, B., Lindholm, A. and Thornton, J. (1980) Histochemical properties of muscle fibre types and enzyme activities in skeletal muscles of standardbred trotters of different ages. *Equine Veterinary Journal*, **12**, 175–80.

Evans, C. (1995) *The homeopathic treatment of horses*. Thesis, Diploma, Equine Studies, University College, Dublin.

Evans, D.L. and Rose, R.J. (1988) Determination and repeatability of maximum oxygen uptake and other cardiorespiratory measurements in the exercising horse. *Equine Veterinary Journal*, **20**, 94–8.

Eyre, P. (1972) Equine pulmonary emphysema: a bronchopulmonary mould allergy. *Veterinary Record*, **91**, 134–40.

Fagan, T.W. (1928) Factors that influence the chemical composition of hay. *Welsh Journal of Agriculture*, **4**, 92.

Feldman, J.F. (1987) Hypocalcemia associated with colic in a horse. *Equine Practice*, **9**, 7–10.

Ferrante, P.L., Kronfeld, D.S., Taylor, L.E. and Meacham, T.N. (1994a) Plasma [H⁺] responses to exercise in horses fed a high-fat diet and given sodium bicarbonate. *Journal of Nutrition*, **124**, 2736S–7S.

Ferrante, P.L., Taylor, L.E., Kronfeld, D.S. and Meacham, T.N. (1994b) Blood lactate concentration during exercise in horses fed a high-fat diet and administered sodium bicarbonate. *Journal of Nutrition*, **124**, 2738S–9S.

Ferrante, P.L., Taylor, L.E., Meacham, T.N., Kronfeld, D.S. and Tiegs, W. (1993) Evaluation of acid–base status and strong ion difference (SID) in exercising horses. *Proceedings of the 13th Equine Nutrition and Physiology Society*, University of Florida, Gainesville, 21–23 January 1993, No. 504, pp. 123–4.

Ferraro, J. and Cote, J.F. (1984) Broodmare management techniques improve conception rates. *Standardbred*, **12**, 56–8.

Fonnesbeck, P.V. (1981) Estimating digestible energy and TDN for horses with chemical analysis of feeds. *Journal of Animal Science*, **53** (Suppl. 1), 241, abstract 290.

Fonnesbeck, P.V. and Symons, L.D. (1967) Utilization of the carotene of hay by horses. *Journal of Animal Science*, **26**, 1030–8.

Forbes, T.J., Dibb, C., Green, J.O., Hopkins, A. and Peel, S. (1980) *Factors Affecting Productivity of Permanent Grassland. A National Farm Survey*. Grassland Research Institute, Hurley.

Ford, C.W., Morrison, I.M. and Wilson, J.R. (1979) Temperature effects on lignin hemicellulose and cellulose in tropical and temperate grasses. *Australian Journal of Agricultural Research*, **30**, 621–33.

Ford, E.J.H. and Evans, J. (1982) Glucose utilization in the horse. *British Journal of Nutrition*, **48**, 111–17.

Ford, E.J.H. and Simmons, H.A. (1985) Gluconeogenesis from caecal propionate in the horse. *British Journal of Nutrition*, **53**, 55–60.

Ford, J. and Lokai, M.D. (1979) Complications of sand impaction colic. *Veterinary Medicine and Small Animal Clinician*, **74**, 573–8.

Foster, C.V.L. and Harris, R.C. (1989) Plasma carnitine concentrations in the horse following oral supplementation using a triple dose regime. *Equine Veterinary Journal*, **21**, 376–7.

Foster, C.V.L. and Harris, R.C. (1992) Total carnitine content of the middle gluteal muscle of Thoroughbred horses: normal values, variability and effect of acute exercise. *Equine Veterinary Journal*, **24**, 52–7.

Foster, C.V.L., Harris, R.C. and Pouret, E.J.M. (1989) Effect of oral L-carnitine on its concentration in the plasma of yearling thoroughbred horses. *Veterinary Record*, **125**, 125–8.

Foster, C.V.L., Harris, R.C. and Snow, D.H. (1988) The effect of oral L-carnitine supplementation on the muscle and plasma concentrations in the thoroughbred horse. *Comparative Biochemistry and Physiology*, **91A**, 827–35.

Fowden, A.L., Comline, R.S. and Silver, M. (1984) Insulin secretion and carbohydrate metabolism during pregnancy in the mare. *Equine Veterinary Journal*, **16**, 239–46.

Franceso, L.L., Saurin, L.L. and Dibana, G.F. (1981) Mechanism of negative potassium balance in the magnesium deficient rat. *Proceedings of the Society for Experimental Biology and Medicine*, **168**, 382–8.

Francis-Smith, K. and Wood-Gush, D.G.M. (1977) Coprophagia as seen in Thoroughbred foals. *Equine Veterinary Journal*, **9**, 155–7.

Frank, C.J. (1970) Equine colic – a routine modern approach. *Veterinary Record*, **87**, 497–8.

Frank, N.B., Meacham, T.N. and Fontenot, J.P. (1987) Effect of feeding two levels of protein on performance and nutrition of exercising horses. *Proceedings of the 10th Equine Nutrition and Physiology Society*, Colorado State University, Fort Collins, 11–13 June 1987, pp. 579–83.

Frape, D.L. (1975) Recent research into the nutrition of the horse. *Equine Veterinary Journal*, **7**, 120–30.

Frape, D.L. (1980) Facts about feeding horses. *In Practice*, **2**, 14–21.

Frape, D.L. (1981) Digestibility studies in horses and ponies. *Proceedings of the 32nd Annual Meeting of the European Association of Animal Production*, Zagreb, Yugoslavia, 31 August–3 September, IVb.

Frape, D.L. (1983) Nutrition of the horse. In: *Pharmacological Basis of Large Animal Medicine* (eds J.A. Bogan, P. Lees and A.T. Yoxall). Blackwell Science, Oxford.

Frape, D.L. (1984a) Straw in the diet of other ruminants and non-ruminant herbivores. In: *Straw and Other Fibrous Byproducts for Food. Developments in Animal and Veterinary Sciences*, 14 (eds E. Owen and F. Sundstl), pp. 487–532. Elsevier Science, Amsterdam.

Frape, D.L. (1984b) The relevance of red cell potassium in diagnosis. *Equine Veterinary Journal*, **16**, 401–2.

Frape, D.L. (1985) Toxicology and diet. *Equine Veterinary Journal*, **17**, 426–7.

Frape, D. L. (1986) *Equine Nutrition and Feeding*, pp. 1–373. Longman Scientific & Technical, Harlow.

Frape, D.L. (1987) Calcium balance and dietary protein content. *Equine Veterinary Journal*, **19**, 265–70.

Frape, D.L. (1988) Dietary requirements and athletic performance of horses. *Equine Veterinary Journal*, **20**, 163–72.

Frape, D.L. (1989) Nutrition and the growth and racing performance of thoroughbred horses. *Proceedings of the Nutrition Society*, **48**, 141–52.

Frape, D.L. (1993) Arterio-venous differences of NEFA during exercise. *Equine Veterinary Journal*, **25**, 4–5.

Frape, D.L. (1994) Diet and exercise performance in the horse. *Proceedings of the Nutrition Society*, **53**, 189–206.

Frape, D.L. (1996) Sherlock Holmes and chemical poisons. *Equine Veterinary Journal*, **28**, 89–91.

Frape, D.L. and Boxall, R.C. (1974) Some nutritional problems of the horse and their possible relationship to those of other herbivores. *Equine Veterinary Journal*, **6**, 59–68.

Frape, D.L., Cash, R.S.G. and Ricketts, S.W. (1983) Panda food allergy. *Lancet*, **i**, 870–1.

Frape, D.L., Peace, C.K. and Ellis, P.M. (1979) Some physiological changes in a fit and an unfit horse associated with a long distance ride. *Proceedings of the 30th Annual Meeting of the European Association of Animal Production*, H-VI-5.

Frape, D.L. and Pringle, J.D. (1984) Toxic manifestations in a dairy herd consuming haylage contaminated by lead. *Veterinary Record*, **114**, 615–16.

Frape, D.L. and Tuck, M.G. (1977) Determining the energy values of ingredients in pig feeds. *Proceedings of the Nutrition Society*, **36**, 179–87.

Frape, D.L., Tuck, M.G., Sutcliffe, N.H. and Jones, D.B. (1982) The use of inert markers in the measurement of the digestibility of cubed concentrates and of hay given in several proportions to the pony, horse and white rhinoceros (*Diceros simus*). *Comparative Biochemistry and Physiology*, **72A**, 77–83.

Freeman, D.W., Potter, G.D., Schelling, G.T. and Kreider, J.L. (1988) Nitrogen metabolism in mature horses at varying levels of work. *Journal of Animal Science*, **66**, 407–12.

Freeman, K.P., Roszel, J.F., McClure, J.M., Mannsman, R., Patton, P.E. and Naile, S. (1993) A review of cytological specimens from horses with and without clinical signs of respiratory disease. *Equine Veterinary Journal*, **25**, 523–6.

Fricker Ch, Riek, W. and Hugelshofer, J. (1982) Occlusion of the digital arteries – a model for pathogenesis of navicular disease. *Equine Veterinary Journal*, **14**, 203–7.

Gabal, M.A., Awad, Y.L., Morcos, M.B., Barakat, A.M. and Malik, G. (1986) Fusariotoxicoses of farm animals and mycotoxic leucoencephalomalacia of the equine associated with the finding of trichothecenes in feedstuffs. *Veterinary and Human Toxicology*, **28**, 207.

Gallagher, J.R. and McMeniman, N.P. (1988) The nutritional status of pregnant and non-pregnant mares grazing South East Queensland pastures. *Equine Veterinary Journal*, **20**, 414–16.

Gallagher, K., Leech, J. and Stowe, H. (1992a) Protein energy and dry matter consumption by racing standardbreds: a field survey. *Journal of Equine Veterinary Science*, **12**, 382–8.

Gallagher, K., Leech, J. and Stowe, H. (1992b) Protein energy and dry matter consumption by racing Thoroughbreds: a field survey. *Journal of Equine Veterinary Science*, **12**, 43–8.

Garner, H.E., Coffman, J.R., Hahn A.W., Hutcheson, D.P. and Tumbleson, M.E. (1975) Equine laminitis of alimentary origin: an experimental model. *American Journal of Veterinary Research*, **36**, 441–4.

Garner, H.E., Hutcheson, D.P., Coffman, J.R., Hahn, A.W. and Salem, C. (1977) Lactic acidosis: a factor associated with equine laminitis. *Journal of Animal Science*, **45**, 1037–41.

Garner, H.E., Moore, J.N., Johnson, J.H., *et al.* (1978) Changes in the caecal flora associated with the onset of laminitis. *Equine Veterinary Journal*, **10**, 249–52.

Gartner, R.J.W., Blaney, B.J. and McKenzie, R.A. (1981) Supplements to correct oxalate induced negative calcium and phosphorus balances in horses fed tropical grass hays. *Journal of Agricultural Science, Cambridge*, **97**, 581–9.

Genetzky, R.M., Loparco, F.V. and Ledet, A.E. (1987) Clinical pathologic alterations in horses during a water deprivation test. *American Journal of Veterinary Research*, **48**, 1007–11.

Gibbs, P.G., Potter, G.D., Blake, R.W. and McMullan, W.C. (1982) Milk production of quarterhorse mares during 150 days of lactation. *Journal of Animal Science*, **54**, 496–9.

Gibbs, P.G., Potter, G.D., Schelling, G.T., Kreider, J.L. and Boyd, C.L. (1988) Digestion of hay protein in different segments of the equine digestive tract. *Journal of Animal Science*, **66**, 400–6.

Gibbs, P.G., Potter, G.D., Schelling, G.T., Kreider, J.L. and Boyd, C.L. (1996) The significance of small vs. large intestinal digestion of cereal grain and oilseed protein in the equine. *Journal of Equine Veterinary Science*, **16**, 60–5.

Gibbs, P.G., Sigler, D.H. and Goehring, T.B. (1987) Influence of diet on growth and development of yearling horses. *Proceedings of the 10th Equine Nutrition and Physiology Society*, Colorado State University, Fort Collins, 11–13 June 1987, pp. 37–42.

Gibson, W. (1726) *The True Method of Dieting Horses*. Osborn and Longman, London.

Giddings, R.F., Argenzio, R.A. and Stevens, C.E. (1974) Sodium and chloride transport across the equine cecal mucosa. *American Journal of Veterinary Research*, **35**, 1511–14.

Giles, C.J. (1983) Outbreak of ragwort (*Senecio jacobea*) poisoning in horses. *Equine Veterinary Journal*, **15**, 248–50.

Gillespie, J.R., Kauffman, A., Steere, J. and White, L. (1975) Arterial blood gases and pH during long distance running in the horse. *Proceedings of the 1st International Symposium on Equine Hematology*, 450–68.

Gilmour, J.S. (1973a) Grass sickness: the paths of research. *Equine Veterinary Journal*, **5**, 102–4.

Gilmour, J.S. (1973b) Observations on neuronal changes in grass sickness of horses. *Research in Veterinary Science*, **15**, 197–200.

Gilmour, J.S., Brown, R. and Johnson, P. (1981) A negative serological relationship between cases of grass sickness in Scotland and *Clostridium perfringens* type A enterotoxin. *Equine Veterinary Journal*, **13**, 56–8.

Gilmour, J.S. and Jolly, G.M. (1974) Some aspects of the epidemiology of equine grass sickness. *Veterinary Record*, **95**, 77–81.

Glade, M.J. (1983a) Nutrition and performance of racing thoroughbreds. *Equine Veterinary Journal*, **15**, 31–6.

Glade, M.J. (1983b) Nitrogen partitioning along the equine digestive tract. *Journal of Animal Science*, **57**, 943–53.

Glade, M.J. (1984) The influence of dietary fiber digestibility on the nitrogen requirements of mature horses. *Journal of Animal Science*, **58**, 638–46.

Glade, M.J. (1986) The control of cartilage growth in osteochondrosis: a review. *Journal of Equine Veterinary Science*, **6**, 175–87.

Glade, M.J. and Bell, P.I. (1981) Nitrogen partitioning along the equine digestive tract. *Journal of Animal Science*, **53** (Suppl. 1), 243, abstract 294.

Glade, M.J., Beller, D., Bergen, J., *et al.* (1985) Dietary protein in excess of requirements inhibits renal calcium and phosphorus reabsorption in young horses. *Nutrition Reports International*, **31**, 649–59.

Glade, M.J. and Biesik, L.M. (1986) Enhanced nitrogen retention in yearling horses supplemented with yeast culture. *Journal of Animal Science*, **62**, 1635–40.

Glade, M.J., Gupta, S. and Reimers, T.J. (1984) Hormonal responses to high and low planes of nutrition in weanling Thoroughbreds. *Journal of Animal Science*, **59**, 658–65.

Glade, M.J., Krook, L., Schryver, H.F. and Hintz, H.F. (1982) Calcium metabolism in glucocorticoid-treated pony foals. *Journal of Nutrition*, **112**, 77–86.

Glade, M.J. and Luba, N.K., (1987a) Serum triiodothyronine and thyroxine concentrations in weanling horses fed carbohydrate by direct gastric infusion. *American Journal of Veterinary Research*, **48**, 578–82.

Glade, M.J. and Luba, N.K. (1987b) Benefits to foals of feeding soybean meal to lactating broodmares. *Proceedings of the 10th Equine Nutrition and Physiology Society*, Colorado State University, Fort Collins, 11–13 June 1987, pp. 593–4.

Glade, M.J., Luba, N.K. and Schryver, H.F. (1986) Effects of age and diet on the development of mechanical strength by the third metacarpal and metatarsal bones of young horses. *Journal of Animal Science*, **63**, 1432–44.

Glade, M.J. and Sist, M.D. (1990) Supplemental yeast culture alters the plasma amino acid profiles of nursling and weanling horses. Equine Nutrition and Physiology Society, 11th Symposium, *Journal of Equine Veterinary Science*, September/October, p. 8.

Gleeson, M., Greenhaff, P.L. and Maughan, R.J. (1987) Diet, acid–base status and the metabolic response to exercise in man. *Proceedings of the Physiological Society*, 108P.

Glendinning, S.A. (1974) A system of rearing foals on an automatic calf feeding machine. *Equine Veterinary Journal*, **6**, 12–16.

Glinsky, M.J., Smith, R.M., Spires, H.R. and Davis, C.L. (1976) Measurement of volatile fatty acid production rates in the cecum of the pony. *Journal of Animal Science*, **42**, 1465–71.

Goater, L.E., Meacham, T.H., Gwazdauskas, F.C. & Fontenot, J.P. (1981) Effect of dietary energy level in mares during gestation. *Journal of Animal Science*, **53** (Suppl. 1), 243, abstract 295.

Goater, L.E., Snyder, J.L., Huff, A.N. and Meacham, T.N. (1982) A review of recurrent problems in feeding horses. *Journal of Equine Veterinary Science*, **2**, 58–61.

Godbee, R.G. and Slade, L.M. (1979) Range blocks with urea for broodmares increase the nutritional value of pasture feeding. *Feedstuffs*, **51**(16), 34–5.

Godbee, R.G. and Slade, L.M. (1981) The effect of urea or soybean meal on the growth and protein status of young horses. *Journal of Animal Science*, **53**, 670–6.

Godbee, R.G., Slade, L.M. and Lawrence, L.M. (1979) Use of protein blocks containing urea for minimally managed broodmares. *Journal of Animal Science*, **48**, 459–63.

Goering, H.K. and Van Soest, P.J. (1975) Forage fiber analyses (apparatus, reagents, procedures and some applications). Agricultural Handbook, 379, ARS, USDA, Washington DC.

Goodman, H.M., van der Noot, G.W., Trout, J.R. and Squibb, R.L. (1973) Determination of energy source utilized by the light horse. *Journal of Animal Science*, **37**, 56–62.

Gossett, K.A., Cleghorn, B., Martin, G.S. and Church, G.E. (1987) Correlation between anion gap, blood L-lactate concentration and survival in horses. *Equine Veterinary Journal*, **19**, 29–30.

Gotte, J.O. (1972) In: *Handbuch Tierernährung, Vol. 2* (eds W. Lenkeit, K. Breirem and E. Crasemann), pp. 393–8. Paul Parey, Berlin/Hamburg.

Gottlieb, M. (1989) Muscle glycogen depletion patterns during draught work in standardbred horses. *Equine Veterinary Journal*, **21**, 110–15.

Graham, P.M., Ott, E.A., Brendermuhl, J.H. and TenBroeck, S.H. (1993) The effect of supplemental lysine and threonine on growth and development of yearling horses. *Proceedings of the 13th Equine Nutrition and Physiology Society*, 21–23 January 1993, University of Florida, Gainesville, pp. 80–1.

Graham, P.M., Ott, E.A., Brendermuhl, J.H. and TenBroeck, S.H. (1994) The effect of supplemental lysine and threonine on growth and development of yearling horses. *Journal of Animal Science*, **72**, 380–6.

Green, D.A. (1961) A review of studies on the growth rate of the horse. *British Veterinary Journal*, **117**, 181.

Green, D.A. (1969) A study of growth rate in Thoroughbred foals. *British Veterinary Journal*, **125**, 539–46.

Green, D.A. (1976) Growth rate in Thoroughbred yearlings and two-year-olds. *Equine Veterinary Journal*, **8**, 133–4.

Green, J.O. (1982) *A Sample Survey of Grassland in England and Wales 1970–1972*. Grassland Research Institute, Hurley, Maidenhead.

Greene, H.J. and Oehme, F.W. (1976) A possible case of equine aflatoxicosis. *Clinical Toxicology*, **9**, 251–4.

Greenhaff, P.L., Hanak, J., Harris, R.C., *et al.* (1991a) Metabolic alkalosis and exercise performance in the thoroughbred horse. *Equine Exercise Physiology*, **3**, 353–60.

Greenhaff, P.L., Harris, R.C. and Snow, D.H. (1990a) The effect of sodium bicarbonate (NaHCO$_3$) administration upon exercise metabolism in the thoroughbred horse. *Journal of Physiology*, **420**, 69P.

Greenhaff, P.L., Harris, R.C., Snow, D.H., Sewell, D.A. and Dunnett, M. (1991b) The influence of metabolic alkalosis upon exercise metabolism in the thoroughbred horse. *European Journal of Applied Physiology*, **63**, 129–34.

Greenhaff, P.L., Snow, D.H., Harris, R.C. and Roberts, C.A. (1990b) Bicarbonate loading in the Thoroughbred: dose, method of administration and acid–base changes. *Equine Veterinary Journal*, (Suppl. 9), 83–5.

Greet, T.R.C. (1982) Observations on the potential role of oesophageal radiography in the horse. *Equine Veterinary Journal*, **14**, 73–9.

Greiwe-Crandell, K.M., Meacham, T.N., Fregin, G.F. and Walberg, J.L. (1989) Effect of added dietary fat on exercising horses. *Proceedings of the 11th Equine Nutrition Physiology Society*, Oklahoma State University, 18–20 May 1989, pp. 101–6.

Greiwe-Crandell, K.M., Kronfeld, D.S., Gay, L.A. and Sklan, D. (1995) Seasonal vitamin A depletion in grazing horses is assessed better by the relative dose response test than by serum retinol concentration. *Journal of Nutrition*, **125**, 2711–16.

Greppi, G.F., Casini, L., Gatta, D., Orlandi, M. and Pasquini, M. (1996) Daily fluctuations of haematology and blood biochemistry in horses fed varying levels of protein. *Equine Veterinary Journal*, **28**, 350–3.

Griffiths, I.R., Kyriakides, E., Smith, S., Howie, F. and Deary, A.W. (1993) Immunocytochemical and lectin histochemical study of neuronal lesions in autonomic ganglia of horses with grass sickness. *Equine Veterinary Journal*, **25**, 446–52.

Groenendyk, S., English, P.B. and Abetz, I. (1988) External balance of water and electrolytes in the horse. *Equine Veterinary Journal*, **20**, 189–93.

Gronwall, R. (1975) Effects of fasting on hepatic function in ponies. *American Journal of Veterinary Research*, **36**, 145–8.

Gronwall, R., Engelking, L.R., Answer, M.S., Erichsen, D.F. and Klentz, R.D. (1975) Bile secretion in ponies with biliary fistulas. *American Journal of Veterinary Research*, **36**, 653–8.

Gronwall, R., Engelking, L.R. and Noonan, N. (1980) Direct measurement of biliary bilirubin excretion in ponies during fasting. *American Journal of Veterinary Research*, **41**, 125–6.

Güllner, H., Gill, J.R. and Bartler, F.C. (1981) Correction of hypokalemia by magnesium repletion in familial hypokalemic alkalosis with tubulopathy. *American Journal of Medicine*, **71**, 578–82.

Guy, P.S. and Snow, D.H. (1977a) The effect of training and detraining on muscle composition in the horse. *Journal of Physiology*, **269**, 33–51.

Guy, P.S. and Snow, D.H. (1977b) The effect of training and detraining on lactate dehydrogenase isoenzymes in the horse. *Biochemical and Biophysical Research Communications*, **75**, 863–9.

Haenlein, G.F.W. (1969) Nutritive value of a pelleted horse ration. *Feedstuffs*, **41**(26), 19–20.

Haggar, R.J. and Jones, D. (1989) Increasing flora diversity in grassland swards. *Proceedings of the 16th International Grassland Congress*, Nice, pp. 1633–4.

Hails, M.R. and Crane, T.D. (1982) Plant poisoning in animals. A Bibliography From the World Literature, 1960–1979. *Veterinary Bulletin*, **52**, No. 8, 557–1048. Also produced (1983) by Commonwealth Agricultural Bureaux of Animal Health, Slough.

Hall, G.M., Adrian, T.E., Bloom, S.R. and Lucke, J.N. (1982) Changes in circulating gut hormones in the horse during long distance exercise. *Equine Veterinary Journal*, **14**, 209–12.

Hambleton, P.L., Slade, L.M., Hamar, D.W., Kienholz, E.W. and Lewis, L.D. (1980) Dietary fat and exercise conditioning effect on metabolic parameters in the horse. *Journal of Animal Science*, **51**, 1330–9.

Hammond, C.J., Mason, D.K. and Watkins, K.L. (1986) Gastric ulceration in mature Thoroughbred horses. *Equine Veterinary Journal*, **18**, 284–7.

Hanna, C.J., Eyre, P., Wells, P.W. and McBeath, D.G. (1982) Equine immunology 2: immunopharmacology–biochemical basis of hypersensitivity. *Equine Veterinary Journal*, **14**, 16–24.

Hansen, T.O. and White, N.A., (II) (1988) Total parenteral nutrition in four healthy adult horses. *American Journal of Veterinary Research*, **49**, 122–4.

Hanson, C.M., Kline, K.H., Foreman, J.H. and Frey, L.P. (1993) Effect of sodium bicarbonate on plasma volume, electrolytes and blood gases in resting quarter horses. *Proceedings of the 13th Equine Nutrition and Physiology Society*, University of Florida, Gainesville, 21–23 January 1993, No. 504, 115–20.

Harbers, L.H., McNally, L.K. and Smith, W.H. (1981) Digestibility of three grass hays by the horse and scanning electron microscopy of undigested leaf remnants. *Journal of Animal Science*, **53**, 1671–7.

Harkins, J.D., Beadle, R.E. and Kamerling, S.G. (1993) The correlation of running ability and physiological variables in Thoroughbred racehorses. *Equine Veterinary Journal*, **25**, 53–60.

Harkins, J.D. and Kamerling, S.G. (1992) Effects of induced alkalosis on performance in thoroughbreds during a 1600-m race. *Equine Veterinary Journal*, **24**, 94–8.

Harkins, J.D., Morris, G.S., Tulley, R.T., Nelson, A.G. and Kamerling, S.G. (1992) Effect of added dietary fat on racing performance in thoroughbred horses. *Journal of Equine Veterinary Science*, **12**, 123–9.

Harmeyer, J., Twehues, R., Schlumbohm, C., Stadermann, B. and Meyer, H. (1992) The role of vitamin D on calcium metabolism in horses. *First Europäische Konferenz über die Ernährung des Pferdes*, Institut für Tierernährung, Tierärzliche Hochschule, Hannover, 3–4 September 1992, pp. 81–5.

Harmon, D.L., Britton, R.A. and Prior, R.L. (1983) Rates of utilization of L(+) and D(−) lactate in bovine tissues. *Federation Proceedings*, **42**, 815, abstract 3065.

Harper, O.F. and Noot, G.W.V. (1974) Protein requirement of mature maintenance horses. *Journal of Animal Science*, **39**, 183, abstract 181.

Harrington, D.D. (1974) Pathologic features of magnesium deficiency in young horses fed purified rations. *American Journal of Veterinary Research*, **35**, 503–13.

Harrington, D.D. (1975) Influence of magnesium deficiency on horse foal tissue concentrations of Mg, calcium and phosphorus. *British Journal of Nutrition*, **34**, 45–57.

Harrington, D.D. (1982) Acute vitamin D_2 (ergocalciferol) toxicosis in horses: case report and experimental studies. *Journal American Veterinary Medical Association*, **180**, 867–73.

Harrington, D.D. and Walsh, J.J. (1980) Equine magnesium supplements: evaluation of magnesium oxide, magnesium sulphate and magnesium carbonate in foals fed purified diets. *Equine Veterinary Journal*, **12**, 32–3.

Harris, R.C. (1987) Carbonic anhydrase isoenzymes – enigmatic variations. *Equine Veterinary Journal*, **19**, 489–93.

Harris, P. and Snow, D.H. (1988) The effects of high intensity exercise on the plasma concentration of lactate, potassium and other electrolytes. *Equine Veterinary Journal*, **20**, 109–13.

Harris, P. and Snow, D.H. (1992) Plasma potassium and lactate concentrations in thoroughbred horses during exercise of varying intensity. *Equine Veterinary Journal*, **24**, 220–5.

Harris, R.C., Dunnett, M. and Snow, D.H. (1991a) Muscle carnosine content is unchanged during maximal intermittent exercise. *Equine Exercise Physiology*, **3**, 257–61.

Harris, R.C. and Foster, C.V.L. (1990) Changes in muscle free carnitine and acetylcarnitine with increasing work intensity in the Thoroughbred horse. *European Journal of Applied Physiology*, **60**, 81–5.

Harris, R.C. and Hultman, E. (1992a) Muscle phosphagen status studied by needle biopsy. In: *Energy Metabolism: Tissue Determinants and Cellular Corollaries* (eds J.M. Kinney and H.N. Tucker), pp. 367–79. Raven Press, New York.

Harris, R.C. and Hultman, E. (1992b) Nutritional strategies for enhanced performance in the racing camel: lessons learned from man and horse. *Proceedings of the 1st International Camel Conference*, Dubai, United Arab Emirates, 2–6 February 1992, pp. 243–6.

Harris, R.C., Marlin, D.J., Dunnett, M., Snow, D.H. and Hultman, E. (1990) Muscle buffering capacity and dipeptide content in the thoroughbred horse, greyhound dog and man. *Comparative Biochemistry and Physiology*, **97A**, 249–51.

Harris, R.C., Marlin, D.J. and Snow, D.H. (1987) Metabolic response to maximal exercise of 800 and 2000 m in the thoroughbred horse. *Journal of Applied Physiology*, **63**, 12–19.

Harris, R.C., Marlin, D.J. and Snow, D.H. (1991b) Lactate kinetics, plasma ammonia and performance following repeated bouts of maximal exercise. *Equine Exercise Physiology*, **3**, 173–8.

Harris, R.C., Marlin, D.J., Snow, D.H. and Harkness, R.A. (1991c) Muscle ATP loss and lactate accumulation at different work intensities in the exercising Thoroughbred horse. *European Journal of Applied Physiology*, **62**, 235–44.

Harris, R.C., Söderlund, K. and Hultman, E. (1992) Elevation of creatine in resting and exercised muscle of normal subjects by creatine supplementation. *Clinical Science*, **83**, 367–74.

Harrison, R.J. (1974) Vitamin B$_{12}$ content in erythrocytes in horse and sheep. *Research Veterinary Science*, **17**, 259–60.

Hart, G.H., Goss, H. and Guilbert, H.R. (1943) Vitamin A deficiency not the cause of joint lesions in horses. *American Journal of Veterinary Research*, **4**, 162–8.

Harthoorn, A.M. and Young, E. (1974) A relationship between acid–base balance and capture myopathy in zebra (*Equus burchelli*) and an apparent therapy. *Veterinary Record*, **95**, 337–42.

Hatak, J. (1977) Effect of the degree of training and the physiological condition of horses on the dynamics of some urine metabolites. *Veterinarstvi*, **27**(7), 301–2.

Hatfull, R.S., Milner, I. and Stanway, V. (1980) Determination of theobromine in animal feeding stuffs. *Journal of the Association of Public Analysts*, **18**, 19–22.

Hathaway, R.L., Oldfield, J.E. and Buettner, M.R. (1981) Effect of selenium in a mineral salt mixture on heifers grazing tall fescue and quack grass pastures. *Proceedings of the Western Section of the American Society of Animal Science*, Vancouver, British Columbia, 23–25 June 1981, **32**, 32–3.

Hawkes, J., Hedges, M., Daniluk, P., Hintz, H.F. and Schryver, H.F. (1985) Feed preferences of ponies. *Equine Veterinary Journal*, **17**, 20–2.

Haywood, P.E., Teale, P. and Moss, M.S. (1990) The excretion of theobromine in Thoroughbred racehorses after feeding compounded cubes containing cocoa husk – establishment of a threshold value in horse urine. *Equine Veterinary Journal*, **22**, 244–6.

Heat, S.E., Bell, R.J. and Harland, R.J. (1988) Botulinum type C intoxication in a mare. *Canadian Veterinary Journal*, **29**, 530–1.

Hemken, R.W., Boling, J.A., Bull, L.S., Hatton, R.H., Buckner, R.C. and Bush, L.P. (1981) Interaction of environmental temperature and anti-quality factors on the severity of summer fescue toxicosis. *Journal of Animal Science*, **52**, 710–14.

Hemken, R.W., Jackson, J.A., Jr and Boling, J.A. (1984) Toxic factors in tall fescue. *Journal of Animal Science*, **58**, 1011–16.

Heneke, D.R., Potter, G.D. and Kreider, J.L. (1981) Rebreeding efficiency in mares fed different levels of energy during late gestation. *Proceeding of the 7th Equine Nutrition and Physiology Society*, Airlie House, Warrenton, Virginia, 30 April–2 May 1981, pp. 101–4.

Henry, M.M. and Moore, J.N. (1991) Whole blood re-calcification time in equine colic. *Equine Veterinary Journal*, **23**, 303–8.

Herd, R.P. (1986a) Epidemiology and control of equine strongylosis at Newmarket. *Equine Veterinary Journal*, **18**, 447–52.

Herd, R.P. (1986b) Pasture hygiene: a nonchemical approach to equine endoparasite control. *Modern Veterinary Practice*, **67**, 36–8.

Herd, R.P. and Willardson, K.L. (1985) Seasonal distribution of infective strongyle larvae on horse pastures. *Equine Veterinary Journal*, **17**, 235–7.

Heusner, G.L., Albert, W.W. and Norton, H.W. (1976) Energy and systems of feeding for weanlings. *Journal of Animal Science*, **43**, 253, abstract 170.

Higgins, A.J. and Wright, I.M. (eds) (1995) *The Equine Manual*. W.B. Saunders, London.

Hill, K.R. (1959) Discussion on seneciosis in man and animals. *Proceedings of the Royal Society of Medicine*, **53**, 281–2.

Hinchcliff, K.W., McKeever, K.H., Muir, W.W. and Sams, R. (1993) Effect of oral sodium loading on acid:base responses of horses to intense exertion. *Proceedings of the 13th Equine Nutrition and Physiology Society*, University of Florida, Gainesville, 21–23 January 1993, No. 504, pp. 121–2.

Hinckley, K.A., Fearn, S., Howard, B.R. and Henderson, I.W. (1996) Nitric oxide donors as treatment for grass induced acute laminitis in ponies. *Equine Veterinary Journal*, **28**, 17–28.

Hiney, K.M. and Potter, G.D. (1996) A review of recent research on nutrition and metabolism in the athletic horse. *Nutrition Research Reviews*, **9**, 149–73.

Hinton, M. (1978) On the watering of horses: a review. *Equine Veterinary Journal*, **10**, 27–31.

Hintz, H.F. (1980a) Growth in the horse. In: *Stud Manager's Handbook*, Vol. 16, pp. 59–66. Agriservices Foundation, Clovis, California.

Hintz, H.F. (1980b) Diagnosis of nutritional status. In: *Stud Manager's Handbook*, Vol. 16, pp. 185–7. Agriservices Foundation, Clovis, California.

Hintz, H.F. (1983) Nutritional requirements of the exercising horse – a review. In: *Equine Exercise Physiology* (eds D.H. Snow, S.G.B. Persson and R.J. Rose), pp. 275–90. Granta Editions, Cambridge.

Hintz, H.F. (1994) Nutrition and equine performance. *Journal of Nutrition*, **124**, 2723S–9S.

Hintz, H.F., Hintz, R.L. and Van Vleck, L.D. (1979a) Growth rate of Thoroughbreds. Effect of age of dam, year and month of birth and sex of foal. *Journal of Animal Science*, **48**, 480–7.

Hintz, H.F. and Kallfelz, F.A. (1981) Some nutritional problems of horses. *Equine Veterinary Journal*, **13**, 183–6.

Hintz, H.F., Lowe, J.E. and Schryver, H.F. (1969) Protein sources for horses. *Proceedings of the Cornell Nutrition Conference for Feed Manufacturers*, 65–8. Cornell University, Ithaca, New York.

Hintz, H.F. and Meakim, D.W. (1981) A comparison of the 1978 National Research Council's recommendations of nutrient requirements of horses with recent studies. *Equine Veterinary Journal*, **13**, 187–91.

Hintz, H.F., Ross, M.W., Lesser, F.R., *et al.* (1978a) The value of dietary fat for working horses. I. Biochemical and hematological evaluations. *Journal of Equine Medicine and Surgery*, **2**, 483–8.

Hintz, H.F., Ross, M.W., Lesser, F.R., *et al.* (1978b) Value of supplemental fat in horse rations. *Feedstuffs*, **50**(12), 27–8.

Hintz, H.F. and Schryver, H.F. (1972) Magnesium metabolism in the horse. *Journal of Animal Science*, **35**, 755–9.

Hintz, H.F. and Schryver, H.F. (1973) Magnesium, calcium and phosphorus metabolism in ponies fed varying levels of magnesium. *Journal of Animal Science*, **37**, 927–30.

Hintz, H.F. and Schryver, H.F. (1975a) The evolution of commercial horse feeds. *Feedstuffs*, **47**(36), 26–8.

Hintz, H.F. and Schryver, H.F. (1975b) Recent developments in equine nutrition. *Proceedings of the Cornell Nutrition Conference for Feed Manufacturers*, 95–8. Cornell University, Ithaca, New York.

Hintz, H.F. and Schryver, H.F. (1976a) Corrugated paper boxes can be ingredients in complete pelleted diets for horses. *Feedstuffs*, **48**(44), 51–3.

Hintz, H.F. and Schryver, H.F. (1976b) Current status of mineral nutrition in the equine. *Proceedings of the University of Maryland Nutrition Conference for Feed Manufacturers*, University of Maryland Agriculture Experimental Station, College Park, Maryland, pp. 42–4.

Hintz, H.F. and Schryver, H.F. (1976c) Nutrition and bone development in horses. *Journal–American Veterinary Medical Association*, **168**, 39–44.

Hintz, H.F. and Schryver, H.F. (1976d) Potassium metabolism in ponies. *Journal of Animal Science*, **42**, 637–43.

Hintz, H.F., Schryver, H.F. and Cymbaluk, N.F. (1979b) Feeding dehydrated forages to horses. In: *Proceedings of the 2nd International Green Crop Drying Congress* (ed. R.E. Howarth), pp. 314–18. University of Saskatchewan, Saskatoon, Canada.

Hintz, H.F., Schryver, H.F., Doty, J., Lakin, C. and Zimmerman, R.A. (1984) Oxalic acid content of alfalfa hays and its influence on the availability of calcium, phosphorus and magnesium. *Journal of Animal Science*, **58**, 939–42.

Hintz, H.F., Schryver, H.F. and Lowe, J.E. (1973) Digestion in the horse. *Feedstuffs*, **45**(27), 25–31.

Hintz, H.F., Schryver, H.F. and Lowe, J.E. (1976) Delayed growth response and limb conformation in young horses. *Proceedings of the Cornell Nutrition Conference for Feed Manufacturers*, 94–6. Cornell University, Ithaca, New York.

Hintz, H.F., Sedgewick, C.J. and Schryver, H.F. (1976) Some observations on digestion of a pelleted diet by ruminants and non-ruminants. *International Zoology Yearbook*, **16**, 54–7.

Hintz, H.F., White, K., Short, C., Lowe, J. and Ross, M. (1980) Effects of protein levels on endurance horses. *Journal of Animal Science*, **51** (Suppl.), 202–3.

Hintz, H.F., Williams, A.J., Rogoff, J. and Schryver, H.F. (1973) Availability of phosphorus in wheat bran when fed to ponies. *Journal of Animal Science*, **36**, 522–5.

Hintz, R.L., Hintz, H.F. and Van Vleck, L.D. (1978c) Estimation of heritabilities for weight, height and front cannon bone circumference of Thoroughbreds. *Journal of Animal Science*, **47**, 1243–5.

Hodge, S.L., Kreider, J.L., Potter, G.D. and Harms, P.G. (1981) Influence of photoperiod on the pregnant and postpartum mare. *Journal of Animal Science*, **330** (Suppl. 1), abstract 505.

Hodgson, D.R. (1993) Exercise-associated myopathy: is calcium the culprit? *Equine Veterinary Journal*, **25**, 1–3.

Hodgson, D.R., Rose, R.J., Allen, J.R. and Dimauro, J. (1984) Glycogen depletion patterns in horses performing maximal exercise. *Research in Veterinary Science*, **36**, 169–73.

Holland, J.L., Kronfeld, D.S. and Meacham, T.N. (1996) Behavior of horses is affected by soy lecithin and corn oil in the diet. *Journal of Animal Science*, **74**, 1252–5.

Hollands, T. and Cuddeford, D. (1992) Effect of supplementary soya oil on the digestibility of nutrients contained in a 40:60 roughage/concentrate diet fed to horses. *First Europäische Konferenz über die Ernährung des Pferdes*, Institut für Tierernahrüng, Tierrnahrüng Hochschule, Hannover, 3–4 September 1992, pp. 128–32.

Holmes, J.R. (1982) A superb transport system. The circulation. *Equine Veterinary Journal*, **14**, 267–76.

Holmes, P.H. (1993) Interactions between parasites and animal nutrition: the veterinary consequences. *Proceedings of the Nutrition Society*, **52**, 113–20.

van der Holst, W., Tjalsma, E.J. and Wonder, C.I. (1984) Experiences with oral administration of β-carotene to pony mares in early spring. *35th Annual Meeting of the European Association for Animal Production*, the Hague, pp. 2–15.

Holt, P.E. and Pearson, H. (1984) Urolithiasis in the horse – a review of 13 cases. *Equine Veterinary Journal*, **16**, 31–4.

Holton, D.W., Garrett, B.J. and Cheeke, P.R. (1983) Effects of dietary supplementation with BHA, cysteine and B vitamins on tansy ragwort toxicity in horses. *Proceedings of the Western Section of the American Society of Animal Science*, Washington State University, Pullman, 26–29 July, **34**, 183–6.

Hood, D.M., Hightower, D., Amoss, M.S., Jr, *et al.* (1987) Thyroid function in horses affected with laminitis. *Southwestern Veterinarian*, **38**, 85–91.

Hopkins, A. (1986) Botanical composition of permanent grassland in England and Wales in relation to soil, environment and management factors. *Grass and Forage Science*, **41**, 237–46.

Hopkins, A., Martyn, T.M. and Bowling, P.J. (1994) Companion species to improve seasonality of production and nutrient uptake in grass/clover swards. *Proceedings of the 15th General Meeting, European Grassland Federation* (ed. L. Mannetje), Wageningen, the Netherlands, pp. 73–6.

Houpt, K.A. (1991) Investigating equine ingestive, maternal, and sexual behavior in the field and in the laboratory. *Journal of Animal Science*, **69**, 4161–6.

Houpt, K.A. (1983) Taste preferences in horses. *Equine Practice*, **5**, 22–5.

Houpt, K.A., Hintz, H.F. and Pagan, J.D. (1981) The response of mares and their foals to brief separation. *Journal of Animal Science*, **53** (Suppl. 1), 129, abstract 6.

Houpt, K.A. and Houpt, T.R. (1988) Social and illumination preferences of mares. *Journal of Animal Science*. **66**, 2159–64.

Houpt, K.A., Parsons, M.S. and Hintz, H.F. (1982) Learning ability of orphan foals, of normal foals and of their mothers. *Journal of Animal Science*, **55**, 1027–32.

Houpt, T.R. and Houpt, K.A. (1971) Nitrogen conservation by ponies fed a low protein ration. *American Journal of Veterinary Research*, **32**, 579–88.

Householder, D.D., Potter, G.D., Lichtenwainer, R.E. and Hesby, J.H. (1976) Growth and digestion in horses fed sorghum or oats. *Journal of Animal Science*, **43**, 254.

Hoven, R. Van Den., Breuknk, H.J., Vaandrager-Verduin, M.H.V. and Scholte, H.R. (1989) Normal resting values of plasma free carnitine and acylcarnitine in horses predisposed to exertional rhabdomyolysis. *Equine Veterinary Journal*, **21**, 307–8.

Hoyt, J.K., Potter, G.D., Greene, L.W. and Anderson, J.G., Jr (1995a) Mineral balance in resting and exercised miniature horses. *Journal of Equine Veterinary Science*, **15**, 310–14.

Hoyt, J.K., Potter, G.D., Greene, L.W. and Anderson, J.G., Jr (1995b) Copper balance in miniature horses fed varying amounts of zinc. *Journal of Equine Veterinary Science*, **15**, 357–9.

Huchton, J.D., Potter, G.D., Sorensen, A.M., Jr and Orts, F.A. (1976) Foal development related to prepartum nutrition. *Journal of Animal Science*, **43**, 253, abstract 171.

Huisman, J., Heinz, Th., Van der Poel, A.F.B., Van Leeuwen, P., Souffrant, W.B. and Verstegen, M.W.A. (1992) True protein digestibility and amounts of endogenous protein measured with the ^{15}N-dilution technique in piglets fed on peas (*Pisum sativum*) and common beans (*Phaseolus vulgaris*). *British Journal of Nutrition*, **68**, 101–10.

Hunt, R.J. (1993) A retrospective evaluation of laminitis in horses. *Equine Veterinary Journal*, **25**, 61–4.

Hunter, L. and Houpt, K.A. (1989) Bedding material preferences of ponies. *Journal of Animal Science*, **67**, 1986–91.

Hyslop, J.J., Jessop, N.S., Stefansdottir, G.J. and Cuddeford, D. (1997) Comparative protein and fibre degradation measured *in situ* in the caecum of ponies and in the rumen of steers. *Proceedings of the British Society of Animal Science*, Scarborough, March 1997, p. 121.

International Union of Biochemistry and Molecular Biology on the Nomenclature and Classification of Enzymes (1992) *Enzyme Nomenclature Recommendations*. Academic Press, San Diego, California.

INRA (1990) *L'Alimentation des Chevaux* (ed. W. Martin-Rosset). INRA Publications, Versailles.

Jablonska, E.M., Ziolkowska, S.M., Gill, J., Szykula, R. and Faff, J. (1991) Changes in some haematological and metabolic indices in young horses during the first year of jump-training. *Equine Veterinary Journal*, **23**, 309–11.

Jacobs, I., (1981) Lactate, muscle glycogen and exercise performance in man. *Acta Physiologica Scandinavica,* **495** (Suppl.), 3–35.

Jacobs, K.A. and Bolton, J.R. (1982) Effect of diet on the oral glucose tolerance test in the horse. *Journal – American Veterinary Medical Association*, **180**, 884–6.

Jacobs, K.A., Norman, P, Hodgson, D.R.G. and Cymbaluk, N. (1982) Effect of diet on oral D-xylose absorption test in the horse. *American Journal of Veterinary Research*, **43**, 1856–8.

Jaeschke, G. and Keller, H. (1978) The ascorbic acid status in horses. 2. Clinical aspects and deficiency symptoms. *Berlin und München Tierärztliche Wochenschrift*, **91**, 375–9.

James, L.F., Panter, K.E., Nielsen, D.B. and Molyneux, R.J. (1992) The effect of natural toxins on reproduction in livestock. *Journal of Animal Science*, **70**, 1573–9.

Jarrett, S.H. and Schurg, W.A. (1987) Use of a modified relative dose response test for determination of vitamin A status in horses. *Nutrition Reports International*, **35**, 733–42.

Jarridge, R. and Tisserand, J.L. (1984) Métabolisme besoins et alimentation azotée du cheval, In: *Le Cheval. Reproduction, Sélection, Alimentation, Exploitation* (R. Jarridge and W. Martin-Rossett), pp. 277–302. INRA, Paris.

Jeffcott, L.B. (1974a) Some practical aspects of the transfer of passive immunity to newborn foals. *Equine Veterinary Journal*, **6**, 109–15.

Jeffcott, L.B. (1974b) Studies on passive immunity in the foal. I. γ-Globulin and antibody variations associated with the maternal transfer of immunity and the onset of active immunity. *Journal of Comparative Pathology*, **84**, 93–101.

Jeffcott, L.B. (1974c) Studies on passive immunity in the foal. II. The absorption of ^{125}I-labelled PVP (polyvinyl pyrrolidone) by the neonatal intestine. *Journal of Comparative Pathology*, **84**, 279–89.

Jeffcott, L.B. (1974d) Studies on passive immunity in the foal. III. The characterization and significance of neonatal proteinuria. *Journal of Comparative Pathology*, **84**, 455–65.

Jeffcott, L.B. (1991) Osteochondrosis in the horse – searching for the key to pathogenesis. *Equine Veterinary Journal*, **23**, 331–8.

Jeffcott, L.B., Dalin, G., Drevemo, S., Fredricson, I., Björne, K. and Bergquist, A. (1982a) Effect of induced back pain on gait and performance of trotting horses. *Equine Veterinary Journal*, **14**, 129–33.

Jeffcott, L.B., Field, J.R., McLean, J.G. and O'Dea, K. (1986) Glucose tolerance and insulin sensitivity in ponies and standardbred horses. *Equine Veterinary Journal*, **18**, 97–101.

Jeffcott, L.B. and Kold, S.E. (1982) Stifle lameness in the horse: a survey of 86 referred cases. *Equine Veterinary Journal*, **14**, 31–9.

Jeffcott, L.B., Rossdale, P.D., Freestone, J., Frank, C.J. and Towers-Clark, P.F. (1982b) An assessment of wastage in Thoroughbred racing from conception to 4 years of age. *Equine Veterinary Journal*, **14**, 185–98.

Jeffcott, L.B., Rossdale, P.D. and Leadon, D.P. (1982c) Haematological changes in the neonatal period of normal and induced premature foals. *Journal of Reproduction and Fertility* Suppl. **32**, 537–44.

Ji, L.L., Dillon, D.A., Bump, K.D. and Lawrence, L.M. (1990) Antioxidant enzymes response to exercise in equine erythrocytes. *Journal of Equine Veterinary Science*, **10**, 380–3.

Johnson, A.L. (1986) Serum concentrations of prolactin, thyroxine and triiodothyronine relative to season and the estrous cycle in the mare. *Journal of Animal Science*, **62**, 1012–20.

Johnson, A.L. and Malinowski, K. (1986) Daily rhythm of cortisol, and evidence for a photo-inducible phase for prolactin secretion in non-pregnant mares housed under non-interrupted and skeleton photoperiods. *Journal of Animal Science*, **63**, 169–75.

Johnson, B.L., Stover, S.M., Daft, B.M., *et al.* (1994) Causes of death in racehorses over a two year period. *Equine Veterinary Journal*, **26**, 327–30.

Johnson, K.A., Sigler, D.H. and Gibbs, P.G. (1988) Nitrogen utilization and metabolic responses of ponies to intense anaerobic exercise. *Journal of Equine Veterinary Science*, **8**, 249–54.

Johnson, R.J. and Hart, J.W. (1974a) Influence of feeding and fasting on plasma free amino acids in the equine. *Journal of Animal Science*, **38**, 790–98.

Johnson, R.J. and Hart, J.W. (1974b) Utilization of nitrogen from soybean meal, biuret and urea by equine. *Nutrition Reports International*, **9**, 209–15.

Johnson, R.J. and Hughes, I.M. (1974) Alfalfa cubes for horses. *Feedstuffs*, **46**(43), 31.

Jones, D.G.C., Greatorex, J.C., Stockman, M.J.R. and Harris, C.P.J. (1972) Gastric impaction in a pony: relief via laparotomy. *Equine Veterinary Journal*, **4**, 98–9.

Jones, D.L., Potter, G.D., Greene, L.W. and Odom, T.W. (1992) Muscle glycogen in exercised miniature horses at various body conditions and fed a control or fat supplemented diet. *Journal of Equine Veterinary Science*, **12**, 287–91.

Jones, S. and Blackmore, D.J. (1982) Observations on the isoenzymes of aspartate aminotransferase in equine tissues and serum. *Equine Veterinary Journal*, **14**, 311–16.

Jordan, R.M. (1979) Effect of corn silage and turkey litter on the performance of gestating pony mares and weanlings. *Journal of Animal Science*, **49**, 651–3.

Jordan, R.M. (1982) Effect of weight loss of gestating mares on subsequent production. *Journal of Animal Science*, **55** (Suppl. 1), 208.

Jordan, R.M. and Marten, G.C. (1975) Effect of three pasture grasses on yearling pony weight gains and pasture carrying capacity. *Journal of Animal Science*, **40**, 86–9.

Jordan, R.M., Myers, V.S., Yoho, B. and Spurell, F.A. (1975) Effect of calcium and phosphorus levels on growth, reproduction and bone development of ponies. *Journal of Animal Science*, **40**, 78–84.

Josseck, H., Zenker, W. and Geyer, H. (1995) Hoof horn abnormalities in Lipizzaner horses and the effect of dietary biotin on macroscopic aspects of hoof horn quality. *Equine Veterinary Journal*, **27**, 175–82.

Judson, G.J. and Mooney, G.J. (1983) Body water and water turnover rate in Thoroughbred horses in training. In: *Equine Exercise Physiology* (eds D.H. Snow, S.G.B. Persson and R.J. Rose), pp. 354–61. Granta Editions, Cambridge.

Julen, T.R., Potter, G.D., Greene, L.W. and Stott, G.G. (1995) Adaptation to a fat-supplemented diet by cutting horses. *Journal of Equine Veterinary Science*, **15**, 436–40.

Julliand, V. (1992) Microbiology of the equine hindgut. *First Europäische Konferenz über die Ernährung des Pferdes*, Institut für Tierernährung, Tierärzliche Hochschule, Hannover, 3–4 September 1992, pp. 42–7.

Kamphues, J., Denell, S. and Radicke, S. (1992) Lipopolysaccharid-Konzentrationen im Magen-Darm-Trakt von Ponys nach Aufnahme von Heu bzw. einer kraftfutterreichen Ration. *First Europäische Konferenz über die Ernährung des Pferdes*, Institut für Tierernährung, Tierärzliche Hochschule, Hannover, 3–4 September 1992, pp. 59–63.

Kamphues, J. and Schad, D. (1992) Verstopfungskoliken bei Pferden nach Fütterung von Windhalm-(*A. peraspica venti*)-Heu. *First Europäische Konferenz über die Ernährung des Pferdes*, Institut für Tierernährung, Tierärzliche Hochschule, Hannover, 3–4 September 1992, pp. 213–15.

Kane, E., Baker, J.P. and Bull, L.S. (1979) Utilization of a corn oil supplemented diet by the pony. *Journal of Animal Science*, **48**, 1379–84.

Kaup, F.J., Drommer, W., Damsch, S. and Deegen, E. (1990) Ultrastructural findings in horses with chronic obstructive pulmonary disease (COPD) II: pathomorphological changes of the terminal airways and the alveolar region. *Equine Veterinary Journal*, **22**, 349–55.

Kaup, F.J., Drommer, W. and Deegen, E. (1990) Ultrastructural findings in horses with chronic obstructive pulmonary disease (COPD) I: alterations of the larger conducting airways. *Equine Veterinary Journal*, **22**, 343–8.

Keenan, D.M. (1978) Changes of plasma uric acid levels in horses after galloping. *Research in Veterinary Science*, **25**, 127–8.

Kellock EM (1982) The origins of the Thoroughbred. *Equi*, No. 13, November/December, 4–5

Kelly, A.P., Jones, R.T., Gillick, J.C. and Simms, L.D. (1984) Outbreak of botulism in horses. *Equine Veterinary Journal*, **16**, 519–21.

Kelly, W.R. and Lambert, M.B. (1978) The use of cocoa-bean meal in the diets of horses: pharmacology and pharmacokinetics of theobromine. *British Veterinary Journal*, **134**, 171–80.

Kempson, S.A. (1987) Scanning electron microscope observations of hoof horn from horses with brittle feet. *Veterinary Record*, **120**, 568–70.

Kempson, S.A., Currie, R.J.W. and Johnston, A.M. (1989) Influence of biotin supplementation on pig claw horn: a scanning electron microscopic study. *Veterinary Record*, **124**, 37–40.

Kennedy, L.G. and Hershberger, T.V. (1974) Protein quality for the non-ruminant herbivore. *Journal of Animal Science*, **39**, 506–511.

Kienzle, E., Radicke, S., Wilke, S., Landes, E. and Meyer, H. (1992) Praeileale Stärkeverdauung in Abhängigheit von Stärkeart und -zubereitung. *First Europäische Konferenz über die Ernährung des Pferdes*, Institut für Tierernahrüng, Tierärzliche Hochschule, Hannover, 3–4 September 1992, pp. 103–6.

Kilshaw, P.J. and Sissons, J.W. (1979) Gastrointestinal allergy to soyabean protein in preruminant calves. Allergenic constituents of soyabean products. *Research in Veterinary Science*, **27**, 366–71.

Kim, H.L., Herrig, B.W., Anderson, A.C., Jones, L.P. and Calhoun, M.C. (1983) Elimination of adverse effects of ethoxyquin (EQ) by methionine hydroxy analog (MHA). Protective effects of EQ and MHA for bitterweed poisoning in sheep. *Toxicology Letters*, **16**, 23–9.

Kinde, H., Hietala, S.K., Bolin, C.A. and Dowe, J.T. (1996) Leptospiral abortion in horses following a flooding incident. *Equine Veterinary Journal*, **28**, 327–30.

King, J.N. and Gerring, E.L. (1991) The action of low dose endotoxin on equine bowel motility. *Equine Veterinary Journal*, **23**, 11–17.

Klein, H.-J., Schulze, E., Deegen, E. and Giese, W. (1988) Metabolism of naturally occurring [^{13}C] glucose given orally to horses. *American Journal of Veterinary Research*, **49**, 1259–62.

Klendshoj, C., Potter, G.D., Lichtenwalner, R.E. and Householder, D.D. (1979) Nitrogen digestion in the small intestine of horses fed crimped or micronized sorghum grain or oats. In: *Proceedings of the 6th Equine Nutrition and Physiology Society*, A. and M. University, Texas, pp. 91–4.

Knaus, E. (1981) Diseases of the foal in the first days of life. *Proceedings of the 32nd Annual Meeting of the European Association of Animal Production*, Zagreb, Yugoslavia, 31 August–3 September, IIIa.

Knight, D.A. and Tyznik, W.J. (1985) The effect of artificial rearing on the growth of foals. *Journal of Animal Science*, **60**, 1–5.

Kohn, C.W., Muir, W.W. and Sams, R. (1978) Plasma volume and extracellular fluid volume in horses at rest and following exercise. *American Journal of Veterinary Research*, **39**, 871–4.

Kold, S. (1992) Is it possible to accelerate the restoration of a deficient skeleton? *Equine Veterinary Journal*, **24**, 335.

Kollarczik, B., Enders, C., Friedrich, M. and Gedek, B. (1992) Effect of diet composition on microbial spectrum in the jejunum of horses. *First Europäische Konferenz über die Ernährung des Pferdes*, Institut für Tierernahrüng, Tierärzliche Hochschule, Hannover, 3–4 September 1992, pp. 49–54.

Körber, H.-D. (1971) Zur Kolikstatistik des Pferdes. *Berlin und München Tierärztliche Wochenschrift*, **84**, 75–7.

Kosiniak, K. (1981) Some properties of semen from young stallions and its value for preservation in liquid nitrogen. *Proceedings of the 32nd Annual Meeting of the European Association of Animal Production*, Zagreb, Yugoslavia, 31 August–3 September, I-10.

Kossila, V. and Ljung, G. (1976) Value of whole oat plant pellets in horse feeding. *Annales Agriculturae Fenniae*, **15**, 316–21.

Kossila, V., Tanhuanpaa, E., Virtanen, E. and Luoma, E. (1972) Hb value, blood glucose, cholesterol, minerals and trace elements in saddle horses. 1. Differences due to age and maintenance. *Journal of the Scientific Agricultural Society of Finland*, **44**, 249–57.

Kownacki, M. (1983) The development of type in the Polish Konik. *Rocz. Nauk rol.* **B82-1**, 71–104.

Kozak, A. and Bickel, H. (1981) Die Verdaulichkeit von Gras dreier Weidetypen Ostafrikas. *Proceedings of the 32nd Annual Meeting of the European Association of Animal Production*, Zagreb, Yugoslavia, 31 August–3 September, I-11.

Kronauer, M. and Bickel, H. (1981) Estimation of the energy value of East African pasture grass. *Proceedings of the 32nd Annual Meeting of the European Association of Animal Production*, Zagreb, Yugoslavia, 31 August–3 September, I-12.

Kronfeld, D.S., Ferrante, P.L. and Grandjean, D. (1994) Optimal nutrition for athletic performance, with emphasis on fat adaptation in dogs and horses. *Journal of Nutrition*, **124**, 2745S–53S.

Krook, L. (1968) Dietary calcium–phosphorus and lameness in the horse. *Cornell Veterinarian*, **58**, 59–73.

Krook, L., Bélanger, L.F., Henrikson, P., Lutwak, L. and Sheffy, B.E. (1970) Bone flow. *Rev. Can. Biol.*, **29**, 157–67.

Kruczynska, H., Ponikiewska, T. and Berthold, St. (1981) Einfluss der mit Gülle gedüngten Weide auf die Milchleistung und Mineralbilanz bei Kühen. *Proceedings of the 32nd Annual Meeting of the European Association of Animal Production*, Zagreb, Yugoslavia, 31 August–3 September, IV-14.

Krzywanek, H. (1974) Lactic acid concentrations and pH values in trotters after racing. *Journal of the South African Veterinary Association*, **45**, 355–60.

Kumar, U., Sareen, V.K. and Singh, S. (1994) Effect of *Saccharomyces cerevisiae* yeast culture supplement on ruminal metabolism in buffalo calves given a high concentrate diet. *Animal Production*, **59**, 209–215.

Langlands, J.P. and Cohen, R.D.H. (1978) The nutrition of ruminants grazing native and improved pastures. III. Mineral composition of bones and selected organs from grazing cattle. *Australian Journal of Agricultural Research*, **29**, 1301–11.

Lapointe, J.-M., Vrins, A. and McCarvill, E. (1994) A survey of exercise-induced pulmonary haemorrhage in Quebec Standardbred racehorses. *Equine Veterinary Journal*, **26**, 482–5.

Lavoie, J.P. and Teuscher, E. (1993) Massive iron overload and liver fibrosis resembling haemochromatosis in a racing pony. *Equine Veterinary Journal*, **25**, 552–4.

Lawrence, L., Kline, K., Miller-Graber, P., *et al.* (1987a) Effect of sodium bicarbonate on racing standardbreds. *Proceedings of the 10th Equine Nutrition and Physiology Society*, Colorado State University, Fort Collins, 11–13 June 1987, pp. 499–503.

Lawrence, L., Kline, K., Miller-Graber, P., *et al.* (1990) Effect of sodium bicarbonate on racing standardbreds. *Journal of Animal Science*, **68**, 673–7.

Lawrence, L.A., Ott, E.A., Asquith, R.L. and Miller, G.J. (1987b) Influence of dietary iron on growth, tissue mineral composition, apparent phosphorus absorption and chemical and mechanical properties of bone in ponies. *Proceedings of the 10th Equine Nutrition and Physiology Society*, Colorado State University, Fort Collins, 11–13 June 1987, pp. 563–71.

Lawrence, L.A., Ott, E.A., Miller, G.J., Poulos, P.W., Piotrowski, G. and Asquith, R.L. (1994) The mechanical properties of equine third metacarpals as affected by age. *Journal of Animal Science*, **72**, 2617–23.

Lawrence, L.M., Slade, L.M., Nockels, C.F. and Shideler, R.K. (1978) Physiologic effects of vitamin E supplementation on exercised horses. *Proceedings of the Western Section of the American Society of Animal Science*, Vancouver, British Columbia, 23–25 June 1981, **29**, 173–7.

Lawrence, L.M., Soderholm, L.V., Roberts, A.M., Williams, J. and Hintz, H.F. (1993) Feeding status affects glucose metabolism in exercising hoses. *Journal of Nutrition*, **123**, 2152–7.

Lawrence, L.M., Williams, J., Soderholm, L.V., Roberts, A.M. and Hintz, H.F. (1995) Effect of feeding state on the response of horses to repeated bouts of intense exercise. *Equine Veterinary Journal*, **27**, 27–30.

Lawson, G.H.K., McPherson, E.A., Murphy, J.R. *et al.* (1979) The presence of precipitating antibodies in the sera of horses with chronic obstructive pulmonary disease (COPD). *Equine Veterinary Journal*, **11**, 172–6.

Ledgard, S.F., Steele, K.W. and Saunders, W.H.M. (1982) Effects of cow urine and its major constituents on pasture properties. *New Zealand Journal of Agricultural Research*, **25**, 61–8.

Lee, J., McAllister, E.S. and Scholz, R.W. (1995) Assessment of selenium status in mares and foals under practical management conditions. *Journal of Equine Veterinary Science*, **15**, 240–5.

Lees, P., Dawson, J. and Sedgwick, A.D. (1986) Eicosanoids and equine leucocyte locomotion *in vitro*. *Equine Veterinary Journal*, **18**, 493–7.

Leonard, T.M., Baker, J.P. and Willard, J.G. (1974) Effect of dehydrated alfalfa on equine digestion. *Journal of Animal Science*, **39**, 184, abstract 188.

Leonard, T.M., Baker, J.P. and Willard, J.G. (1975) Influence of distillers' feeds on digestion in the equine. *Journal of Animal Science*, **40**, 1086–90.

Levine, S.B., Myhre, G.D., Smith, G.L., Burns, J.G. and Erb, H. (1982) Effect of a nutritional supplement containing N, N-dimethylglycine (DMG) on the racing standardbred. *Equine Practice*, **4**, 17–20.

Lewis, L.D. (1982) *Feeding and Care of the Horse*. Lea and Febiger, Philadelphia.

Lindholm, A. (1974) Glycogen depletion pattern and the biochemical response to varying exercise intensities in standardbred trotters. *Journal of the South African Veterinary Association*, **45**, 341–3.

Lindholm, A., Bjerneld, H. and Saltin, B. (1974) Glycogen depletion pattern in muscle fibres of trotting horses. *Acta Physiologica Scandinavica*, **90**, 475–84.

Lindholm, A., Johansson, H.-E. and Kjaersgaard, P. (1974) Acute rhabdomyolysis ('tying-up') in standardbred horses. A morphological and biochemical study. *Acta Veterinaria Scandinavica*, **15**, 325–39.

Lindholm, A. and Piehl, K. (1974) Fibre composition, enzyme activity and concentrations of metabolites and electrolytes in muscles of standardbred horses. *Acta Veterinaria Scandinavica*, **15**, 287–309.

Lindholm, A. and Saltin, B. (1974) The physiological and biochemical response of standardbred horses to exercise of varying speed and duration. *Acta Veterinaria Scandinavica*, **15**, 310–24.

Lindner, A., von Wittke, P., Bendig, M. and Sommer, H. (1991) Effect of an energy enriched electrolyte fluid concentrate on heart rate and lactate concentration of ponies during and after exercise. *Proceedings of the 12th Equine Nutrition and Physiology Society*, University of Calgary, 6–8 June 1991, pp. 93–4.

Lindner, A., von Wittke, P. and Frigg, M. (1992) Effect of biotin supplementation on the V_{LA4} of thoroughbred horses. *Journal of Equine Veterinary Science*, **12**, 149–52.

Littlejohn, A., Kruger, J.M. and Bowles, F. (1977) Exercise studies in horses: 2. The cardiac response to exercise in normal horses and in horses with chronic obstructive pulmonary disease. *Equine Veterinary Journal*, **9**, 75–83.

Lloyd, D.R., Evans, D.L., Hodgson, D.R., Suann, C.J. and Rose, R.J. (1993) Effects of sodium bicarbonate on cardiorespiratory measurements and exercise capacity in Thoroughbred horses. *Equine Veterinary Journal*, **25**, 125–9.

Lloyd, K.C.K. (1988) Alternative diagnosis in the colic patient. *Veterinary Clinics of North America: Equine Practice*, **4**, 17–34.

Loew, F.M. (1973) Thiamin and equine laryngeal hemiplegia. *Veterinary Record*, **92**, 372–3.

Loew, F.M. and Bettany, J.M. (1973) Thiamine concentrations in the blood of standardbred horses. *American Journal of Veterinary Research*, **34**, 1207–1208.

Lohrey, E., Tapper, B. and Hove, E.L. (1974) Photosensitization of albino rats fed on lucerne–protein concentrate. *British Journal of Nutrition*, **31**, 159–67.

Lolas, G.M. and Markakis, P. (1975) Phytic acid and other phosphorus compounds of beans (*Phaseolus vulgaris*), *Journal of Agricultural and Food Chemistry*, **23**, 13–15.

Lopez, N.E., Baker, J.P. and Jackson, S.G. (1987) Effect of cutting and vacuum cleaning on the digestibility of oats by horses. *Proceedings of the 10th Equine Nutrition and Physiology Society*, Colorado State University, Fort Collins, 11–13 June 1987, pp. 611–13.

López, S., Hovell, F.D.DeB., Manyuchi, B. and Smart, R.I. (1995) Comparison of sample preparation methods for the determination of the rumen degradation characteristics of fresh and ensiled forages by the nylon bag technique. *Animal Science*, **60**, 439–50.

Lopez-Rivero, J.L., Morales-Lopez, J.L., Galisteo, A.M. and Aguera, E. (1991) Muscle fibre type composition in untrained and endurance-trained Andalusian and Arab horses. *Equine Veterinary Journal*, **23**, 91–3.

Lord, K.A. and Lacey, J. (1978) Chemicals to prevent the moulding of hay and other crops. *Journal of the Science of Food and Agriculture*, **29**, 574–5.

Löscher, W., Jaeschke, G. and Keller, H. (1984) Pharmacokinetics of ascorbic acid in horses. *Equine Veterinary Journal*, **16**, 59–65.

Lucke, J.N. and Hall, G.M. (1978) Biochemical changes in horses during a 50-mile endurance ride. *Veterinary Record*, **102**, 356–8.

Lucke, J.N. and Hall, G.M. (1980) Further studies on the metabolic effects of long distance riding: Golden Horseshoe Ride 1979. *Equine Veterinary Journal*, **12**, 189–92.

Lunn, P.G. and Austin, S. (1983) Dietary manipulation of plasma albumin concentration. *Journal of Nutrition*, **113**, 1791–802.

Lyons, E.T., Drudge, J.H. and Tolliver, S.C. (1980) Antiparasitic activity of parbendazole in critical tests in horses. *American Journal of Veterinary Research*, **41**, 123–4.

McCall, C.A. (1990) A review of learning behavior in horses and its application in horse training. *Journal of Animal Science*, **68**, 75–81.

McCall, C.A. Potter, G.D., Friend, T.H. and Ingram, R.S. (1981) Learning abilities in yearling horses using the Hebb–Williams closed field maze. *Journal of Animal Science*, **53**, 928–33.

MacCallum, F.J., Brown, M.P. and Goyal, H.O. (1978) An assessment of ossification and radiological interpretation in limbs of growing horses. *British Veterinary Journal*, **134**, 366–73.

McCann, J.S., Caudle, A.B., Thompson, F.N., Stuedemann, J.A., Heusner, G.L. and Thompson, D.L., Jr (1992) Influence of endophyte-infected tall fescue on serum prolactin and progesterone in gravid mares. *Journal of Animal Science*, **70**, 217–23.

McCann, J.S., Meacham, T.N. and Fontenot, J.P. (1987) Energy utilization and blood traits of ponies fed fat-supplemented diets. *Journal of Animal Science*, **65**, 1019–26.

MacCarthy, D.D., Spillane, T.A. and O'Moore, L.B. (1976) Type and quality of oats used for bloodstock feeding in Ireland. *Irish Journal of Agricultural Research*, **15**, 47–54.

McDaniel, A.L., Martin, S.A., McCann, J.S. and Parks, A.H. (1993) Effects of *Aspergillus oryzae* fermentation extract on *in vitro* equine cecal fermentation. *Journal of Animal Science*, **71**, 2164–72.

McDonald, P., Edwards, R.A. and Greenhalgh, J.F.D. (1981) *Animal Nutrition*. Longman, London and New York.

McGavin, M.D. and Knake, R. (1977) Hepatic midzonal necrosis in a pig fed aflatoxin and a horse fed moldy hay. *Veterinary Pathology*, **14**, 182–7.

McGorum, B.C. and Dixon, P.M. (1993) Evaluation of local endobronchial antigen challenges in the investigation of equine chronic obstructive pulmonary disease. *Equine Veterinary Journal*, **25**, 269–72.

McGorum, B.C., Dixon, P.M. and Halliwell, R.E.W. (1993a) Evaluation of intradermal mould antigen testing in the diagnosis of equine chronic obstructive pulmonary disease. *Equine Veterinary Journal*, **25**, 273–5.

McGorum, B.C., Dixon, P.M. and Halliwell, R.E.W. (1993b) Responses of horses affected with chronic obstructive pulmonary disease to inhalation challenges with mould antigens. *Equine Veterinary Journal*, **25**, 261–7.

McGuinness, E.E., Morgan, R.G.H., Levison, D.A., Frape, D.L., Hopwood, D. and Wormsley, K.G. (1980) The effect of long-term feeding of soya flour on the rat pancreas. *Scandinavian Journal of Gastroenteritis*, **15**, 497–502.

McKeever, K.H., Hinchcliff, K.W., Reed, S.M. and Robertson, J.T. (1993) Plasma constituents during incremental treadmill exercise in intact and splenectomised horses. *Equine Veterinary Journal*, **25**, 233–6.

McKenzie, R.A., Blaney, B.J., Gartner, R.J.W., Dillon, R.D. and Standfast, N.F. (1979) A technique for the conduct of nutritional balance experiments in horses. *Equine Veterinary Journal*, **11**, 232–4.

McLaughlin, C.L. (1982) Role of peptides from gastrointestinal cells in food intake regulation. *Journal of Animal Science*, **55**, 1515–27.

McLean, L.M., Hall, M.E. and Bederka, J.P., Jr. (1987) Plasma amino acids/intermediary metabolites in the racing horse. *Proceedings of the 10th Equine Nutrition and Physiology Society*, Colorado State University, Fort Collins, 11–13 June 1987, pp. 437–42.

McMeniman, N.P. and Hintz, H.F. (1992) Effect of vitamin E status on lipid peroxidation in exercised horses. *Equine Veterinary Journal*, **24**, 482–4.

McMiken, D.F. (1983) An energetic basis of equine performance. *Equine Veterinary Journal*, **15**, 123–33.

McPherson, E.A., Lawson, G.H.K., Murphy, J.R., Nicholson, J.M., Breeze, R.G. and Pirie, H.M. (1979a) Chronic obstructive pulmonary disease (COPD) in horses: aetiological studies: responses to intradermal and inhalation antigenic challenge. *Equine Veterinary Journal*, **11**, 159–66.

McPherson, E.A., Lawson, G.H.K., Murphy, J.R., Nicholson, J.M., Breeze, R.G. and Pirie, H.M. (1979b) Chronic obstructive pulmonary disease (COPD): factors influencing the occurrence. *Equine Veterinary Journal*, **11**, 167–71.

McPherson, E.A. and Thomson, J.R. (1983) Chronic obstructive pulmonary disease in the horse: nature of the disease. *Equine Veterinary Journal*, **15**, 203–206.

McPherson, R. (1978) Selenium deficiency. *Veterinary Record*, **103**, 60.

Madelin, T.M., Clarke, A.F. and Mair, T.S. (1991) Prevalence of serum precipitating antibodies in horses to fungal and thermophilic actinomycete antigens: effects of environmental challenge. *Equine Veterinary Journal*, **23**, 247–52.

Madigan, J.E. and Evans, J.W. (1973) Insulin turnover and irreversible loss rate in horses. *Journal of Animal Science*, **36**, 730–33.

Mäenpää, P.H., Koskinen, T. and Koskinen, E. (1988a) Serum profiles of vitamins A, E and D in mares and foals during different seasons. *Journal of Animal Science*, **66**, 1418–23.

Mäenpää, P.H., Pirhonen, A. and Koskinen, E. (1988b) Vitamin A, E, and D nutrition in mares and foals during the winter season: effect of feeding two different vitamin mineral concentrates. *Journal of Animal Science*, **66**, 1424–9.

Mair, T.S., Hillyer, M.H., Taylor, F.G.R. and Pearson, G.R. (1991) Small intestinal malabsorption in the horse: an assessment of the specificity of the oral glucose tolerance test. *Equine Veterinary Journal*, **23**, 344–6.

Mair, T.S. and Osborn, R.S. (1990) The crystalline composition of normal equine urine deposits. *Equine Veterinary Journal*, **22**, 364–5.

Malinowski, K., Johnson, A.L. and Scanes, C.G. (1985) Effects of interrupted photoperiods on the induction of ovulation in anestrous mares. *Journal of Animal Science*, **61**, 951–5.

Maloiy, G.M.O. (1970) Water economy of the Somali donkey. *American Journal of Physiology*, **219**, 1522–7.

Malone, J.C. (1969) Hazards to domestic pet animals from common toxic agents. *Veterinary Record*, **84**, 161–5.

Markham, G. (1936) *Markhams Maister-Peece. In Two Bookes.* Nicholas and John Okes, London.

Marlin, D.J., Harris, R.C., Gash, S.P. and Snow, D.H. (1989) Carnosine content of the middle gluteal muscle in thoroughbred horses with relation to age, sex and training. *Comparative Biochemistry and Physiology*, **93A**, 629–32.

Marquardt, R.R. McKirdy, J.A., Ward, T. and Campbell, L.D. (1975) Amino acid, hemagglutinin and trypsin inhibitor levels, and proximate analyses of faba beans (*Vicia faba*) and faba bean fractions. *Canadian Journal of Animal Science*, **55**, 421–9.

Marti, E., Gerber, H., Essich, G., Oulehla, J. and Lazary, S. (1991) The genetic basis of equine allergic diseases. 1. Chronic hypersensitivity bronchitis. *Equine Veterinary Journal*, **23**, 457–60.

Martin, B., Robinson, S. and Robertshaw, D. (1978) Influence of diet on leg uptake of glucose during heavy exercise. *American Journal of Clinical Nutrition*, **31**, 62–7.

Martin, K.L., Hoffman, R.M., Kronfeld, D.S., Ley, W.B. and Warnick, L.D. (1996a) Calcium decreases and parathyroid hormone increases in serum of periparturient mares. *Journal of Animal Science*, **74**, 834–9.

Martin, R.G., McMeniman, N.P. and Dowsett, K.F. (1991) Effects of a protein deficient diet and urea supplementation on lactating mares. *Journal of Reproductive Fertility*, (Suppl. 44), 543–50.

Martin, R.G., McMeniman, N.P. and Dowsett, K.F. (1992) Milk and water intakes of foals suckling grazing mares. *Equine Veterinary Journal*, **24**, 295–9.

Martin, R.G., McMeniman, N.P., Norton, B.W. and Dowsett, K.F. (1996b) Utilization of endogenous and dietary urea in the large intestine of the mature horse. *British Journal of Nutrition*, **76**, 373–86.

Martin-Rosset, W. (1990) L'alimentation des chevaux, techniques et pratiques, pp 1–232 [Ed. committee], 75007 Paris, INRA.

Martin-Rosset, W., Andrieu, J. and Jestin, M. (1996a) Prediction of organic matter digestibility (OMD) of forages in horses from the chemical composition. *47th European Association of Animal Production Meeting*, Lillehammer, Norway, 25–29th August. Horse Commission Session-H4: Nutrition, pp 1–5.

Martin-Rosset, W., Andrieu, J. and Jestin, M. (1996b) Prediction of the organic matter digestibility (OMD) of forages in horses by the pepsine cellulase method. *47th European Association of Animal Production Meeting*, Lillehammer, Norway, 25–29th August. Horse Commission Session-H4: Nutrition, pp 1–6.

Martin-Rosset, W., Andrieu, J. and Vermorel, M. (1996c) Routine methods for predicting the net energy value of feeds for horses. *47th European Association of Animal Production Meeting*, Lillehammer, Norway, 25–29th August. Horse Commission Session-H4: Nutrition, pp 1–15.

Martin-Rosset, W., Boccard, R. and Robelin, J. (1979) Relative growth of different organs, tissues and body regions in the foal from birth to 30 months. *Proceedings of the 30th Annual Meeting of the European Association of Animal Production*, Zagreb, Yugoslavia, 31 August–3 September, H6.1.

Martin-Rosset, W., Doreau, M. and Cloix, J. (1978) Grazing behaviour of a herd of heavy brood mares and their foals. *Annals of Zootechnology*, **27**, 33–45.

Martin-Rosset, W., Doreau, M. and Thivend, P. (1987) Digestion of diets based on hay or maize silage in growing horses. *Reproduction Nutritional Development*, **27**, 291–2.

Martin-Rosset, W. and Dulphy, J.P. (1987) Digestibility interactions between forages and concentrates in horses: influence of feeding level – comparison with sheep. *Livestock Production Science*, **17**, 263–76.

Martin-Rosset, W., Vermorel, M., Doreau, M., Tisserand, J.L. and Andrieu, J. (1994) The French horse feed evaluation systems and recommended allowances for energy and protein. *Livestock Production Science*, **40**, 37–56.

Mason, D.K. and Kwok, H.W. (1977) Some haematological and biochemical parameters in racehorses in Hong Kong. *Equine Veterinary Journal*, **9**, 96–9.

Masri, M.D., Merritt, A.M., Gronwall, R. and Burrows, C.F. (1986) Faecal composition in foal heat diarrhoea. *Equine Veterinary Journal*, **18**, 301–306.

Matsuoka, T., Novilla, M.N., Thomson, T.D. and Donoho, A.L. (1996) Review of monensin toxicosis in horses. *Journal of Equine Veterinary Science*, **16**, 8–15.

Matthews, H. and Thornton, I. (1982) Seasonal and species variation in the content of cadmium and associated metals in pasture plants at Shipham. *Plant and Soil*, **66**, 181–93.

Mawdsley, A. (1993) *Linear assessment of the Thoroughbred horse*. MEqS thesis, Faculties of Agriculture and Veterinary Medicine, National University of Ireland, Dublin.

Mayhew, I.G. (1994) Odds and SODs of equine motor neuron disease. *Equine Veterinary Journal*, **26**, 342–3.

Mayhew, I.G., Brown, C.M., Stowe, H.D., Trapp, A.L., Derksen, F.J. and Clement, S.F. (1987) Equine degenerative myeloencephalopathy: a vitamin E deficiency that may be familial. *Journal of Veterinary Internal Medicine*, **1**, 45–50.

Meadows, D.G. (1979) Utilization of dietary protein or non-protein nitrogen by lactating mares fed soybean meal or urea. *Dissertation Abstracts International B*, **40**, 999.

Meakin, D.W., Ott, E.A., Asquith, R.L. and Feaster, J.P. (1981) Estimation of mineral content of the equine third metacarpal by radiographic photometry. *Journal of Animal Science*, **53**, 1019–26.

Mehring, J.S. and Tyznik, W.J. (1970) Equine glucose tolerance. *Journal of Animal Science*, **30**, 764–6.

Merkt, H. and Günzel, A.-R. (1979) A survey of early pregnancy losses in West German Thoroughbred mares. *Equine Veterinary Journal*, **11**, 256–8.

Merritt, A.M. (1975) Treatment of diarrhoea in the horse. *Journal of the South African Veterinary Association*, **46**, 89–90.

Merritt, J.B. and Pearson, R.A. (1989) Voluntary food intake and digestion of hay and straw diets by donkeys and ponies. *Proceedings of the Nutrition Society*, **48**, 169A.

Merritt, T., Mallonee, P.G. and Merritt, A.M. (1986) D-xylose absorption in the growing foal. *Equine Veterinary Journal*, **18**, 298–300.

van der Merwe, J.A. (1975) Dietary value of cubes in equine nutrition. *Journal of the South African Veterinary Association*, **46**, 29–37.

Meyer, H. (1982) *Contributions to Digestive Physiology of the Horse*. Paul Parey, Hamburg and Berlin.

Meyer, H. (1983a) The pathogenesis of disturbances in the alimentary tract of the horse in the light of newer knowledge of digestive physiology. *Proceedings of the Horse Nutrition Society*, Uppsala, Sweden, 1983 C, 95–109.

Meyer, H. (1983b) Protein metabolism and protein requirement in horses. *Proceedings of the 4th International Symposium on Protein Metabolism and Nutrition*, Clermont-Ferrand, France, 5–9 September, 1983, No. 16. INRA, Paris.

Meyer, H. (1987) Nutrition of the equine athlete. In: *Equine Exercise Physiology 2*, pp. 644–73. ICEEP Publications, Davis, California.

Meyer, H. (1990) Contributions to water and mineral metabolism of the horse. In: *Advances in Animal Physiology and Animal Nutrition, Journal of Animal Physiology and Animal Nutrition*, pp. 1–102. (Suppl. 21), Paul Parey, Berlin and Hamburg.

Meyer, H. (1992) Intestinaler Wasser- und Elektrolytstoffwechsel Pferdes. *First Europäische Konferenz über die Ernährung des Pferdes*, Institut für Tierernähring, Tierärztliche Hochschule, Hannover, 3–4 September 1992, pp. 67–72.

Meyer, H. and Ahlswede, L. (1976) Intrauterine growth and the body composition of foals, and the nutrient requirements of pregnant mares. *Ubersichten zur Tierernährung*, **4**, 263–92.

Meyer, H. and Ahlswede, L. (1977) Studies on Mg metabolism in the horse. *Zentralblatt für Veterinärmedizin*, **24A** (2), 128–39.

Meyer, H., Ahlswede, L. and Pferdekamp, M. (1975a) Untersuchungen über Magenentleerung und Zusammensetzung des Mageninhaltes beim Pferd. *Deutsche Tierärztliche Wochenschrift*, **87**, 43–7.

Meyer, H., Ahlswede, L. and Reinhardt, H.J. (1975b) Untersuchungen über Freßdauer, Kaufrequenz und Futterzerkleinerung beim Pferd. *Deutsche Tierärztliche Wochenschrift*, **82**, 49–96.

Meyer, H., Coenen, M. and Stadermann, B. (1993) The influence of size on the weight of the gastrointestinal tract and the liver of horses and ponies. *Proceedings of the 13th Equine Nutrition and Physiology Society*, University of Florida, Gainesville, 21–23 January 1993, No. 504, 18–23.

Meyer, H., Heckotter, E., Merkt, M., Bernoth, E.M., Kienzle, E. and Kamphus, J. (1986) Current problems in veterinary advice on feeding. 6. Adverse effects of feeds in horses. *Deutsche Tierärztliche Wochenschrift*, **93**, 486–90.

Meyer, H., Lindemann, G. and Schmidt, M. (1982a) Einfluss unterschiedlicher Mischfuttergaben pro Mahlzeit auf praecaecale- und postileale Verdauungsvorgänge beim Pferd. In: *Contributions to Digestive Physiology of the Horse. Advances in Animal Physiology and Animal Nutrition. Supplement to Journal of Animal Physiology and Animal Nutrition*, **13**, 32–9. Paul Parey, Berlin and Hamburg.

Meyer, H., Muuss, H., Güldenhaupt, V. and Schmidt, M. (1982b) Intestinaler Wasser-, Natrium- und Kaliumstoffwechsel beim Pferd. In: *Contributions to Digestive Physiology of the Horse. Advances in Animal Physiology and Animal Nutrition. Supplement to Journal of Animal Physiology and Animal Nutrition*, **13**, 52–60. Paul Parey, Berlin and Hamburg.

Meyer, H., Pferdekamp, M. and Huskamp, B. (1979) Studies on the digestibility and tolerance of different feeds by typhlectomized ponies. *Deutsche Tierärztliche Wochenschrift*, **86**, 384–90.

Meyer, H., Radicke, S., Kienzle, E., Wilke, S., Kleffken, D. and Illenseer, M. (1995) Investigations on preileal digestion of starch from grain, potato and manioc in horses. *Journal of Veterinary Medicine*, **42**, 371–81.

Meyer, H. and Sallmann, H.-P. (1996) Fettfütterung beim Pferd. *Übers Tierernährung*, **24**, 199–227.

Meyer, H., Schmidt, M. and Guldenhaupt, V. (1981) Untersuchungen über Mischfutter für Pferde. *Deutsche Tierärztliche Wochenschrift*, **88**, 2–5.

Meyer, H., Schmidt, M., Lindemann, G. and Muuss, H. (1982c) Praecaecale und postileale Verdaulichkeit von Mengen- (Ca, P, Mg) und Spurenelementen (Cu, Zn, Mn) beim Pferd. In: *Contributions to Digestive Physiology of the Horse. Advances in Animal Physiology and Animal Nutrition.* Supplement to *Journal of Animal Physiology and Animal Nutrition*, **13**, 61–9. Paul Parey, Berlin and Hamburg.

Meyer, H. and Stadermann, B. (1990) Energie- und Nährstoffbedarf hochtragender Stuten. *Effem-Forschung für Heimtiernahrung*, International Stockmen School, Houston, Texas, February 1990, Waltham Report No. 31, 1–14.

Meyer, H., Winkel, C., Ahlswede, L. and Weidenhaupt, C. (1978) Untersuchungen über Schweissmenge und Schweisszusammensetzung beim Pferd. *Tierärztliche Umschulung*, **33**, 330–36.

Meyers, M.C., Potter, G.D., Evans, J.W., Greene, L.W. and Crouse, S.F. (1989) Physiologic and metabolic response of exercising horses to added dietary fat. *Journal of Equine Veterinary Science*, **9**, 218–23.

Meyers, M.C., Potter, G.D., Greene, L.W., Crouse, S.F. and Evans, J.W. (1987) Physiological and metabolic response of exercising horses to added dietary fat. *Proceedings of the 10th Equine Nutrition and Physiology Society*, Colorado State University, Fort Collins, 11–13 June 1987, pp. 107–113.

Micol, D. and Martin-Rosset, W. (1995) Feeding systems for horses on high forage diets in the temperate zone. In: *Proceedings of the IVth International Symposium on the Nutrition of Herbivores*, pp. 569–84; Clermont-Ferrand, France, 11–15 September, *Recent Developments in the Nutrition of Herbivores*, (eds M. Journet, E. Grenet, M.-H. Farce et al.), INRA Editions, Paris.

Milić, B.Lj. (1972) Lucerne tannins. I. Content and composition during growth. *Journal of the Science of Food and Agriculture*, **23**, 1151–6.

Milić, B.Lj. and Stojanović, S. (1972) Lucerne tannins. III. Metabolic fate of lucerne tannins in mice. *Journal of the Science of Food and Agriculture*, **23**, 1163–7.

Milić, B.Lj., Stojanović, S. and Vučurević, N. (1972) Lucerne tannins. II. Isolation of tannins from lucerne, their nature and influence on the digestive enzymes *in vitro*. *Journal of the Science of Food and Agriculture*, **23**, 1157–62.

Miller-Graber, P.A. and Lawrence, L.M. (1988) The effect of dietary protein level on exercising horses. *Journal of Animal Science*, **66**, 2185–92.

Miller-Graber, P.A., Lawrence, L.M., Foreman, J.H., Bump, K.D., Fisher, M.G. and Kurcz, E.V. (1991) Dietary protein level and energy metabolism during treadmill exercise in horses. *Journal of Nutrition*, **121**, 1462–9.

Miller-Graber, P.A., Lawrence, L.M., Kline, K., et al. (1987) Plasma ammonia and other metabolites in the racing standardbred. *Proceedings of the 10th Equine Nutrition and Physiology Society*, Colorado State University, Fort Collins, 11–13 June 1987, pp. 397–402.

Millward, D.J., Davies, C.T.M., Halliday, D., Wolman, S.L., Matthews, D. and Rennie, M. (1982) Effect of exercise on protein metabolism in humans as explored with stable isotope. *Federation Proceedings*, **41**, 2686–91.

Milne, D.W. (1974) Blood gases, acid–base balance and electrolyte and enzyme changes in exercising horses. *Journal of the South African Veterinary Association*, **45**, 345–54.

Milne, D.W., Skarda, R.J., Gabel, A.A., Smith, L.G. and Ault, K. (1976) Effects of training on biochemical values in standardbred horses. *American Journal of Veterinary Research*, **37**, 285–90.

Milner, J. and Hewitt, D. (1969) Weight of horses: improved estimate based on girth and length. *Canadian Veterinary Journal*, **10**, 314–17.

Ministère de l'Agriculture (1980) *Aménagement et Équipement des Centres Équestres. Section Technique des Équipements Hippiques*. Fiche Nos CE.E.4, 5, 13 and 14. Service des Haras et de l'Equitation, Institut de Cheval, Le Lion-d'Angèrs.

Ministry of Agriculture, Fisheries and Food (1977) *Drainage of Grassland*. ADAS Leaflet No. 7. HMSO, London.

Ministry of Agriculture, Fisheries and Food (1978) *Drainage Maintenance*. ADAS Leaflet No. 7. HMSO, London.

Ministry of Agriculture, Fisheries and Food (1979) *Lime and Fertiliser Recommendations. 1. Arable Crops and Grassland*. ADAS Booklet No. 2191. HMSO, London.

Ministry of Agriculture, Fisheries and Food (1984) *Poisonous Plants in Britain and Their Effects on Animals and Man* (Reference Book 161). HMSO. London.

Minnick, P.D., Brown, C.M., Braselton, W.E., Meerdink, G.L. and Slanker, M.R. (1987) The induction of equine laminitis with an aqueous extract of the heartwood of black walnut (*Juglans nigra*). *Veteri-*

nary and Human Toxicology, **29**, 230–33.

Mishra, P.C. (1988) Ultrastructural changes in an alimentary model of equine laminitis and the comparative vascular changes induced by histamine and endotoxin – including an hypothesis as to the pathogenesis of the lesions in the foot. *Dissertation Abstracts International*, **48**, 1910B.

Mitten, L.A., Hinchcliff, K.W., Holcombe, S.J. and Reed, S.M. (1994) Mechanical ventilation and management of botulism secondary to an injection abscess in an adult horse. *Equine Veterinary Journal*, **26**, 420–33.

Moffitt, P.G., Potter, G.D., Kreider, J.L. and Moritani, T.M. (1985) Venous lactic acid levels in exercising horses fed N,N-dimethylglycine. *Proceedings of the 9th Equine Nutrition and Physiology Society*, Michigan State University, 23–25 May 1985, pp. 248–53.

Moise, L.L. and Wysocki, A.A. (1981) The effect of cottonseed meal on growth of young horses. *Journal of Animal Science*, **53**, 409–413.

Moore, B.E. and Dehority, B.A. (1993) Effects of diet and hindgut defaunation on diet digestibility and microbial concentrations in the cecum and colon of the horse. *Journal of Animal Science*, **71**, 3350–58.

Moore, J.N. (1988) Recognition and treatment of endotoxemia. *Veterinary Clinics of North America: Equine Practice*, **4**, 105–113.

Moore, J.N. (1991) Rethinking endotoxaemia in 1991. *Equine Veterinary Journal*, **23**, 3–4.

Moore, J.N., Garner, H.E., Berg, J.N. and Sprouse, R.F. (1979) Intracecal endotoxin and lactate during the onset of equine laminitis: a preliminary report. *American Journal of Veterinary Research*, **40**, 722–3.

Moore, J.N., Garner, H.E. and Coffman, J.R. (1981) Haematological changes during development of acute laminitis hypertension. *Equine Veterinary Journal*, **13**, 240–42.

Moore, J.N., Garner, H.E., Shapland, J.E. and Hatfield, D.G. (1980) Lactic acidosis and arterial hypoxemia during sublethal endotoxemia in conscious ponies. *American Journal of Veterinary Research*, **41**, 1696–8.

Moore, J.N., Garner, H.E., Shapland, J.E. and Hatfield, D.G. (1981) Prevention of endotoxin-induced arterial hypoxaemia and lactic acidosis with flunixin meglumine in the conscious pony. *Equine Veterinary Journal*, **13**, 95–8.

Moore, J.N., Owen, R. ap R. and Lumsden, J.H. (1976) Clinical evaluation of blood lactate levels in equine colic. *Equine Veterinary Journal*, **8**, 49–54.

Moraillon, R., De Faucompret, P. and Cloche, D. (1978) Results of the long-term administration of a flaked or granulated complete feed into saddle horses. *Recueil de Médecine Vétérinaire*, **154**, 999–1007.

Morris, E.A. and Seeherman, H.J. (1991) Clinical evaluation of poor performance in the racehorse: the results of 275 evaluations. *Equine Veterinary Journal*, **23**, 169–74.

Morse, E.V., Duncan Margo, A., Page, E.A. and Fessler, J.F. (1976) Salmonellosis in Equidae: a study of 23 cases. *Cornell Veterinarian*, **66**, 198–213.

Moss, M.S. (1975) Recent advances in the field of doping detection. *Equine Veterinary Journal*, **7**, 173–4.

Moss, M.S. and Clarke, E.G.C. (1977) A review of drug 'clearance times' in racehorses. *Equine Veterinary Journal*, **9**, 53–6.

Moss, M.S. and Haywood, P.E. (1984) Survey of positive results from racecourse antidoping samples received at Racecourse Security Services' Laboratories. *Equine Veterinary Journal*, **16**, 39–42.

Moyer, W., Spencer, P.A. and Kallish, M. (1991) Relative incidence of dorsal metacarpal disease in young Thoroughbred racehorses training on two different surfaces. *Equine Veterinary Journal*, **23**, 166–8.

Mullen, P.A. (1970) Variations in the albumin content of blood serum in Thoroughbred horses. *Equine Veterinary Journal*, **2**, 118–20.

Mullen, P.A., Hopes, R. and Sewell, J. (1979) The biochemistry, haematology, nutrition and racing performance of two-year-old Thoroughbreds throughout their training and racing season. *Veterinary Record*, **104**, 90–95.

Mundt, H.C. (1978) Untersuchungen über die Verdaulichkeit von aufgeschlossenem Stroh beim Pferd. Published thesis, Tierarztliche Hochschule, Hannover.

Murphy, J.R., McPherson, E.A. and Dixon, P.M. (1980) Chronic obstructive pulmonary disease (COPD): effects of bronchodilator drugs on normal and affected horses. *Equine Veterinary Journal*, **12**, 10–14.

Murray, A. (1993) *The intake of a molassed mineral block by a group of horses at pasture.* MEqS thesis, Faculties of Agriculture and Veterinary Medicine, National University of Ireland, Dublin.

Murray, M. (1985) Hepatic lipidosis in a post parturient mare. *Equine Veterinary Journal*, **17**, 68–9.

Murray, M.J. and Mahaffey, E.A. (1993) Age-related characteristics of gastric squamous epithelial mucosa in foals. *Equine Veterinary Journal*, **25**, 514–17.

Murray, M.J., Murray, C.M., Sweeney, H.J., Weld, J., Digby, N.J.W. and Stoneham, S.J. (1990) Prevalence of gastric lesions in foals without signs of gastric disease: an endoscopic survey. *Equine Veterinary Journal*, **22**, 6–8.

Murray, M.J. and Schusser, G.F. (1993) Measurement of 24-h gastric pH using an indwelling pH electrode in horses unfed, fed and treated with ranitidine. *Equine Veterinary Journal*, **25**, 417–21.

Murray, M.J., Schusser, G.F., Pipers, F.S. and Gross, S.J. (1996) Factors associated with gastric lesions in Thoroughbred racehorses. *Equine Veterinary Journal*, **28**, 368–74.

Mussman, H.C. and Rubiano, A. (1970) Serum protein electrophoregram in the Thoroughbred in Bogota, Colombia. *British Veterinary Journal*, **126**, 574–8.

Muuss, H., Meyer, H. and Schmidt, M. (1982) Entleerung und Zusammensetzung des Ileumchymus beim Pferd. In: *Contributions to Digestive Physiology of the Horse. Advances in Animal Physiology and Animal Nutrition*. Supplement to *Journal of Animal Physiology and Animal Nutrition*, **13**, 13–23. Paul Parey, Berlin and Hamburg.

Muylle, E. and van den Hende, C. (1983) The concept of osmolality. *Equine Veterinary Journal*, **15**, 80–81.

Muylle, E., van den Hende, C., Deprez, P., Nuytten, J. and Oyaert, W. (1986) Non-insulin dependent diabetes mellitus in a horse. *Equine Veterinary Journal*, **18**, 145–6.

Muylle, E., van den Hende, C., Nuytten, J., Deprez, P., Vlaminck, K. and Oyaert, W. (1984b) Potassium concentration in equine red blood cells: normal values and correlation with potassium levels in plasma. *Equine Veterinary Journal*, **16**, 447–9.

Muylle, E., Nuytten, J., van den Hende, C., Deprez, P., Vlaminck, K. and Oyaert, W. (1984a) Determination of red cell potassium content in horses with diarrhoea. A practical approach for therapy. *Equine Veterinary Journal*, **16**, 450–52.

Muylle, E., van den Hende, C., Nuytten, J., Oyaert, W. and Vlaminck, K. (1983) Preliminary studies on the relationship of red blood cell potassium concentration and performance. In: *Equine Exercise Physiology* (eds D.H. Snow, S.G.B. Persson and R.J. Rose), pp. 366–70. Granta Editions, Cambridge.

Muylle, E., van den Hende, C., Oyaert, W., Thoonen, H. and Vlaminck, K. (1981) Delayed monensin sodium toxicity in horses. *Equine Veterinary Journal*, **13**, 107–108.

Muylle, E., Oyaert, W., de Roose, P. and van den Hende, C. (1973) Hypocalcaemia in the horse. *Vlaams Diergeneeskundig Tijdschrift*, **42**, 44–51 [in Dutch].

Nagata, Y., Takagi, S. and Kubo, K. (1972a) Studies on gas metabolism in light horses fed a complete pelleted ration. I. Gas metabolism at rest (the effects of diets and season on gas metabolism at rest). *Experimental Reports of the Equine Health Laboratory*, No. 9, 84–9.

Nagata, Y., Takagi, S. and Kubo, K. (1972b) Studies on gas metabolism in light horses fed a complete pelleted ration. II. Gas metabolism at excitement (the effect of epinephrine infusion on gas metabolism). *Experimental Reports of the Equine Health Laboratory*, No. 9, 90–5.

Nahani, F. and Atiabt, N. (1977) Electrophoretic analysis of blood serum protein of normal horses. *Journal of the Veterinary Faculty of the University of Tehran*, **33**, 75–9.

Nahapetian, A. and Bassiri, A. (1975) Changes in concentrations and interrelationships of phytate, phosphorus, magnesium, calcium and zinc in wheat during maturation. *Journal of Agricultural and Food Chemistry*, **23**, 1179–83.

National Institute of Agricultural Botany (1983–4a) *Grasses and Legumes for Conservation*. Technical Leaflet No. 2. NIAB, Cambridge.

National Institute of Agricultural Botany (1983–4b) *Recommended Varieties of Herbage Legumes*. Farmers Leaflet No. 4. NIAB, Cambridge.

National Institute of Agricultural Botany (1983–4c) *Recommended Varieties of Grasses*. Farmers Leaflet No. 16. NIAB, Cambridge.

National Research Council (1978) *Nutrient Requirements of Domestic Animals. No. 6. Nutrient Requirements of Horses*, 4th edn revised. National Academy of Sciences, Washington DC.

National Research Council (1989) *Nutrient Requirements of Domestic Animals. Nutrient Requirements of Horses*, 5th edn revised. National Academy of Sciences, Washington DC.

Naylor, J.M., Kronfield, D.S. and Acland, H. (1980) Hyperlipemia in horses: effects of undernutrition and diseases. *American Journal of Veterinary Research*, **41**, 899–905.

Naylor, J.M., Jones, V. and Berry, S.-L. (1993) Clinical syndrome and diagnosis of hyperkalaemic periodic paralysis in quarter horses. *Equine Veterinary Journal*, **25**, 227–32.

Neave, R.M.S. and Callear, J.F.F. (1973) Further clinical studies on the uses of mebendazole (R17635) as an anthelmintic in horses. *British Veterinary Journal*, **129**, 79–82.

Nielsen, B.D., Potter, G.D., Morris, E.L., *et al.* (1993) Training distance to failure in young racing quarter horses fed sodium zeolite A. *Proceedings of the 13th Equine Nutrition and Physiology Society*, University of Florida, Gainesville, 21–23 January 1993, No. 504, 5–10.

Nimmo, M.A., Snow, D.H. and Munro, C.D. (1982) Effects of nandrolone phenylpropionate in the horse: (3) skeletal muscle composition in the exercising animal. *Equine Veterinary Journal*, **14**, 229–33.

Noot, G.W.V., Symons, L.D., Lydman, R.K. and Fonnesbeck, P.V. (1967) Rate of passage of various feedstuffs through the digestive tract of horses. *Journal of Animal Science*, **26**, 1309–1311.

Ödberg, F.O. and Francis Smith, K. (1976) A study on eliminative and grazing behaviour – the use of the field by captive horses. *Equine Veterinary Journal*, **8**, 147–9.

O'Donohue, D.D. (1991) *A study of the feeding, management and some skeletal problems of growing Thoroughbred horses in Ireland*. MVM thesis, Faculty of Veterinary Medicine, University College, Dublin.

O'Donohue, D.D., Smith, F.H. and Strickland, K.L. (1992) The incidence of abnormal limb development in the Irish Thoroughbred from birth to 18 months. *Equine Veterinary Journal*, **24**, 305–309.

Oftedal, O.T., Hintz, H.F. and Schryver, H.F. (1983) Lactation in the horse: milk composition and intake by foals. *Journal of Nutrition*, **113**, 2096–2106.

Oldham, S.L., Potter, G.D., Evans, J.W., Smith, S.B., Taylor, T.S. and Barnes, W. (1990) Storage and mobilization of muscle glycogen in exercising horses fed a fat-supplemented diet. *Journal of Equine Veterinary Science*, **10**, 353–9.

O'Moore, L.B. (1972) Nutritional factors in the rearing of the young Thoroughbred horses. *Equine Veterinary Journal*, **4**, 9–16.

Ordidge, R.M., Schubert, F.K. and Stoker, J.W. (1979) Death of horses after accidental feeding of monensin. *Veterinary Record*, **104**, 375.

Orr, J.A., Bisgard, G.E., Forster, H.V., Rawlings, C.A., Buss, D.D. and Will, J.A. (1975) Cardiopulmonary measurements in nonanesthetized, resting normal ponies. *American Journal of Veterinary Research*, **36**, 1667–70.

Orskov, E.R. and Hovell, F.D. de B. (1981) Principles and appropriate technology for improving the nutritive value of tropical feeds. *Proceedings of the 32nd Annual Meeting of the European Association of Animal Production*, Zagreb, Yugoslavia, 31 August–3 September, I–6.

Orton, R.G. (1978) Biochemical changes in horses during endurance rides. *Veterinary Record*, **102**, 469.

Orton, R.K., Hume, I.D. and Leng, R.A. (1985a) Effects of level of dietary protein and exercise on growth rates of horses. *Equine Veterinary Journal*, **17**, 381–5.

Orton, R.K., Hume, I.D. and Leng, R.A. (1985b) Effects of exercise and level of dietary protein on digestive function in horses. *Equine Veterinary Journal*, **17**, 386–90.

Osborn, T.G., Schmidt, S.P., Marple, D.N., Rahe, C.H. and Steenstra, J.R. (1992) Effect of consuming fungus-infected and fungus-free tall fescue and ergotamine tartrate on selected physiological variables of cattle in environmentally controlled conditions. *Journal of Animal Science*, **70**, 2501–2509.

Osborne, M. (1981) Rearing the orphan foal. *Proceedings of the 32nd Annual Meeting of the European Association of Animal Production*, Zagreb, Yugoslavia, 31 August–3 September, IIIa–1.

Østblom, L.C., Lund, C. and Melsen, F. (1982) Histological study of navicular bone disease. *Equine Veterinary Journal*, **14**, 199–202.

Østblom, L.C., Lund, C. and Melsen, F. (1984) Navicular bone disease: results of treatment using egg-bar shoeing technique. *Equine Veterinary Journal*, **16**, 203–206.

Osweiler, G.D., van Gelder, G.D. and Buck, G.A. (1978) Epidemiology of lead poisoning in animals. In: *Toxicity of Heavy Metals in the Environment* (ed. F.W. Oehme), Part 1, pp. 143–71. Marcel Dekker, New York.

Ott, E.A. (1981) Influence of level of feeding on digestive efficiency of the horse. *Proceedings of the 7th Equine Nutrition and Physiology Society*, Airlie House, Warrenton, Virginia, 30 April–2 May 1981, pp. 37–43.

Ott, E.A. and Asquith, R.L. (1989) The influence of mineral supplementation on growth and skeletal development of yearling horses. *Journal of Animal Science*, **67**, 2831–40.

Ott, E.A. and Asquith, R.L. (1994) Trace mineral supplementation of broodmares. *Journal of Equine Veterinary Science*, **14**, 93–101.

Ott, E.A. and Asquith, R.L. (1995) Trace mineral supplementation of yearling horses. *Journal of Animal Science*, **73**, 466–71.

Ott, E.A., Asquith, R.L. and Feaster, J.P. (1981) Lysine supplementation of diets for yearling horses. *Journal of Animal Science*, **53**, 1496–503.

Ott, E.A., Asquith, R.L., Feaster, J.P. and Martin, F.G. (1979a) Influence of protein level and quality on the growth and development of yearling foals. *Journal of Animal Science*, **49**, 620–28.

Ott, E.A., Feaster, J.P. and Lieb, S. (1979b) Acceptability and digestibility of dried citrus pulp by horses. *Journal of Animal Science*, **49**, 983–7.

Owen, J.M. (1975) Abnormal flexion of the corono-pedal joint or 'contracted tendons' in unweaned foals. *Equine Veterinary Journal*, **7**, 40–45.

Owen, J.M. (1977) Liver fluke infection in horses and ponies. *Equine Veterinary Journal*, **9**, 31.

Owen, J.M., McCullagh, K.G., Crook, D.H. and Hinton, M. (1978) Seasonal variations in the nutrition of horses at grass. *Equine Veterinary Journal*, **10**, 260–66.

Owen, R. ap R. (1985) Potato poisoning in a horse. *Veterinary Record*, **117**, 246.

Pagan, J.D. (1989) Calcium, hindgut function affect phophorus needs. *Feedstuffs*, **61**, No 35, 21 August, pp. 1–2.

Pagan, J.D., Essén-Gustavsson, B., Lindholm, A. and Thornton, J. (1987a) The effect of exercise and diet on muscle and liver glycogen repletion in standardbred horses. *Proceedings of the 10th Equine Nutrition and Physiology Society*, Colorado State University, Fort Collins, 11–13 June 1987, pp. 431–6.

Pagan, J.D., Essén-Gustavsson, B., Lindholm, A. and Thornton, J. (1987b) The effect of dietary energy source on exercise performance in standardbred horses. In: *Equine Exercise Physiology 2*, pp. 686–700. Granta Editions, Cambridge.

Pagan, J.D., Essén-Gustavsson, B., Lindholm, A. and Thornton, J. (1987c) The effect of dietary energy source on blood metabolites in standardbred horses during exercise. *Proceedings of the 10th Equine Nutrition and Physiology Society*, Colorado State University, Fort Collins, 11–13 June 1987, pp. 425–30.

Pagan, J.D. and Hintz, H.F. (1986a) Equine energetics. 1. Relationship between body weight and energy requirements in horses. *Journal of Animal Science*, **63**, 815–21.

Pagan, J.D. and Hintz, H.F. (1986b) Equine energetics. 11. Energy expenditure in horses during submaximal exercise. *Journal of Animal Science*, **63**, 822–30.

Pagan, J.D., Hintz, H.F. and Rounsaville, T.R. (1984) The digestible energy requirements of lactating pony mares. *Journal of Animal Science*, **58**, 1382–7.

Pagan, J.D., Jackson, S.G. and DeGregorio, R.M. (1993a) The effect of early weaning on growth and development in Thoroughbred foals. *Proceedings of the 13th Equine Nutrition and Physiology Society*, University of Florida, Gainesville, 21–23 January 1993, No. 504, 76–9.

Pagan, J.D., Tiegs, W., Jackson, S.G. and Murphy, H.Q. (1993b) The effect of different fat sources on exercise performance in Thoroughbred racehorses. *Proceedings of the 13th Equine Nutrition and Physiology Society*, University of Florida, Gainesville, 21–23 January 1993, No. 504, 125–9.

Palmer, E. and Driancourt, M.A. (1981) Consequences of foaling at different seasons and under different photoperiods. *Proceedings of the 32nd Annual Meeting of the European Association of Animal Production*, Zagreb, Yugoslavia, 31 August–3 September, I–1.

Palmer, J.L. and Bertone, A.L. (1994) Joint structure, biochemistry and biochemical disequilibrium in synovitis and equine joint disease. *Equine Veterinary Journal*, **26**, 263–77.

Parry, B.W. (1983) Survey of 79 referral colic cases. *Equine Veterinary Journal*, **15**, 345–8.

Parry, B.W., Anderson, G.A. and Gay, C.C. (1983) Prognosis in equine colic: a study of individual variables used in case assessment. *Equine Veterinary Journal*, **15**, 337–44.

Parsons, A.J., Johnson, I.R. and Harvey, A. (1988) Use of a model to optimize the interaction between frequency and severity of intermittent defoliation and to provide a fundamental conparison of the continuous and intermittent defoliation of grass. *Grass and Forage Science*, **43**, 49–59.

Pashen, R.L. and Allen, W.R. (1976) Genuine anoestrus in mares. *Veterinary Record*, **99**, 362–3.

Patience, J.F. (1990) A review of the role of acid–base balance in amino acid nutrition. *Journal of Animal Science*, **68**, 398–408.

Patterson, P.H., Coon, C.N. and Hughes, I.M. (1985) Protein requirements of mature working horses. *Journal of Animal Science*, **61**, 187–96.

Pearson, R.A., Cuddeford, D., Archibald, R.F. and Muirhead, R.H. (1992) Digestibility of diets containing different proportions of alfalfa and oat straw in thoroughbreds, Shetland ponies, highland ponies and donkeys. *First Europäische Konferenz über die Ernährung des Pferdes*, Institut für Tierernahrüng, Tierärzliche Hochschule, Hannover, 3–4 September 1992, pp. 153–7.

Pearson, R.A. and Merritt, J.B. (1991) Intake, digestion and gastrointestinal transit time in resting donkeys and ponies and exercised donkeys given *ad libitum* hay and straw diets. *Equine Veterinary Journal*, **23**, 339–43.

Pedersen, E.J.N. and Møller, E. (1976) Perennial ryegrass and clover in pure stand and in mixture. The influence of mixture, nitrogen fertilization and number of cuts on yield and quality. *Beretning fra Faellesudvalget for Statens Planteavls-og Husdyrbrugsforsøg, København*, No. 6, 1–27.

Peel, S., Mayne, C.S., Titchen, N.M. and Huckle, C.A. (1987) Beef production from grass/white clover swards. In: *Efficient Beef Production* (ed. J. Frame), pp. 97–104. British Grassland Society, Occasional Symposium, No. 22.

Peek, S.F., Divers, T.J. and Jackson, C.J. (1997) Hyperammonaemia associated with encephalopathy and abdominal pain without evidence of liver disease in four mature horses. *Equine Veterinary Journal*, **29**, 70–74.

Pérez, R., Valenzuela, S., Merino, V., *et al.* (1996) Energetic requirements and physiological adaptation of draught horses to ploughing work. *Animal Science*, **63**, 343–51.

Persson, S.G.B. (1983) Evaluation of exercise tolerance and fitness in the performance horse. In: *Equine Exercise Physiology* (eds D.H. Snow, S.G.B. Persson and R.J. Rose), pp. 441–57. Granta Editions, Cambridge.

Peterson, A.J., Bass, J.J. and Byford, M.J. (1978) Decreased plasma testosterone concentrations in rams affected by ryegrass staggers. *Research in Veterinary Science*, **25**, 266–8.

Pethick, D.W., Harman, N. and Chong, J.K. (1987) Non-esterified long-chain fatty acid metabolism in fed sheep at rest and during exercise. *Australian Journal of Biological Sciences*, **40**, 221–34.

Pethick, D.W., Rose, R.J., Bryden, W.L. and Gooden, J.M. (1993) Nutrition utilisation by the hindlimb of Thoroughbred horses at rest. *Equine Veterinary Journal*, **25**, 41–4.

Platt, D. and Bayliss, M.T. (1994) An investigation of the proteoglycan metabolism of mature equine articular cartilage and its regulation by interleukin – 1. *Equine Veterinary Journal*, **26**, 297–303.

Platt, H. (1978) Growth and maturity in the equine foetus. *Journal of the Royal Society of Medicine*, **71**, 658–61.

Platt, H. (1982) Sudden and unexpected deaths in horses: a review of 69 cases. *British Veterinary Journal*, **138**, 417–29.

Plummer, C., Knight, P.K., Ray, S.P. and Rose, R.J. (1991) Cardiorespiratory and metabolic effects of propranolol during maximal exercise. In: *Equine Exercise Physiology*, 3 (eds S.G.B. Persson, A. Lindholm and L. Jeffcott), pp. 465–74. ICEEP Publications, Davis, California

Podoll, K.L., Bernard, J.B., Ullrey, D.E., DeBar, S.R., Ku, P.K. and Magee, W.T. (1992) Dietary selenate versus selenite for cattle, sheep, and horses. *Journal of Animal Science*, **70**, 1965–70.

Poggenpoel, D.G. (1988) Measurements of heart rate and riding speed on a horse during a training programme for endurance rides. *Equine Veterinary Journal*, **20**, 224.

Pohlenz, J., Stockhofe-Zurwieden, N. and Rudat, R. (1992) Pathology and potential pathogenesis of typhlocolitis in horses. *First Europäische Konferenz über die Ernährung des Pferdes*, Institut für Tierernahrüng, Tierärzliche Hochschule, Hannover, 3–4 September 1992, pp. 201–206.

Pollitt, C.C. (1990) An autoradiographic study of equine hoof growth. *Equine Veterinary Journal*, **22**, 366–8.

Popplewell, J.C., Topliff, D.R., Freeman, D.W. and Breazile, J.E. (1993) Effects of dietary cation–anion balance on acid–base balance and blood parameters in anaerobically exercised horses. *Proceedings of the 13th Equine Nutrition and Physiology Society*, University of Florida, Gainesville, 21–23 January 1993, No. 504, 191–6.

Porter, J.K. (1995) Analysis of endophyte toxins: fescue and other grasses toxic to livestock. *Journal of Animal Science*, **73**, 871–80.

Porter, J.K. and Thompson, F.N. Jr (1992) Effects of fescue toxicosis on reproduction in livestock. *Journal of Animal Science*, **70**, 1594–603.

Potter, G.D., Arnold, F.F., Householder, D.D., Hansen, D.H. and Brown, K.M. (1992a) Digestion of starch in the small or large intestine of the equine. *First Europäische Konferenz über die Ernährung des Pferdes*, Institut für Tierernahrüng, Tierärzliche Hochschule, Hannover, 3–4 September 1992, pp. 107–111.

Potter, G.D., Evans, J.W., Webb, G.W. and Webb, S.P. (1987) Digestible energy requirements of Belgian and Percheron horses. *Proceedings of the 10th Equine Nutrition and Physiology Society*, Colorado State University, Fort Collins, 11–13 June 1987, pp. 133–8.

Potter, G.D., Gibbs, P.G., Haley, R.G. and Klendshoj, C. (1992c) Digestion of protein in the small and large intestines of equines fed mixed diets. *First Europäische Konferenz über die Ernährung des Pferdes*, Institut für Tierernahrüng, Tierärzliche Hochschule, Hannover, 3–4 September 1992, pp. 140–3.

Potter, G.D., Hughes, S.L., Jullen, T.R. and Swinney, D.L. (1992b) A review of research on digestion and utilization of fat by the equine. *First Europäische Konferenz über die Ernährung des Pferdes*, Institut für Tierernahrüng, Tierärzliche Hochschule, Hannover, 3–4 September 1992, pp. 119–23.

Potter, G.D., Webb, S.P., Evans, J.W. and Webb, G.W. (1990) Digestible energy requirements for work and maintenance of horses fed conventional and fat supplemented diets. *Journal of Equine Veterinary Science*, **10**, 214–18.

Prescott, J.F., Staempfli, H.R., Barker, I.K., Bettoni, R. and Delaney, K. (1988) A method for reproducing fatal idiopathic colitis (colotis X) in ponies and isolation of a clostridium as a possible agent. *Equine Veterinary Journal*, **20**, 417–20.

Price, J.S., Jackson, B., Eastell, R., *et al.* (1995) Age related changes in biochemical markers of bone metabolism in horses. *Equine Veterinary Journal*, **27**, 201–207.

Prinz, K. (1978) Effect of vitamin A–E emulsion on stallion semen. *Tierartz Umschau*, **33**, 27–30.

Prior, R.L., Hintz, H.F., Lowe, J.E. and Visek, W.J. (1974) Urea recycling and metabolism of ponies. *Journal of Animal Science*, **38**, 565–71.

Proudman, C.J. and Edwards, G.B. (1993) Are tapeworms associated with equine colic? A case control study. *Equine Veterinary Journal*, **25**, 224–6.

Pusztai, A., Clarke, E.M.W. and King, T.P. (1979) The nutritional toxicity of *Phaseolus vulgaris* lectins. *Proceedings of the Nutrition Society*, **38**, 115–20.

Putnam, M. (1973) Micronization – a new feed processing technique. *Flour Animal Feed Milling*, June, 40–41.

Quinn, P.J., Baker, K.P. and Morrow, A.N. (1983) Sweet itch: responses of clinically normal and affected horses to intradermal challenge with extracts of biting insects. *Equine Veterinary Journal*, **15**, 266–72.

Raisz, L.G. and Bingham, P.J. (1972) Effect of hormones on bone development. *Annual Review of Pharmacology*, **12**, 337–52.

Ralston, S.L. (1984) Controls of feeding in horses. *Journal of Animal Science*, **59**, 1354–61.

Ralston, S.L. (1988) Nutritional management of horses competing in 160 km races. *Cornell Veterinarian*, **78**, 53–61.

Ralston, S.L. (1992) Effect of soluble carbohydrate content of pelleted diets on postprandial glucose and insulin profiles in horses. *First Europäische Konferenz über die Ernährung des Pferdes*, Institut für Tierernahrüng, Tierärzliche Hochschule, Hannover, 3–4 September 1992, pp. 112–15.

Ralston, S.L. and Baile, C.A. (1981) Feeding behaviour of ponies after intragastric nutrient and intravenous glucose infusion. *Journal of Animal Science*, **53** (Suppl. 1), 131, abstract 12.

Ralston, S.L. and Baile, C.A. (1982a) Plasma glucose and insulin concentrations and feeding behavior in ponies. *Journal of Animal Science*, **54**, 1132–7.

Ralston, S.L. and Baile, C.A. (1982b) Gastrointestinal stimuli in the control of feed intake in ponies. *Journal of Animal Science*, **55**, 243–53.

Ralston, S.L. and Baile, C.A. (1983) Effects of intragastric loads of xylose, sodium chloride and corn oil on feeding behavior of ponies. *Journal of Animal Science*, **56**, 302–308.

Ralston, S.L. and Breuer, L.H. (1996) Field evaluation of a feed formulated for geriatric horses. *Journal of Equine Veterinary Science*, **16**, 334–8.

Ralston, S.L., Freeman, D.E. and Baile, C.A. (1983) Volatile fatty acids and the role of the large intestine in the control of feed intake in ponies. *Journal of Animal Science*, **57**, 815–25.

Ralston, S.L., Van den Broek, G. and Baile, C.A. (1979) Feed intake patterns and associated blood glucose, free fatty acid and insulin changes in ponies. *Journal of Animal Science*, **49**, 838–45.

Randall, R.P. and Pulse, R.E. (1974) Taste reactions in the immature horse. *Journal of Animal Science*, **38**, 1330, abstract 45.

Randall, R.P., Schurg, W.A. and Church, D.C. (1978) Response of horses to sweet, salty, sour and bitter solutions. *Journal of Animal Science*, **47**, 51–5.

Rasmussen, R.A., Cole, C.L. and Miller, M.S. (1944) Carotene, vitamin A and ascorbic acid in mare's milk. *Journal of Animal Science*, **3**, 346–52.

Raub, R.H., Jackson, S.G. and Baker, J.P. (1989) The effect of exercise on bone growth and development in weanling horses. *Journal of Animal Science*, **67**, 2508–14.

Reed, S.M. and Andrews, F.M. (1986) The biochemical evaluation of liver function in the horse. *Proceedings of the American Association of Equine Practitioners*, **32**, 81–93.

Reid, J.T. and Tyrrell, H.F. (1964) Effect of level of intake on energetic efficiency of animals. *Proceedings of the Cornell Nutrition Conference for Feed Manufacturers*, 25–38. Cornell University, Ithaca, New York.

Reid, R.L. and Horvath, D.J. (1980) Soil chemistry and mineral problems in farm livestock. A review. *Animal Feed Science and Technology*, **5**, 95–167.

Reitnour, C.M. (1978) Response to dietary nitrogen in ponies. *Equine Veterinary Journal*, **10**, 65–8.

Reitnour, C.M. (1979) Effect of cecal administration of corn starch on nitrogen metabolism in ponies. *Journal of Animal Science*, **49**, 988–91.

Reitnour, C.M. (1982) Protein utilization in response to caecal corn starch in ponies. *Equine Veterinary Journal*, **14**, 149–52.

Reitnour, C.M., Baker, J.P., Mitchell, G.E. Jr, Little, C.O. and Kratzer, D.D. (1970) Amino acids in equine cecal contents, cecal bacteria and serum. *Journal of Nutrition*, **100**, 349–54.

Reitnour, C.M. and Salsbury, R.L. (1972) Digestion and utilization of cecally infused protein by the equine. *Journal of Animal Science*, **35**, 1190–93.

Reitnour, C.M. and Salsbury, R.L. (1975) Effect of oral or caecal administration of protein supplements on equine plasma amino acids. *British Veterinary Journal*, **131**, 466–71.

Reitnour, C.M. and Salsbury, R.L. (1976) Utilization of proteins by the equine species. *American Journal of Veterinary Research*, **37**, 1065–7.

Rej, R., Rudofsky, U. and Magro, A. (1990) Effects of exercise on serum amino-transferase activity and pyridoxal phosphate saturation in Thoroughbred racehorses. *Equine Veterinary Journal*, **22**, 205–208.

Rerat, A. (1978) Digestion and absorption of carbohydrates and nitrogenous matters in the hindgut of the omnivorous nonruminant animal. *Journal of Animal Science*, **46**, 1808–837.

Revington, M. (1983a) Haematology of the racing Thoroughbred in Australia. 1. Reference values and the effect of excitement. *Equine Veterinary Journal*, **15**, 141–4.

Revington, M. (1983b) Haematology of the racing Thoroughbred in Australia. 2. Haematological values compared to performance. *Equine Veterinary Journal*, **15**, 145–8.

Reynolds, J.A., Potter, G.D., Odom, T.W., *et al.* (1993) Physiological responses to training in racing two-year old quarter horses fed sodium zeolite A. *Proceedings of the 13th Equine Nutrition and Physiology Society*, University of Florida, Gainesville, 21–23 January 1993, No. 504, 197–202.

Ribeiro, J.M.C.R., MacRae, J.C. and Webster, A.J.F. (1981) An attempt to explain differences in the nutritive value of spring and autumn harvested dried grass. *Proceedings of the Nutrition Society*, **40**, 12A.

Rice, L., Ott, E.A., Beede, D.K., *et al.* (1992) Use of oral tolerance tests to investigate disaccharide digestion in neonatal foals. *Journal of Animal Science*, **70**, 1175–81.

Ricketts, S.W. and Frape, D.L. (1990) Big bale silage as a horse feed. *Veterinary Record*, **118**, 55.

Ricketts, S.W., Greet, T.R.C., Glyn, P.J., *et al.* (1984) Thirteen cases of botulism in horses fed big bale silage. *Equine Veterinary Journal*, **16**, 515–18.

Roberts, M.C. (1974a) The D(+) xylose absorption test in the horse. *Equine Veterinary Journal*, **6**, 28–30.

Roberts, M.C. (1974b) Total serum cholesterol levels in the horse. *British Veterinary Journal*, **130**, xvi–xviii.

Roberts, M.C. (1974c) The development and distribution of alkaline phosphatase activity in the small intestine of the horse. *Research in Veterinary Science*, **16**, 110–11.

Roberts, M.C. (1974d) Amylase activity in the small intestine of the horse. *Research in Veterinary Science*, **17**, 400–401.

Roberts, M.C. (1975) Carbohydrate digestion and absorption studies in the horse. *Research in Veterinary Science*, **18**, 64–9.

Roberts, M.C. (1983) Serum and red cell folate and serum vitamin B_{12} levels in horses. *Australian Veterinary Journal*, **60**, 101–105.

Roberts, M.C. and Hill, F.W.G. (1973) The oral glucose tolerance test in the horse. *Equine Veterinary Journal*, **5**, 171–3.

Roberts, M.C., Hill, F.W.G. and Kidder, D.E. (1974) The development and distribution of small intestinal disaccharidases in the horse. *Research in Veterinary Science*, **17**, 42–8.

Roberts, M.C., Kidder, D.E. and Hill, F.W.G. (1973) Small intestinal beta-galactosidase activity in the horse. *Gut*, **14**, 535–40.

Roberts, M.C. and Norman, P. (1979) A re-evaluation of the D(+) xylose absorption test in the horse. *Equine Veterinary Journal*, **11**, 239–43.

Roberts, M.C. and Seawright, A.A. (1983) Experimental studies of drug-induced impaction colic in the horse. *Equine Veterinary Journal*, **15**, 222–8.

Robinson, D.W. and Slade, L.M. (1974) The current status of knowledge on the nutrition of equines. *Journal of Animal Science*, **39**, 1045–66.

Robinson, J.A., Allen, G.K., Green, E.M., Fales, W.H., Loch, W.E. and Wilkerson, C.G. (1993) A prospective study of septicaemia in colostrum-deprived foals. *Equine Veterinary Journal*, **25**, 214–19.

Rodd, J.G. (1979) *Exercise: a factor to be considered in the determination of the thiamin requirement*. MSc thesis, Cornell University, Ithaca, New York.

Rollinson, J., Taylor, F.G.R. and Chesney, J. (1987) Salinomycin poisoning in horses. *Veterinary Record*, **121**, 126–8.

Romić, S. (1974) Changes in some blood properties of horses during fattening. *Poljoprivredna Znanstvena Smotra*, **33**, 17–24.

Romić, S. (1978) Dietary and protective value of lucerne in the feeding of horses. *Poljoprivredna Znanstvena Smotra*, **45**, 5–17.

Ronéus, B.O., Hakkarainen, R.V.J., Lindholm, C.A. and Työppönen, J.T. (1986) Vitamin E requirements of adult standardbred horses evaluated by tissue depletion and repletion. *Equine Veterinary Journal*, **18**, 50–58.

Ronéus, M. and Lindholm, A. (1991) Muscle characteristics in Thoroughbreds of different ages and sexes. *Equine Veterinary Journal*, **23**, 207–210.

Rooney, J.R. (1968) Biomechanics of equine lameness. *Cornell Veterinarian*, **58**, 49–58.

Rose, R.J. (1979) Studies on some aspects of intravenous fluid infusion in the dog. PhD thesis, University of Sydney.

Rose, R.J. (1981) A physiological approach to fluid and electrolyte therapy in the horse. *Equine Veterinary Journal*, **13**, 7–14.

Rose, R.J., Allen, J.R., Hodgson, D.R. and Kohnke, J.R. (1983) Studies on isoxsuprine hydrochloride for the treatment of navicular disease. *Equine Veterinary Journal*, **15**, 238–43.

Rose, R.J., Arnold, K.S., Church, S. and Paris, R. (1980a) Plasma and sweat electrolyte concentrations in the horse during long distance exercise. *Equine Veterinary Journal*, **12**, 19–22.

Rose, R.J. and Hodgson, D.R. (1982) Haematological and plasma biochemical parameters in endurance horses during training. *Equine Veterinary Journal*, **14**, 144–8.

Rose, R.J., Ilkiw, J.E., Arnold, K.S., Backhouse, J.W. and Sampson, D. (1980) Plasma biochemistry in the horse during three-day event competition. *Equine Veterinary Journal*, **12**, 132–6.

Rose, R.J., Ilkiw, J.E. and Martin, I.C.A. (1979) Blood-gas, acid–base and haematological values in horses during an endurance ride. *Equine Veterinary Journal*, **11**, 56–9.

Rose, R.J., Ilkiw, J.E., Sampson, D. and Backhouse, J.W. (1980) Changes in blood-gas, acid–base and metabolic parameters in horses during three-day event competition. *Research in Veterinary Science*, **28**, 393–5.

Rose, R.J., Purdue, R.A. and Hensley, W. (1977) Plasma biochemistry alterations in horses during an endurance ride. *Equine Veterinary Journal*, **9**, 122–6.

Rose, R.J. and Sampson, D. (1982) Changes in certain metabolic parameters in horses associated with food deprivation and endurance exercise. *Research in Veterinary Science*, **32**, 198–202.

Ross, M.W., Lowe, J.E., Cooper, B.J., Reimers, T.J. and Froscher, B.A. (1983) Hypoglycemic seizures in a Shetland pony. *Cornell Veterinarian*, **73**, 151–69.

Rossdale, P.D. (1971) Experiences in the use of corticosteroids in horse practice. In: *The Application of Corticosteroids in Veterinary Medicine*. Symposium of the Royal Society of Medicine, London, pp. 29–31. Glaxo Laboratories, Greenford.

Rossdale, P.D. (1972) Modern concepts of neonatal disease in foals. *Equine Veterinary Journal*, **4**, 117–28.

Rossdale, R.D., Burguez, P.N. and Cash, R.S.G. (1982) Changes in blood neutrophil/lymphocyte ratio related to adrenocortical function in the horse. *Equine Veterinary Journal*, **14**, 293–8.

Rossdale, P.D. and Ricketts, S.W. (1980) *Equine Stud Farm Medicine*. Cassell (Baillière Tindall), London.

Roughan, P.G. and Slack, C.R. (1973) Simple methods for routine screening and quantitative estimation of oxalate content of tropical grasses. *Journal of the Science of Food and Agriculture*, **24**, 803–811.

Round, M.C. (1968a) The diagnosis of helminthiasis in horses. *Veterinary Record*, **82**, 39–43.

Round, M.C. (1968b) Experiences with thiabendazole as an anthelmintic for horses. *British Veterinary Journal*, **124**, 248–58.

Rowe, J.B., Lees, M.J. and Pethick, D.W. (1994) Prevention of acidosis and laminitis associated with grain feeding in horses. *American Institute of Nutrition. Journal of Nutrition*, **124**, 2742S–4S.

Roždestvenskaja, G.A. (1961) Growth of the bony tissue in skeleton of extremities in foals from birth to one year of age. *Trudy vses Inst Konevodstva*, **23**, 321–30.

Rudolph, W.G. and Corvalan, E.O. (1992) Urinary and serum gamma glutamyl transpeptidase in relation to urinary pH and proteinuria in healthy Thoroughbred horses in training. *Equine Veterinary Journal*, **24**, 316–17.

Rudra, M.M. (1946) Vitamin A in the horse. *Biochemistry Journal*, **40**, 500.

Russell, M.A., Rodiek, A.V. and Lawrence, L.M. (1986) Effect of meal schedules and fasting on selected plasma free amino acids in horses. *Journal of Animal Science*, **63**, 1428–31.

Saastamoinen, M.T. (1990) Factors affecting growth and development of foals and young horses. *Acta Agriculturae Scandinavica*, **40**, 387–96.

Saastamoinen, M.T. and Koskinen, E. (1993) Influence of quality of dietary protein supplement and anabolic steroids on muscular and skeletal growth of foals. *Animal Production*, **56**, 135–44.

Saastamoinen, M.T., Lähdekorpi, M. and Hyyppä, S. (1990) Copper and zinc levels in the diet of pregnant and lactating mares. *Proceedings of the 41st Annual Meeting of the European Association for Animal Production*, pp. 1–9.

Saba, N., Symons, A.M. and Drane, H.M. (1974) The effects of feeding white clover pellets and red clover hay on teat length, plasma gonadotrophins and pituitary function in wethers. *Journal of Agricultural Science, Cambridge*, **82**, 357–61.

Sainsbury, D.W.B. (1981) Ventilation and environment in relation to equine respiratory disease. *Equine Veterinary Journal*, **13**, 167–70.

St John, Sir Paulet (1780) *Every Man His Own Farrier*. J. Wilkes for S. Crowder, Winton.

Salimem, K. (1975) Cobalt metabolism in horse serum levels and biosynthesis of vitamin B_{12}. *Acta Physiologica Scandinavica*, **16**, 84–94.

Sallmann, H.P., Kienzlle, E., Fuhrmann, H., Grunwald, D., Eilmans, I. and Meyer, H. (1992) Einfluss einer marginalen Fettversorgung auf Fettverdaulichkeit, Lipidgehalt und- Zusammensetzung von Chymus, Gewebe und Blut. *First Europäische Konferenz über die Ernährung des Pferdes*, Institut für Tierernährung, Tierärztliche Hochschule, Hannover, 3–4 September 1992, pp. 124–7.

von Sandersleben, J. and Schlotke, B. (1977) Muscular dystrophy (white muscle disease) in foals, apparently a disease of increasing prevalence. *Deutsche Tierärztliche Wochenschrift*, **84**, 105–107.

Sandford, J. and Aitken, M.M. (1975) Effects of some drugs on the physiological changes during exercise in the horse. *Equine Veterinary Journal*, **7**, 198–202.

Sandgren, B. (1993) *Osteochondrosis in the tarsocrural joint and osteochondral fragments in the metacarpo/metatarsophalangeal joints in young standardbreds*. PhD thesis, Swedish University of Agricultural Sciences, Uppsala.

Santidrian, S. (1981) Intestinal absorption of D-glucose, D-galactose and L-leucine in male growing rats fed raw field bean (*Vicia faba*) diet. *Journal of Animal Science*, **53**, 414–19.

Sato, T., Oda, K. and Kubo, M. (1978) Hematological and biochemical values of Thoroughbred foals in the first six months of life. *Cornell Veterinarian*, **69**, 3–19.

Savage, C.J. (1991) *The influence of nutrition on skeletal growth and induction of osteochondrosis (dyschondroplasia) in horses*. PhD thesis, University of Melbourne, Australia.

Savage, C.J., McCarthy, R.N. and Jeffcott, L.B. (1993a) Effects of dietary energy and protein on induction of dyschondroplasia in foals. *Equine Veterinary Journal* (Suppl. 16), 74–9.

Savage, C.J., McCarthy, R.N. and Jeffcott, L.B. (1993b) Effects of dietary phosphorus and calcium on induction of dyschondroplasia in foals. *Equine Veterinary Journal* (Suppl. 16), 80–83.

Schell, T.C., Lindemann, M.D., Kornegay, E.T., Blodgett, D.J. and Doerr, J.A. (1993) Effectiveness of different types of clay for reducing the detrimental effects of aflatoxin-contaminated diets on performance and serum profiles of weanling pigs. *Journal of Animal Science*, **71**, 1226–31.

Schmidt, M., Lindemann, G. and Meyer, H. (1982) Intestinaler N-Umsatz beim Pferd. In: *Contributions to Digestive Physiology of the Horse. Advances in Animal Physiology and Animal Nutrition.* Supplement to *Journal of Animal Physiology and Animal Nutrition*, **13**, 40–51. Paul Parey, Berlin and Hamburg.

Schmidt, S.P., Hoveland, C.S., Clark, E.M., *et al.* (1983) Association of an endophytic fungus with fescue toxicity in steers fed Kentucky 31 tall fescue seed or hay. *Journal of Animal Science*, **55**, 1259–63.

Schoental, R. (1960) The chemical aspects of seneciosis. *Proceedings of the Royal Society of Medicine*, **53**, 284–8.

Scholefield, D. and Matkin, E.A. (1987) Amelioration of poached swards. In: *Grassland for the 90s. Publication and Environmental Research*, p. 9.3. British Grassland Society, Winter Meeting, Hurley, Maidenhead. *Proceedings of the 13th Equine Nutrition and Physiology Society*, University of Florida, Gainesville, 21–23 January 1993, No. 504, 18–23.

Scholte, H.R., Verduin, M.H.M., Ross, J.D., *et al.* (1991) Equine exertional rhabdomyolysis: activity of the mitochondrial respiratory chain and the carnite system in skeletal muscle. *Equine Veterinary Journal*, **23**, 142–4.

Schryver, H.F. (1975) Intestinal absorption of calcium and phosphorus by horses. *Journal of the South African Veterinary Association*, **46**, 39–45.

Schryver, H.F., Bartel, D.L., Langrana, N. and Lowe, J.E. (1978b) Locomotion in the horse: kinematics and external and internal forces in the normal equine digit in the walk and trot. *American Journal of Veterinary Research*, **39**, 1728–33.

Schryver, H.F., Craig, P.H., Hintz, H.F., Hogue, D.E. and Lowe, J.E. (1970b) The site of calcium absorption in the horse. *Journal of Nutrition*, **100**, 1127–31.

Schryver, H.F., Craig, P.H. and Hintz, H.F. (1970a) Calcium metabolism in ponies fed varying levels of calcium. *Journal of Nutrition*, **100**, 955–64.

Schryver, H.F. and Hintz, H.F. (1972) Calcium and phosphorus requirements of the horse: a review. *Feedstuffs*, **44**(28), 35–8.

Schryver, H.F., Hintz, H.F. and Craig, P.H. (1971a) Calcium metabolism in ponies fed a high phosphorus diet. *Journal of Nutrition*, **101**, 259–64.

Schryver, H.F., Hintz, H.F. and Lowe, J.E. (1971b) Calcium and phosphorus interrelationships in horse nutrition. *Equine Veterinary Journal*, **3**, 102–109.

Schryver, H.F., Hintz, H.F. and Lowe, J.E. (1974a) Calcium and phosphorus in the nutrition of the horse. *Cornell Veterinarian*, **64**, 493–515.

Schryver, H.F., Hintz, H.F. and Lowe, J.E. (1975) The effect of exercise on calcium metabolism in horses. *Proceedings of the Cornell Nutrition Conference for Feed Manufacturers* 99–101. Cornell University, Ithaca, New York.

Schryver, H.F., Hintz, H.F. and Lowe, J.E. (1978a) Calcium metabolism, body composition and sweat losses of exercised horses. *American Journal of Veterinary Research*, **39**, 245–8.

Schryver, H.F., Hintz, H.F., Lowe, J.E., Hintz, R.L., Harper, R.B. and Reid, J.T. (1974b) Mineral composition of the whole body, liver and bone of young horses. *Journal of Nutrition*, **104**, 126–32.

Schryver, H.F., Meakim, D.W., Lowe, J.E., Williams, J., Soderholm, L.V. and Hintz, H.F. (1987) Growth and calcium metabolism in horses fed varying levels of protein. *Equine Veterinary Journal*, **19**, 280–87.

Schryver, H.F., Oftedal, O.T., Williams, J., Soderholm, L.V. and Hintz, H.F. (1986) Lactation in the horse: the mineral composition of mare milk. *Journal of Nutrition*, **116**, 2142–7.

Schryver, H.F., Parker, M.T., Daniluk, P.D., *et al.* (1987) Salt consumption and the effect of salt on mineral metabolism in horses. *Cornell Veterinarian*, **77**, 122–31.

Schryver, H.F., Van Wie S., Daniluk, P. and Hintz, H.F. (1978c) The voluntary intake of calcium by horses and ponies fed a calcium deficient diet. *Journal of Equine Medicine and Surgery*, **2**, 337–40.

Schubert, R. (1990) Zusätzliche Gaben von Vitamin E verbessern die Rennleistung. *Vollblut, Zucht und Rennen*, **121**, 189–90. Dt. Sportverl. K. Stoof, Köln, Sommer.

Schubert, R. (1991) Nutrition of the performance horse. Influence of high vitamin E doses on performance of racehorse. Session II: horse production. *Proceedings of the 42nd Annual Meeting of the European Association for Animal Production*, 8–12 September, Berlin, p. 538.

Schuh, J.C.L., Ross, C. and Meschter, C. (1988) Concurrent mercuric blister and dimethyl sulphoxide (DMSO) application as a cause of mercury toxicity in two horses. *Equine Veterinary Journal*, **20**, 68–71.

Schulz, E. and Peterson, U. (1978) Evaluation of horse beans (*Vicia faba* L. *minor*), sweet lupins (*Lupinus lutens* L.) and solvent-extracted rapeseed oil meal. *Landwirtschrift Forschung*, **31**, 218–32.

Schurg, W.A., Frei, D.L., Cheeke, P.R. and Holtan, D.W. (1977) Utilization of whole corn plant pellets by horses and rabbits. *Journal of Animal Science*, **45**, 1317–21.

Schurg, W.A. and Pulse, R.E. (1974) Grass straw: an alternative roughage for horses. *Journal of Animal Science*, **38**, 1330, abstract 46.

Schwabenbauer, K., Meyer, H. and Lindemann, G. (1982) Gehalt an flüchtigen Fettsäuren und Ammoniak in Caecuminhalt des Pferdes in Abhängigkeit von Futterart, Futterreihenfolge und Fütterungszeitpunkt. In: *Contributions to Digestive Physiology of the Horse. Advances in Animal Physiology and Animal Nutrition.* Supplement to *Journal of Animal Physiology and Animal Nutrition*, **13**, 24–31. Paul Parey, Berlin and Hamburg.

Scott, B.D., Potter, G.D., Evans, J.W., Reagor, J.C., Webb, G.W. and Webb, S.P. (1987) Growth and feed utilization by yearling horses fed added dietary fat. *Proceedings of the 10th Equine Nutrition and Physiology Society*, Colorado State University, Fort Collins, 11–13 June 1987, pp. 101–106.

Scott, B.D., Potter, G.D., Greene, L.W., Hargis, P.S. and Anderson, J.G. (1992) Efficacy of a fat-supplemented diet on muscle glycogen concentrations in exercising thoroughbred horses maintained in varying body conditions. *Journal of Equine Veterinary Science*, **12**, 109–113.

Scott, B.D., Potter, G.D., Greene, L.W., Vogelsang, M.M. and Anderson, J.G. (1993) Efficacy of a fat-supplemented diet to reduce thermal stress in exercising Thoroughbred horses. *Proceedings of the 13th*

Equine Nutrition and Physiology Society, University of Florida, Gainesville, 21–23 January 1993, No. 504, 66–71.

Seawright, A.A., Groenendyk, S. and Silva, K.I.N.G. (1970) An outbreak of oxalate poisoning in cattle grazing *Setaria sphacelata. Australian Veterinary Journal*, **46**, 293–6.

Seeherman, H.J. and Morris, E.A. (1991) Comparison of yearling, two-year-old and adult Thoroughbreds using a standardised exercise test. *Equine Veterinary Journal*, **23**, 175–84.

Sellers, A.F. and Lowe, J.E. (1986) Review of large intestinal motility and mechanisms of impaction in the horse. *Equine Veterinary Journal*, **18**, 261–3.

Sewell, D.A. and Harris, R.C. (1991) Lactate and ammonia appearance in relation to exercise duration in the thoroughbred horse. *Journal of Physiology*, **434**, 43P.

Sewell, D.A. and Harris, R.C. (1992) Adenine nucleotide degradation in the thoroughbred horse with increasing exercise duration. *European Journal of Applied Physiology*, **65**, 271–7.

Sewell, D.A., Harris, R.C. and Dunnett, M. (1991a) Carnosine accounts for most of the variation in physico-chemical buffering in equine muscle. *Equine Exercise Physiology*, **3**, 276–80.

Sewell, D.A., Harris, R.C., Hanak, J. and Jahn, P. (1992a) Muscle adenine nucleotide degradation in the thoroughbred horse as a consequence of racing. *Comparative Biochemistry and Physiology*, **101B**, 375–81.

Sewell, D.A., Harris, R.C., Marlin, D.J. and Dunnett, M. (1991b) Muscle fibre characteristics and carnosine content of race-trained thoroughbred horses. *Journal of Physiology*, **435**, 79P.

Sewell, D.A., Harris, R.C., Marlin, D.J. and Dunnett, M. (1992b) Estimation of the carnosine content of different fibre types in the middle gluteal muscle of the thoroughbred horse. *Journal of Physiology*, **455**, 447–53.

Sheldrick, R.D., Lavender, R.H. and Martyn, T.M. (1991) A comparison of methods to rejuvenate grass–clover swards – strip-seeding or sward management? In: *Grazing September Grassland Renovation and Weed Control, Europe, European Grassland Federation Symposium*, Graz, 18–21 September 1991, pp. 19–21.

Sheldrick, R.D., Lavender, R.H., Martyn, T.M. and Deschard, G. (1990) *Rates and frequencies of superphosphate fertiliser application for grass–clover swards*. Session 1: Poster 2, pp. 1–2. British Grassland Society, Research Meeting No. 2, Scottish Agricultural College, Ayr.

Shupe, J.L., Eanes, E.D. and Leone, N.C. (1981) Effect of excessive exposure to sodium fluoride on composition and crystallinity of equine bone tumors. *American Journal of Veterinary Research*, **42**, 1040–42.

Siciliano, P.D. and Wood, C.H. (1993) The effect of added dietary soybean oil on vitamin E status of the horse. *Journal of Animal Science*, **71**, 3399–402.

Silva, C.A.M., Merkt, H., Bergamo, P.N.L., *et al.* (1987) Consequence of excess iodine supply in a Thoroughbred stud in southern Brazil. *Journal of Reproduction Fertilization* (Suppl. 35), 529–33.

Sissons, S. and Grossman, J.D. (1961) *The Anatomy of the Domestic Animals*. W.B. Saunders, Philadelphia and London.

Skarda, R.T., Muir, W.W., Milne, D.W. and Gabel, A.A. (1976) Effects of training on resting and postexercise ECG in standardbred horses, using a standardized exercise test. *American Journal of Veterinary Research*, **37**, 1485–8.

Sklan, D. and Donoghue, S. (1982) Serum and intracellular retinol transport in the equine. *British Journal of Nutrition*, **47**, 273–80.

Slade, L.M. (1987) Effects of feeds on racing performance of quarter horses. *Proceedings of the 10th Equine Nutrition and Physiology Society*, Colorado State University, Fort Collins, 11–13 June 1987, pp. 585–91.

Slade, L.M., Bishop, R., Morris, J.G. and Robinson, D.W. (1971) Digestion and the absorption of ^{15}N-labelled microbial protein in the large intestine of the horse. *British Veterinary Journal*, **127**, xi–xiii.

Slade, L.M., Lewis, L.D., Quinn, C.R. and Chandler, M.L. (1975) Nutritional adaptations of horses for endurance performance. *Proceedings of the Equine Nutrition and Physiology Society*, 114–28.

Slade, L.M., Robinson, D.W. and Casey, K.E. (1970) Nitrogen metabolism in nonruminant herbivores. I. The influence of nonprotein nitrogen and protein quality on the nitrogen retention of adult mares. *Journal of Animal Science*, **30**, 753–60.

Slagsvold, P., Hintz, H.F. and Schryver, H.F. (1979) Digestibility by ponies of oat straw treated with anhydrous ammonia. *Animal Production*, **28**, 347–52.

Slater, M.R. and Hood, D.M. (1997) A cross-sectional epidemiological study of equine hoof wall problems and associated factors. *Equine Veterinary Journal*, **29**, 67–9.

Slater, M.R., Hood, D.M. and Carter, G.K. (1995) Descriptive epidemiological study of equine laminitis. *Equine Veterinary Journal*, **27**, 364–7.

Slocombe, R.F., Huntington, P.J., Friend, S.C.E., Jeffcott, L.B., Luff, A.R. and Finkelstein, D.K. (1992) Pathological aspects of Australian stringhalt. *Equine Veterinary Journal*, **24**, 174–83.

Smith, A., Allcock, P.J., Cooper, E.M. and Forbes, T.J. (1982) *Permanent Grassland Studies. 4. An Investigation Into the Influence of Sward Composition and Environment on Stocking Rate, Using Census and Survey Data*, pp. 1–51. The GRI–ADAS Joint Permanent Pasture Group. Grassland Research Institute, Hurley.

Smith, B.L. and O'Hara, P.J. (1978) Bovine photosensitization in New Zealand. *New Zealand Veterinary Journal*, **26**, 2–5.

Smith, B.S.W. and Wright, H. (1984) 25-Hydroxyvitamin D concentrations in equine serum. *Veterinary Record*, **115**, 579.

Smith, J.D., Jordan, R.M. and Nelson, M.L. (1975) Tolerance of ponies to high levels of dietary copper. *Journal of Animal Science*, **41**, 1645–9.

Smith, J.E., Cipriano, J.E., DeBowes, R. and Moore, K. (1986) Iron deficiency and pseudo-iron deficiency in hospitalized horses. *Journal of American Veterinary Medical Association*, **188**, 285–7.

Smith, J.E., Erickson, H.H., DeBowes, R.M. and Clark, M. (1989) Changes in circulating equine erythrocytes induced by brief, high-speed exercise. *Equine Veterinary Journal*, **21**, 444–6.

Smith, J.E., Moore, K., Cipriano, J.E. and Morris, P.G. (1984) Serum ferritin as a measure of stored iron in horses. *Journal of Nutrition*, **114**, 677–81.

Smith, J.F., Jagusch, K.T., Brumswick, L.F.C. and Kelly, R.W. (1979) Coumestans in lucerne and ovulation in ewes. *New Zealand Journal of Agricultural Research*, **22**, 411–16.

Smith, T.K., James, L.J. and Carson, M.S. (1980) Nutritional implications of *Fusarium* mycotoxins. *Proceedings of the Cornell Nutrition Conference for Feed Manufacturers*, 35–42. Cornell University, Ithaca, New York.

Snow, D.H. (1977) Identification of the receptor involved in adrenaline mediated sweating in the horse. *Research in Veterinary Science*, **23**, 246–7.

Snow, D.H. (1985) The horse and dog, elite athletes – why and how? *Proceedings of the Nutrition Society*, **44**, 267–72.

Snow, D.H. (1987) Assessment of fitness in the horse. *In Practice*, **9**, January, 26–30.

Snow, D.H. (1994) Ergogenic aids to performance in the race horse: nutrients or drugs. *Journal of Nutrition*, **124**, 2730S–5S.

Snow, D.H., Baxter, P.B. and Rose, R.J. (1981) Muscle fibre composition and glycogen depletion in horses competing in an endurance ride. *Veterinary Record*, **108**, 374–8.

Snow, D.H. and Frigg, M. (1987a) Oral administration of different formulation of ascorbic acid to the horse. *Proceedings of the 10th Equine Nutrition and Physiology Society*, Colorado State University, Fort Collins, 11–13 June 1987, pp. 617–19.

Snow, D.H. and Frigg, M. (1987b) Plasma concentration at monthly intervals of ascorbic acid, retinol, β-carotene and α-tocopherol in two thoroughbred racing stables and the effects of supplementation. *Proceedings of the 10th Equine Nutrition and Physiology Society*, Colorado State University, Fort Collins, 11–13 June 1987, pp. 55–60.

Snow, D.H., Gash, S.P. and Cornelius, J. (1987) Oral administration of ascorbic acid to horses. *Equine Veterinary Journal*, **19**, 520–23.

Snow, D.H. and Guy, P.S. (1980) Muscle fibre type composition of a number of limb muscles in different types of horse. *Research in Veterinary Science*, **28**, 137–44.

Snow, D.H. and Harris, R.C. (1989) The use of conventional and unconventional supplements in the thoroughbred horse. *Proceedings of the Nutrition Society*, **48**, 135–9.

Snow, D.H., Harris, R.C., Macdonald, I.A., Forster, C.D. and Marlin, D.J. (1992) Effects of high-intensity exercise on plasma catecholamines in the Thoroughbred horse. *Equine Veterinary Journal*, **24**, 462–7.

Snow, D.H., Kerr, M.G., Nimmo, M.A. and Abbott, E.M. (1982) Alterations in blood, sweat, urine and muscle composition during prolonged exercise in the horse. *Veterinary Record*, **110**, 377–84.

Snow, D.H. and Mackenzie, G. (1977a) Some metabolic effects of maximal exercise in the horse and adaptations with training. *Equine Veterinary Journal*, **9**(3), 134–40.

Snow, D.H. and MacKenzie, G. (1977b) Effect of training on some metabolic changes associated with submaximal endurance exercise in the horse. *Equine Veterinary Journal*, **9**, 226–30.

Snow, D.H., Persson, S.G.B. and Rose, R.J. (eds) (1983) *Equine Exercise Physiology*. Granta Editions, Cambridge.

Snow, D.H., Ricketts, S.W. and Mason, D.K. (1983) Haematological response to racing and training exercise in Thoroughbred horses, with particular reference to the leucocyte response. *Equine Veterinary Journal*, **15**, 149–54.

Snow, D.H. and Rose, R.J. (1981) Hormonal changes associated with long-distance exercise. *Equine Veterinary Journal*, **13**, 195–7.

Snow, D.H. and Summers, R.J. (1977) The actons of the β-adrenoceptor blocking agents propranolol and metoprolol in the maximally exercised horse. *Journal of Physiology*, **271**, 39–40P.

Sobel, A.E. (1955) Composition of bones and teeth in relation to blood and diet. *Voeding*, **16**, 567–75.

Soldevila, M. and Irizarry, R. (1977) Complete diets for horses. *Journal of the Agricultural University of Puerto Rico*, **61**, 413–15.

Solleysel, S. de (1771) *The Compleat Horseman: or, Perfect Farrier. In Two Parts*. R. Bonwicke and others, London.

Sommer, H. and Felbinger, U. (1983) The influence of racing on selected serum enzymes, electrolytes and other constituents in Thoroughbred horses. In: *Equine Exercise Physiology* (eds D.H. Snow, S.G.B. Persson and R.J. Rose), pp. 362–5. Granta Editions, Cambridge.

Spais, A.G., Papasteriadis, A., Roubiës, N., Agiannidis, A., Yantzis, N. and Argyroudis, S. (1977) Studies on iron, manganese, zinc, copper and selenium retention and interaction in horses. *Proceedings of the 3rd International Symposium on Trace Element Metabolism in Man and Animals*, Greece, pp. 501–5. Arbeitskreis für Tierernährungsforschung Weihenstephan, Freising-Weihenstephan, Germany.

Spencer, H., Kramer, L. and Osis, D. (1988) Do protein and phosphorus cause calcium loss? *Journal of Nutrition*, **118**, 657–60.

Spiers, S., May, S.A., Bennett, D. and Edwards, G.B. (1994) Cellular sources of proteolytic enzymes in equine joints. *Equine Veterinary Journal*, **26**, 43–7.

Sprouse, R.F., Garner, H.E. and Green, E.M. (1987) Plasma endotoxin levels in horses subjected to carbohydrate induced laminitis. *Equine Veterinary Journal*, **19**, 25–8.

Srivastava, V.K. and Hill, D.C. (1976) Effect of mild heat treatment on the nutritive value of low glucosinolate–low erucic acid rapeseed meals. *Journal of the Science of Food and Agriculture*, **27**, 953–8.

Stadermann, B., Nehring, T. and Meyer, H. (1992) Ca and Mg absorption with roughage or mixed feed. *First Europäische Konferenz über die Ernährung des Pferdes*, Institut für Tierarnahrüng, Tierärzliche Hochschule, Hannover, pp. 77–80.

Staley, T.E., Jones, E.W., Corley, L.D. and Anderson, I.L. (1970) Intestinal permeability to *Escherichia coli* in the foal. *American Journal of Veterinary Research*, **31**, 1481–3.

Stanek, Ch. (1981) Conservative therapy for correction of poor limb conformation in foals. *Proceedings of the 32nd Annual Meeting of the European Association of Animal Production*, Zagreb, Yugoslavia, 31 August–3 September, IIIa–4.

Steiss, J.E., Traber, M.G., Williams, M.A., Kayden, H.J. and Wright, J.C. (1994) Alpha tocopherol concentrations in clinically normal adult horses. *Equine Veterinary Journal*, **26**, 417–19.

Stick, J.A., Robinson, N.E. and Krehbiel, J.D. (1981) Acid–base and electrolyte alterations associated with salivary loss in the pony. *American Journal of Veterinary Research*, **42**, 733–7.

Sticker, L.S., Thompson, D.L., Jr, Bunting, L.D., Fernandez, J.M. and DePew, C.L. (1995) Dietary protein and (or) energy restriction in mares: plasma glucose, insulin, nonesterified fatty acid, and urea nitrogen responses to feeding, glucose, and epinephrine. *Journal of Animal Science*, **73**, 136–44.

Sticker, L.S., Thompson, D.L., Jr, Bunting, L.D. and Fernandez, J.M. (1996) Dietary protein and energy restriction in mares: rapid changes in plasma metabolite and hormone concentrations during dietary alteration. *Journal of Animal Science*, **74**, 1326–35.

Sticker, L.S., Thompson, D.L., Jr, Fernandez, J.M., Bunting, L.D. and DePew, C.L. (1985) Dietary protein and (or) energy restriction in mares: plasma growth hormone, IGF-I, prolactin, cortisol, and thyroid hormone responses to feeding, glucose, and epinephrine. *Journal of Animal Science*, **73**, 1424–32.

Stoker, J.W. (1975) Monensin sodium in horses. *Veterinary Record*, **97**, 137–8.

Stowe, H.D. (1968a) Effects of age and impending parturition upon serum copper of thoroughbred mares. *Journal of Nutrition*, **95**, 179–84.

Stowe, H.D. (1968b) Alpha–tocopherol requirements for equine erythrocyte stability. *American Journal of Clinical Nutrition*, **21**, 135–42.

Stowe, H.D. (1982) Vitamin A profiles of equine serum and milk. *Journal of Animal Science*, **54**, 76–81.

Strickland, K., Smith, F., Woods, M. and Mason, J. (1987) Dietary molybdenum as a putative copper antagonist in the horse. *Equine Veterinary Journal*, **19**, 50–54.

Stubley, D., Campbell, C., Dant, C. and Blackmore, D.J. (1983) Copper and zinc levels in the blood of Thoroughbreds in training in the United Kingdom. *Equine Veterinary Journal*, **15**, 253–6.

Stull, C.L. and Rodiek, A.V. (1988) Responses of blood glucose, insulin and cortisol concentrations to common equine diets. *Journal of Nutrition*, **118**, 206–213.

Stull, C.L., Rodiek, A.V. and Arana, M.J. (1987) The effects of common equine feeds on blood levels of glucose, insulin, and cortisol. *Proceedings of the 10th Equine Nutrition and Physiology Society*, Colorado State University, Fort Collins, 11–13 June 1987, pp. 61–6.

Stutz, W.A., Topliff, D.R., Freeman, D.W., Tucker, W.B., Breazile, J.W. and Wall, D.L. (1992) Effect of dietary cation–anion balance on blood parameters in exercising horses. *Journal of Equine Veterinary Science*, **12**, 164–7.

Sufit, E., Houpt, K.A. and Sweeting, M. (1985) Physiological stimuli of thirst and drinking patterns in ponies. *Equine Veterinary Journal*, **17**, 12–16.

Suttle, N.F. (1983) The nutritional basis for trace element deficiencies in ruminant livestock. In: *Trace Elements in Animal Production and Veterinary Practice* (eds N.F. Suttle, R.G. Gunn, W.M. Allen, K.A. Linklater and G. Wiener, pp. 19–25. British Society of Animal Production, Occasional Publication No. 7, Edinburgh.

Suttle, N.F., Gunn, R.G., Allen, W.M., Linklater, K.A. and Wiener, G. (eds) (1983) *Trace Elements in Animal Production and Veterinary Practice*. British Society of Animal Production, Occasional Publication No. 7, Edinburgh.

Suttle, N.F., Small, J.N.W., Collins, E.A., Mason, D.K. and Watkins, K.L. (1996) Serum and hepatic copper concentrations used to define normal, marginal and deficient copper status in horses. *Equine Veterinary Journal*, **28**, 497–9.

Sutton, E.I., Bowland, J.P. and McCarthy, J.F. (1977) Studies with horses comparing 4 N-HCl insoluble ash as an index material with total fecal collection in the determination of apparent digestibilities. *Canadian Journal of Animal Science*, **57**, 543–9.

Sutton, E.I., Bowland, J.P. and Ratcliff, W.D. (1977) Influence of level of energy and nutrient intake by mares on reproductive performance and on blood serum composition of the mares and foals. *Canadian Journal of Animal Science*, **57**, 551–8.

Swartzman, J.A., Hintz, H.F. and Schryver, H.F. (1978) Inhibition of calcium absorption in ponies fed diets containing oxalic acid. *American Journal of Veterinary Research*, **39**, 1621–3.

Sweeting, M.P., Houpt, C.E. and Houpt, K.A. (1985) Social facilitation of feeding and time budgets in stabled ponies. *Journal of Animal Science*, **60**, 369–74.

Tallowin, J.R.B. and Brookman, S.K.E. (1988) Herbage potassium levels in a permanent pasture under grazing. *Grass and Forage Science*, **43**, 209–12.

Talukdar, A.H., Calhoun, M.L. and Stinson, A.W. (1970) Sensory end organs in the upper lip of the horse. *American Journal of Veterinary Research*, **31**, 1751–4.

Tasker, J.B. (1967) Fluid and electrolyte studies in the horse. III: intake and output of water, sodium and potassium in normal horses. *Cornell Veterinarian*, **57**, 649–57.

Taylor, L.E., Ferrante, P.L., Kronfeld, D.S. and Meacham, T.N. (1995) Acid–base variables during incremental exercise in sprint-trained horses fed a high-fat diet. *Journal of Animal Science*, **73**, 2009–2018.

Taylor, L.E., Ferrante, P.L., Meacham, T.N., Kronfeld, D.S. and Tiegs, W. (1993) Acid–base responses to exercise in horses trained on a diet containing added fat. *Proceedings of the 13th Equine Nutrition and Physiology Society*, University of Florida, Gainesville, 21–23 January 1993, No. 504, 185–90.

Taylor, M.C., Loch, W.E., Heimann, E.D. and Morris, J.S. (1981) Effect of nitrogen fertilization and selenium supplementation on the hair selenium concentration in pregnant pony mares grazing fescue. *Journal of Animal Science*, **53** (Suppl. 1), 266, abstract 350.

Teeter, S.M., Stillions, M.C. and Nelson, W.E. (1967) Maintenance levels of calcium and phosphorus in horses. *Journal – American Veterinary Medical Association*, **151**, 1625–8.

Thijssen, H.H.W., Van der Bogaard, A.E.J.M., Wetzel, J.M., Maes, J.H.J. and Muller, A.P. (1983) Warfarin pharmacokinetics in the horse. *American Journal of Veterinary Research*, **44**, 1192–6.

Thomas, B., Thompson, A., Oyenuga, V.A. and Armstrong, R.H. (1952) The ash constituents of some herbage plants at different stages of maturity. *Empire Journal of Experimental Agriculture*, **20**, 10–13.

Thomas, P.T. (1963) Breeding herbage plants for animal production and well-being. *Proc Br Vet Ass A Congr*, 1–4.

Thompson, J.P., Casey, P.B. and Vale, J.A. (1995) Pesticide incidents reported to the Health and Safety Executive 1989/90–1991/92. *Human and Experimental Toxicology*, **14**, 630–603.

Thompson, K.N. (1995) Skeletal growth rates of weanling and yearling Thoroughbred horses. *Journal of Animal Science*, **73**, 2513–17.

Thompson, K.N., Jackson, S.G. and Baker, J.P. (1988) The influence of high planes of nutrition on skeletal growth and development of weanling horses. *Journal of Animal Science*, **66**, 2459–67.

Thomson, J.R. and McPherson, E.A. (1981) Prophylactic effects of sodium cromoglycate on chronic obstructive pulmonary disease in the horse. *Equine Veterinary Journal*, **13**, 243–6.

Thomson, J.R. and McPherson, E.A. (1983) Chronic obstructive pulmonary disease in the horse: therapy. *Equine Veterinary Journal*, **15**, 207–210.

Thomson, J.R. and McPherson, E.A. (1984) Effects of environmental control on pulmonary function of horses affected with chronic obstructive pulmonary disease. *Equine Veterinary Journal*, **16**, 35–8.

Thorén-Tolling, K. (1988) Serum alkaline phosphatase isoenzymes in the horse – variation with age, training and in different pathological conditions. *Journal of Veterinary Medical Association*, **35**, 13–23.

Thornton, I. (1983) Soil–plant–animal interactions in relation to the incidence of trace element disorders in grazing livestock. In: *Trace Elements in Animal Production and Veterinary Practice* (eds N.F. Suttle, R.G. Gunn, W.M. Allen, K.A. Linklater and G. Wiener), pp. 39–49. British Society of Animal Production, Occasional Publication No. 7, Edinburgh.

Thornton, J.R. and Lohni, M.D. (1979) Tissue and plasma activity of lactic dehydrogenase and creatine kinase in the horse. *Equine Veterinary Journal*, **11**, 235–8.

Tisserand, J.L., Candau, M., Houiste, A. and Masson, C. (1977) Evolution de quelques paramètres physico-chimiques du contenu caecal d'un poney au cours du nycthémère. *Annals of Zootechnology*, **26**, 429–34.

Tisserand, J.-L. and Martin-Rosset, W. (1996) Evaluation of nitrogen value of feeds in the horse in the MADC system. *47th European Association of Animal Production Meeting*, Lillehammer, Norway, 25–29th August. Horse Commission Session-H4: Nutrition, pp. 1–14.

Tisserand, J.L., Masson, C., Ottin-Pecchio, M. and Creusot, A. (1977) Mesure du pH et la concentration en AGV dans le caecum et le colon du poney. *Ann. Biol. anim. Bioch. Biophys.*, **17**, 533–7.

Tobin, T. and Combie, J. (1984) Some reflections on positive results from medication control tests in the USA. *Equine Veterinary Journal*, **16**, 43–6.

Todhunter, R.J., Erb, H.N. and Roth, L. (1986) Gastric rupture in horses: a review of 54 cases. *Equine Veterinary Journal*, **18**, 288–93.

Topliff, D.R., Lee, S.F. and Freeman, D.W. (1987) Muscle glycogen, plasma glucose and free fatty acids in exercising horses fed varying levels of starch. *Proceedings of the 10th Equine Nutrition and Physiology Society*, Colorado State University, Fort Collins, 11–13 June 1987, pp. 421–4.

Topliff, D.R., Potter, G.D., Kreider, J.L. and Cregan, C.R. (1981) Thiamin supplementation for exercising horses. *Proceedings of the 7th Equine Nutrition and Physiology Society*, Airlie House, Warrenton, Virginia, 30 April–2 May 1981, pp. 167–72.

Topliff, D.R., Potter, G.D., Krieder, J.L., Dutson, T.R. and Jessup, G.T. (1985) Diet manipulation, muscle glycogen metabolism and anaerobic work performance in the equine. *Proceedings of the 9th Equine Nutrition and Physiology Society*, Airlie House, Warrenton, Virginia, 30 April–2 May 1981, pp. 167–72.

Torún, B., Scrimshaw, N.S. and Young, V.R. (1977) Effect of isometric exercises on body potassium and dietary protein requirements of young men. *American Journal of Clinical Nutrition*, **30**, 1983–93.

Townley, P., Baker, K.P. and Quinn, P.J. (1984) Preferential landing and engorging sites of *Culicoides* species landing on a horse in Ireland. *Equine Veterinary Journal*, **16**, 117–20.

Townson, J. (1992) *A survey and assessment of racehorse stables in Ireland*. MEqS thesis, Faculties of Agriculture and Veterinary Medicine, National University of Ireland, Dublin.

Treacher, T.T., Orr, R.J. and Parsons, A.J. (1986) Direct measurement of the seasonal pattern of production on continuously stocked swards (ed. J. Frame), pp. 204–5. *Proceedings of Conference of British Grassland Society*, Malvern, *Occasional Symposium*, No. 19.

Tyler, C.M., Hodgson, D.R. and Rose, R.J. (1996) Effect of a warm-up on energy supply during high intensity exercise in horses. *Equine Veterinary Journal*, **28**, 117–20.

Tyznik, W.K. (1968) Nutrition. In: *Care and Training of the Trotter and Pacer* (ed. J.C. Harrison), USTA, Columbus, Ohio.

Tyznik, W.K. (1975) Recent advances in horse nutrition. *Proc 35th Semi-annual meeting of the AFMA Nutrition Council*, Kansas City, Missouri, 12–13 November 1975, 32–8. American Feed Manufacturers Association, Arlington, Virginia.

Ullrey, D.E., Ely, W.T. and Covert, R.L. (1974) Iron, zinc and copper in mare's milk. *Journal of Animal Science*, **38**, 1276–7.

Ullrey, D.E., Struthers, R.D., Hendricks, D.G. and Brent, B.E. (1966) Composition of mare's milk. *Journal of Animal Science*, **25**, 217–22.

Underwood, E.J. (1977) *Trace Elements in Human and Animal Nutrition*, 14th edn. Academic Press, New York and London.

United Kingdom Agricultural Supply Trade Association (1984) *Code of Practice for Cross-contamination in Animal Feeding Stuffs Manufacture*. Amended Code, June 1984. UKASTA, London.

Urch, D.L. and Allen, W.R. (1980) Studies on fenbendazole for treating lung and intestinal parasites in horses and donkeys. *Equine Veterinary Journal*, **12**, 74–7.

Valberg, S. (1986) Glycogen depletion patterns in the muscle of standard trotters after exercise of varying intensities and durations. *Equine Veterinary Journal*, **18**, 479–84.

Valberg, S., Essén-Gustavsson, B., Lindholm, A. and Persson, S.G.B. (1985) Energy metabolism in relation to skeletal muscle fibre properties during treadmill exercise. *Equine Veterinary Journal*, **17**, 439–43.

Valberg, S., Essén-Gustavsson, B., Lindholm, A. and Persson, S.G.B. (1989) Blood chemistry and skeletal muscle responses during and after different speeds and durations of trotting. *Equine Veterinary Journal*, **21**, 91–5.

Valberg, S., Haggendal, J. and Lindholm, A. (1993) Blood chemistry and skeletal muscle metabolic responses to exercise in horses with recurrent exertional rhabdomyolysis. *Equine Veterinary Journal*, **25**, 17–22.

Valberg, S., Jönsson, L., Lindholm, A. and Holmgren, N. (1993) Muscle histopathology and plasma aspartate amino transferase, creatine kinase and myoglobin changes with exercise in horses with recurrent exertional rhabdomyolysis. *Equine Veterinary Journal*, **25**, 11–16.

Van Dam, B. (1978) Vitamins and sport. *British Journal of Sports Medicine*, **12**, 74–9.

Veira, D.M. (1986) The role of ciliate protozoa in nutrition of the ruminant. *Journal of Animal Science*, **63**, 1547–60.

Verberne, L.R.M. and Mirck, M.H. (1976) A practical health programme for prevention of parasitic and infectious diseases in horses and ponies. *Equine Veterinary Journal*, **8**, 123–5.

Vermorel, M. and Martin-Rosset, W. (1997) Concepts, scientific bases, sructures and validation of the French horse net energy system (UFC). *Livestock Production Science*, **47**, 261–75.

Vermorel, M., Martin-Rosset, W. and Vernet, J. (1997) Energy utilization of twelve forages or mixed diets for maintenance by sport horses. *Livestock Production Science*, **47**, 157–67.

Vernet, J., Vermorel, M. and Martin-Rosset, W. (1995) Energy cost of eating long hay, straw and pelleted food in sport horses. *Animal Science*, **61**, 581–8.

Vogel, C. (1984) Navicular disease and equine insurance. *Veterinary Record*, **115**, 89.

Waite, R. and Sastry, K.N.S. (1949) The composition of timothy (*Phleum pratense*) and some other grasses during seasonal growth. *Emp J exp Agric*, **17**, 179–82.

Walker, D. and Knight, D. (1972) The anthelmintic activity of mebendazole: a field trial in horses. *Veterinary Record*, **90**, 58–65.

Wall, D.L., Topliff, D.R., Freeman, D.W., Breazile, J.E., Wagner, D.G. and Stutz, W.A. (1993) The effect of dietary cation–anion balance on mineral balance in the anaerobically exercised horse. *Proceedings of the 13th Equine Nutrition and Physiology Society*, University of Florida, Gainesville, 21–23 January 1993, No. 504, 50–53.

Wallace, W.M. and Hastings, A.B. (1942) The distribution of the bicarbonate ion in mammalian muscle. *Journal of Biological Chemistry*, **144**, 637–49.

Wang, E., Ross, P.F., Wilson, T.M., Riley, R.T. and Merrill, A.H., Jr (1992) Increases in serum sphingosine and sphinganine and decreases in complex sphingolipids in ponies given feed containing fumonisins, mycotoxins produced by *Fusarium moniliforme*. *Journal of Nutrition*, **122**, 1706–1716.

Ward, D.S., Fessler, J.F., Bottoms, G.D. and Turek, J. (1987) Equine endotoxaemia: cardiovascular, eicosanoid, hematologic, blood chemical, and plasma enzyme alterations. *American Journal of Veterinary Research*, **48**, 1150–56.

Waterman, A. (1977) A review of the diagnosis and treatment of fluid and electrolyte disorders in the horse. *Equine Veterinary Journal*, **9**, 43–8.

Watson, E.D., Cuddeford, D. and Burger, I. (1996) Failure of β-carotene absorption negates any potential effect on ovarian function in mares. *Equine Veterinary Journal*, **28**, 233–6.

Watson, T.D.G., Burns, L., Love, S., Packard, C.J. and Shepherd, J. (1992) Plasma lipids, lipoproteins and post-heparin lipases in ponies with hyperlipaemia. *Equine Veterinary Journal*, **24**, 341–6.

Webb, J.S., Lowenstein, P.L., Howarth, R.J., Nichol, I. and Foster, R. (1973) *Provisional Geochemical*

Atlas of Northern Ireland. Applied Geochemical Research Group Technical Communication No. 60, Imperial College, London.

Webb, J.S., Thornton, I., Howarth, R.J., Thompson, M. and Lowenstein, P.L. (1978) *The Wolfson Geochemical Atlas of England and Wales.* Oxford University Press, Oxford.

Webb, S.P., Potter, G.D. and Evans, J.W. (1987a) Physiologic and metabolic response of race and cutting horses to added dietary fat. *Proceedings of the 10th Equine Nutrition and Physiology Society*, Colorado State University, Fort Collins, 11–13 June 1987, pp. 115–20.

Webb, S.P., Potter, G.D., Evans, J.W. and Greene, L.W. (1987b) Digestible energy requirements for mature cutting horses. *Proceedings of the 10th Equine Nutrition and Physiology Society*, Colorado State University, Fort Collins, 11–13 June 1987, pp. 139–44.

Webb, S.P., Potter, G.D., Evans, J.W. and Webb, G.W. (1990) Influence of body fat content on digestible energy requirements of exercising horses in temperate and hot environments. *Journal of Equine Veterinary Science*, **10** (2), 116–20.

Webster, A.J.F. (1980) The energetic efficiency of growth. *Livestock Production Science*, **7**, 243–52.

Webster, A.J.F., Osuji, P.O., White, F. and Ingram, J.F. (1975) The influence of food intake on portal blood flow and heat production in the digestive tract of the sheep. *British Journal of Nutrition*, **34**, 125–39.

Webster, A.J., Clarke, M.T.M. and Wathes, C.M. (1987) Effects of stable design, ventilation and management on the concentration of respirable dust. *Equine Veterinary Journal*, **19**, 448–53.

Weiss, T. (1982) Equine nutrition in New Mexico. *Veterinary Medicine and Small Animal Clinician*, **77**, 817–9.

Welch, K.J., Perry, T.W., Adams, S.B. and Battaglia, R.A. (1981) Effect of partial typhlectomy on nutrient utilization in ponies. *Journal of Animal Science*, **53** (Suppl. 1), 92, abstract 58.

Weller, R.F. and Cooper, A. (1995) The effect of the grazing management of mixed swards on herbage production, clover composition and animal performance. In: *Grassland to the 21st Century* (ed. G.E. Pollott), pp. 292–4. *British Grassland Society, Occasional Symposium*, No. 29, 50th Meeting, Harrogate.

Wells, P.W., McBeath, D.G., Eyre, P. and Hanna, C.J. (1981) Equine immunology: an introductory review. *Equine Veterinary Journal*, **13**, 218–22.

West, H.J. (1996) Clinical and pathological studies in horses with hepatic disease. *Equine Veterinary Journal*, **28**, 146–56.

West, J.B. and Mathieu-Costello, O. (1994) Stress failure of pulmonary capillaries as a mechanism for exercise induced pulmonary haemorrhage in the horse. *Equine Veterinary Journal*, **26**, 441–7.

Weston, C.F.M., Cooper, B.T., Davies, J.D. and Levine, D.F. (1987) Veno-occlusive disease of the liver secondary to ingestion of comfrey. *British Medical Journal*, **295**, 183.

Whang, R., Morasi, H.T. and Rogers, D. (1967) The influence of sustained magnesium deficiency on muscle potassium repletion. *Journal of Laboratory Clinical Medicine*, **70**, 895–902.

White, G. (1789) *Natural History and Antiquities of Selborne.* British Veterinary Association Library, London.

White, J. (1823) *A Compendium of the Veterinary Art*, Vol. 3. British Veterinary Association Library, London.

White, K.K., Short, C.E., Hintz, H.F., *et al.* (1978) The value of dietary fat for working horses. II. Physical evaluation. *Journal of Equine Medicine and Surgery*, **2**, 525–30.

White, M.G. and Snow, D.H. (1987) Quantitative histochemical study of glycogen depletion in the maximally exercised Thoroughbred. *Equine Veterinary Journal*, **19**, 67–9.

White, N.A., Moore, J.N. and Douglas, M. (1983) SEM study of *Strongylus vulgaris* larva-induced arteritis in the pony. *Equine Veterinary Journal*, **15**, 349–53.

Whitehead, C.C. and Bannister, D.W. (1980) Biotin status, blood pyruvate carboxylase (EC 6.4.1.1) activity and performance in broilers under different conditions of bird husbandry and diet processing. *British Journal of Nutrition*, **43**, 541–9.

Willard, J.G. (1976) Feeding behavior in the equine fed concentrate versus roughage diets. *Dissertation Abstracts International*, **36**, 4772-B-3-B.

Willard, J.G., Bull, L.S. and Baker, J.P. (1978) Digestible energy requirements of the light horse at two levels of work. *Proc 70th A Meet Am Soc Anim Sci*, 324.

Willard, J.G., Willard, J.C., Wolfram, S.A. and Baker, J.P. (1977) Effect of diet on cecal pH and feeding behavior of horses. *Journal of Animal Science*, **45**, 87–93.

Williams, M. (1974) The effect of artificial rearing on the social behaviour of foals. *Equine Veterinary Journal*, **6**, 17–18.

Williams, N.R., Rajput-Williams, J., West, J.A., Nigdikar, S.V., Foote, J.W. and Howard, A.N. (1995) Plasma, granulocyte and mononuclear cell copper and zinc in patients with diabetes mellitus. *Analyst*, **120**, 887–90.

Williamson, H.M. (1974) Normal and abnormal electrolyte levels in the racing horse and their effect on performance. *Journal of the South African Veterinary Association*, **45**, 335–40.

Willoughby, R.A., MacDonald, E. and McSherry, B.J. (1972a) The interaction of toxic amounts of lead and zinc fed to young growing horses. *Veterinary Record*, **91**, 382–3.

Willoughby, R.A., MacDonald, E., McSherry, B.J. and Brown, G. (1972b) Lead and zinc poisoning and the interaction between Pb and Zn poisoning in the foal. *Canadian Journal of Comparative Medicine*, **36**, 348–52.

Wilsdorf, G., Berschneider, F. and Mill, J. (1976) Oriented determination of enzyme activities in race-horses and studies of selenium levels in their feed. *Monatschrift Veterinaermedizin*, **31**, 741–6.

Wilson, T.M., Morrison, H.A., Palmer, N.C., Finley, G.G. and van Dreumel, A.A. (1976) Myodegeneration and suspected selenium/vitamin E deficiency in horses. *Journal – American Veterinary Medical Association*, **169**, 213–17.

Winter, L. (1980) *A survey of feeding practices at two Thoroughbred race tracks*. MSc thesis, Cornell University, Ithaca, New York.

Wintzer, H.J. (1986) Influence of supplementary vitamin H (biotin) on growth and quality of hoof horn in horses. *Tierärzliche Praxis*, **14**, 495–500.

Wirth, B.L., Potter, G.D. and Broderick, G.A. (1976) Cottonseed meal and lysine for weanling foals. *Journal of Animal Science*, **43**, 261, abstract 200.

Wiseman, A., Dawson, C.O., Pirie, H.M., Breeze, R.G. and Selman, I.E. (1973) The incidence of precipitins to *Micropolyspora faeni* in cattle fed hay treated with an additive to suppress bacterial and mould growth. *Journal of Agricultural Science, Cambridge*, **81**, 61–4.

Witherspoon, D.M. (1971) The oestrous cycle of the mare. *Equine Veterinary Journal*, **3**, 114–17.

Wolfram, S.A., Willard, J.C., Willard, J.G., Bull, L.S. and Baker, J.P. (1976) Determining the energy requirements of horses. *Journal of Animal Science*, **43**, 261, abstract 201.

Wolter, R., Durix, A. and Letourneau, J.C. (1974) Influence du mode de présentation du fourrage sur la vitesse du transit digestif chez le poney. *Annals Zootechnologie*, **23**, 293–300.

Wolter, R., Durix, A. and Letourneau, J.C. (1975) Influence du mode de présentation du fourrage sur la digestibilité chez le poney. *Annals Zootechnologie*, **24**, 237–42.

Wolter, R., Gouy, D., Durix, A., Letourneau, J.C., Carcelen, M. and Landreau, J. (1978) Digestibilité et activité biochimique intracaecale chez le poney recevant une même aliment complet présenté sous forme granulée, expansée ou semi-expansée. *Annals Zootechnologie*, **27**, 47–60.

Wolter, R., Meunier, B., de Faucompret, R., Durix, A. and Landreau, J. (1977) Assai d'un aliment complet, granulé ou expansé, en comparaison avec le régime traditionnel chez des chevaux de sport. *Revue de Médecine Vétérinaire*, **128**, 71–81.

Wolter, R., Moraillon, R. and Taulat, B. (1971) Aliments complets pour chevaux: nouveaux essais. *Recueil de Médecine Vétérinaire*, **147**, 565–76.

Wolter, R. and Wehrle, P. (1977) Appreciation of the principal mineral complements destined for horses. *Revue de Médecine Vétérinaire*, **128**, 467–8, 473–83.

Wooden, G.R., Knox, K.L. and Wild, C.L. (1970) Energy metabolism in light horses. *Journal of Animal Science*, **30**, 544–8.

Wootten, J.F. and Argenzio, R.A. (1975) Nitrogen utilization within equine large intestine. *American Journal of Physiology*, **229**, 1062–7.

Worden, A.N., Sellers, K.C. and Tribe, D.E. (1963) *Animal Health, Production and Pasture*. Longmans Green, London.

Worth, M.J., Fontenot, J.P. and Meacham, T.N. (1987) Physiological effects of exercise and diet on metabolism in the equine. *Proceedings of the 10th Equine Nutrition and Physiology Society*, Colorado State University, Fort Collins, 11–13 June 1987, pp. 145–51.

Wright, I.M. (1993) A study of 118 cases of navicular disease: treatment by navicular suspensory desmotomy. *Equine Veterinary Journal*, **25**, 501–509.

Yamamoto, M., Tanaka, Y. and Sugano, M. (1978) Serum and liver lipid composition and lecithin:cholesterol acyltransferase in horses, *Equus caballus*. *Comparative Biochemistry and Physiology*, B **62**, 185–93.

Yoakam, S.C., Kirkham, W.W. and Beeson, W.M. (1978) Effect of protein level on growth in young ponies. *Journal of Animal Science*, **46**, 983–91.

Zenker, W., Josseck, H. and Geyer, H. (1995) Histological and physical assessment of poor hoof horn quality in Lipizzaner horses and a therapeutic trial with biotin and a placebo. *Equine Veterinary Journal*, **27**, 183–91.

Zentek, J., Nyari, A. and Meyer, H. (1992) Untersuchungen zur postprandialen H_2- und CH_4-Exhalation beim Pferd. *First Europäische Konferenz über die Ernährung des Pferdes*, Institut für Tierernahrüng, Tierärzliche Hochschule, Hannover, 3–4 September 1992, pp. 64–6.

Zimmerman, N.I., Wickler, S.J., Rodiek, A.V. and Howler, M.A. (1992) Free fatty acids in exercising Arabian horses fed two common diets. *Journal of Nutrition*, **122**, 145–50.

Conclusion

Progress in nutritional knowledge and in commercial feed quality has been made during the last decade. This progress has been based on the considerable amount of recently published experimental work, but many gaps in our understanding of dietary needs still exist. It is hoped that the present text will have provided some useful conclusions and advice in summarizing the mass of evidence.

Several aspects of equine husbandry require particular nutritional attention. Whereas there is much more knowledge today concerning the physiology and endocrinology of equine reproduction, nutritional information to accompany this is still relatively scarce. Abnormalities in bone growth and development in young horses have received increasing attention, but the prevalence of these abnormalities remains relatively constant. More is known now about respiratory disease related to housing and feeds, although in many cases housing design has not reflected this. Nevertheless, a wider use has been made in the UK of haylage and silage that has had a considerable impact on the problem and which has given a further lease of life to poorly designed stables. The assessment of roughage quality in stable and stud is a continuing problem making precise definitions of nutrient and energy needs somewhat premature. This is an issue to be surmounted, as roughage will undoubtedly remain a backbone to the feeding of the horse.

Index

therapeutic, prophylactic, 383, 385
anticoagulant, warfarin, 416
antihistamines, 423
   see also cromoglycate, sodium
antioxidants, 82, 83
   natural, 82, 83
   synthetic, 82, 83
antiprotease, *see* toxins, plant
appetite, 33, 34, 157, 158, 202, 203
articular cartilage, 254
ascorbic acid, 89
   administration, 89
ash, dietary, 107, 108
   see also declaration of analysis
aspartate amino transferase, *see* enzymes
ataxia, *see* neurological disorders
autonomic nervous system, 306
   also see dysautonomia
azoturia, 410–12, 432
   also see myoglobinuria, urine

bacteria
   as soil-borne pathogens, 344
   endotoxins from, 381–5
   enterotoxin, 385
   exotoxin, 399
   fermentation by, 9, 14
   in feed spoilage, 380, 381, 390
   large intestinal, 17–24, 310, 381, 384
   pathogenic (intestinal), 135, 215, 381, 385, 400
   rumen, 18
   tuberculosis, 409
   also see silage and haylage
beans
   field horse (*V. faba*), 133, 134
   hyacinth (*Dolichos lablab*), 134
   kidney (*P. vulgaris*), 134
   navy, lima (*P. lunatus*), 134
bedding, 421
bentonite, sodium, 107
bicarbonate, 288–94
bit, 2
bile, 10, 12
   acids, 435, 436
   duct obstruction, 436
biuret, 38
bleeders, 271, 272
blood, 274–81
   albumin, 252, 432, 433
   amino acids in, 252, 309
   ammonia in, 236, 263, 287, 288, 308–10, 407
   anaemia, *see* anaemia
   analyses of, 55, 223
   base excess of, 290
   bilirubin in, 408, 436
   buffers in, 278
   B vitamins in, 84, 85, 88

calcium and phosphorus in, 44, 45, 433
clotting of, 407
creatinine in, 410
electrolytes in, 283, 433
enzymes in, *see* enzymes
enzyme activity coefficients of
   see enzyme
free fatty acids in, 263
gases in, 274, 276
glucose in, 236, 278, 279
haematocrit, *see* packed cell volume
haemoglobin, 216, 252
haemolysis of, 407
haemolytic icterus of, 216
hyperlipidaemia, 236
insulin activity of, *see* insulin
ketones (3-hydroxybutyrate) in, 264
lactate in, 265–7, 286
leucocytes in, 274
osmotic pressure of, 284, 312, 432
packed cell volume (haematocrit) of, 274–6
pH of, 281, 283, 293–5, *see also* pH
plasma globulins of, *see* disease resistance and
     immune response
plasma protein patterns of disease in, 441
prothrombin time of, 407
serum composition of, 54, 55, 283
spleen as source of, 279
total plasma protein of, 252, 308, 312
trace elements in, 433
urea in, 308–10
viscosity of, 279, 280
volume of, 274
bodyweight
   changes in, 280
   estimates of, 158–61
   excessive, 271
   fat reserves in (depot fat), 265, 271, 300–302
   loss, 409
bone
   abnormalities of, 244, 251
   density and exercise, 256, 272
   disorders, 44, 224–9, 244, 251, 252, 254–8, 358,
     361, 389, 398, 415, 416, 441
   epiphysitis, *see under* epiphysis
   flour, meal, 48, 135
   fractures of, 44, 57
   growth of, 243–8, 250, 254, 431
   in leg, 243–8
   metabolism, measurement of, 440–2
   minerals, 43–7, 56, 59–61
   mobilization, resorption of, 222
   structure of, 59, 245, 246, 272
   see also Vitamin D
boron, *see* trace elements
botulism, 344, 399, 400 *and see* haylage
breed, 279